ROUTLEDGE HANDBOOK OF WATER AND DEVELOPMENT

Edited by Sofie Hellberg, Fredrik Söderbaum, Ashok Swain, and Joakim Öjendal

Routledge
Taylor & Francis Group
LONDON AND NEW YORK

Designed cover image: © Getty Images

First published 2024
by Routledge
4 Park Square, Milton Park, Abingdon, Oxon OX14 4RN

and by Routledge
605 Third Avenue, New York, NY 10158

Routledge is an imprint of the Taylor & Francis Group, an informa business

British Library Cataloguing-in-Publication Data
A catalogue record for this book is available from the British Library

ISBN: 978-0-367-55876-5 (hbk)
ISBN: 978-0-367-55877-2 (pbk)
ISBN: 978-1-003-09554-5 (ebk)

DOI: 10.4324/9781003095545

Typeset in Times New Roman
by KnowledgeWorks Global Ltd.

ROUTLEDGE HANDBOOK OF WATER AND DEVELOPMENT

Water is essential for human life and at the centre of political, economic, and socio-cultural development. This *Routledge Handbook of Water and Development* offers a systematic, wide-ranging, and state-of-the-art guide to the diverse links between water and development across the globe. It is organized into four parts:

- **Part I** explores the most significant theories and approaches to the relationship between water and development.
- **Part II** consists of carefully selected in-depth case studies, revealing how water utilization and management are deeply intertwined with historical development paths and economic and socio-cultural structures.
- **Part III** analyses the role of governance in the management of water and development.
- **Part IV** covers the most urgent themes and issues pertaining to water and development in the contemporary world, ranging from climate change and water stress to agriculture and migration.

The 32 chapters by leading experts are meant to stimulate researchers and students in a wide range of disciplines in the social and natural sciences, including Geography, Environmental Studies, Development Studies, and Political Science. The *Handbook* will also be of great value to policymakers and practitioners.

Sofie Hellberg is an Associate Professor of Peace and Development Research at the School of Global Studies, University of Gothenburg, Sweden, and a Research Associate at the Department of Anthropology and Development Studies at the University of Johannesburg, South Africa.

Fredrik Söderbaum is a Professor of Peace and Development Research at the School of Global Studies, University of Gothenburg, Sweden, and an Associate Research Fellow at the United Nations University Institute of Comparative Regional Integration Studies (UNU-CRIS), Bruges, Belgium.

Ashok Swain is Head of the Department of Peace and Conflict Research, UNESCO Chair on International Water Cooperation and Director of the Research School of International Water Cooperation at Uppsala University, Sweden. He is also the founding Editor-in-Chief of the journal *Environment and Security*, published by SAGE and the Environmental Peacebuilding Association.

Joakim Öjendal is a Professor in Peace and Development Research since 2006 at the School of Global Studies, Gothenburg University, Sweden. He has worked on *resource politics*, *peacebuilding*, and *post-war democratization* for three decades in research, policy, and education.

CONTENTS

FIGURES

TABLES

ABOUT THE CONTRIBUTORS

Editors

Sofie Hellberg is an Associate Professor of Peace and Development Research at the School of Global Studies, University of Gothenburg, Sweden, and a Research Associate at the Department of Anthropology and Development Studies at the University of Johannesburg, South Africa. She studies, and teaches, water politics, environmental, climate governance, and theories of power and agency. Hellberg has published in leading journals and with international publishers on topics ranging from Education for Sustainable Development (ESD) to research methodology. Her previous work on water appears in international journals including *Geoforum*, *Water Alternatives, Environment and Planning E: Nature and Space* and *Local Environment*, as well as in a monograph on *The Biopolitics of Water* (Routledge, 2018).

Fredrik Söderbaum is a Professor of Peace and Development Research at the School of Global Studies, University of Gothenburg, and an Associate Research Fellow at the United Nations University Institute of Comparative Regional Integration Studies (UNU-CRIS), Bruges, Belgium. Söderbaum has published extensively in leading journals on comparative regionalism, global and regional governance, development research, security studies, and African politics. His most recent books include *Contestations of the Liberal International Order: A Populist Script of Regional Cooperation* (Cambridge University Press, 2021), *Rethinking Regionalism* (Palgrave Macmillan, 2016), and *Intersecting Interregionalism: Regions, Global Governance and the EU* (Springer, 2014).

Ashok Swain is Head of the Department of Peace and Conflict Research, UNESCO Chair on International Water Cooperation and Director of the Research School of International Water Cooperation at Uppsala University, Sweden. He is also the founding Editor-in-Chief of the journal *Environment and Security*, published by SAGE and the Environmental Peacebuilding Association. He has written extensively on new security challenges, water-sharing issues, environment, conflict and peace, and democratic development issues. His most recent publications include *Handbook of Security and the Environment* (Edward Elgar, 2021) coedited with Joakim Öjendal and Anders Jägerskog.

Joakim Öjendal is a Professor in Peace and Development Research since 2006 at the School of Global Studies, Gothenburg University, Sweden. He has worked on *resource politics*, *peacebuilding*, and *post-war democratization* for three decades in research, policy and education. He has published widely in leading journals and with international publishers, for instance, being the co-editor of *Water Security, a Four Volume Set of SAGE Major Works* as well as *Transboundary Water Management and the Climate Change Debate,* published with Earthscan, both in 2014. His most recent publications include *Handbook of Security and the Environment* (Edward Elgar, 2021) coedited with Ashok Swain and Anders Jägerskog.

Contributors

Dr Zafar Adeel serves as the Executive Director of the Pacific Water Research Centre, and as a faculty at the School of Sustainable Energy Engineering at Simon Fraser University. He has over 25 years of experience in a broad range of environmental science and policy issues, including 18 years of service as a UN official. He served as the Director, United Nations University Institute for Water, Environment and Health (2006–2016). He also served in a number of international leadership roles, including chairing a group of 60 organizations called UN-Water (2010–2012), and co-chairing the Millennium Ecosystem Assessment team that produced the global desertification synthesis (2005).

Evelyn Arriagada Oyarzún is an Anthropologist and MA in social sciences (Universidad de Chile), a MA in Political and Social Sciences (Universitat Pompeu Fabra), and a PhD candidate at the Institute for Resources, Environment and Sustainability (University of British Columbia). Her dissertation focuses on shifting subjectivities and hydrosocial relationships in contemporary water justice movements in Chile. Her PhD is supported by a Vanier CGS Award, Killam Doctoral Scholarship, Public Scholar Initiative, and the National Agency for Research and Development (ANID – Chile) – among other funding. Her most recent co-authored publication explores the experiences of "lived environmental citizenship" in female rural leaders in central Chile.

Karen Bakker is a Canadian Author, Researcher, and Entrepreneur known for her work on digital transformation, environmental governance, and sustainability. A Rhodes Scholar with a PhD from Oxford, Bakker is a Professor at the University of British Columbia, and currently the Matina S. Horner Distinguished Visiting Professor at the Radcliffe Institute for Advanced Study at Harvard University. Bakker's current research focuses on the intersection of digital technologies and environmental governance, digital environmental humanities, digital geographies, political ecology, and political economy. Her most recent book, The Sounds of Life: How Digital Technology Is Bringing Us Closer to the Worlds of Animals and Plants, was published by Princeton University Press in October 2022.

Anik Bhaduri is the Director of the Sustainable Water Future Programme (Water Future) of Future Earth. Water Future is a global platform facilitating international scientific collaboration to drive solutions to the world's water security problems. Anik coordinates this large network of more than 400 researchers from Water Future's 15 international and interdisciplinary working groups and facilitates integration and synthesis exercises in collaboration with working group members. Anik is also an Associate Professor at the Australian River Institute, Griffith University, Australia where he works on several topics and projects, ranging from transboundary water sharing to adaptive water management under climate change.

Rutgerd Boelens is a Professor "Political Ecology of Water" with CEDLA, University of Amsterdam, and a Professor "Water Governance and Social Justice" at Wageningen University. He is Visiting Professor at the Catholic University of Peru and the Central University of Ecuador. He coordinates the Justicia Hídrica/Water Justice alliance and the international research/action programmes Riverhood and River Commons (www.movingrivers.org). His research focuses on political ecology, water rights, legal pluralism, water cultural politics, governmentality, hydrosocial territories, environmental justice and social mobilization. Recent books: "*Water, Power and Identity. The Cultural Politics of Water in the Andes*," and "*Water Justice*," Cambridge University Press (with T. Perreault & J. Vos).

Jessica Budds is a Professor of Development Geography at the University of Bonn. Her research explores shifting water-society relations in the context of economic change in the Global South, particularly Latin America. Her main foci have comprised the socio-ecological implications of the application of neoliberal principles to water policy (water privatization, water markets, payments for watershed ecosystem services), and of increased demand for water by industrial sectors as part of growth strategies (agriculture, mining, hydropower). Her current work is examining how responses to water scarcity may serve to legitimize and sustain existing unequal power relations and forms of capital accumulation.

Deliang Chen holds the prestigious August Röhss Chair at the University of Gothenburg and is a world-renowned climate researcher. He studies regional climate changes and their impacts on water, ecosystem, and the environment from an Earth System Science perspective. Currently, he focuses on climate and environmental changes over the Tibetan Plateau and its surrounding. He is a coordinating lead author of the sixth assessment report of the Intergovernmental Panel on Climate Change (IPCC).

Patricia Urteaga Crovetto is a Professor and Researcher at the Pontificia Universidad Catolica del Peru. She earned a law degree from the Pontificia Universidad Catolica del Peru, and a Master's and Ph.D. in Anthropology from the University of California, Berkeley. Her research topics focus on natural resources, indigenous rights, extractive industries, and water. Some of her last publications are: (2021) "Transforming nature, crafting irrelevance: The commodification of marginal land for sugarcane and cocoa agroindustry in Peru." In: Abubakari Ahmed & Alexandros Gasparatos (editors). *Political Ecology of Industrial Crops*, Routledge, Taylor and Francis Group. Political Ecology of Industrial Crops – 1st Edition, and (2019) *Derecho Humano al Agua, Petróleo y Pueblos Indígenas en la Amazonía*. Lima: Departmento Academico de Derecho, CICAJ.

Gaylean Davies has a BA in Psychology and an MA in Sociology and International Development. She is currently a PhD student at the Institute for Resources, Environment and Sustainability (IRES) at the University of British Columbia (UBC). Gaylean is an interdisciplinary social scientist whose current research sits at the intersection of gender equality and water governance, access, use, and knowledge. Her PhD research focuses on utilizing a critical feminist and development lens to understand water projects designed to empower women. Gaylean's research is supported by a Killam Doctoral Scholarship and through the Public Scholar Initiative.

Nelson Ekane is a Research Fellow at the Stockholm Environment Institute (SEI). His research on sanitation governance disentangles how sanitation policies articulate at multiple governance levels and translate into investment and behaviour change at household and community levels.

His research on risk governance involves measuring the nature and variability of public beliefs, attitudes, and policy preferences for wastewater reuse and land application of treated sewage sludge. Nelson's recent publication in this field in Frontiers in Sustainable Food Systems is entitled "Resources and Risks: Perceptions on the Application of Sewage Sludge on Agricultural Land in Sweden, a Case Study."

Aiko Endo received MSc in Coastal and Ocean Policy course from the University of Plymouth, UK, in 2003 and PhD in the Graduate School of Biosphere Science of Hiroshima University, Japan in 2008. She led the project on water-energy-food nexus funded by RIHN, Japan, and has published nexus articles focusing on methods and methodology including "Methodology for Nexus Approach Toward Sustainable Use of Geothermal Hot Spring Resources" in 2021. She joined the Graduate School of Fisheries and Environmental Sciences of Nagasaki University, Japan as a Professor in 2022. Her current research interest is plastic marine debris on the seabed.

Jo Geere is a Director of the Water Security Research Centre and physiotherapy lecturer in the School of Health Sciences, University of East Anglia, Norfolk U.K. Her current research focusses on health, water security and how WaSH access can support community resilience, health, and wellbeing of vulnerable people. Recent most relevant publications include Venkataramanan, V., et al., (2020): "In pursuit of 'safe' water: the burden of personal injury from water fetching in 21 low-income and middle-income countries" in *BMJ Global Health* and J. L. Geere and P. R. Hunter. (2020): "The impacts of water carriage and sanitation provision on maternal and child health. A combined analysis of 49 Multiple Indicator Cluster Surveys from 41 countries" in *International Journal of Hygiene and Environmental Health.*

Dr Peter Gleick is co-founder and Senior Fellow of the Pacific Institute, Oakland, California. Gleick is a hydrologist and climatologist working on broad issues of freshwater sustainability, the human right to water, and water conflicts. He is a member of the US National Academy of Sciences, a MacArthur Fellow, and most recently author of "The Three Ages of Water: Prehistoric Past, Imperiled Present, and a Hope for the Future" (PublicAffairs, 2023).

Leila M. Harris is a Professor with the Institute for Resources, Environment, and Sustainability and with the Institute for Gender, Race, Sexuality, and Social Justice at the University of British Columbia. Her work analyses a range of governance and political considerations from feminist, equity, and sustainability perspectives – with work that has focused on water politics, governance, and justice in a variety of contexts (from Turkey to Ghana, South Africa, and Canada). Recent projects consider lived experiences and equity concerns related to the uneven implementation of the human right to water, ongoing water governance shifts in varied contexts, as well as focus on social, affective, and political dimensions of the uneven geography of water access, quality, and infrastructures. Recent publications include a focus on issues of trustworthiness around drinking water and the use of narrative inquiry in political ecology research (*Environment and Planning E*).

Juan Pablo Hidalgo-Bastidas is an Agronomist, specialist in water and land management, and holds a PhD in social sciences. He is currently a postdoctoral researcher in the Water Resources Management group at Wageningen University, the Netherlands. His work especially covers the study of water and environmental justice, the political ecology of water, hydrocracies and social movements, and the role of water technologies in society-nature relations. Juan Pablo is a member-researcher of the Water Justice Alliance (http://justiciahidrica.org/), the

WaterLat-Gobacit Network (http://waterlat.org/es/) and the National Water Forum of Ecuador (https://camaren.org).

Jaime Hoogesteger is an Associate Professor in the field of water commons and collectives at the Water Resources Management Group of Wageningen University. His research interests revolve around the themes of commons, water collectives, social movements and co-governance arrangements, especially in contexts of rapid agrarian change and broader rural transformations. His approach is interdisciplinary and is greatly inspired by notions of political ecology and socio-environmental justice as evidenced by his publications.

Paul R. Hunter is a Professor of Medicine at the Norwich Medical School. Paul's key research interests include the epidemiology of emerging infectious diseases especially that linked to environmental factors; the spread of infection by drinking water, but also recreational water contact and food; zoonotic diseases; risk assessment and risk communication; and in conducting case-control and other epidemiological studies in the UK, Europe, and the developing world. Recent publications include Nik Hassan, N. M. N., Hunter, P. R., & Lake, I. R. (2022): "Risk perception from the consumption of untreated drinking water in a small island community" in *Journal of Water and Health* and de França Doria, M., Segurado, P., Korc, M., Heller, L., Jimenez Cisneros, B., Hunter, P. R., & Forde, M. (2021): "Preliminary assessment of COVID-19 implications for the water and sanitation sector in Latin America and the Caribbean." in *International Journal of Environmental Research and Public Health.*

Dr Inga Jacobs is the IWMI Regional Representative for Southern Africa. Her expertise focuses on water governance, the water-energy-food nexus and its governance frameworks, water stewardship, community-based approaches to water services, hybrid water law, and environmental migration. Before joining IWMI Inga was the research group leader for the integrated water solutions group at the Council for Scientific and Industrial Research (CSIR), South Africa, where she managed a transdisciplinary team of researchers working on the co-development of integrated water solutions in South Africa and throughout the Southern African Development Community (SADC) region. Inga was also the Executive Manager: Business Development, Marketing, and Communications at the Water Research Commission, South Africa. Inga is an active leader in the water sector having served on the board of directors for the Water Institute of Southern Africa and the International Water Association.

Anders Jägerskog (PhD) is the Programme Manager of the Cooperation on International Waters in Africa (CIWA) Trust Fund (ciwaprogram.org) at the World Bank. He is also the focal point at the World Bank for Transboundary Waters. He was Counsellor for regional water resources in the MENA region at the Swedish Embassy in Amman, Jordan; Director, Knowledge Services, at the Stockholm International Water Institute (SIWI) where he headed the Transboundary Water Management Unit. He managed the UNDP Shared Waters Partnership and is an Associate Professor at Peace and Development Research, School of Global Studies, University of Gothenburg He has published over 100 scientific articles, book chapters, debate articles, and reports on global water issues.

Kurt Mørck Jensen is a Chief Adviser in the Ministry of Foreign Affairs of Denmark. He was a senior analyst at the Danish Institute for International Studies from 2010 to 2013 leading a research programme on Water Governance and Water Development. In 2005–2006, he was a research associate at Sydney University's Mekong Research Centre. He has worked in international development for 40 years for Danish and Swedish aid and for the World Bank in Washington D.C. A special focus by him has been 20 years of attention to the politics and transboundary water management in the Mekong River.

Henrik Josefsson is an Associate Professor of Environmental Law and Senior Lecturer of Commercial Law at the Department of Business Studies, Uppsala University. His research interests range from surface and groundwater law, to chemical and nuclear law. Among his publications are "Opportunities to tackle antibiotic resistance development in the aquatic environment through the Water Framework Directive" (*Ambio*, 2023) co-authored with M. Ågerstrand, A.S. Wernersson and D. G. J. Larsson, and "The potential of the EU Water Framework Directive for reducing emissions of pollutants is limited: a case study on river basin specific pollutants in Swedish environmental permitting processes" (*Environ. Sci. Eur.* 2022) co-authored with E. Undeman and H. Ågerstrand.

Aditya Kaushik is the Director for the US-India Business Council's digital economy, portfolio. Prior to joining USIBC, Aditya served as a fellow at the New Emerging Strategic Technologies (NEST) Division at the Ministry of External Affairs, India. Aditya worked as an Associate Director of the Water Solutions Lab (WSL) at Divecha Centre for Climate Change, IISc. Kaushik holds a Master of Science degree in EE from USC and a Master of Arts degree in Law and Diplomacy from the Fletcher School at Tufts University with an emphasis on Science and Technology Diplomacy. His research interest includes AI and regulation.

Rosemary Knight is a Professor of Geophysics and Earth System Science at Stanford University as well as a Senior Fellow at the Stanford Woods Institute for Environment. She is currently working with students to find ways of using geophysical methods to understand the processes occurring in the top 100 metres of Earth, with a specific interest in the use of geophysical imaging for the evaluation and management of our groundwater resources. Her recent publications include the Development and Application of a 1D Compaction Model to Understand 65 Years of Subsidence in the San Joaquin Valley (Water Resources Research, 2022) with Matthew Lees and Ryan Smith.

Hui-Wen Lai, Department of Earth Sciences, University of Gothenburg, focuses on processes of physical interactions between large-scale circulation, mesoscale convective systems, and regional climate changes over the complex terrain region, and is interested in improving our understanding of how precipitation has changed in this region using numerical weather and climate models. She is a contributing author of the sixth assessment report of the Intergovernmental Panel on Climate Change (IPCC).

Bruce A. Lankford is an Emeritus Professor of Water and Irrigation Policy at the School of International Development, University of East Anglia, UK. He has nearly 40 years of experience in agriculture, irrigation and water resources management. His research covers; irrigation policy in Sub-Saharan Africa; serious games in natural resource management; design for irrigation manageability; irrigation efficiency and the paracommons; water accounting; river basin management and water allocation; and irrigation resilience. Recent publications include Lankford, B., et al (2023): "Irrigation area, efficiency and water storage mediate the drought resilience of irrigated agriculture in a semi-arid catchment" in *Science of the Total Environment* and Lankford, B., & Orr, S. (2022): "Exploring the Critical Role of Water in Regenerative Agriculture; Building Promises and Avoiding Pitfalls" in *Frontiers in Sustainable Food Systems.*

Alex Loftus is a Professor of Political Ecology at King's College London. His research focuses on inequalities in relation to water access and everyday struggles to address these inequalities. In recent years, Alex has become increasingly interested in the political economy of water infrastructure, in processes of financialization, and in efforts to achieve human rights to water and sanitation. He is the author of *Everyday Environmentalism* (2012), co-editor of *Water Politics* (2019), and co-author of *Water: A Critical Introduction* (2023).

Alan Mackworth is a Professor Emeritus in the Department of Computer Science at the University of British Columbia. His work focuses on constraint-based artificial intelligence with applications in vision, robotics, situated agents, assistive technology and sustainability. He is known as a pioneer in the areas of constraint satisfaction, robot soccer, hybrid systems, and constraint-based agents. He served as the founding Director of the UBC Laboratory for Computational Intelligence. His most recently published book is Artificial Intelligence: Foundations of Computational Agents (Cambridge University Press, 2010) with co-author David L Poole.

Deborah McGregor is an Anishinabe and Associate Professor and Canada Research Chair in Indigenous Environmental Justice, Osgoode Hall Law School and Faculty of Environmental and Urban Change, York University. Her research has focused on Indigenous knowledge systems and their various applications in diverse contexts including water security and governance, environmental and climate justice, health, and environment, Indigenous legal and intellectual traditions. She remains actively involved in a variety of Indigenous communities, serving as an advisor and continuing to engage in community-based research and initiatives. Professor McGregor has been at the forefront of Indigenous environmental justice and Indigenous research theory and practice.

Lyla Mehta is a Professor at the Institute of Development Studies, UK, and a Visiting Professor at Noragric, Norwegian University of Life Sciences. She uses the case of water and sanitation to focus on the politics of scarcity, gender, human rights, and access to resources, resource grabbing, and power and policy processes in rural, peri-urban, and urban contexts. Her work also focusses on climate change, uncertainty and transformation and forced displacement. She has extensive research and field experience in India and southern Africa and is currently leading a Belmont/Norface/EU/ISC project on "Transformations as praxis" in South Asia. Her most recent books are Water, Food Security, Nutrition and Social Justice and *The Politics of Climate Change and Uncertainty in India.*

Richard Meissner is an Associate Professor at the University of South Africa's Department of Political Sciences. Before joining UNISA he was a Senior Researcher at the Council for Scientific and Industrial Research's Water Centre between 2010 and 2021. He holds a Doctoral degree in International Politics from the University of Pretoria. Richard specializes in several study areas ranging from water security at the local government level to the analysis of transboundary river basin governance and politics. His research focuses on the complexities and interactions between and among non-state actors, international organizations, and state/government organs.

Erik Mostert is a Lecturer in water management and law at the Delft University of Technology. His research interests include European and Dutch water law, the history of Dutch water management, and the link between water and culture. In addition, he has conducted internationally-comparative research on water management in the Netherlands, France, the UK, Germany, and Portugal. His publications include "Water and national identity in the Netherlands; the history of an idea" (*Water History* 12(3): 311–329) and "Law and politics in river basin management; The Implementation of the Water Framework Directive in the Netherlands" (*Water* 12(2): 1–12).

Raymond T. Ng is a Professor of Computer Science and Scientific Director of the Data Science Institute at the University of British Columbia. Since 2016, he has held the Canadian Research Chair on Data Science and Analytics. Raymond's main research area is data mining, with a focus on health informatics and text mining. Since the inception of the PROOF Centre of Excellence, he has held the position of the Chief Informatics Officer of the Centre. His recent publications include Predicting the Survival of Patients With Cancer From Their Initial Oncology Consultation Document Using Natural Language Processing (JAMA Network Open, 2023) with co-authors John-Jose Nunez, Bonnie Leung, Cheryl Ho, and Alan T. Bates.

Dr Karen Nortje is a Senior Researcher (Gender and Social Inclusion) at the International Water Management Institute (IWMI). At IWMI, Karen is responsible for driving efforts to deliver a step change in IWMI's understanding of gender, youth and inclusion challenges in sustainable agriculture and natural resource management in Africa. She brings socio-economic and socio-cultural insights to the oftentimes technical challenges that her interdisciplinary team works on. She has twenty years' of experience working as an Anthropologist, and this has afforded her the opportunity to engage with stakeholder groups, in particular communities, relying on local knowledges to help understand and solve environmental, socio-cultural, and development issues. She has conducted fieldwork throughout South Africa, as well as, amongst others, in Mozambique, Zambia, India, Cost Rica, Vietnam, and the United States.

Jens Christian Refsgaard is a Professor emeritus in water resources at the Geological Survey of Denmark and Greenland (GEUS). Co-ordinated several major EU-funded and Danish-funded research projects. Co-coordinator/team leader of several multi-disciplinary international research and capacity-building projects in Europe, Asia, and Africa – working experience from more than 40 countries. Published more than 160 journal papers with more than 8000 ISI citation (h index 42 in ISI Web of Science) and 17000 citations (h-index 57) on Google Scholar. Major research interests: water resources management including good modelling practice and uncertainty assessments. One of Refsgaard's most recent publication is "Hydrological process knowledge in catchment modelling – lessons and perspectives from 60 years development" in *Hydrological Processes* (2022).

Max Ritts is an Environmental Geographer with a PhD from the University of British Columbia (2018). His research explores intersections of social power, sensory practice, and ecological transformation. His in-process book, A Resonant Ecology (under contract with Duke UP), examines the material, affective, and conceptual force of industrial development through situated engagements with sound culture (including eco-acoustics, whalesong, industrial noise, and Indigenous heavy metal). His recent publications include Reworking the P in digital forests: The cosmopolitics of socio-technical worlds (Progress in Environmental Geography, 2022) with Jennifer Gabrys, Michelle Westerlaken, Danilo Urzedo, and Trishant Simlai.

Dr Stephen Rule has managed quantitative surveys and policy research on education, drought response, financial literacy, youth development, religion, migration, gambling behaviour and urban settlements, in South Africa, Nigeria, Kenya, Rwanda, Namibia, Lesotho, and the UK. He is a Research Fellow in the Department of Geography and Environmental Studies at Stellenbosch

University and has held posts in the Scottish Centre for Social Research, Human Sciences Research Council, and the SA Monitoring and Evaluation Association. A related publication, together with Parker, S., Majikijela, Y. & Lunga, W, is: "Neighbourhood responses to drought in the Western Cape" in *GeoJournal* (2021).

Maria Rusca is a Lecturer at the Global Development Institute, University of Manchester. Her work focuses on political ecologies of water and hydroclimatic extremes, critical disaster studies, and experimental political ecologies. Although her work is firmly rooted in human geography, she is committed to developing research at the nexus between social and natural sciences to further the fields of political ecology and nature/society. Recent publications include work on the urban metabolism of waterborne diseases (*Annals of the American Association of Geographers*), everyday urbanism and uneven waterscapes (*Wires Water*), and urban climate extremes (*Nature Climate Change*).

Dr Elisa Savelli, affiliated with the Department of Earth Sciences, Air, Water and Landscape Science, Uppsala University, Uppsala, Sweden, and the Centre of Natural Hazards and Disaster Science, CNDS, Uppsala, Sweden is an experienced academic and practitioner in the water sector specialized in the provision of water services in developing countries. She has led several pieces of research on pro-poor water services and policy implementation in Southern and Eastern Africa, as well as coordinated humanitarian water, sanitation, and hygiene interventions in West Africa and the Middle East. Presently Elisa's field of work concerns the intersection between politics and the production and distribution of hydrological risk. Her work combines political ecologies of land, water, and vulnerability with socio-hydrological insights around human-water systems coevolution

Klaas Schwartz is an Associate Professor of Urban Water Governance in the Department of Water Governance at the IHE Delft Institute for Water Education. Mr. Schwartz has 25 years of experience in education and training relating to the field of water services governance. His main area of expertise concerns water services management, informality and water services and reform of public water utilities. In his research activities, he seeks to develop a better understanding of the management and governance practices of practitioners in service providers. He has been involved in research, education, and training activities in various countries in Asia, Africa, and Latin America.

Mahisha Sritharan is a Research Associate at York University and works with the Toronto Environmental Alliance. Her research focuses on environmental justice, climate change, social and environmental determinants of health, and Indigenous knowledge. Mahisha has co-authored a number of research publications focused on Indigenous environmental justice and climate justice including "Indigenous environmental justice and sustainability" with Professor Deborah McGregor and Steven Whitaker published in the Current Opinion in Environmental Sustainability journal. She holds a Master's in Environmental Studies from York University and a Bachelor of Arts in Environmental Geography and Health Studies from the University of Toronto.

Diana Suhardiman is the director of the Royal Netherlands Institute of Southeast Asian and Caribbean Studies, an institute of the Royal Netherlands Academy of Arts and Sciences. She is a Special Professor of Natural Resource Governance, Climate, and Equity at Leiden University, the Netherlands. Putting power and politics central in the contemporary struggles of natural resource governance, her research focuses on unpacking messy processes (re)shaping water, land, environmental, and climate governance in Southeast Asia. It places knowledge (re)production central in unpacking actors' and institutions' views, strategies, and responses to agrarian change, climatic change, and processes of globalization including how capitalist intrusions affect livelihood (re)making.

Larry A. Swatuk is a Professor in the School of Environment, Enterprise and Development (SEED) at the University of Waterloo, Canada. He is also Extraordinary Professor in the Institute for Water Studies, University of the Western Cape, South Africa and External Researcher, Bonn International Centre for Conflict Studies, Germany. His recent research focuses on the unintended negative consequences of state-led climate action leading to potential "boomerang effects." Among his most recent publications are the co-edited collections (with Dr Corrine Cash) entitled *The Political Economy of Urban Water Security under Climate Change* (Palgrave, 2022) and *The Political Economy of Climate Finance: Lessons from International Development* (Palgrave, 2022).

Alexandre Teixeira is a PhD Student at Australian Rivers Institute/Griffith University, also affiliated with Brazilian Water and Sanitation Agency – ANA with research interests in water security, green-grey infrastructure investment, dynamic optimization modelling and climate change. Graduated in Civil Engineer at the Federal University of Rio de Janeiro (Brazil) and also Master's Degree in Civil Engineering Program from the Federal University of Rio de Janeiro/Brazil (COPPE/UFRJ). Since 2004, has been working as a water resources specialist at the National Water and Sanitation Agency of Brazil (ANA), where coordinated water resources planning studies.

Dr Mireia Tutusaus is a Monitoring, Evaluation and Learning advisor at VEI Dutch Water Operators. In her role at VEI, she monitors the progress and impact of Water Operators Partnership (WOP) projects implemented with partners in Africa, Asia, and Latin America. She also initiates research and provides advice to the strategic development of the project portfolio of VEI as WOP implementer. Mireia is also a research affiliate at IHE Delft where she lectures and supervises MSc students. Mireia obtained her PhD from the University of Amsterdam (the Netherlands) researching the "practiced" commercialization reforms of public water utilities. Her main area of expertise is the development and analysis of business and governance models for drinking water services in urban areas and small towns, with a specific interest on financial sustainability.

Susann Baez Ullberg is a Senior Lecturer of Cultural Anthropology at Uppsala University. Her research interests have since long focused on hydrosocial relations. She has carried out ethnographic research in Argentina to study how social memory and oblivion forge stratified flood risk; in Peru to inquire the temporalities of water megaprojects; and more recently, focusing on aquifers to understand how people and organizations engage with groundwater. Among her publications are "Making the megaproject: Water infrastructure and hydrocracy at the public-private interface in Peru" (*Water Alternatives*, 2019) and "Forgetting Flooding? Post-disaster economy and embedded remembrance in suburban Santa Fe, Argentina" (*Nature and Culture*, 2017).

Dr Susan van de Meene is a Lecturer with at the Monash Sustainable Development Institute, Monash University, Australia. With an interdisciplinary background in human geography and environmental engineering, Susan's research has focused on urban sustainability, particularly in urban water governance and climate policy. She has investigated sustainability perspectives of urban water professionals, policy learning among cities and realizing urban co-benefits through climate mitigation policies. This research has drawn on theories of governance, collaboration, network learning, and has published in *Sustainability*, *Global Environmental Change*, *Journal of Cleaner Production* and *Policy Science*. Susan has also published more broadly for the industry through reports and industry magazines.

Gert Jan Veldwisch is an Associate Professor in the field of Water, Society & Technology at Wageningen University. His research and education aim to bridge the gap between science and implementation practices in relation to societal and technological transformations. Working on cases across the Global South his work spans themes such as drinking water supply, irrigation, practices and politics of modelling, rural-urban water transfers, and river (co-)management.

Dr Jeroen Warner teaches, trains and publishes on Crisis and Disaster Studies at Wageningen University, where he is a senior Associate Professor and Research Coordinator. A founder member of the London Water Research Group, Jeroen works on domestic and transboundary water conflict and cooperation, multi-stakeholder participation in water governance. He won a CAPES scholarship as Special Visiting Professor at the University of Sao Paulo and coordinated the European Horizon 2020 project EDUCEN, on cities, cultures, and catastrophes. Jeroen is Editor-in-Chief of the International Journal of Water Governance, and associate editor of *Natural Hazards* and *Regions and Cohesion.* He (co)published seven books and well over 100 academic and professional articles

Steven Whitaker, B.Sc., MES, B.Ed., is a Teacher and Research Assistant. His research is focused on topics relating to forestry, climate, water, and environmental and sustainability issues. He is a teacher at the Shawanoswe School in Whitefish River First Nation. He co-authored many Indigenous community-based reports and co-authored, Indigenous Environmental Justice and Sustainability In the journal *Current Opinion in Environmental Sustainability in 2020.*

Neda Zawahri is a Professor in the Department of Political Science at Cleveland State University. For more than two decades, she has focused on water politics and security, managing transboundary rivers, and adapting to climate change in the Middle East and South Asia. She has carried out extensive field research throughout these regions. Her work appears in international relations and environmental politics journals, including *International Studies Quarter*, *Journal of Peace Research*, and *Global Environmental Politics*. Bringing together exerts in the field, she has co-edited several special issues and a book on the management of transboundary water resources.

Margreet Zwarteveen is a Professor of Water Governance at IHE Delft Institute for Water Education in the Netherlands, with an affiliation with the University of Amsterdam. Originally trained as an engineer, Margreet is now a critical (feminist) social scientist with an interest in water. She has over 30 years of experience in researching how institutions, technologies and markets shape water allocation and regulate water flows, often with highly uneven outcomes. Beyond exposing the structural power relations that help explain water-based inequities, Margreet is interested in noticing and supporting potentially more equitable and sustainable ways of doing and relating to water. Her recent publications include a co-edited (with Jenniver Sehring and Rozemarijn ter Horst) book *Gender Dynamics in Transboundary Water Governance. Feminist Perspectives on Water Conflict and Cooperation* (Routledge, 2022), and a special issue of the Journal *Water Alternatives*on groundwater governance.

PREFACE AND ACKNOWLEDGEMENTS

The drive to compile this book derives from the recognition that water and the way it is managed and used is absolutely fundamental for understanding human development and the societies we inhabit. Across the globe, we are undoubtedly facing a number of water challenges. This is not least due to the effects of climate change, which are reshaping water availability, temporally and spatially, and which will alter the relationship between water and development.

This *Handbook* offers a systematic, wide-ranging and state-of-the-art guide to the diverse links between water and development – in time, space, and across disciplines. The 32 chapters by leading experts are meant to stimulate discussion among researchers and students in a wide range of disciplines in the social and natural sciences.

We have deliberately avoided a narrow or pre-defined framework in order to give the contributors the opportunity to explore all the different dimensions of the relationship between water and development. Since development is an essentially contested concept, we have encouraged the contributors to work with different understandings (and theories) of development. However, the book has been guided by the notion that *all* societies – not simply "poor countries" – are in the process of development. Although a universal approach to development is not a new idea, it has resurfaced in the context of the discussion of global development and the UN Sustainable Development Goals. We believe that this approach is particularly relevant to this book because every society must deal with the management and governance of its water resources.

In developing this *Handbook*, we have tried to uphold a balance between breadth and specialization as well as between geographical and disciplinary representation. We have also searched for gender balance and to involve authors at different levels of their career. While we may not have succeeded in all our ambitions, we have gathered a diverse team of contributors who are at the very forefront of research on water and development.

While Ashok Swain and Joakim Öjendal are responsible for picking up the idea to do a *Handbook* in the first place, the project would not have been completed without Sofie Hellberg, who has been the main driver of the book and she has done a large part of the editorial work. As a team, we are indebted to Sara Löwgren for helping us to assess the chapters during a particularly hectic moment in time, and to Philip Wade who prepared the manuscript before submission. We also want to thank Andrew Mould, Egle Zigaite, Claire Maloney, and Megan Smith at Taylor and Francis/Routledge for their professionalism and assistance in taking the book from idea to finished text. Finally, we want to thank our families for their patience during the years of working with this project.

Sofie Hellberg, Fredrik Söderbaum, Ashok Swain, and Joakim Öjendal

1

INTRODUCTION

The Water–Development Nexus

Sofie Hellberg, Fredrik Söderbaum, Ashok Swain, and Joakim Öjendal

Rationale and Aim

Water is essential for human life and the making of human society. Water and development are thus inextricably linked: development is impossible without water, and sustainable use of water requires development. The relationship between water and development has become ever more urgent as the world moves towards a range of severe water challenges, such as water scarcity in megacities and for food production, backlogs in water and sanitation services, financial gaps in the provision of water infrastructure and growing competition over water resources at all levels. Water is also central for dealing with global climate change, which is bound to amplify all types of societal cleavages and increase the problems involved in managing water for development purposes. In essence, there is a need for much greater awareness and debate about how water and development hang together.

This *Handbook of Water and Development* offers a systematic, wide-ranging and state-of-the-art guide to the diverse links between water and development – in time, space and across disciplines. In the widest sense, it explores the past, present and future role of water for human and societal development. The *Handbook* offers an original and much-needed reference point for researchers, students, practitioners, policymakers and anyone concerned with the role of water in societal development.

This field of study has expanded rapidly during the past two decades, and in our understanding it will continue to grow in urgency in the foreseeable future. A central justification for this *Handbook* is that it fills a void in the existing literature. Currently, there is no other comparable volume that systematically and comprehensively deals with the relationship between water and development. Other handbooks and research monographs in the field approach water from specific standpoints and perspectives, and rarely with a concern for development. For instance, there are several handbooks on water and security, the most important arguably being *Handbook on Water Security*, edited by Pahl-Wostl et al. (2016), and the four-volume mini-series on *Water Security*, edited by Jägerskog et al. (2015). The influential *Water Security: Principles, Perspectives and Practices*, edited by Lankford et al. (2013), could also be mentioned in this context.

The Oxford Handbook on Water Politics and Policy, edited by Ken Conca and Erika Wienthal (2018), is highly relevant for development, but it focuses specifically on politics, policy and water

DOI: 10.4324/9781003095545-1

governance. While a few other publications specifically deal with development and development-related aspects of water, these usually focus on a particular theme or issue area related to development, as illustrated by *Water Resources and Development*, edited by Clive Agnew and Philip Woodhouse (2011), and the *Routledge Handbook of Water and Health*, edited by Jamie Bartram (2015). In other words, the existing literature is rather different from our wide-ranging exploration of all dimensions of the water–development nexus.

In developing this *Handbook*, we have sought to achieve a balance between, on the one hand, breadth and an open-minded approach, and, on the other, more specialized and theoretically guided framings. The *Handbook* is comprehensive in terms of geographical and disciplinary representation, the gender mix of the authors, and the fact that the contributors include scholars at different stages of their academic careers. We cannot claim that we have succeeded fully with our ambitions, given the realities of the hierarchies and the meritocracy within academia. In addition, the pandemic made the process of putting together this volume considerably more difficult to manage. However, at the end of the day, this *Handbook* includes representations from a wide theoretical spectrum and most parts of the globe, gathering some of the most sophisticated and recent research available.

The *Handbook* is organized into four parts. The first part explores the most significant theories and approaches to the relationship between water and development. The second consists of carefully selected in-depth case studies, shedding light on how water utilization and governance are deeply intertwined with historical development paths and economic and societal structures in different contexts. The third part analyses the role of governance in the management of water and development. Finally, the fourth part deals with the most pressing themes and issues pertaining to water and development in the contemporary world, ranging from climate change and water stress to agriculture and migration. Before outlining our general approach to water and development and situating the chapters within the overall structure, however, we need to clarify what we mean by development.

Development

Development is an essentially contested concept, and there is no consensus on the meaning of the term. Development is a moving target, and there is a 'co-existence of strategies and theories with deeply diverging views on what development is and should be' (Bull and Bøås, 2012: 320; see also Hettne, 1995; Hydén, 2014). These divergences are often closely related to contending understandings of other related concepts, such as 'good change', 'progress', 'transformation', 'emancipation', 'growth' and 'justice'.

The varying conceptions of development are not by themselves problematic. After all, development is complex, multi-dimensional and shifting over time (Hettne, 2009). Conceptual and theoretical pluralism is both necessary (to reflect diversity) and productive (to generate creative debate) (Scholte and Söderbaum, 2017). Rather, the problem lies in the tendency among both researchers and practitioners to talk past each other (Bull and Bøås, 2012). The resulting misunderstandings and frictions often tend to result from a weak understanding of the conditions under which different theories and perspectives were developed and how to deal with development in different contexts. In a number of cases, such frictions have resulted in scholars avoiding or trying to do away with development altogether. In contrast, our position is that development as a concept, idea and ambition is crucial. However, to retain development, one has to appreciate the diversity of theories and practices. The development literature is rich, and for this purpose we distinguish between three main theoretical orientations, here termed 'classical', 'post-development' and 'global development'.[1]

'Classical' or 'traditional' perspectives view development research as a new social science discipline that was established after World War II. Classical development research includes theories of modernization, structuralism, dependency and 'another development' (Hettne, 1995). These approaches regard development as a particular set of problems of 'poor countries' in the 'developing world' or 'the South' (previously the 'Third World'). In general, scholars working within these perspectives have little interest in complicating or questioning the concept of development, because the main attention goes to studying the (usually material) hardships of poor countries and poor people. Development in this sense was also closely linked to the nation-state and state-building, expanding, as it were, in an era of decolonization and state-building in the 1960s and 1970s.

Within classical approaches, one can distinguish between those who define development 'as a long-term process of structural societal transformation' and those who define it 'as a short-to-medium term outcome of desirable targets' (Sumner and Tribe, 2008: 11). Long-term views are typically more accentuated in academic circles, while short-to-medium-term views are more prevalent in policy circles and often involve a focus on evaluations and indicators (Sumner and Tribe, 2008: 11–14). Policy-led approaches also have a close connection to international development cooperation, since considerable attention is directed towards finding concrete solutions to poverty and (the lack of) development.

A second category of theoretical perspectives, often characterized as 'post-development', questions the whole concept and project of development (e.g. Escobar, 1995; Rahnema and Bawtree, 1997). This line of thinking originated out of the 'development impasse' that arose in the wake of emerging globalization and was fed by the obvious over-belief in the state as the agent and referent in the development process, and started out as a broad post-colonial critique of development as an instrument of colonial power and the portrayal of the 'Third World' as homogeneous (Schuurman, 1993, 2000). The mainstream conceptualization of development was also fiercely criticized because it failed to recognize the intricacies of local conditions. The critique of the grand idea of development was further deepened by the post-modernist/post-colonialist pulse, according to which development is based on orthodox Eurocentric scientific discourses. Post-development conceptions often link development theory and practices with colonialism and associated paternalism, racism and arrogance. 'Development' is therefore seen as viewing the local as passive, inferior and as the object rather than the subject of development. From a radical post-development perspective, 'development' is an imposed colonialist 'meta-narrative'. Seen from this perspective, 'the way out of development … [is] about unmaking the knowledge-power effects of the development discourse (including established scientific discourses) and carving out space for a variety and a mixture of localised and pluralistic histories, cultures, identities and knowledge systems' (Buch-Hansen and Lauridsen, 2012: 295).

The third perspective, 'global development', reassesses development in the light of globalization. It emphasizes the need to address global challenges and provide global and regional public goods in the context of issues that range from climate change and financial stability to peace and global health. The 'global' perspective straddles neat distinctions between rich and poor countries and also rejects the state-centrism, national developmentalism and methodological nationalism inherent in much classical development thinking (Buch-Hansen and Lauridsen, 2012; Hettne, 2009: 293). According to this perspective, development studies should not be restricted to 'poor countries', regardless of how these are defined: all countries are seen as being in the process of – and in need of – 'development'. The broadened scope of such an approach has two important implications. First, it makes it necessary to transcend the binary between developed and developing countries. Second, it implies the rejection of both methodological nationalism and the view that development was 'invented' after World War II.

The focus on *sustainable* development has become increasingly important over time and in relation to the changing views on whether to focus only on poor societies or on all societies. The World Commission on Environment and Development was pathbreaking in establishing and promoting the concept of sustainable development already in the late 1980s. However, most of the early advocates of sustainable development maintained the dichotomy between developed/poor and developing/rich countries because they often saw poverty as one of the main reasons for environmental degradation (WCED, 1987). The result was that sustainable development was conveniently married to a classical development agenda. However, more recently, *sustainability* has been integrated within and become central to more or less all development discourses and perspectives. While we have not done away with power relations between those described as developed and developing countries, the binary categorization has been both smoothened and challenged. It is, for instance, clearly stated in the UN's 2030 Agenda for Sustainable Development that development is a *universal* concern: *all* societies (or countries) are undergoing development. Undeniably, climate change turns sustainability into a global concern with deep consequences for development efforts in every corner of the world.

In line with the global development approach, we treat development as a universal and 'global' phenomenon in order to escape the problematic binary between developed and developing countries as well as the obsolete view that development is only linked to the nation-state. We follow the post-development approach in the sense that we are wary of the simplification of the concept of development and its 'subjects', and because it is important to be sensitive to the power it un/intentionally projects. Indeed, awareness of cultural diversity in contemporary globalization has encouraged the spread of post-development critiques, a perspective that certainly holds relevance and is advocated by some of the contributors to this book. However, the foundations of the classical approach are also of relevance given its focus on material conditions and societal progress. In this context, it is relevant to mention that some classical approaches also emphasized that the concept of development was of universal relevance (e.g. liberal development economics) or was global in the sense that it depended on world-systemic dynamics (e.g. structuralism and dependency theory). However, a genuinely global approach is broader than this and emphasizes the need to address global challenges and provide global and regional public goods in the context of issues that range from climate change and financial stability to peace and global health, which are all included in the Sustainable Development Goals (SDGs) and Agenda 2030. Hence, all countries are seen as being in the process of – and in need of – 'development'.

Given the breadth of this volume, different contributors may draw on different understandings of development. For example, the chapter by McGregor, Sritharan and Whitaker on Indigenous perspectives provides anti-colonial critiques related to questions of water and development, while Mehta's chapter on water for human development has its own take on the concept of development. However, while it is possible to enact competing approaches to development, this *Handbook* is based on the idea that the different approaches at least in some respects complement, refine and deepen each other. Their interaction and mingling is, we believe, stimulating and creative. Having outlined the main approaches to development, we now turn to the most important approaches that focus on water–society interactions.

Water and Society

Water is exceptional in that it is a prerequisite for all life. It is also completely integrated into our daily lives and thus 'so normal that it retreats into the background of awareness' (Sofoulis, 2005: 448). In terms of its importance, water is thus both hyper-visible and invisible, and, curiously, both life-generating and neglectable. Water is at bottom often described in terms of its chemical composition and its necessity for all life-forms. It is an essential component for food production, as well as our health, and a crucial component in many economic activities; for example, more than 40% of employment worldwide is 'heavily water-dependent' (UNESCO, 2016). Water also cuts across many other issues, global in character, that include (but are not limited to) natural disasters, such as floods and droughts, conflicts, migration and (gender) inequality. It is, furthermore, connected to questions of energy and inextricably linked to climate change, since the latter's impact will largely be felt through water-related aspects and events.

However, in interaction with humans and human life, water is a contested concept. Water is 'what we make of it' (Linton, 2010: 3). What water 'is' and 'does' is thus not as self-evident as it might first seem (Linton, 2010; Hellberg, 2018; see also Franco-Torres, 2021). The naming of water as *the* scarce resource, in contrast to seeing it as a living entity (Wilson and Inkster, 2018), can illustrate the ontological politics of defining water and the possibility of a range of different viewpoints on the use, governance and ownership of water.

Contemporary scholarship on the social aspects of water is rich in addressing the questions of what water 'is' and 'does' and what it has meant for the construction of human societies – in other words, development. In this section, we identify a number of different approaches, in both the social and the natural sciences, that focus on the connections between water and development. These approaches represent different research strands and bodies of literature, ranging from hydrosocial research, hydro-sociology and socio-hydrology to hydropolitics, water security and policy-oriented approaches.

In the classical development discourse, water was the quintessential resource (together with land and labour) in the rural areas where the bulk of the population lived, sustaining themselves from agriculture. *Rural development* was a term under which entire research programmes were pursued and university departments labelled (see Molden et al., 2001), primarily focused on the rural poor and improving material living conditions. A core conception of rural development was often centred on water and irrigation, and how distribution and access should be governed (Uphoff, 1996). Under modernity, social change and the imperative of increasing output, the organization, sharing and governance of water resources became crucial. With the Green Revolution – and increasing global dependence on it – reliable water access became even more important (Öjendal and Hellberg, this volume). During the 1980s and 1990s, the discourse on reliable water access was one of the dominant discourses in the development literature (Chambers, 1983; Öjendal, 2000).

There are several other bodies of literature that are key in addressing water–society relations and thereby provide insights into water–development linkages. *Hydrosocial research* is one influential approach that is mainly based in human geography. It focuses broadly on issues of power in 'human–water or social–nature systems' (Wesselink et al., 2017: 6). The so-called hydrosocial cycle is a central concept that attends to the social and political aspects of water and questions the understanding of water as a mere resource, a natural fact (Linton, 2010). Hydrosocial research foregrounds questions of inequalities and injustices in water use and access, especially as a result of neoliberal forms of governing, as well as how the interactions between water users and technologies are constitutive of identities and subjectivities (see Hellberg, this volume; Budds and Loftus, this volume).

There are a number of other fields that have established themselves as addressing the connections and relations between humans and water. Orienting more towards the hydrological terrain, and thereby natural science, society–water interactions have been studied through the lens of *hydro-sociology*, originally developed by Malin Falkenmark (1979). The focus of Falkenmark's pioneering research was, among other things, to establish links between human activities and the water cycle and to make predictions regarding the future of water scarcity (see Falkenmark, 1997). Closely connected to this field – regardless of whether one considers the two to be separate 'sciences' or not (see Sivukumar, 2012) – is *socio-hydrology*, which also has its base in hydrology and addresses the relations between water and human societies. The emphasis here lies on the multiple 'feedbacks between humans and the water environment' that, in turn, make it possible to account for 'puzzles, paradoxes and unintended consequences' in water management (Pande and Sivapalan, 2016: 1; see also Di Baldassarre et al., 2015).

While hydrosocial research has its main base in human geography, and hydro-sociology and socio-hydrology in hydrology, the vast literature on *hydropolitics* draws largely on concepts and theories from political science. Hydropolitics is concerned with the questions of 'who gets what [water], when, where and how' (Turton, 2002: 16). It focuses on issues of governance, power relations between stakeholders, and the ways in which processes of conflict and cooperation matter in the context of water, often with a predominant focus on the state as a central actor. While water has always been political, this discourse came to openly acknowledge that 'high politics' in transboundary settings had a direct bearing on subnational water access and allocation. Hence, while the water wars thesis (Starr, 1991) was thoroughly rejected (Wolf, 1999), transboundary issues nevertheless profoundly affected the possibility of pursuing efficient development at the regional level (see Waterbury, 1979). For instance, the evolution of the highly charged relationship between Israel and Palestine determined how much water would be available for smallholders on the West Bank (Wolf, 1995), and the relationship between Ethiopia and Egypt determined how water could be accessed for millions of small farmers in the Nile valley (Waterbury, 1979). The list is endless, but in this field it was soon clear that hydropolitics had a major significance for development prospects (Ohlsson, 1995; Sandström et al., 2016).

The focus on conflict in a water context invokes the concept of *water security*, which is another central concept within debates on the social aspects of water. However, academics, policy institutions and organizations employ multiple definitions of this concept. While water security can be understood in a narrow sense as addressing questions of intrastate conflict and military security, it is often understood in a broader sense. It is in this broad application that the concept has become one of the most influential in global water policy discussions during the past decade. Cook and Bakker (2012) identified four main themes that dominate contemporary approaches to water security: water availability, human vulnerability to hazards, human needs and sustainability. Contemporary approaches to water security embrace a broad understanding and thereby include most, if not all, of these themes in their definitions. Effectively, water security in this interpretation has had little reference to military security, the so-called water wars and the like (see above).

Integrated water resources management (IWRM) has received considerable attention in policy-oriented research. Its policy orientation implies that IWRM is not so much a separate approach in itself, but rather an integrated part of the other above-mentioned fields. IWRM is commonly defined as 'the co-ordinated development and management of water, land and related resources, in order to maximize the resultant economic and social welfare in an equitable manner without compromising the sustainability of vital ecosystems' (GWP, 2000: 22). The approach has been promoted at the global policy level from the so-called Dublin Principles in 1992 up until the more recent UN Sustainable Development Goal 6. In spite of its hegemonic position in policy circles (or possibly because of it), it has not escaped critique (Jeffrey and Gearey, 2006).

In academic writings, IWRM has been labelled a 'nirvana concept' (Molle, 2008) that can allow for 'business as usual' as powerful actors appropriate the IWRM language to promote their own agendas and attract financial resources. In such an understanding, real effects in terms of a transition to a more sustainable water management are lacking. There are also vast challenges involved in integrating the factors and knowledge needed to support a truly integrated water management (Cook and Spray, 2012; Savenije and Van der Zaag, 2008). Critique has been voiced both in relation to the dominance of engineers and physical science within the approach (see, e.g. McDonnell, 2008) *and* about the lack of adherence to scientific principles within the framework (see Cook and Spray, 2012; Jeffrey and Gearey, 2006; Petit and Baron, 2009). Here we can identify a tension that lies at the heart of water management and governance in general, and at the heart of many of the chapters in this *Handbook*, namely, the relationship between water science/scholarship and the politics of water (see Conca, 2005).

Moreover, what we learn from the above-mentioned literature, which the various chapters of this *Handbook* will also illustrate, is that how we define and understand water and its connection to human societies is crucial for how we conceive of appropriate ways of *managing and using* water. This is true regardless of whether we understand water to be in abundance, readily available for human consumption and for building our societies, or as a scarce resource under threat from overuse, pollution and the effects of climate change, whose availability can and should be amended through technological innovation.

As will be obvious throughout this *Handbook*, two circumstances lie at its core. *First*, as we have seen above, water is an increasingly scarce resource that is needed in all walks of life, and one that is becoming increasingly critical both for human and for planetary well-being. *Second*, it is our conviction that *how* water is being brought in to foster human development and the progress of societies is critical. How water has, can and should be used for developmental purposes is what we turn to below, outlining the structure of this *Handbook*.

Structure of the Book

Based on the above, and as an organizing tool, we explore the water–development nexus through a division into four parts, namely, in terms of *theories*, *case studies*, *governance mechanisms*, and a wide-ranging number of *themes and issues*.

Theories and approaches to water and development

The links that we can identify between water and development depend on how we conceptualize both development and water. The first part of this *Handbook* deals with the most significant theories and approaches to water and development. The first chapter in this part, by Joakim Öjendal and Sofie Hellberg, investigates the role of water–society connections in modernization theory and for the creation of modern societies. The second chapter, by Larry Swatuk, outlines liberal institutionalist approaches and argues for the need for deeper reflection of the epistemological and ontological foundations of these perspectives. Lyla Mehta's chapter on water and human development takes yet another perspective on development and explores both the entitlements approach and the capabilities approach to the water domain, which emphasize well-being, the multifaceted nature of water and water's role in the realization of freedoms, rather than merely access to water. Taken together, as argued by Mehta, the two approaches strengthen the case for the human right to water. Sofie Hellberg's chapter on critical and post-structural approaches outlines how we can understand the allocation, distribution and management of water as issues of power and rule, and as related to the creation of particular societies and subjects. While insights from critical and post-structural approaches are in many ways shared in feminist perspectives, Margreet Zwarteveen's contribution, which focuses on feminist scholarship, deepens perspectives on the politics of water by discussing the division between nature and society and the importance of being aware of the situatedness of all water knowledges, including scientific ones. The chapter by Deborah McGregor, Mahisha Sritharan and Steve Whitaker on Indigenous perspectives in turn presents anti-colonial critiques of and insights on the 'water-as-resource/commodity' perspective currently held by global, regional and local governance authorities, and simultaneously offers perspectives on what Indigenous perspectives can contribute to attempts to redress the global water crisis.

Case studies on water and development

The second part of the book is made up of carefully selected in-depth case studies, which explore how water utilization and governance are deeply intertwined with historical development paths and the political, economic and societal structures of different contexts. Although there is no ambition to be representative of each region, we have selected some of the most intriguing cases in different parts of the world. In East Asia, Joakim Öjendal explores the case of Cambodia, while in Southern Africa Richard Meissner, Nikki Funke, Stephen Rule, Karen Nortje and Inga Jacobs-Mata investigate South Africa. In South America, Patricia Urteaga Crovetto analyses the case of Peru, and, in the Middle East, Neda Zawahri provides insights into Jordan. Finally, Erik Mostert provides a European example through the case of the Netherlands.

All these case studies demonstrate in various ways that water has been (and remains) crucial for the formation of our societies in its simultaneously hyper-visible and invisible characteristics. The cases show that the water–development nexus needs to be understood in historical and contextual perspective and that it has always shaped societies. Even if our cases are context-bound, water is not simply viewed as a 'national' resource. The basic idea is rather that history and context matter, and that the nation-state is merely a point of departure for context-sensitive explorations showing how water and development also depend on the larger regional contexts and their associated resource regimes.

Governing water and development

Since water is essential for human life, it must be carefully managed and governed. The third part puts the spotlight on the role of governance and politics in the use of water for development. Again, since water is multi-dimensional, there are many meanings of 'governance' in this context, and we emphasize the most prominent ideas and governance mechanisms and approaches in the field.

The *Handbook* covers key approaches and themes in water governance. The first chapter in this part, by Klaas Schwarz and Mireia Tutusaus, focuses on the coexistence of the two dominant discourses on water services, namely, those around the commercialization of water provision and the human right to water. The focus on the application of economic and market principles to water management and governance is also a theme in the contribution by Jessica Budds and Alex Loftus. This chapter provides an overview of the lessons learned from the study of the neoliberalization of water, in particular the tensions between commodification and enabling access, and how water has become a site of capital accumulation. In the next chapter, Peter Gleick provides a historical and judicial perspective on the development of the idea of the human right to water, arguing that the application of this idea is still in an early phase and that there is limited agreement as to how it should be applied and enforced.

The chapter by Kurt Mörck Jensen and Jens Christian Refsgaard explores five decades of water management approaches and their implementation challenges in the context of both developing and developed countries. The chapter by Jeroen Warner and Richard Meissner zooms in on one of the cornerstones of contemporary water governance approaches, namely, the adoption of the concept of participation in terms of the promise of more inclusive and informed decision-making in the water sector. In the chapter on transboundary waters, Jeroen Warner finds that Wolf's (1995) rejection of the water wars thesis still stands, but that we should be wary of the different forms of cooperation that exist in transboundary settings. The last chapter in this part of the book, by Anik Bhaduri, Alexandre Teixeira and Aditya Kaushik, discusses the role of water for Agenda 2030 and summarizes progress and actions needed for the implementation of the specific water and sanitation goal (SDG 6).

Themes and issues

The fourth part of the book focuses on a wide range of current and emerging themes and issues pertaining to water and development in the contemporary world. These are either well-known sectors and issues that do not have (and may never have) a final solution or emerging issues of some significance whose full ramifications we do not yet know. The issues and themes have been selected on the basis that they involve a certain degree of urgency and significance and contribute to, or are related to, societal change and, as we frame it here, development.

The first chapter in this part, by Jaime Hoogesteger, Diana Suhardiman, Gert Jan, Veldwisch, Jean Pablo Hidalgo-Bastidas and Rutgerd Boelens, outlines the importance of irrigated agriculture for food production, as well as its socio-environmental challenges. One of these challenges is the effects of irrigation on groundwater, which is an issue that is thoroughly examined by Susann Baez Ullberg and Henrik Josefsson. Their chapter explores institutional, legal, political, social and cultural aspects of groundwater research and suggests lines for future research in order to broaden understanding of the relation between groundwater and development. Continuing the theme of water resource availability, the next chapter, by Zafar Adeel, provides an overview of various notions of water stress and scarcity. As such, it both problematizes these notions and provides insights into how they intersect with the policy domain. One aspect of water scarcity is that it is often assumed to have a linear and almost causal relationship to migration. In the chapter on migration, Anders Jägerskog and Ashok Swain problematize the relationship between water, migration and development, and propose ideas for policy measures based on their analysis.

The next chapter, by Deliang Chen and Hui-Wen Lai, focuses on the links between water, sustainable development and climate change. It introduces key concepts and methods for the projection of the future climate as well as presenting the state-of-the-art of emission and climate change scenarios according to the Intergovernmental Panel on Climate Change (IPCC). The subsequent contribution by Elisa Savelli addresses a phenomenon that is projected to increase owing to climate change, namely, drought. Savelli engages fruitfully with both natural and social sciences in order to account for the socio-environmental complexity of droughts. In the same vein, the next chapter on the water–energy nexus, by Aiko Endo, outlines challenges of applying interdisciplinary and transdisciplinary approaches to the complexity and uncertainty of the linkages between water and energy.

In the subsequent chapter on inequality, Maria Rusca draws on critical water studies to examine the production and politics of water inequalities in relation to both overarching governance structures and everyday practices. In terms of access to, control of and roles related to water, one of the key dimensions of inequality is gender. Gaylean Davis, Evelyn Arriagada and Leila Harris adopt a critical perspective to unpack the discourses and practices at the intersection of gender, water and development.

In the chapter on urban water, Susan van de Meene explores both conventional approaches to urban water management and possible future scenarios in which small-scale solutions and alternative governance approaches can represent a possibility for cities in the Global South to 'leapfrog' and avoid the negative impacts of conventional water management approaches. In their chapter on health, Jo Geere, Paul Hunter and Bruce Lankford take a broad perspective on water, sanitation and hygiene (WaSH) services, including the multiple impacts on health following unsafe access to such services. Nelson Ekane's chapter deepens the discussion of one of the aspects of WaSH, namely, sanitation, focusing on the obstacles to the achievement of the SDG targets on sanitation and the possibilities for overcoming them. The last chapter of the *Handbook*, on digital water – by Karen Bakker, Rosemary Knight, Raymond Ng, Alan Mackworth and Max Ritts – explores digital solutions in terms of their potential to address many of the water challenges that have been discussed in the previous chapters, namely, as a way of increasing sustainability in the water sector, mitigating risk and health hazards, addressing water insecurity, and providing new avenues for involving citizens in water issues. The chapter also discusses how such solutions raise a number of ethical issues in relation to data management and power dynamics between actors, including that of uneven development globally.

The Future of Water and Development

The inventory of the water–development interactions in this *Handbook* reveals numerous water challenges for the future of society, human and non-human life, and the well-being of ecosystems. In this last section of this introduction, we paint a broad picture of what we see as the main challenges to come and how we see the role of this *Handbook* in relation to those challenges. Our list of challenges is by no means exhaustive, but, as we interpret the insightful contributions to this volume, these are among the most pressing and urgent ones.

There is little doubt that *climate change* will have a severe impact on water and development in the future. It has already been affecting precipitation patterns and will continue to reshape water availability both temporally and spatially. Climate change has such profound effects on water that it is often referred to primarily as a 'water crisis' (UN-Water, 2023). We can expect to see an increase in water-related events, such as floods and droughts, as well as rising sea levels, all of which will have significant social effects on livelihoods and food production, as well as migration and urbanization patterns and dynamics. Although the way we govern, manage and use water might be seen as part of the solution in terms of how we address climate change, the scaling up of efforts to mitigate carbon emissions will also involve challenges for water sources. As they are currently being implemented, these efforts mainly involve a shift from fossil fuels to the use of (clean) electricity, which requires massive amounts of metals for batteries. This, in turn, poses threats to water sources in both old and new mining areas. As part of the move away from fossil fuels, increased pressure on damming and developing hydropower will also have severe effects on water access, distribution and ecosystems.

The availability of freshwater resources is challenged not only by climate change but also by processes of industrialization and irrigated agriculture, as well as population dynamics, such as migration, urbanization and an expanding middle class. *Water stress and scarcity* are projected to increase alarmingly. We can expect to see more countries, regions and cities falling into the category of being 'water-stressed'. Several large cities are running out of water, Cape Town being one of the most notable examples in the past years, which might be followed by cities, such as São Paulo, Beijing and many others. However, and this is also true for the projections of the effects of climate change – the forecasts of water-scarcity dynamics are highly uncertain, making *uncertainty* in water management a challenge of its own.

As scarcity increases, the demand will increase for *technological solutions for freshwater availability*, such as wastewater and desalination plants and huge water-transfer schemes. Although more efficient now than in the past, wastewater and desalination plants use massive amounts of energy, and some of them – like one of the world's largest, Ras Al Khair Desalination Plant in Saudi-Arabia, which produces both potable water and electricity – run on fossil fuels, namely, natural gas (Power-Technology, 2023). These dynamics further complicate the relations between water and climate change and point to the unavoidable link between water and energy. Moreover, although technical fixes may solve local drinking water challenges and enable small-scale production of high-value products, they cannot meet the massive water demands of agriculture and food production.

Relatedly, all the dynamics mentioned above – effects of climate change, water scarcity and pollution of water sources, along with the rising cost of water because of expensive investment in new technology – have the potential to increase one of the most explosive challenges of our time, namely, *inequality*. The ones who are likely to be most affected because of changing precipitation patterns, droughts and increasing water scarcity are the poorest, including smallholders and informal dwellers in cities where water services are lacking and prices of water tend to be high. Moreover, indigenous people will face increasing demands on their land and water, as well as an increase in polluted water sources due to rising demand for metals.

These pressures on water resources, together with the vast inequalities in access and usage, make it abundantly clear that maintenance of the economic growth paradigm – which continues to increase the demand for water for the same time as it has been estimated that 1.6 billion people will lack safely managed drinking water by 2030 (United Nations, 2022: 13) – is simply unsustainable, both ecologically and socially, but also in the long run economically. As many water scholars have emphasized during the past decades, our relationship to water needs to change. Or, put in the words of this *Handbook*: current relationships and interactions between water and development need to be altered and a new water paradigm construed.

Taken together, the challenges mentioned above demonstrate the need for insightful and critical water scholarship that can account for and problematize dynamics, not only in the hydrological cycle, but also in relation to the social aspects of these challenges in all their complexities. Given the many water challenges we face, the ambition of this *Handbook* has been to build on, and add to, water and development scholarship that addresses the role of water in society in ways that are both comprehensive and pertinent, as well as innovative and constructive.

Note

1 This description draws freely on Scholte and Söderbaum (2017).

References

Agnew, C., and Woodhouse, P. (2011). *Water Resources and Development*. London, UK: Routledge.

Bartram, J. (ed.) (2015). *Routledge Handbook of Water and Health*. London, UK: Routledge.

Buch-Hansen, M., and Lauridsen, L. S. (2012). "The past, present and future of development studies." *Forum for Development Studies*, 39(3), 293–300.

Bull, B., and Bøås, M. (2012). "Between ruptures and continuity: Modernisation, dependency and the evolution of development theory." *Forum for Development Studies*, 39(3), 319–336.

Chambers, R. (1983). *Rural Development: Putting the Last First*. Harlow, UK: Longman.

Conca, K. (2005). *Governing Water: Contentious Transnational Politics and Global Institution Building*. Cambridge, MA: MIT Press.

Conca, K., and Wienthal, E. (eds.) (2018). *The Oxford Handbook on Water Politics and Policy*. Oxford, UK: Oxford University Press.

Cook, B. R., and Spray, C. J. (2012). "Ecosystem services and integrated water resource management: Different paths to the same end?" *Journal of Environmental Management*, 109, 93–100.

Cook, C., and Bakker, K. (2012). "Water security: Debating an emerging paradigm." *Global Environmental Change*, 22(1), 94–102.
Di Baldassarre, G., Viglione, A., Carr, G., Kuil, L., Yan, K., Brandimarte, L., and Bloeschl, G. (2015). "Debates – Perspectives on socio-hydrology: Capturing feedbacks between physical and social processes." *Water Resources*, 51, 4770–4781.
Escobar, A. (1995). *Encountering Development: The Making and Unmaking of the Third World*. Princeton, NJ: Princeton University Press.
Falkenmark, M. (1979). "Main problems of water use and transfer of technology." *GeoJournal*, 3(5), 435–443.
Falkenmark, M. (1997). "Society's interaction with the water cycle: A conceptual framework for a more holistic approach." *Hydrological Sciences Journal*, 42(4), 451–466.
Franco-Torres, M. (2021). "The path to the new urban water paradigm: From modernity to metamodernism." *Water Alternatives*, 14(3), 820–840.
Global Water Partnership (GWP). (2000). *TAC Background Paper No. 4: Integrated Water Resources Management*. Stockholm: Global Water Partnership.
Hellberg, S. (2018). *The Biopolitics of Water: Governance Scarcity and Populations*. London, UK: Routledge.
Hettne, B. (1995). *Development Theory and the Three Worlds*. Harlow, UK: Longman.
Hettne, B. (2009). *Thinking About Development*. London, UK: Zed Books.
Hydén, G. (2014). "The ever-changing nature of development research: The newest challenge." *Forum for Development Studies*, 41(3), 501–511.
Jägerskog, A., Swain, A., and Öjendal, J. (eds.) (2015). *Water Security*. Vols I–IV. London, UK: Sage.
Jeffrey, P., and Gearey, M. (2006). "Integrated water resources management: Lost on the road from ambition to realisation?" *Water Science and Technology*, 53(1), 1–8.
Lankford, B., Bakker, K., Zeitoun, M., and Conway, D. (eds.) (2013). *Water Security: Principles, Perspectives and Practices*. London, UK: Routledge.
Linton, J. (2010). *What Is Water? The History of a Modern Abstraction*. Vancouver, BC: UBS Press.
McDonnell, R. A. (2008). "Challenges for integrated water resources management: How do we provide the knowledge to support truly integrated thinking?" *International Journal of Water Resources Development*, 24, 131–143.
Molden, D., Amarasinghe, U., and Hussain, I. (2001). "Water for rural development: Background paper on water for rural development prepared for the World Bank". *IWMI Working Paper 32*. Colombo: IWMI.
Molle, F. (2008). "Nirvana concepts, narratives and policy models: Insights from the water sector." *Water Alternatives*, 1(1), 131–156.
Ohlsson, L. (ed.) (1995). *Hydropolitics: Conflicts over Water as a Development Constraint*. London, UK: Zed Books.
Öjendal, J. (2000). *Sharing the Good: Modes of Managing Water Resources in the Lower Mekong River Basin*. Göteborg: Padrigu Papers.
Pahl-Wostl, C., Bhaduri, A., and Gupta, J. (eds.) (2016). *Handbook on Water Security*. Cheltenham, UK: Edward Elgar.
Pande, S., and Sivapalan, M. (2016). "Progress in socio-hydrology: A meta-analysis of challenges and opportunities." *Wires Water*, 4(4), 1–18.
Petit, O., and Baron, C. (2009). "Integrated water resources management: From general principles to its implementation by the state. The case of Burkina Faso." *Natural Resources Forum*, 33, 49–59.
Power-Technology. (2023). "Ras Al-Khair IWPP, Saudi Arabia." [Viewed 1 February 2023]. Available from: https://www.power-technology.com/marketdata/ras-al-khair-iwpp-saudi-arabia/
Rahnema, M., and Bawtree, V. (eds.) (1997). *The Post-Development Reader*. London, UK: Zed Books.
Sandström, E., Jägerskog, A., and Oestigaard, T. (eds.) (2016). *Land and Hydropolitics in the Nile River Basin: Challenges and New Investments*. London, UK: Routledge.
Savenije, H. H. G., and Van der Zaag, P. (2008). "Integrated water resources management: Concepts and issues." *Physics and Chemistry of the Earth*, 33, 290–297.
Scholte, J. A., and Söderbaum, F. (2017). "A changing global development agenda?" *Forum for Development Studies*, 44(1), 1–12.
Schuurman, F. J. (ed.) (1993). *Beyond the Impasse: New Directions in Development Theory*. London, UK: Zed Books.
Schuurman, F. J. (2000). "Paradigms lost, paradigms regained? Development studies in the twenty-first century." *Third World Quarterly*, 21(1), 7–20.

Sivukumar, B. (2012). “Socio-hydrology: Not a new science, but a recycled and re-worded hydrosociology.” *Hydrological Processes*, 26(24), 3788–3790.
Sofoulis, Z. (2005). “Big water, everyday water: A sociotechnical perspective.” *Continuum: Journal of Media and Cultural Studies*, 19(4), 445–463.
Starr, J. R. (1991). “Water wars.” *Foreign Policy*, 82, 17–36.
Sumner, A., and Tribe, M. (eds.) (2008). *International Development Studies: Theories and Methods in Research and Practice*. London, UK: Sage.
Turton, A. R. (2002). “Hydropolitics: The concept and its limitations. In: A. Turton and R. Henwood (eds.). *Hydropolitics in the Developing World: A Southern African Perspective*. Pretoria: African Water Research Unit, pp. 13–22.
UNESCO. (2016). *World Water Development Report 2016: Water and Jobs* [Viewed 31 January 2023]. Available from: https://unesdoc.unesco.org/ark:/48223/pf0000243938
United Nations. (2022). *The Sustainable Development Goals Report 2022* [Viewed 31 January 2023]. Available from: https://unstats.un.org/sdgs/report/2022/The-Sustainable-Development-Goals-Report-2022.pdf
UN-Water. (2023). *Water and Climate Change* [Viewed 31 January 2023]. Available from: https://www.unwater.org/water-facts/water-and-climate-change
Uphoff, N. (1996). *Learning from Gal Oya: Possibilities for Participatory Development and Post-Newtonian Social Science*. Rugby: Practical Action Publishing.
Waterbury, J. (1979). *Hydropolitics of the Nile Valley*. Syracuse, NY: Syracuse University Press.
Wesselink, A., Kooy, M., and Warner, J. (2017). “Socio-hydrology and hydrosocial analysis: Toward dialogues across disciplines.” *WIREs Water*, 4(2), e1196.
Wilson, N. J., and Inkster, J. (2018). “Respecting water: Indigenous water, governance, ontologies and the politics of kinship on the ground.” *Environment and Planning E: Nature and Space*, 1(4), 516–538.
Wolf, A. T. (1995). *Hydropolitics Along the Jordan River: Scarce Water and Its Impact on the Arab–Israeli Conflict*. Tokyo: United Nations University Press.
Wolf, A. T. (1999). “‘Water wars’ and water reality: Conflict and cooperation along international waterways.” In: S. Lonergan (ed.), *Environmental Change, Adaptation, and Human Security*. Dordrecht: Kluwer Academic, pp. 251–265.
World Commission on Environment and Development (WCED). (1987). *Our Common Future*. Oxford, UK: Oxford University Press.

PART I

Theories and Approaches to Water and Development

2

WATER AS A TOOL FOR MODERNITY

Joakim Öjendal and Sofie Hellberg

Introduction

Ever since humans turned agriculturalists in the plains between the Euphrates and Tigris rivers (and elsewhere), water has been a central resource to master; it has been the source of food, development, growth and political power (Caponera, 1992: 1). This chapter sets out to trace the role of water in the building of societies in general, and of the modern project in particular. It is critical to the sustainability of modernity and concludes by reflecting on the imperative of a new water paradigm transgressing classic development ideas on water.

Historical Background – Water as the Foundation

When agriculture increasingly started to out-compete cultures based on herding, hunting and gathering, settlements were established, societies were growing, and proto-states emerged. Most famously some of the earliest state formations have been described by the German historian Karl Wittfogel (1957) who coined them "hydraulic civilizations." In this grand theory, he attributed the rise of the earliest state formations to the elites' control of water in a centralized state. The "dynamic core" of the societies in question was engaged in the construction and management of large-scale irrigation and "[…]there can be no doubt that in all these cases the resulting regime was decisively shaped by the leadership and social control required by hydraulic agriculture" (Ibid.: 27). Sophisticated water management was the origin and a necessary condition for early societal development, he argued.

Although the "hydraulic civilization" hypothesis may have been over-stating the hegemonic role of water in early state formations, the significance of access, control and distribution of water were central for the creation and durability of these states. There are no doubts that the earliest and most sophisticated early civilizations were born in valleys with ample access to water and/or in hydraulic regimes where irrigation were sophisticated and extensive (Caponera, 1992). This is striking from some authoritative historical documents, such as the Hammurabi code, the Hindi legal system and the ancient Chinese water regulations where water, and the order under which it was used, were dealt with extensively (Ibid.; cf. Wittfogel, 1957). While water is the cradle of many civilizations, it is also the source for social and political degeneration would the hydraulic regime falter (Caponera, 1978).

DOI: 10.4324/9781003095545-3

Judging from the above, a few crucial points emerge. *Firstly*, since ancient times, water has been a critical resource and the handling of it has been determining whether a society would bloom or falter. *Secondly*, water was early on the subject to grand schemes with the ability to amass power with those controlling the water systems, central for the creation of bureaucracies and early states. Water carries political significance. *Thirdly*, the management, techniques, and resources various regimes possessed, determined how efficiently water could be used and how powerful the regime could be. To fail, on the other hand, could be the beginning of the end for great civilizations and empires alike. In a generic sense, these still apply.

Water has continued throughout history to be a central resource for human development, but it was with 20th century "high modernity" (cf. Giddens, 1990) and its technological progress that water (again) turned into a commodity of huge importance, subject to major interventions and a central part of the enforced change and societal development. This era was characterized by a conviction of scientific and technological progress, relying on technicians, engineers and experts to "conquer," "tame" and "order" nature. It was through major interventions in cultural, historical and physical structures that progress could be achieved. Moreover, this mindset was emerged in capitalist (although often state-led) and industrial development. "Featuring historically as the premier development project, delivering electricity, water and the spectacle of modernity, dams were frequently central to state-building missions" (Dye, 2022). This took place irrespective of regime and ideology, irrespective of whether to uphold colonialism or to support independence and nation-building, and irrespective of rich or poor countries.

It was, however, not just technological improvements that drove water to centre stage for development interventions. "Modernity" has metaphorically been understood as an unstoppable machine, categorizing and ordering, counting and measuring, objectifying humans and resources, and controlling and taming nature. This is pursued through dam-building, canal-works and major drainage projects, but also through more spectacular interventions, such as the reversing of rivers, making the desert bloom, or towing icebergs to the shores of Australia and the Middle East. At the bottom, as Anthony Giddens tells us (1991), modernity – with its scientific progress and its technical abilities – shaped a mindset and an identity that (ironically) went beyond simple development rationality, often embracing megalomaniac ideas.

James Scott (1998) has seen these inherent features of modernity as a result of the state formation and the intense frenzy of states to harness nature, pursue big schemes and push modernist development at all costs (cf. Li, 2005; Menga & Swyngedouw, 2018). Development (in its classic development paradigm see Introduction this volume) has often been pursued through the rationale of "seeing like a state," in a technical and industrial manner. No natural resource is as prone to be caught up in this way of thinking and ordering of nature as water is. Thus, technological progress, intensive state-building under competition, the emerging development imperative, and water's ability to be harnessed and put to use for development purposes on a large scale, provided the basis for a perfect storm in the 20th century. Water became a resource through which development and state-building should be achieved, often in a highly visionary and romanticized manner. Dams figured on the postal stamps of young states alongside satellites and futuristic cities.

Below we will observe three interpretations of water and development in the 20th century, all carrying the key features of modernity. Each one being major water-driven ideas which have each come to alter human development during the 20th century and onwards. Notably, they are based in, and reinforced by, the classic development paradigm (see Introduction, this volume). The *first* is the birth, export and regime of major infrastructure works for water harnessing, in particular the prevalence of mega-dams for energy, irrigation and flood prevention. *Secondly*, we will assess the "green revolution" and its reliance on irrigated (and industrialized) agriculture, with remarkable results from the 1950s and onwards. *Thirdly*, we will describe how water was/is the foundation for the functioning of major cities through massive and highly sophisticated water and sanitation systems, and which significance this still has. Each of these narratives is at the very core of the 20th-century development debate and practice (and especially for the second half of the century), illuminating the role and the overwhelming importance of water in the development process under a certain form of modernity. We will end the chapter by discussing how this form of modernity has been challenged at the beginning of the 21st century, which in contrast is marked by a paradigm that emphasises sustainability generally and demand management specifically, but also how its core assumptions still continue to steer development in the water sector.

The Evolution of Mega Dams and Its Demise (and Its Return)

A jolly good dam here! A pity there is no water.

(British Colonial Officer, Rhodesia)

Few interventions have triggered the modernist visionaries as much as the mega-dams and their potential effects. For instance, Jawaharlal Nehru said, famously, in his vision for independent India that dams are "the temples of modern India" and the US-led mission to the Mekong in 1958 said that damming the Mekong River "[...] will contribute more towards improving human welfare in this area than any other undertaking" (UN, 1958 in Öjendal, 2000). In 1935, the 221-metre-tall Hoover Dam on the Colorado River was completed, opening a new era of mega-dam constructions with 35,000 major dams constructed world-wide between 1950 and the late 1980s (Öjendal, 2000). These mega-dams embody modernity, making possible the taming, ordering and using of nature for the benefit of humankind. As the necessary infrastructure for major hydropower as well for irrigation, the dams appear as the perfect machine delivering electricity 24 hours a day, 365 days a year, with only minor maintenance requirements (or so it can be understood). Currently there are some 57,000 major dams in the world and 300 mega dams, more than half of them in China and the United States. Currently, some 20% of the global energy generation comes from hydropower facilitated by dams (IRN, 2022a, 2022b).

The almost religious conviction of the benefits of major dams is imbedded in modernity itself. The origin of the global dam-building frenzy, however, could be traced back to the creation of the Tennessee Valley Authority (TVA) and the major initiative to use the state's water resources as a means to enforce modernization. A grand masterplan for dam construction and energy generation was constructed in the early 1930s, including the building of an omnipotent river basin authority, and came gradually to direct the development of the valley for half a century. The last dam in that scheme was only finished in 1979. It was federally financed and constituted, at the time, a great experiment for the potential of harnessing water resources on an industrial scale and its developmental effects. At least 38 major multipurpose dams (irrigation, hydropower, and flood management) were constructed (and many minor ones) constituting a backbone of the region's development, water being the "engine of growth" in the area. Dams became symbols of progress, wealth, modernity and power.

The initiative was early on regarded as a huge success, and it was broadly agreed that a model had been found that "worked" and could be repeated and exported (Öjendal, 2000: 59). As such it came to have major influence in the post-war development scene.

> The TVA's international influence was immense. Thousands of foreign planners, engineers and politicians visited the Tennessee at the expense of the US government and returned

home to spread the gospel that one or more multipurpose dam built by a central authority could quickly transform any regional economy from subsistence agriculture to agribusiness and industry.

(McCully, 1996: 245)

Throughout the 1950s, basically any major dam in the developing world – whether on Sao Francisco River in Brazil, Papaloapan River in Mexico, the Cuca in Colombia, the Dez in Iran, or the Damodar in India – was presented as a carbon copy of the TVA-experience (Ibid.). There was a virtual army of engineers and planners conquering the "Third World" to deliver the competence for constructing these modern "temples," and aid-driven funding followed suit. While the TVA was the role model, work on the Columbia River, also in the United States, followed a similar development path controlled by the US Army Corps of Engineers. Water, dams and hydropower-driven regional development became a major export industry. In the 1950s, approximately 1,000 major dams were built every year world-wide and there are few major rivers in the world which have escaped either dam structures or plans for them to be built, often in megalomaniac plans of "cascades of dams" fully regulating river flows, "saving" nations and "guaranteeing" development.

However, after four decades of intensive dam-building and large-scale water harnessing, critique built into an avalanche. The dams became increasingly expensive to build than anticipated, externalities were always disregarded, environmental and livelihood destructions amassed, protests from people that were forcibly moved (and not properly compensated) grew, and the cleanliness of the energy and the durability of the structures were increasingly questioned (IRN, 2022a, 2022b). In the 1990s, the World Bank – the largest funder of dams in the developing world – established a de facto moratorium for the construction of new dams. Subsequently, in cooperation with IUCN, a process of re-assessment of effects of dams for development was initiated, eventually resulting in a global process with a multitude of studies, stakeholder consultations, and high-level policy meetings. It constituted a much-needed reality check and resulted in a set of comprehensive guidelines for dam-building, all the way from planning to management (WCD, 2000). Under these guidelines – endorsed by all major stakeholders, including the industry – dam-building was far less attractive than hitherto and new constructions remained very limited for a decade.

However, the allure of (late) modernity anything but waned. Rather large-scale water usage and major dam-building moved to other places and a new set of actors being in a phase of development fitting classic modernity. Modern "temples" have continued to be planned and constructed also in this period in the name of development, growth and modernization. For instance, in China, on the Yangtse Kiang River, and Ethiopia on the Blue Nile. Water scarcity/competition pushes actors to harness water; climate change drives the search for non-fossil energy; and the looming global food-crisis motivates extensive irrigation structures. In fact, in the last decade the mega-dams are returning to the development agenda, especially in countries with restricted transparency and limited public scrutiny (Pearse-Smith, 2014), or as Dye notes, "in politically illiberal countries" (2022: 3).

For instance, amongst many others, China is currently building some of the largest dams in the world, Laos is already pursuing (and Cambodia has just announced renewed interest in) the damming of the Mekong mainstream, and in Brazil an aggressive hydropower agenda is set in motion for the Amazon River Basin (Ibid.). In these new times, the World Bank has gradually returned to funding major dams. Under pressure from the global water/energy/food crisis and embedded in the urge for growth and development, modernity's reflex of taming the nature for human's benefit is coming back. Learnings and insights, from the decades-long critique, and the guidelines from the World Commission on Dams (WCD) have however, had an impact. Dams are coming back, but in "an evolved high modernist ideology" (Dye, 2018: 2), at least in some places. In others, the same mistakes as were committed in the 20th century are bound to be repeated.

The Green Revolution Saving (Temporarily)

Paradoxically, the initiation of the *Club of Rome* and the consolidation of *the green revolution* took place simultaneously in the first years of the 1970s: the former predicting the inability of mankind to survive sustainably due to the finite amount of global resources in relation to perpetual population growth and the endless need of economic expansion (Meadows et al., 1972). The latter, in contradistinction, presenting the most ambitious solution ever to the problem of feeding a rapidly growing global population. Modern research, modernized agricultural techniques in combination with certain development beliefs at the time, created from the 1950s and onwards a virtual revolution in agriculture which global ramifications. The green revolution rested on three pillars: newly developed High Yielding Varieties (HYV); the extensive use of fertilizers and pesticides; and, most importantly here, a reliable and endless source of water for irrigation. Importantly, each of these were needed for success, requiring "industrial" precision. Norman Borlaug, the father of the green revolution, praises the role of modern techniques for the green revolution "Never before in the history of agriculture has a transplantation of high-yielding varieties coupled with an entirely new technology and strategy been achieved on such a massive scale [...]" (Borlaug, 1970).

Even more paradoxical up to now, history has proved them both "right" in their assumptions and analyses. The original assessments of Club of Rome have at large been repeatedly validated for instance by the original authors (Meadows et al., 2004) as well as independently at a later stage (Turner & Alexander, 2014). The bottom-line argument of the Club is that *either civilization or growth must end, and soon.* The global resources and its ecological system is, they argue, unable to sustain both. Advocates of modernity, on the other hand, operate on the belief that such "limits" can be avoided and overcome with progress and innovations. The green revolution, on its part, has (at least in a narrow sense) been a tremendous modern success, credited with saving billions of lives, so far (cf. Gollin et al., 2021). Below, we will disentangle the benefits of the green revolution, its sustainability and its dependence on the dwindling water source.

The origin of the Green Revolution is traced back to a range of experimental practices in Mexico drawing on cutting edge research in the United States. During the 1950/1960s considerable success was experienced and it has been described as a rural revolution on par with the country's political revolutions (Cotter, 2003). The geopolitical power of this evolution was that pressure on land-redistribution lessened, minimizing the risk for communist revolutions. Hence, the green revolution, just like hydropower expansion reviewed above, became a favoured US export-item. It was, however, when it was exported to India and Pakistan that it took on a global significance. In discussions with Indian authorities, Punjab in northeastern India was chosen – both Ford and Rockefeller foundations were highly active in this process – as the site for the next phase of the revolution, due to its fertile soil and reliable access to water. Its subsequent success, volume and efficiency came to permanently alter global agriculture. Within three decades, in India, the output of rice per hectare tripled and the price of rice was reduced to a third during the same time period (Barta, 2007, July 28). Once globally acknowledged, it was repeated worldwide: famines and starvation were avoided and bottomless urban poverty was escaped through higher food production and lowered food prices.

As it were, the conditions for the green revolution (as above) were best applied under a large-scale approach and with (at the time) high-tech mechanization. When genetically generated, the HYVs were readily available and the right amount of fertilizers and pesticides were rather simple to apply. However, the third pillar, sufficient water and reliable irrigation services, required strict organization and a competent bureaucracy (a wink at Wittfogel is warranted here), conducive topography, and a water resource itself in ample and regulated amounts. For decades this threw water access, irrigation systems and social organization to the forefront as well as reservoir-building, canal construction and groundwater pumping. This may constitute the grandest project of modernity of them all, with the well-being of billions of people as well as the ecological survival of the entire planet at stake.

In the same way as mega-dams are now irreversibly integrated in the global energy infrastructure, the green revolution is a pillar of global food production. However, there is a wide critique of the sustainability of the green revolution, which forebodes (the beginning of) a global crisis. As Borlaug acknowledges: "There are no miracles in agricultural production" (Borlaug, 1970), and while overall defending the green revolution, he also acknowledges that it is not a magic bullet, just a temporary relief requiring constant research and progress. Critique and worries abound. *Firstly*, we have the "Malthusian argument": that with higher food production, there is higher population growth, and at the end of the day we have the same problem again, but on a larger scale. This argument is however, empirically refuted: food security, development, and wealth-creation are typically seen to reduce population growth in the long-term (Gollin et al., 2021).

Secondly, and more seriously, are the agricultural practices sustainable? Overall, they are not. Irrigated land is subject to, inter alia, erosion, top-soil depletion, salinization and dysfunctional reservoirs serving the irrigated fields. Lack of control of water distribution – as in climate change or water scarcity – makes it harder yet to fine-tune the possible damages done to the soil by irrigation (cf. Khan et al., 2006). *Thirdly*, without a paradigmatic shift in irrigation practices, there is not enough water on the planet to keep up with the need for extended irrigation (and erratic rainfall in the wake of global climate change will worsen this dilemma). *Fourthly*, the enforced mono-crop ideal is crowding out variety, and the extensive use of pesticides reduces biodiversity (including other food stuff in the fields) (Kilusang et al., 2007). This also introduces vulnerability towards new forms of pests and crop malfunctions. *Finally*, the commercialization of agriculture makes it dependent on external trade, value-chains, specialization, and financial flows (and on poor people's purchasing power). In total this constitutes a chain of vulnerabilities of which some is bound to undermine the contemporary success at some point, and deeply questioning the viability of Green Revolution 2.0.

At bottom, we have a dwindling water resource which, at least in some regions, are impossible to solve in the long-term, and the sheer volume of water for agricultural use makes any simple substitute unrealistic. Hence, in spite of a half-a-century long success for the project of the green revolution, in the longer run, the Club of Rome may be right after all, juxtapositioning "growth" and "civilization." Nobody has an answer to how to make water suffice for food production, except, possibly, the grand old lady of research on global water scarcity, Professor Malin Falkenmark, stating that within a generation, we will all have to be vegetarians (2012). I is as simple as that.

Urban Water and Sanitation (an Imperative for Functional Modernity)

When understanding the connection between water and development through the lens of modernity as aimed at in this chapter, urban water and sanitation systems attain a central role. These systems, alongside dams, are seen as the ultimate symbols of modernity. Historically, they were intrinsic to economic and social well-being in cities of the North and the elites in colonial contexts. They have been essential for constructing (the image of) the modern state, and modern (hygienic and clean) citizens that could form part of a productive and healthy labour, and military, force (Gandy, 2006a; Scott, 1998). Water and sanitation systems have played a central role in promoting public health and conveniency, and it is impossible to imagine major cities, as we know them, without these systems. They have however arguably also created infrastructural ideals that have not been suitable everywhere, especially in post-colonial contexts where systems implemented in the colonial era have involved malfunctioning and produced inequalities (Kooy & Bakker, 2008; Nilsson, 2016). They have certainly also come with environmental and resource related costs.

Zeroing in on the origin of the grand idea – an idea was unthinkable that before modernity – of supplying entire cities with water and sanitation, we notice that Western cities have stood as an urban ideal as they developed from the second half of the 19th century and onwards. A particular form of water infrastructure is central to this ideal in the form of unified and integrated water (and later) sewage systems (Gandy, 2006a). As in the cases observed above, this form of modernity was exported to (and adopted by) the rest of the world.

This era, when large and integrated public water systems were installed, has been termed the "modern city," the "hydraulic city," or the "sanitary city" (Ibid.). Matthew Gandy has described this shift in urban governance through the notion of the "bacteriological city" (Gandy, 2006a; 2014). The creation of the bacteriological city, according to Gandy, was stimulated by a range of factors, public health being one, as urban living conditions in the pre-bacteriological era caused outbreaks of disease. At this time, ill-health was understood as related to individual morals and to "stagnant air and water" (Gandy, 2006a: 15). This understanding transformed along with scientific progress in the mid-1800s, especially through John Snow's demonstration, in 1854 that contaminated water was the cause of cholera (Ibid.: 16). Public health increasingly became understood as a social problem (Nilsson, 2016, based on Melosi, 2000).

During this era, better living conditions and public health was however not be seen as an end in and of itself. According to Gandy, its significance should not be overestimated since the concern for urban governance was first and foremost focused on establishing an urban environment that facilitated political control as well as stimulated the cities as spaces for capital accumulation (Gandy, 2006a). Insofar as it was a factor, public health increasingly became a *strategic* concern because of military conscription and the risk of unrest in deprived areas (Gandy, 2004; 2006a; 2014). The flow of clean and readily available water – and thereby the physical restructuring of the urban environment – was needed for growing industries and against the constant threat of fire not least, as Gandy notes, because of demands put forward by the insurance companies (2006a).

In this shift from dispersed water arrangements to integrated and large-scale public ones, a shift in urban governance simultaneously took place. Water became incorporated into newly created municipal administrations and into an urban form that increasingly was rationalized and scientifically" managed (Gandy, 2006a: 14). Another factor that was majorly important was the financial instruments, the municipal bonds, which made large scale projects possible without having to resort to high tax increases. When Berlin constructed its water systems in the mid-19th century, it used financial resources placed on the London capital markets. In the beginning of the 20th century, water infrastructural projects made up the largest part of municipal debts for American cities. As a result, the connection between international capital markets and urbanization became increasingly stronger (Ibid.: 17).

During the era of the bacteriological city, the human/society relationship was rearranged and distinctions between inside/outside, private/public and bad/good water were created (Gandy, 2014; Kaika, 2005; Sultana, 2013). The private home, for example, became especially important, during the 1900s, as *the* place for the good and purified water in contrast to the uncontrolled water outside of the house, which, in turn, was associated with threat and danger (Kaika, 2005). Human waste became something which was supposed to quickly be removed from human settlements rather than a valuable resource and fertilizer (Borowy, 2021).

In the Western context, class and gender differences played a role in terms of who got what water services (Kaika, 2005), but not with as stark contrast as in the colonies. Here the implementation of western technology became a symbol for superiority, progress and civilization and hence functioned to legitimise colonization and occupation (Nilsson, 2016). The implementation of water supply systems came to create a way of life that resembled European standards for the ruling elites (Ibid., based on Headrick, 1988). Local populations, such as for example those in Jakarta, were denied access and their practises was portrayed as unclean and backward (Kooy & Bakker, 2008; Nilsson, 2016). Water infrastructure systems, in these contexts, were hence a means to demarcate difference between citizen and subject (Mamdani, 1996, referred to in Gandy, 2004), and thus between those understood as human and less than human, constituting a hydraulic racism (Mbembe, 2019) in both the literal and symbolic sense.

The European urban model of water and sanitation systems was not only discriminatory during the colonial era but had effect long after independence. In colonial cities, such as Kampala (Uganda) and Lagos (Nigeria), these systems have only been partially completed which has created not only devastating outbreaks of disease but have also created a legacy "incomplete modernity" (Gandy, 2006b; Nilsson, 2016). After independence such unfinished infrastructural systems proved not to be easily remedied. Previously racially segregated systems increasingly became segregated in terms of class as this large scale and capital-intensive systems created a need to recover cost from the more well-off (Nilsson, 2016). The *mental model of modernity and progress* through western infrastructural ideals, stressed by the elite and by donors, has arguably held back more locally adapted innovation and technological development (Ibid.).

Modernity 2.0 – A Shifting Water Paradigm

This modern hydraulic ideal is that the assumption that abundant water resources exist, readily available for industrialization, urbanization and irrigated agriculture is now questioned globally. In the 1990s, the so-called hydraulic mission came to an impasse as governments were unable or unwilling to pay for large infrastructural projects and environmental and resource related concerns became evident, especially concerning large dams (Bakker, 2010; Molle et al., 2009). Peter Gleick's edited volume *Water in Crisis* (1993) became seminal in the way that it portrayed the state of water and its governance being at a dead-end. In developing countries, "state failure" became the concept that was used to describe the fact that universal access had not been achieved (Bakker, 2010: 87). Because of this, the water sector became subject to a reform agenda, which included a move away from supply oriented solutions, and instead stressed demand management, including participatory processes, human rights approaches, as well as reforms of privatization, commercialization and liberalization (Bakker, 2010).

To which extent this new paradigm is challenging the rationality of modernity, or simply a reformation at gunpoint by pragmatic forces, can be debated. According to Jamie Linton, the way that the so-called global water crisis was portrayed, and which gave rise to these new ways of understanding water was in itself framed within the terms of modern water (2010: 93). As such it is still constructed in terms of an unbalanced relationship between the availability of water, on the one hand, and existing and predicted future demands, based on current uses and trends, on the other. If the latter be the case we are not really entering into a new paradigm, and sustainability is questionable.

There are indeed examples in abundance of the support for the claim that we have not challenged the core assumptions of the paradigm of modern water. As we have observed above, we have recently seen a (re)turn to big water infrastructure, such as dams, inter-basin transfers and desalination plants (Crow-Miller et al., 2017). While softer "demand management" approaches became the order of the day, the World Bank has "quietly returned to financing large dams since the mid-2000s" (Ibid.: 196). The differences compared to this previous era, however, consist of a new language being used to legitimize these large-scale projects and that new players have entered the field (Ibid.). According to Crow-Miller et al. (2017), discourses about "green development" and "sustainability" now encircle these projects and private finance as well as Chinese banks are new important actors financing them (Ibid.).

As mentioned above, the paradigm shift has included processes of the liberalization (and indeed the commodification) of water. While these changes rarely involves that ownership of water *per se* is transferred to private companies, often processes of privatization, liberalization, and decentralization has functioned as conditionalities in development cooperation (see Boelens, Hoogesteger, & Baud, 2015; Hansson, 2014). Hence, this new paradigm, has not advocated for "pure" market solutions but has rather proposed a range of interventions; including guaranteeing the human right to water and water for the environment. These interventions can nevertheless be understood as interventions that work within a system that applies economic market principles as the overarching model (Hellberg, 2018).

Following the above, contemporary water governance has not left behind the core logic that we, through technology, can (and should) make more water available for human consumption and processes of capital accumulation. We do not, for instance, seem to want to let go of modernist ideas of flushing toilets regardless of how much pressure they put on water resources, but rather seek solutions, such as wastewater treatment and desalination to meet these wants. In comparison to modernist rationales around water, the way that water is understood has however changed. Contemporary water management discourse has "complicated" (Linton, 2010) the way we think about water. Not least, climate change has affected the way that water is a "natural fact" which can be quantified and measured (Ibid.) since one fundamental aspect of the effect of water resources emanating from climate change is uncertainty. Hence, modernist rationalities exist alongside the view that good governance is needed for water stewardship in which both the public and private sector are focused on supporting a form of governance that allow market rationalities to function, through simultaneous processes of regulation and de-regulation.

Given the sustainability and resource-related challenges we are facing, there need to be alterations in the modernist paradigm in relation to water. Any emerging water governance paradigm needs address the questions of what constitutes reasonable, equitable and sustainable uses of water. In such a paradigm, water is no longer something that is taken for granted; it is no longer seen as an abundant resource that is readily available for modernization, industrialization and urbanization, but something which has ecological, cultural and political dimensions (Linton, 2010: 7). In turn, this means that we need to question ideas around what constitutes a modern city, a modern state or modern ways of life (Hellberg, 2020).

What eventually will be considered "modern" and "developed" water and sanitation solutions are, however, far from given. Modernity has a habit of re-inventing itself when facing obstacles, and may appear in a 2.0-version, with or without classic modernist hallmarks. Either way, we need to invent new mental models for the utilization of water for development purposes, or, phrased in more academic terms, we need a truly new water paradigm.

References

Bakker, K. (2010). *Privatizing Water: Governance Failure and the World's Urban Water Crisis*. Cornell University Press, Ithaca, NY.

Barta, P. (2007). "Feeding billions, a grain at a time." *Wall Street Journal*. 28(7), A1.

Boelens, R., Hoogesteger, J. & Baud, M. (2015). "Water reform governmentality in Ecuador: Neoliberalism, centralization, and the restraining of polycentric authority and community rule-making." *Geoforum*. 64, 281–291.

Borlaug, N. (1970). "The green revolution, peace, and humanity." Nobel laurate speech in 1970 [Viewed 26 January 2023]. Available from: https://www.nobelprize.org/prizes/peace/1970/borlaug/lecture/

Borowy, I. (2021). "Human excreta: Hazardous waste or valuable Resource? Shifting views of modernity." *Journal of World History*. 32(3), 517–545.

Caponera, D. (1978). "Water laws in hydraulic civilisations." In Ulmen.

Caponera, D. (1992). *Principles of Water Law and Administration – National and International*. A. A. Balkema, Rotterdam.

Cotter, J. (2003). *Troubled Harvest: Agronomy and Revolution in Mexico, 1880–2002*. Praeger, Westport, CT.

Crow-Miller, B., Webber, M. & Molle, F. (2017). "The (re)turn to infrastructure for water management?." *Water Alternatives*. 10(2), 195–207.

Dye, B. J. (2018). *The Politics of Dam Resurgence: High Modernist Statebuilding and the Emerging Powers in Africa*. University of Oxford, Oxford, UK.

Dye, B. J. (2022). "Dam building by the illiberal modernisers: Ideological drivers for Rwanda and Tanzania's megawatt mission." *Critical African Studies*. 14(3), 231–249.

Falkenmark, M. (2012). Food Security: Overcoming Water Scarcity Realities. In *Feeding a Thirsty World Challenges and Opportunities for a Water and Food Secure Future*. SIWI, Stockholm.

Gandy, M. (2004). "Rethinking urban metabolism: Water, space and the modern city." *City*. 8(3), 363–379.

Gandy, M. (2006a). "The bacteriological city and its discontents." *Historical Geography*. 34(6), 14–25.

Gandy, M. (2006b). "Planning, anti-planning and the infrastructure crisis facing metropolitan Lagos." *Urban Studies*. 43, 371–396.

Gandy, M. (2014). *The Fabric of Space: Water, Modernity, and the Urban Imagination*. The MIT Press, Cambridge, MA.

Giddens, A. (1990). *The Consequences of Modernity*. Polity Press, Cambridge, UK.

Giddens, A. (1991). *Modernity and Self-Identity: Self and Society in the Late Modern Age*. Stanford University Press, Stanford, CA.

Gleick, P. H. (ed.) (1993). *Water in Crisis: A Guide to the World's Freshwater Resources*. Oxford University Press, New York, NY.

Gollin, D., Hansen, C. W. & Wingender, A. M. (2021). "Two blades of grass: The impact of the green revolution." *Journal of Political Economy*. 129(8), 2344–2384.

Hansson, S. (2014). "Dependence and autonomy: Conditions of possibility for Nigerien ownership in the water sector." *Journal of Intervention and Statebuilding*. 9(2), 256–277.

Headrick, D. R. (1988). *The Tentacles of Progress. Technology Transfer in the Age of Imperialism*. Oxford University Press, Oxford, UK.

Hellberg, S. (2018). *The Biopolitics of Water: Governance Scarcity and Populations*. Routledge, London, UK.

Hellberg, S. (2020). "Scarcity as a means of governing: Challenging the neoliberal governmentality in the context of the South African drought." *Environment and Planning E: Nature and Space*. 3(1), 186–206.

IRN. (2022a). [Viewed 14 December 2022]. Available from: https://archive.internationalrivers.org/questions-and-answers-about-large-dams

IRN. (2022b). [Viewed 27 November 2022]. Available from: https://archive.internationalrivers.org/questions-and-answers-about-large-dams

Kaika, M. (2005). *City of Flows: Modernity, Nature and the City*. Routledge, London, UK.

Khan, S., Tariq, R., Cui, Y. & Blackwell, J. (2006). "Can irrigation be sustainable?" *Agricultural Water Management*. 80, 87–99.

Kilusang, M., Layosa, P. M., Acharya, K., Quijano, R. F., Adapon, S. Q., Pelegrina, W. R. & Cruz, P. A. Z. (eds.) (2007). *The Great Rice Robbery: A Handbook on the Impact of IRRI in Asia*. Pesticide Action Network Asia and the Pacific, Penang, Malaysia. ISBN 978-983-9381-35-1.

Kooy, M. & Bakker, K. (2008). "Technologies of government: Constituting subjectivities, spaces, and infrastructures in colonial and contemporary Jakarta." *International Journal of Urban and Regional Research*. 32(2), 375–391.

Li, T. (2005). "Beyond the 'State' and failed schemes." *American Anthropologist* New Series. 107(3), 383–394.

Linton, J. (2010). *What Is Water? The History of a Modern Abstraction*. UBS Press, Vancouver, BC.

Mamdani, M. (1996). *Citizen and Subject: Contemporary Africa and the Legacy of Late Colonialism*. Princeton University Press, Princeton, NJ.

Mbembe, A. (2019). *Necropolitics*. Duke University Press, Durham, NC.

McCully, P. (1996). *Silenced Rivers: The Ecology and Politics of Large Dams*. Zed Books, London, UK.

Meadows, D., Dennis, H. L., Meadows, J. R. & Behrens, W. W. III (1972). *Limits to Growth*. Universe Books, New York, NY.
Meadows, D. H., Dennis, L. & Meadows, J. R. (2004). *Limits to Growth: The 30-Year Update*. Chelsea Green Publ., White River Junction, VT.
Melosi, M. (2000). *The Sanitary City. Urban Infrastructure in America from Colonial Times to the Present*. John Hopkins University Press, Baltimore, MD.
Menga, F. & Swyngedouw, E. (2018). *Water, Technology and the Nation-State*. Routledge, London, UK.
Molle, F., Mollinga, P. P. & Wester, P. (2009). "Hydraulic bureaucracies and the hydraulic mission: Flows of water, flows of power." *Water Alternatives*. 2(3), 328–349.
Nilsson, D. (2016). "The unseeing state: How ideals of modernity have undermined innovation in Africa's urban water systems." *NTM International Journal of History and Ethics of Natural Sciences Technology and Medicine*. 24(4), 418–510.
Öjendal, J. (2000). *Sharing the Good – Modes of Managing Water Resources in the Lower Mekong River Basin*, PhD thesis, Göteborg University.
Pearse-Smith, S. W. D. (2014). "The return of large dams to the development agenda: A post-development critique." *Consilience – The Journal of Sustainable Development*. 11(1), 123–131.
Scott, J. C. (1998). *Seeing Like a State: How Certain Schemes to Improve the Human Condition Have Failed*. Yale University Press, New Haven, CT.
Sultana, F. (2013). "Water, technology, and development: Transformations of development technonatures in changing waterscapes." *Environment and Planning D: Society and Space*. 31, 337–353.
Turner, G. & Alexander, C. (2014). "Limits to growth was right. New research shows we're nearing collapse." *The Guardian*, Sep. 2, 2014.
UN. (1958). "Programme of studies and investigations for comprehensive development of the Lower Mekong River Basin." UN documents TAA/AFE/3, Apr. 22, 1958.
WCD. (2000). *Dams and Development: A New Framework for Decision-Making*. Earthscan, London, UK.
Wittfogel, K. (1957). *Oriental Despotism: A Comparative Study of Total Power*. Random House, New York, NY.

3
INSTITUTIONAL APPROACHES TO WATER FOR DEVELOPMENT

Larry A. Swatuk

Introduction

Institution building and institutional reform have been at the centre of international development practice for several decades. Following the end of World War II, the Western world placed significant emphasis on a liberal institutionalist approach to development and peace, both within the alliance (e.g. through organizations such as the North Atlantic Treaty Organization, the General Agreement on Tariffs and Trade, and the European Coal and Steel Community) and as a means of recruiting new members (e.g. through mechanisms such as the Colombo Plan, the Lomé Convention and the Caribbean Basin Initiative). At the heart of these initiatives lies the liberal institutionalist belief that functional economic and political cooperation leads to mutually beneficial outcomes for all parties. Supporting this idea are equally foundational beliefs regarding the process: that cooperation can be engineered through institution building; and that appropriately structured institutions can manage stakeholder competition, build capacity, balance interests and minimize conflict.

With the rise of neoliberal globalization, the focus of development has been on limiting the role of the state, building civil society and nurturing a private sector thereby freeing the power of capital and entrepreneurship in development. The rule of law, markets and private property rights have held pride of place in almost all policy documents (Harvey, 2005). Following the end of the Cold War, a concerted effort was placed on bringing the state "back in" through enhanced administrative and regulative capacities, while ensuring transparency, accountability and broader stakeholder participation in governance (World Bank, 1997). For Phillips:

> neoliberal theories understood development as an inherently national process, and the reasons for development failures as lying in internal […] factors – that is, as rooted in "incorrect" government policies, institutional under-development, and excessive state intervention in the workings of national economies.
>
> *(2020: 359)*

As with the broader ideology of modernization, neoliberal institutionalism presumed outside intervention as a necessary, indeed incontestable, aspect of institutional development. It is accurate to

 DOI: 10.4324/9781003095545-4

define the "development industry" as a global institution, consisting of rules, norms and procedures accepted and contested by states, the private sector and civil society organizations (Jolly et al., 2003).

Neoliberal institutional approaches to development share several key characteristics with their post-World War II precursors: they are both Western/modernist in origin, functionalist in nature, reformist in vision and alignment-oriented in emphasis. Where they differ is in relation to the significantly decreased role for the state and increased role for markets and non-state (and non-national) actors in decision-making.

In relation to water resources, contemporary institutional development manifests as an increasingly coordinated response to multiple natural and human-made threats to the security of human communities. Importantly, included among perceived threats are existing institutional arrangements said to be deficient in a variety of ways. We return to this discussion in detail below.

Taking the historical development of water management as known (Dellapenna & Gupta, 2008; Solomon, 2010), the balance of this chapter focuses on the policies, laws and administrative procedures that constitute contemporary institutional arrangements for managing water in development. Beyond this introduction, the next section focuses on institutions, followed by a section on *Water in Development*. The penultimate section focuses on *Water Institutions*, in particular reflecting on existing and emergent policies, laws and administrative procedures. It highlights the role of culture, norms and values in shaping institutions and reflects on the central role of law in determining water resource access and use. The final section briefly reflects on emergent developmental trajectories and possibilities for positive institutional change.

Institutions in Development

In defining institutions, distinguishing them from organizations, and understanding their interrelationship, it is important to quote North at some length:

> Institutions are the humanly devised constraints that structure human interaction. They are made up of formal constraints (rules, laws, constitutions), informal constraints (norms of behavior, conventions, and self-imposed codes of conduct), and their enforcement characteristics. [...] If institutions are the rules of the game, organizations and their entrepreneurs are the players. Organizations are made up of groups of individuals bound together by some common purpose to achieve certain objectives. Organizations include political bodies (political parties, the Senate, a city council, regulatory bodies), economic bodies (firms, trade unions, family farms, cooperatives), social bodies (churches, clubs, athletic associations), and educational bodies (schools, universities, vocational training centers). The organizations that come into existence will reflect the opportunities provided by the institutional matrix.
>
> *(1993)*

For analytical purposes, the institutional matrix may be usefully divided into three key elements: policies; laws; and administrative procedures, each of which may give rise to or privilege the engagement of particular organizational forms (Chopra & Ramachandran, 2021; Haller, 2002). For Jütting (2003: 11), and following North:

> institutions include any form of constraint that human beings devise to shape human interaction. These constraints include both what individuals are prohibited from doing and, sometimes, under what conditions some individuals are permitted to undertake certain activities.
>
> *(1993)*

For Brännmark (2019: 1049), whether formal and explicit, or informal and implicit, institutions understood as "sets of constraints" are productive of particular social arrangements that reflect "collective acceptance"; "an institution [...] exists because we collectively accept the set of constraints that together make up the institution" (Brännmark, 2019: 1049). Wengert emphasizes the socially constructed nature of institutions:

> An institution is the structured outcome of a process by which values, or norms, are articulated, arranged, communicated and applied. It has continuity over time, and influences or controls the behaviour of persons involved with the institution but who may not have participated in the process of institutionalization.
>
> *(1983: 279)*

In a useful review of the literature, Jütting (2003) illustrates three ways of classifying institutions: degree of formality; different levels of hierarchy; and area of analysis (economic, political, legal, social). Each of these forms of classification present a hierarchy of value with informal institutions at the base, and formal, Western approaches at the apex. It should here be noted that while institutions are socially constructed, they simultaneously may be "socially contested" (Brännmark, 2019). Where current water institution building aims to overcome challenges of fragmentation in decision making, Wengert wryly observes, "the present fragmented system reflects deep seated values and historical traditions" (1983: 280). This raises questions not only of the utility and durability of contemporary water institutions within a particular society but of their fungibility across space and time. We return to these issues in relation to water resources in the *Water Institutions* section below.

Water in Development

The centrality of water in social formation and wealth creation is well-chronicled in the literature (Allan, 2006; Solomon, 2010; Worster, 1985). From simple infrastructure such as Zai pits in West Africa (Partey, 2018), to dramatic engineering efforts such as Roman aqueducts, the capture and control of water resources has long been at the centre of human development. Improvements in technological and scientific innovation have supported wealth creation, the rise of empires, and the growth of civilizations around the world and across time: from Mesopotamia to the Aztecs and Incas, from Great Zimbabwe to the many dynasties and kingdoms of present day China. In every case, innovation gave rise to an unending mix of positive and negative effects leading the modern world into what Wright (2004) labels "the progress trap" and humanity towards what Kolbert (2014) describes as "the sixth extinction." Where modernization was argued to be the way forward for "backward" societies, today modernization is recast, most notably in its ecological form (Alier, 2003; Fisher & Freudenburg, 2001), as the pathway to sustainability. Hence notions of "circular economies" aim to rearticulate progress, away from endless growth and biodiversity collapse.

In the context of post-World War II international development, water gained an elevated role in relation to "nation building" and economic growth. Big water infrastructure – for hydropower, irrigation, urban water supply – became an essential cog in the machine of modernization. This is most clearly exhibited in the growth in number of large dams across the Global South. Modernization has long rested on an ontological foundation of fundamental social change where "traditional," agrarian, and other subsistence-oriented societies are regarded as inferior to "modern" Western-industrial ones (Hoogvelt, 2001; Latour, 1993). While the relative weight to be given to the state or the market in development is debated, modernization's value is believed to be its capacity to generate surplus at a greater scale across a wider range of goods and services than any other form of production (Mittelman & Pasha, 1997). As societies grow and consumption increases, traditional institutions – both formal and informal – lose their ability to manage competing demands; hence the perceived importance of modern ways of doing things. Although international development has altered its meta-narrative, focusing on challenges such as poverty alleviation and sustainable development, modernization remains dominant through an institutional apparatus articulated through Western scientific, technological, legal and financial philosophies and practices (Blaney & Tickner, 2017; Escobar, 1995; Ynalvez & Shrum, 2015). Today, the value of modernization as social superiority is rarely stated, instead finding relevance in its hypothesized capacity to deal with the wicked problems arising from late modern socio-ecological complexity (Sachs, 2015).

Water Institutions

Writing in 1976, Fox stated:

> Each nation has an institutional arrangement for managing water resources. This arrangement establishes the conditions under which water resources can be developed and used and provide organizations and individuals with certain resources and authorities to carry out prescribed tasks.
>
> *(Fox, 1976: 743)*

He goes on to say, "[t]he product of this set of interactions should be the preference of society" (Fox, 1976: 745). Wengert (1983: 280) notes, an "institution is thus a way to designate the network of interdependencies or the cluster of relationships among law, agency operations, public finance, political values and influence in the context of the general culture." Implicit in both Fox's and Wengert's definitions is the possibility for variation in institutional form – both in time and space – reflective of societies' particularities. As the "product of a set of interactions," institutions are socially constructed, reflective of particular cultural norms and values which themselves may change over time. The drivers behind the development of water institutions are common to all societies, tied in the first instance to consumptive (food, drinking water) and non-consumptive (spiritual, recreational) needs and wants, and evolving in relation to rights of access and use. As constraints on behaviour, traditional institutions reflect social norms and commonly held values embedded in particular cultures and representative of unique world views.

Norms and values

Like multilateral development institutions with which they are thoroughly enmeshed, contemporary water institutions promote universalist liberal norms and values expressed in a wide variety of national, transboundary/regional and global policies, laws, procedures and organizations. For example, within South Africa's revised water laws, a hierarchy of use value is articulated with "free basic water" for citizens being primary, ecosystem health guaranteed through attention to "environmental flow," beyond which "use" is determined by need and ability to pay. Transboundary water institutions are determined by inter-state agreement, often reflecting key international water law principles such as "equitable and reasonable use" and "the duty to cause no significant harm" (Fatch et al., 2022). They are thus normative in the sense that they seek to establish a particular Western/modernist standard of behaviour. Those who support institutional design believe that water and development challenges are amenable to resolution through the deliberate construction of rules, regulations and procedures. At the same time, the obverse is true: supporters of institutional approaches believe that either the absence or outmoded nature of existing rules and regulations account for current challenges. The focus of development intervention, then, is two-fold: simultaneously creating new structures to guide resource use in a particular direction, what Cleaver (2012) calls "development by design," while encouraging the abandonment/disbandment of non- or weakly-institutional(ized) efforts regarded as yielding sub-optimal outcomes. To accept this narrative requires one to accept the argument that existing ways and means of managing resources are deficient in comparison to those proffered by contemporary institution builders.

In development-speak, this most often translates as Western/modern approaches supplanting traditional ones where access to and use of a resource resulted from particularistic social relations (e.g., kinship, religion and tribalism/ethnicity) or ways of knowing (e.g., traditional ecological knowledge; Indigenous cosmologies). In the modernist imaginary, such culturally derived forms of access and use could yield positive collective outcomes only insofar as pressure on the resource (be it forest, wetland, grassland and lake) was limited. Beyond a certain threshold of complexity, traditional institutions are incapable of yielding the desired, traditionally sustainable outcome. Where the resource is essential and non-substitutable, i.e. water, population increase plus/minus land use change degrades the resource while altering accessibility. In other words, increasing scarcity of key resources demands institutional innovation. Hardin's idea of the "tragedy of the commons" has long served as the conceptual foundation upon which to introduce individual private property rights as the basis for sustainable, equitable and economically efficient resource use outcomes. While it is right to question Hardin's notion of commons being "open access resources" as they were in most instances subject to rules of access and use determined and upheld by the collective, his observation of scarcity driving exclusion, degradation and ultimately collective social loss rings true today (Haller, 2002).

Whether their ecological restoration or efficient management requires the imposition of Western norms such as private property rights and the commoditization of the resource is another question entirely, leading to yet another sense in which contemporary institutional approaches are normative: they presume that Western, capitalist approaches are superior to non-Western, non-market-oriented approaches. The ontological framing here is important. According to the Dublin Principles, water is "scarce and finite" and should be treated as an economic good. This elevates economistic valuations of water above, for example, spiritual, intrinsic and communal valuations (Salmond, 2012; Van der Zaag, 2005).

Traditional approaches to resource access and use are often regarded as non-institutional or non-formal, implying a certain level of economic inefficiency. However, such approaches serve a sub-set of a larger community and may be organized specifically to control access to a key resource such as pastureland or surface water. What Hardin interpreted as open access was in fact a resource managed through non-Western capitalist means, for example according to criteria such as the presence or absence of a totem animal such as a crocodile whose presence may reflect accepted facts about seasonal changes to a surface water resource. Van der Zaag (2005: 25) provides an example where the Manica in Zimbabwe believe that using concrete or cement in weir development "harms the water creatures or water spirits." This taboo against the use of concrete seems inefficient from a Western perspective, for weirs made of simple, local materials leak and collapse. Regarded differently, leakage ensures both flow and infiltration during dry periods, and seasonal weir reconstruction builds social capital. These practices emerge over time and are culturally embedded; as such, they are not easily displaced by Western models devised to increase yields of marketable crops.

The challenge for traditional approaches is that they collapse under acute socio-ecological pressure. Land-use change driven by colonial and neocolonial interests undermined many traditional forms of land management, especially those regarding seasonal mobility where pastoralists and smallholder farmers are everywhere hemmed in by plantation-style private property. This problem has long been the object of study within critical agrarian studies (see Akram-Lohdi et al., 2021).

Culture

In a world water day 2006 press release, the World Health Organization stated:

> Cultural differences play a key role in the way water is perceived, valued and managed in different societies. World health and poverty eradication have cultural connotations; culture has positive and negative health impacts on individual well-being – in particular women's health. Water management practices should be adapted to specific cultures as they constitute distinct systems of knowledge and behavior. Water resources management strategies must take culture fully into account.
>
> *(2006)*

According to Li and Zahra:

> Formal institutions and informal cultural constraints are strongly intertwined. Formal institutions are usually embedded in cultural settings because political, economic and contractual rules are all connected to peoples' conceptions of how things ought to be done.
>
> *(2012: 97)*

Drawing on Hofstede (2001), they state, "culture is the 'software of the mind' and formal institutions are themselves 'products of the dominant cultural value systems'" (Hofstede, 2001 in Li & Zahra, 2012: 97). Citing North (1990), the authors continue: "As a result, the same formal institutions that exist in societies with different cultural values can produce different economic outcomes" (North, 1990, cited in Li & Zahra, 2021).

To be sure, cultures change as shifting situations and conditions demand adaptations from societies. Moreover, interests and ideologies cross cultures so impacting social relations and institutional arrangements. Worster (1985) elegantly chronicled the shifting character of hydraulic societies across time and space, identifying three types: the local subsistence mode; the agrarian state mode; and the capitalist state mode. It would not be inaccurate to characterize each of these modes as simultaneously emerging from and constituting culture. Throughout the world, the capitalist state mode thoroughly dominates current approaches to harnessing water for human use. At the same time, this approach often cohabits alongside each of the earlier modes, particularly in the Global South.

In the capitalist state mode, "water" is simultaneously a commodity and a factor of production. For Worster (1985: 51), in the capitalist state mode:

> there are two roughly equivalent centres of power: a private sector of agriculturalists and a public sector made up of bureaucratic planners and elected representatives. Both need each other, reinforce each other's values, compete for the upper hand without lasting success, and finally agree to work together to achieve a control over nature that is unprecedentedly thorough.
>
> *(1985: 51)*

In pursuit of wealth creation, this ideology is "shared wholeheartedly by agriculturists and water bureaucrats, providing the bond that unites their potentially rival centres of power into a formidable alliance" (Worster, 1985: 53). At their service are the experts, where:

> the contemporary engineer is the best exemplar of that power of expertise. Though not himself [sic] necessarily concerned with profit-making, he reinforces directly and indirectly the rule of instrumentalism and unending economic growth.
>
> *(Worster, 1985: 57)*

The paradox of displacing local subsistence modes with capitalist state modes is that "inefficient" and "unproductive" processes whose ecological footprint was small have been displaced by highly efficient and productive processes that are resource intensive and incredibly destructive (Beck, 1992; Carson, 1962; Wackernagel & Rees, 1996). Institutional reform thus has become more about finding a middle ground than blindly chasing man-over-nature command and control processes (Allan, 2006).

Allan (2006) describes this process as "the new sanctioned discourse," with Sustainable Development Goal 6 (water and sanitation for all) emblematic of this perspective. Importantly, Allan (2006: 38–39) notes that "[p]olicy is not made on the basis of rational science, although well-observed science can play a role [...] Policy is made by agents and policy entrepreneurs operating in complex local discourses, usually at the national level." Interestingly, this era of "reflexive modernity," characterized by uncertainty and globalized risk awareness (Allan, 2006: 47), has opened up spaces for the re-emergence of traditional institutional forms, notably in New Zealand Aotearoa through the passage of the *Te Aw Tupua* (Whangarui River Claims Settlement) Act which recognizes the Whangarui River as "an indivisible and living whole" and confers upon the River "all the rights, power, duties and liabilities of a legal person" (Brierley et al., 2019: 1640).

The rule of law

Legal systems reflect and facilitate social relations (Edelman, 2004, 2007). They may be formal or/and informal, emerge over time, and are subject to minor and/or major change. Laws relating to water are as old as human civilization (Kornfeld, 2009). For Dellapenna and Gupta:

> History shows that water law has developed in a highly contextual manner reflecting the historical, geographical and political contexts of the countries concerned. As a result, today there are 192 different national water law systems, each with country specific characteristics. At the same time, there are a number of unifying forces that have spread common principles of water law to different parts of the world. These include: (a) the spread of civilizations...; (b) the spread of religion...; (c) the impact of conquest and colonization...; (d) the codification of legal principles...; (e) the rise of epistemic communities...; (f) the influence of environmentalism; and (g) the second wave of globalization... These various influences overlapped and often continue to co-exist within a single society, resulting in plural systems of water law competing for application ... although recent decades have seen efforts to integrate different regulations into one comprehensive water law.
>
> *(2009: 10)*

As a result, sovereign states' water resources are subject to a patchwork quilt of legal and administrative rules and directives (Cullet, 2018; Dellapenna & Gupta, 2008; 2009). A quick survey of water laws around the world shows a striking similarity irrespective of regime type. From Australia to China to India and Zimbabwe, the rights of use and access share both foundations (e.g. access to groundwater being a function of land ownership) and evolutions (e.g. water is an economic good; the imposition of user-pay fees). This can be explained by factors (a) through (g) highlighted by Dellapenna and Gupta (2009) above. Put differently, the "rule of law" evolved alongside the interests and capabilities of the most influential (local and global) social actors, in some cases overriding and/or overturning long-entrenched cultural practices. "Rights" to water and land came to be determined by Western/modernist ideas regarding "productive value" measured in terms of crop yield, energy production, percentages of people with access to improved sources of drinking water and sanitation. With the diffusion of (temperate zone) European power around the world, Roman law came to dominate. With its focus on ownership and rights (communal; private; public), expropriation of indigenous resources was codified in colonial law while simultaneously disrupting semi-arid and arid zone norms and practices (Dellapenna & Gupta, 2008). This is epitomized by legal principles such as "first in time/first in right" and "riparian rights" which were extended outward from their points of origin in the West to the rest of the world via colonial/imperial rule. Similarly, vesting groundwater rights in land rights remains the dominant legal practice in much of the world. English groundwater law, founded initially in *Acton v. Blundell* (1843), served as the standard legal principle throughout the British Empire (Grover & Mann, 1991).

At the same time, with the increasing commercialization of water provision, including the involvement of private sector actors, water law increasingly revolves around the rights of access and use contingent upon ability to pay (Cullet, 2018). Turning on the notion of "scarcity," and drawing on the Dublin Principle identifying water as an economic good, legislation concerning how much water may be taken, when and at what price is at the forefront of water reforms worldwide (Barbier, 2019). While, on the one hand, water legislation is permissive guaranteeing the right of an individual, group, or company to use a particular amount of water. On the other hand, it is also punitive i.e., should you not be able to pay, your service will be cut off. With the state having taken "ownership" or acting as "trustee" of the water on behalf of its citizens, this suggests possibilities for equity and sustainability, in line with the emergent human right to water first articulated in UN General Assembly Resolution 64/292 of 28 July 2010. However, it also suggests possibilities for further resource capture and ecological marginalization (Homer-Dixon, 1992).

Contemporary Institutional Arrangements

Many significant actors, forces and factors are driving institutional development and change at a variety of temporal and spatial scales. These drivers are endogenous and exogenous. Irrespective of the driver, all efforts reflect the liberal institutionalist belief that socio-ecological change can be engineered in support of a greater good. As described above, these efforts mirror Western/modernist beliefs, initially in the command and control of nature and, more recently, in pursuit of sustainable development. In this section, the focus is on the other three characteristics of institutional re-engineering: fitting form to function; the focus on reform; and the alignment of laws, policies and procedures through national, regional and global consensus building efforts.

Fitting form to function

Water's institutional architecture reflects changing and often contending perspectives on the value of water, uneven individual and collective capabilities in access and use within and across space and time, technological innovation in the pursuit of particular interests, and ad hoc responses to natural and human-made socio-ecological challenges (Postel, 2017). The institutional and organizational result is generally described as "fragmented," the most serious impacts of which are extensive resource degradation, water-related disease, and increasing resource scarcity (Bakker & Cook, 2011).

Through the twin discourse of crisis and scarcity, often under the spectre of "water wars," water experts and interested parties have settled on the river basin as the appropriate scale for institutional and organizational change and integrated water resources management (IWRM) as the necessary framework for action. River basin organizations (RBOs) reflect a pyramid design, consisting of water user associations at the base, a stakeholder-based council above them, followed by a council executive. As trustee of water for the nation, and ultimate water authority, the national government sits at the apex. Councils are assisted in the planning process by a technical committee. Depending on the human, financial and technical capacity of the country, RBOs will be more or less penetrated by actors exogenous to the basin and often the country. This includes familiar development agencies such as the World Bank, UNDP and FAO, development banks, private contractors and consultants. Participation is intended not only to assist with getting the form right, but in aligning national institutional reform across the country, across transboundary basins, and in line with regional and global principles and practices (Kittikhoun & Schmeier, 2021).

Despite criticisms (Biswas, 2004; Giordano & Shah, 2014a; 2014b), Muller notes:

> IWRM provides a framework within which to consider tradeoffs between different development objectives and, where possible, to identify win-win water investments. By aligning and integrating interests and activities that are traditionally seen as unrelated or that, despite obvious interrelationships, are simply not coordinated, IWRM can foster more efficient and sustainable use of water resources.
>
> *(2006: 1)*

Writing ten years later, Hileman et al. (2016: 6) state:

> A general strength of IWRM is that it emphasizes the importance of the human dimension – water-users, stakeholder groups, and institutions – of water development. IWRM is a participatory, inclusive, and multi-stakeholder process, and addresses conflicts as they arise between parties with diverging goals for water resources management.
>
> *(2016: 6)*

It is generally agreed that for IWRM to be implementable, an enabling environment is necessary. Worldwide there has been focus on repealing outdated water laws and creating new ones wherein appropriate criteria for decision making are set out (see Swatuk, 2002, for an example). Globally, the European Union Water Framework Directive serves as an example for countries and regions to emulate. As stated by the European Commission:

> The best model for a single system of water management is management by river basin – the natural geographical and hydrological unit – instead of according to administrative or political boundaries […] For each river basin district – some of which will traverse national frontiers – a "river basin management plan" needs to be established and updated every six years, and this will provide the context for the co-ordination requirements identified above.
>
> *(EC, n.d.)*

Focus on reform

Institutional change faces a conundrum: existing arrangements are embedded through a long history of development and not easily dislodged. At the same time, they have come to best serve those who have the capacity to influence policies, laws and practices: big agriculture, mining and industry, energy supply, governments at all levels, public and private financial institutions, and beneficiaries of the products and services resulting from this arrangement. In basins, regions and states where water access and use are most inequitable, evidence shows that this constellation of social, economic and political power cannot be dislodged through institutional design. In contrast, meaningful change – if it occurs – tends to result from either a dramatic event or organized pressure from civil society. In the case of the former, reform to California groundwater law emerged out of a multi-year drought; in the case of the latter, decades of organized civil society challenges to treaty law in New Zealand Aotearoa resulted in the investiture of the Whangarui river basin as a legal person.

However, even these innovations do little to alter the high-modern understanding of water as an input to be manipulated by humans for production. Whereas a great deal has been made out of the *Te Awa Tupua* Act, in particular the acknowledgement that the river is "a living being," the actual institutional framework – placing the river in a trust, like a child not old enough to make decisions on its own – reflects liberal understandings of individual rights, rather than Maori cosmologies of humans in nature and nature as indivisible. Similar to Indigenous peoples in North America regarded as "river keepers," they are allotted a familiar resource management role within already established practices. In contrast, the discourse of crisis and scarcity has led to a resurgence of long-established approaches to water in development: damming, diverting, draining (Conca, 2006): assisted by science, technology and public/private finance.

Alignment

From the local to the global, a common language is being spoken, constituting an influential innovation over the last several decades. Policies, laws and procedures are increasingly focused on criteria that go beyond simply serving the wealthy and influential. They provide a basis upon which groups and individuals can hold their states to account in support of broader social interests. At the global level, for example, SDG 6 consists of eight targets and eleven indicators. State performance across these indicators provide entry points for stakeholder groups to press for change.

There is an evolutionary character to not only water institutions but to the discourse enlivening them. From the global Water Forums, regional and basin organizations to national and local legal and institutional systems, there is a discernible alignment regarding what should be done, by whom, how and why. Benchmarking performance through metrics such as the SDGs is useful not only for tracking progress but for facilitating a conversation known to all and providing civil society with the evidence and legitimacy to hold governments (and private sector providers) accountable to their promises.

Conclusion

The proposed methods for meaningful change – multi-level basin organizations, stakeholder engagement, bottom-up decision making, adaptive management, collaborative planning – and the policies, laws and procedures put in place to support action show mixed outcomes. Given water's centrality in economic development, the infiltration of neoliberal, Western philosophies of resource access, use and management should come as no surprise to the reader. The organizational framework of the new water architecture makes space for governments, inter-governmental organizations, international financial institutions, and private companies whose norms reflect Western biases towards viewing water as a commodity. There is little room left for civil society, especially for local groups, less "formal" institutions and Indigenous knowledge. Despite claims that the new water architecture must privilege lowest-appropriate authorities, it is these groups who remain on the periphery of water institutions (Meinzen-Dick, 2007). While critical forms of institutional analysis reveal this, and many norm purveyors argue in support of "stakeholder forums" and "Indigenous knowledge," the fact is that institutions for water in development remain modernizationist in philosophy and largely path-dependent in practice.

References

Akram-Lodhi, H., Dietz, K., Engels, B. and McKay, B. (eds.). (2021). *Handbook of Critical Agrarian Studies*. Cheltenham, UK: Edward Elgar Publishing.

Alier, J. (2003). "Problems of Environmental Degradation: Environmental Justice or Ecological Modernization." *Capitalism Nature Socialism* 14(1), 133–138.

Allan, J. A. (2006). "IWRM: "The New Sanctioned Discourse?" In Mollinga, P., Dixit, A. and Athukorala, K. (eds.). *Integrated Water Resources Management: Global Theory, Emerging Practice and Local Needs*. New Delhi: Sage. pp. 38–63.

Bakker, K. and Cook, C. (2011). "Water Governance in Canada: Innovation and Fragmentation." *International Journal of Water Resources Development* 27(2), 275–289.

Barbier, E. B. (2019). *The Water Paradox: Overcoming the Global Crisis in Water Management*. New Haven, CT: Yale University Press.

Beck, U. (1992). *Risk Society: Towards a New Modernity*. London, UK: Sage.

Biswas, A. K. (2004). "Integrated Water Resources Management: A Reassessment." *Water International* 29(2), 248–256.

Blaney, D. L. and Tickner, A. B. (2017). "Worlding, Ontological Politics and the Possibility of a Decolonial IR." *Millennium: Journal of International Studies* 45(3), 293–311.

Brännmark, J. (2019). "Contested Institutional Facts." *Erkenntnis* 84, 1047–1064.

Brierley, G., Tadaki, M., Hikuroa, D., Blue, B., Sunde, C., Tunnicliffe, J. and Salmond, A. (2019). "A Geomorphic Perspective on the Rights of the River in Aotearoa New Zealand." *River Research and Applications* 35, 1640–1651.

Carson, R. (1962). *Silent Spring*. New York, NY: Houghton Mifflin.

Chopra, A. and Ramachandran, P. (2021). "Understanding Water Institutions and Their Impact on the Performance of the Water Sector in India." *Water Policy* 23, 466–86.

Cleaver, F. (2012). *Development Through Bricolage: Rethinking Institutions for Natural Resource Management*. London, UK: Routledge.

Conca, K. (2006). *Governing Water: Contentious Transnational Politics and Global Institution Building*. Cambridge, MA: MIT Press.

Cullet, P. (2018). "Innovations and Trends in Water Law." In Conca, K. and Weinthal, E. (eds.). *The Oxford Handbook of Water Politics and Policy*. Oxford, UK: Oxford University Press. pp. 327–350.

Dellapenna, J. and Gupta, J. (2008). "Toward Global Law on Water." *Global Governance* 14, 437–453.

Dellapenna, J. and Gupta, J. (2009). "The Evolution of Global Water Law." In Dellapenna, J. and Gupta, J. (eds.). *The Evolution of the Law and Politics of Water*. Dordrecht: Springer Netherlands. pp. 3–19.

EC (European Commission) (n.d.). *Introduction to the EU Water Framework Directive*. [accessed 18 December 2022.] Available at https://ec.europa.eu/environment/water/water-framework/info/intro_en.htm

Edelman, L. B. (2004). "Rivers of Law and Contested Terrain: A Law and Society Approach to Economic Rationality." *Law and Society Review* 38(2), 181–197.

Edelman, L. B. (2007). "Overlapping Fields and Constructed Legalities: The Endogeneity of Law." In O'Brien, J. (ed.). *Private Equity, Corporate Governance and the Dynamics of Capital Market Regulation*. Singapore: World Scientific Publishing. pp. 55–90.

Escobar, A. (1995). *Encountering Development: The Making and Unmaking of the Third World*. Princeton, NJ: Princeton University Press.

Fatch, J., Bolding, A. and Swatuk, L. A. (2022). "Boundaries of Benefit Sharing: Interpretation and Application of Substantive Rules in the Lake Malawi/Niassa/Nyasa Sub-Basin of the Zambezi Watercourse." *Int Environ Agreements*. https://doi.org/10.1007/s10784-022-09585-4

Fisher, D. and Freudenburg, W. (2001). "Ecological Modernization and Its Critics: Assessing the Past and Looking Toward the Future." *Society & Natural Resources* 14(8), 701–709.

Fox, I. K. (1976). "Institutions for Water Management in a Changing World." *Natural Resources Journal* 16(4), 743–758.

Giordano, M. and Shah, T. (2014a). "Non-Integrated Water Resources Management." In Martinez-Santos, P. and Aldaya, M. M. (eds.). *Integrated Water Resources Management in the 21st Century: Revisiting the Paradigm*. London, UK: CRC Press. pp. 37–46.

Giordano, M. and Shah, T. (2014b). "From IWRM Back to Integrated Water Resources Management." *International Journal of Water Resources Development* 30(3), 364–376.

Grover, G. G. and Mann, J. F. Jr (1991). "*Acton v Blundell* Revisited: "Property" in California Groundwater." *Western State University Law Review* 18(2), 589–598.

Haller, T. (2002). "The Understanding of Institutions and their Link to Resource Management from a New Institutionalism Perspective." IP6 Institutional Change and Livelihood Strategies Working Paper No. 1. Zurich: University of Zurich.

Harvey, D. (2005). *A Brief History of Neoliberalism*. Oxford, UK: Oxford University Press.

Hileman, J., Hicks, P. and Jones, R. (2016). "An Alternative Framework for Analysing and Managing Conflicts in Integrated Water Resources Management (IWRM): Linking Theory and Practice." *International Journal of Water Resources Development* 32(5), 675–691.

Hofstede, G. (2001). *Culture's Consequences: Comparing Values, Behaviors, Institutions and Organizations Across Nations*, 2nd Edition. Thousand Oaks, CA: Sage.

Homer-Dixon, T. (1994). "Environmental Scarcities and Violent Conflict: Evidence from Cases." *International Security* 19(1), 5–40.

Hoogvelt, A. (2001). *Globalization and the Postcolonial World: The New Political Economy of Development*. London, UK: Macmillan.

Jolly, R., Boas, M. and McNeill, D. (2003). *Global Institutions and Development: Framing the World?* London, UK: Routledge.

Jütting, J. (2003). *Institutions and Development: A Critical Review. Working Paper No. 210*. Paris, France: OECD Development Centre.

Kittikhoun, A. and Schmeier, S. (eds.) (2021). *River Basin Organizations in Water Diplomacy*, London, UK: Routledge.

Kolbert, E., 2014. *The Sixth Extinction: An Unnatural History*. New York, NY: Henry Holt.

Kornfeld, I. E. (2009). "Mesopotamia: A History of Water and Law." In Dellapenna, J. and Gupta, J. (eds.). *The Evolution of the Law and Politics of Water*. Dordrecht: Springer Netherlands. pp. 26–36.

Latour, B. (1993). *We Have Never Been Modern*. Cambridge, MA: Harvard University Press.

Li, Y. and Zahra, S. A. (2012). "Formal institutions, Culture, and Venture Capital Activity: A Cross-Country Analysis." *Journal of Business Venturing* 27(1), 95–111.

Meinzen-Dick, R. (2007). "Beyond Panaceas in Water Institutions." *PNAS* 104(39), 1–6.

Mittelman, J. H. and Pasha, M. (1997). *Out from Underdevelopment Revisited: Changing Global Structures and the Remaking of the Third World*. New York, NY: St. Martin's.

Muller, M. (2006). *How IWRM Will Contribute to Achieving the MDGs. GWP TEC Policy Brief 4*. Stockholm: GWP.

North, D. C. (1990). *Institutions, Institutional Change and Economic Performance*. Cambridge, UK: Cambridge University Press.

North, D. C. (1993). *Economic Performance through Time*. Prize Lecture. The Sveriges Riksbank Prize in Economic Sciences in Memory of Alfred Nobel 1993. [Accessed 17 March 2022]. Available at https://www.nobelprize.org/prizes/economic-sciences/1993/north/lecture/

Partey, S. T. (2018). "Developing Climate-Smart Agriculture to Face Climate Variability in West Africa: Challenges and Lessons Learnt." *Journal of Cleaner Production* 187, 285–95.

Phillips, N. (2020). "The Political Economy of Development." In J. Ravenhill (ed.). *Global Political Economy*, 6th Edition. Oxford, UK: Oxford University Press. pp. 354–383.

Postel, S. (2017). *Replenish: The Virtuous Cycle of Water and Prosperity*. Washington, DC: Island Press.

Sachs, J. D. (2015). *The Age of Sustainable Development*. New York, NY: Columbia University Press.

Salmond, A. (2012). "Ontological Quarrels: Indigeneity, Exclusion and Citizenship in a Relational World." *Anthropological Theory* 12(2), 115–141.

Solomon, S. (2010). *Water: The Epic Struggle for Wealth, Power and Civilization*. New York, NY: HarperCollins.

Swatuk, L. A. (2002). "The New Water Architecture in Southern Africa: Reflections on Current Trends in Light of Rio +10." *International Affairs* 78(3), 507–530.

Van der Zaag, P. (2005). "Water's Vulnerable Value in Africa." In IHP/HWRP." *Value of Water – Different Approaches in Transboundary Water Management*. Koblenz, Germany: IHP/HWRP-Secretariat. pp. 19–30.

Wackernagel, M. and Rees, W. E. (1996). *Our Ecological Footprint: Reducing Human Impact on the Earth*. Gabriola Island, BC: New Society Publishers.

Wengert, N. (1983). "Water Management Institutions." In Guggiero, E., Rossi, G. and Hendricks, D. (eds.). *Operation of Complex Water Systems*. Boston, The Hague, Dordrecht, Lancaster: Martinus Nijhoff Pub. pp. 277–292.

World Bank. (1997). *World Development Report*. New York, NY: Oxford University Press.

World Health Organization (WHO). (2006). *Water & Culture: The International Decade for Water 2005–2015*. [Accessed 17 March 2022]. Available at https://www.who.int/water_sanitation_health/Water&cultureEnglishv2.pdf?ua=1

Worster, D. (1985). *Rivers of Empire: Water, Aridity, and the Growth of the American West*. Oxford, UK: Oxford University Press.

Wright, R. (2004). *A Short History of Progress*. Toronto, ON: Macmillan.

Ynalvez, M. A. and Shrum, W. M. (2015). "Science and Development." *International Encyclopedia of the Social and Behavioral Sciences*, 2nd Edition, 21, 150–155. http://dx.doi.org/10.1016/B978-0-08-097086-8.85020-5

4

WATER AND HUMAN DEVELOPMENT

Unpacking Scarcity and 'Water Crises'

Lyla Mehta

Introduction

Human development is interested not just in economic growth, but in expanding human capabilities and choices (Anand and Sen, 2000). The concept of human development could perhaps go back to Aristotle who was one of the first defenders of the human good, or human flourishing. In his *Nicomachean Ethics*, Aristotle saw wealth, which is often sought after, as the means for providing the necessities of life, such as food, health and so on (Nussbaum, 1987). The end, for him, was happiness, self-fulfilment and self-realization which led to human flourishing. The Aristotelian notion of the human good links the necessity to 'first ascertain the function of man' and subsequently explores 'life in the sense of activity' (see Nussbaum, 1987; Sen, 1999: 73). Water and sanitation are basic necessities, enabling people and ecosystems to function and human activity to flourish. For poor people, access to water is a prerequisite to achieving a minimum standard of health and to undertake productive activities. Water has multiple cultural and aesthetic values, faces and meanings in the everyday contexts within which people live their lives (HLPE, 2015). Safe drinking water and sanitation are fundamental to the nutrition, health and dignity of all (UNDP, 2006; UNSCN, 2020).

Water is also the lifeblood of most ecosystems, such as forests, lakes or wetlands which are particularly important for poor people, providing them with nutrition and livelihoods. Water is used to grow food both for subsistence and commercial purposes. Water is also fundamental for all other sectors of importance to human well-being, including energy, manufacturing and industrial productivity.

Access to safe and convenient water supplies is also crucial to enhance women's and girls' well-being. Cultural norms dictate that women and girls are responsible for water collection and can spend between three minutes and three hours per day collecting water. This time instead could be used to focus on livelihood and agricultural activities and also improve maternal health and that of infants. Girls, often overburdened by time-consuming water collection activities, should have time to attend school and enjoy a normal childhood. Access to water is deeply unequal and contested around the globe. Poor and marginalized people also lack access to safe and adequate water, either because it is too expensive or because they are excluded due to race, class, caste, ethnicity or gender. They also often do not share the gains of large infrastructure projects such as dams and

DOI: 10.4324/9781003095545-5

irrigation schemes, while often negatively affected by them through displacement as well as loss of livelihoods and land (as discussed later in this chapter, see also WCD, 2000).

Aristotle's mentor, Plato, illuminated the paradox of the value of water and diamonds. Plato in *Euthydemus* (Section 304B) found that what is rare is valuable, while water, considered the best of all, is also the cheapest (see Toye, 2005). Those goods that were rare were useless (like diamonds). By contrast, water was seen to be abundant and useful. Today, however, very few people would consider water to be abundant. Rather the dominant narrative is about the growing water crisis due to scarcity. Despite the progress made in achieving global targets around water and sanitation in the past decades, about three in ten people, or 2.1 billion lack access to safe, readily available water at home, and six in ten, or 4.5 billion, lack safely managed sanitation (WHO and UNICEF, 2017). Accessing water can be particularly challenging for small holders, vulnerable and marginalized populations and women. Even though the human right to water (RTW) and sanitation was globally endorsed by the United Nations in 2010, it is violated every day across the globe. This situation undermines health, nutrition, human wellbeing and dignity and is a global and moral outrage.

This chapter spells out a human development approach to water and explores the application of both the entitlement and capability analysis (EA and CA henceforth) to the different aspects of water. The chapter addresses two challenges: (1) it seeks to advance a human development approach to view the water 'crisis' and water scarcity (cf. Mehta, 2014), and (2) it explores how EA and CA can be applied to water and whether they can help enhance water justice and expand freedoms around water. This is a challenging task because there is no one EA and CA, but instead many approaches (Gasper, 2006). There are also different normative, political and policy implications of these approaches and challenges in application to policy and practical realities (see Robeyns, 2005; Gasper, 2006).

I argue that a human development approach to water scarcity helps capture the multifaceted nature of water as well as challenge dominant and simplistic portrayals of the water 'crisis'. The EA allows us to move away from aggregate views of water scarcity to focus on the structural and institutional arrangements (including market-based mechanisms) that exclude the poor and intensify water-related inequalities. However, merely having access to water is not enough. Instead, a person needs a certain kind of access to water in order to derive certain freedoms or functionings (i.e., capabilities) which in turn depend on a host of factors. The CA thus highlights the importance of the multi-faceted nature of water, its links with wellbeing and other freedoms. Conventionally dominant modes of water management have been characterized by sectoral approaches that separate water and sanitation from water for food, energy, domestic supply, irrigation and floodwater management (Mehta et al., 2020). The chapter shows how both EA and CA help strengthen the case for the human RTW and CA in particular helps break down the false distinction between water for domestic and productive purposes. Thus, despite some limitations, applying the EA and CA help enhance equity considerations in the water sector. This is important because even though water policy rhetoric may be about rights and equity, in practice a focus on volumetric issues of supply and demand as well as considerations of utility and efficiency persist which may not always have the interests of the marginalized upfront.

The Multifaceted Nature of Water

People across the globe value water for both its non-economic and economic roles and it also has deep spiritual significance in many cultures. However, official water resources management discourses (such as those endorsed in the 1992 Dublin principles) largely tend to focus on the economic values of water. Merely viewing water through an economic lens (for example as an economic good) can undermine its embeddedness in the everyday symbolic, cultural and social contexts within which people live their lives (see Hellum et al., 2015). In particular, scholarship on indigeneity and water has highlighted the importance of focussing on the multiple ontologies of water (Wilson and Inkster, 2018).

Water, more than most resources, is highly variable across time and space. Its availability is characterized by the complex interactions of a number of elements which include rainfall, temperature, wind, runoff, evapotranspiration, storage, distribution systems, technologies and water allocation and property regimes as water quality (see Movik, 2012; Mehta et al., 2020).

Due to the fluid nature of water, water rights are usually competing and overlapping and entail a mixture of formal and informal arrangements often shaped by customary law and practices, kinship networks, race, class, gender, caste and patronage (see Movik, 2012; Sultana and Loftus, 2020). However, the multifaceted aspects of water are overlooked in dominant and global portrayals of the resource that tend to largely focus on the volumetric and material aspects of water to which I now turn.

Unpacking Water Scarcity

Water scarcity has emerged as one of the most pressing problems in the twenty-first century. Against a growing alarmism about 'water wars', global agencies, national governments and NGOs have been concerned with emerging water 'crises' and the causality and solutions around water scarcity. Largely, the terms water 'crisis', water shortage, scarcity and stress are used very loosely in conventional debates. In reality, water scarcity is a complex phenomenon and can be analysed differently from social, political, meteorological, hydrological and agricultural perspectives. There has been the tendency though to direct attention to the lack of supply of water due to natural and economic forces rather than look at human-induced land and water use practices and at socio-political considerations (see UNDP, 2006).

There are competing pathways and discourses regarding water security and scarcity. Scarcity of water is typically examined through two lenses – the first looks at the amount of water available per capita in a particular area, taking into account the average volume of water available in rivers, lakes, dams and groundwater aquifers per person. This is considered physical scarcity of water. For example, Swedish hydrologist Malin Falkenmark and colleagues (see Falkenmark and Widstrand, 1992) have classified countries according to a 'water stress index' on the basis of their annual water resources and population. This widely adopted definition proposes a threshold of 1700 cubic metres per person per annum below which countries are said to be water-stressed and water scarcity is less than 1000 cubic metres per person per annum.

However, this lens does not indicate whether that water is available for use or accessed. In some areas there may be a sufficient volume of water available per person, but a lack of infrastructure and investment means that the water is not available where is it needed, or of an appropriate quality for use. This is deemed economic water scarcity (CA, 2007). For example, even though the Democratic Republic of the Congo (DRC) has more than half of Africa's water reserves (UNEP, 2011), currently only 52% of the population has access to an improved water source and 29% has improved sanitation facilities.[1]

But both these lenses obscure other key dimensions of water scarcity. They tend to direct attention to natural and economic forces rather than look at human-induced land and water use practices and at socio-political considerations and how scarcity can be socially mediated or constructed (UNDP, 2006). They also tend to take an aggregate view of populations lacking access to water, rather than breaking down groups due to gender, caste, race and so forth. Thus, it is important to also focus on the socio-political dimensions of scarcity and how access to water is often determined by social difference as well as political will and levels of national economic development. Thus, people's lack of access to water may have little to do with physical scarcity per se but may instead be due to exclusions arising from social positioning, race, power and gender relations or because of the way water is managed, priced and regulated (see Mehta, 2005; UNDP 2006). Finally, scarcity can also be induced through policies and planning that can lead to inefficient use/overexploitation of water resources and also due to market-based mechanisms that promote cut-offs or exclusions.

A Human Development Approach to Water Scarcity

The human development movement has focussed on actual lives lived by a range of people, especially the poor and marginalized. A human development approach to scarcity would thus break down macro and aggregate understandings of scarcity. It would argue that scarcity regarding access to water is unacceptable in the twenty-first century. This is particularly so because scarcity is not 'natural' but generated through socio-political processes, through exclusion, biases and discrimination. For example, in India, so called lower caste women are still denied access to certain wells. In apartheid South Africa, the inequalities based on discriminatory policies were huge. Consequently around 80 percent of the poor in rural areas had no access to water or sanitation in 1994 at the birth of the new South Africa (see Movik, 2012).

A human development approach to scarcity can usefully draw on Amartya Sen's entitlements approach (1981, 1983, 1985, 1999). Sen sees entitlements as the 'the set of alternative commodity bundles that a person can command in a society using the totality of rights and opportunities that he or she faces' (Sen, 1983: 754). EA is largely a descriptive, not normative framework. An entitlement set includes the full range of goods and services that a person acquires by converting her 'endowments' (namely, assets and resources including labour) through 'exchange entitlement mapping' (see Sen, 1981; 1983). These in turn determine the various functionings that a person can achieve, which is directly linked to the capability approach (CA). Capabilities refer to 'the totality of all the alternative functioning vectors a person can choose from' (Sen, 1985: 27) and reflect an individual's well-being (see below). Here the links to Aristotle's notions of the human good or necessities to 'ascertain the function of man' (see Sen, 1999: 73) as outlined earlier are obvious.

In his seminal study of starvation and famines, Sen argued that the fixation with per capita food availability decline (FAD) is a misleading way to look at hunger and famine, since hunger is more about people not having access to food due to wider social and political arrangements as opposed to there not being enough food to eat (Sen, 1981; 1983). Looking at per capita availability of a resource lacks relevant discrimination and is even more gross when applied to the population of the world as a whole (Sen, 1981). Water scarcity is also often misleadingly perceived as per capita water availability rather than inequality in access to water supply. Usually water access is determined by social and political institutions, cultural norms and property rights. Some groups may suffer from lack of water even when there is no decline in water availability in the region. Thus, water shortages (like famines) are entitlement failures. Sen's entitlements framework has been applied to look at water scarcity in India by P. B. Anand to ask why some people suffer from more water deprivation than other and what society can do about this (2007). In particular, the poor may be more vulnerable to health impacts of water supply problems than others. The entitlements lens thus helps us move away from the conventional metrics of water scarcity that were discussed earlier in this chapter.

Poor and landless rural dwellers may lack endowments such as irrigation facilities and wells that can help them have secure water supplies. Poor people who cannot afford to pay for water from taps and handpumps may opt for water from unprotected streams and be exposed to water pollution and disease. Invariably, the poorest have low water endowment. This can negatively affect women's and girls' capabilities and life chances who are culturally required to spend a lot of time on water collection which in turn can also affect the wellbeing of the entire household. Thus, the EA when applied to water would indicate that some people's lack of water does not necessarily imply that water is scarce. Instead, it means that some people lack entitlements to water either because it is too highly priced, due to the lack of infrastructure, or due to social exclusion. Some are seen as 'entitled' by virtue of who they are or are taken to be socially. For example, in many cities of the global South rich people in urban areas have swimming pools while those in peri-urban dwellings are considered to be 'informal' and lack a range of entitlements to water. Furthermore, state sponsored water is usually directed to rich and middle class localities that are connected to the formal water system while the poor and minorities are left to fend for themselves and secure water by informal means (see Anand, 2017). Also water privatization in countries such as South Africa led to thousands of people being cut off from water services which severely restricts citizens' ability to enjoy their constitutional RTW (see Flynn and Chirwa, 2005). Merely having access to water is not enough. Poor quality affects people's entitlements and endowments in different ways. Degradation of local sources forces poor people in the global South to purchase water which in many cases can cost 10–20 times more than what residents in the global North pay (UNDP, 2006). If water is contaminated and causes diarrhoea to different family members, it affects people's capability to engage in productive activities and also leads to rising costs in health care. If people lack a range of entitlements (e.g., technologies to purify water or the finances), they may have no choice but to drink polluted water. Thus, it is important to look at the structural barriers that deny access to water.

The application of the EA to water helps avoid the water sector's conventional focus on utilitarian arguments and notions of the 'common good'. Rather than national and aggregate numerical portrayals of water availability, the focus would be on individual and group entitlements and endowments to water, taking into account factors such as gender, caste, ethnicity, race, occupation and location (e.g., peri-urban, rural or urban). Thus, the EA urges us to go beyond volumetric or per capita measurements of water availability, neo Malthusian understandings of the water 'crisis' and water scarcity to also address the structural and institutional issues concerning water inequalities and injustices.

Water, Capabilities and Well-Being

A human development approach to water would also question the water sector's conventional focus on cost-benefit analyses to evaluate well-being (Anand, 2007). This is where the capability analysis (CA) is most useful. Traditional cost-benefit analyses, which emerged to identify and measure the costs or profits emerging out of infrastructure projects such as dams, failed to capture intangible issues such as changes in socio-cultural identity, the loss of common property resources and geographical space, crucial for a community's well-being (WCD, 2000) due to a conventional one dimensional materialistic evaluative frame that is commonly used for cost-benefit analyses. Rather than focussing on the means of achieving the 'good', Sen prefers to look at the 'actual living that people manage to achieve' (Sen, 1999: 73). Sen argues that even though it is common to 'use incomes and commodities as the material basis of our well-being…what use we can respectively make of a given level of income, depends crucially on a number of contingent circumstances, both personal and social' (1999: 70). Hence, well-being is firmly anchored in a particular social and personal context. This is why Sen advocates for evaluative purposes, the CA as a means to measure well-being. This approach focuses on 'substantive freedoms – the capabilities – to choose a life one has reason to value' (1985, 1993, 1999: 74). Thus, at the heart of this approach one must look at the freedoms that an individual can enjoy. Thus, development, according to Sen, is a process of expanding the real freedoms that people enjoy (1999: 3).

In this broader sense, well-being should increasingly be understood as a multidimensional phenomenon ranging from income to the public provision of goods and services, access to common property resources and other intangible dimensions such as clean air, water, dignity, self-respect and autonomy (Razavi, 1999). Conventional ways of evaluating water and well-being, say through the Sustainable Development Goal (SDG) six focuses on addressing universality and inequality, and also sets out a 'service ladder' a 'service ladder' from 'safely managed' to 'basic', 'unimproved' and surface water (see Cumming and Slaymaker, 2018). While there is greater awareness of gender inequalities in the SDG framework compared with earlier approaches, sex-disaggregated data is seldom collected, making it impossible to monitor progress or devise gender sensitive policies around water. Until recently a source that could be accessed within 30 minutes (round-trip collection time) was considered an 'improved water source' as per the SDG indicators tracking progress on drinking water, sanitation and hygiene. While the new global SDG indication states that the source must be located on premises, in reality many poor women and girls in especially in rural areas continue to make multiple trips of 30 minutes or more with heavy water pots because the service that brings water to their home is unreliable and often breaks down. There is thus a critical need to ensure that more women are part of the decision-making processes when such standards and targets are being set. Finally, as argued by Mitra and Rao (2019) in the enthusiasm to meet the global targets, governments tend to ignore the interconnectedness between water services and food production and their wider embeddedness in ecology (i.e., land use and water flows), gendered labour patterns and how these are determined by power relations. Finally, a more multidimensional approach would also focus on issues such as autonomy, gender equality, links with identity and the freedom to choose. We cannot assume that once water supply is provided within a certain distance (usually about 1000 metres), access to water has been created.

Take the case of Vasava and Tadvi, *Adivasi* (indigenous/so called 'tribal people') in Gujarat, India who were displaced from their ancestral homes from the banks of the River Narmada due to reservoir flooding in the early 1990s. They were resettled 200 kilometres away against their will in villages in the plains where they were considered 'primitive' and 'backward' by the dominant caste Hindu groups who inhabited the villages. The officials thought that the women were now 'developed' because taps were installed a few metres away from their home. But the women preferred water of the river to the tap provided close to the house by the government because the taste of the river water was better. They also wanted the freedom to collect water when it suited them, not when the officials turned on the tap (Mehta and Punja, 2006). Displacement had robbed them of their access to a river they considered 'Mother' and holy, reduced their sense of autonomy and independence and the interactions with friends that they enjoyed through daily water collection. In addition to all the problems around water access, the displaced communities also struggled to make a livelihood out of the poor farm land that was allocated to them and also missed access to forest and common property resources which were very crucial for their livelihood security (Mehta and Punja, 2006).

How does the CA help us to understand displaced people's ill-being after displacement, something that is often neglected or even wilfully ignored by policy makers and resettlement officials? Before displacement, the tribal villagers were entitled to a near unlimited and constant supply of good water from the river. They were able to convert that river water to food by growing crops on the river bed and through fishing. All the villagers had the freedom to go to the river and fetch water whenever it suited them. The ability to extract the water will have varied across households, as may the freedoms and opportunities realized on the basis of the water, but few if any social limitations were placed on its use, and the endowment was unrestricted. After displacement by contrast the supply was intermittent, the water poor, and access limited by the pecking order at the spout. The entitlement was now restricted, the endowment severely cut back, and the poor water impossible to convert into the well-being and greater agency freedom of using good river water. Displacement also reduced their capability to convert resources such as the river, grazing lands and the forest to various freedoms and well-being that promoted their livelihood security, health and provided an additional source of income (e.g., the forest provided herbs, wood and grazing lands for their cattle). Yet by the metric of the engineers and donors, upon displacement, a spout if closer than the bank of the river will count as an improvement, though in this case it contributes to the immiseration of displacement. Thus, the CA is able to account for the deprivation and marginalization of the villagers in terms of water use, sense of place and identities while conventional aggregate framings hide them. Furthermore, the CA focuses on both process and opportunity aspects of substantive freedoms. It allows us to argue that tribal communities suffered both to problems in water quality and access and were also deprived of all the opportunities to convert river water into further capabilities. Finally, they also lacked the freedom to participate in decision-making processes about their displacement and the social and economic aspects of life after relocation, which is linked to what Martha Nussbaum calls process freedom (see Nussbaum, 2003).

Human Development and the Human Right to Safe Drinking Water and Sanitation (HRSDSW)

Even though water is fundamental to the nutrition, health and dignity of all, the RTW was not explicitly acknowledged in the 1948 Universal Declaration of Human Rights. Until the turn of the twenty-first century, there remained considerable resistance to the RTW on the part of some nations and corporations (see Mehta et al., 2020). While the RTW is an important galvanizing call in the global struggle for water justice (cf. Boelens et al., 2018) and several countries, such as South Africa, Ecuador, Bolivia, Gambia, Tanzania and Uruguay, have recognized the human RTW, there still remains a considerable gap between human rights talk and human rights practice.

Despite the long over-due global recognition, the RTW remains conceptually ambiguous. There is sometimes a clear tension between a government's commitment to rights and to market-based mechanisms, with the latter tending to prevail having consequences for poor people's RTW. It is also unclear what constitutes the RTW, i.e., in terms of the actual amount but also whether its narrow scope should be expanded to also look at wider livelihood needs beyond domestic issues (Mehta et al., 2020).

Here the EA and CA come to our help to break down false divisions in the water domain. As the 2006 Human Development Report puts it these are: 'water for life' (i.e., domestic and drinking purposes) and 'water for production' (i.e., water for food, irrigation and industrial activities) (UNDP, 2006). This distinction, however, is highly problematic from the perspective of local users whose daily activities encompass both the domestic and productive elements of water and for whom there is little sense in separating water for drinking and washing from water for small-scale productive activities that are so crucial for survival. Empirical research demonstrates the critical role played by water for productive purposes and livelihoods especially for poor women in both rural and peri-urban areas (Van Koppen et al., 2017), for example, watering crops and livestock, brickmaking/

This is why there are growing calls for an elaboration of a human rights perspective to water and livelihoods that encompasses the productive uses of water while being more interconnected than the current RTW (e.g., Hellum et al., 2015; HLPE, 2015; Van Koppen et al., 2017; Mehta et al., 2020). This broader conceptualization of the RTW is truer to how water is understood and embedded in the daily lives of local women and men around the world. Local communities rarely distinguish water for domestic and subsistence purposes. It is thus important to break down these siloes in the water sector, including in the rights domain, in order to embrace a more holistic definition of water, well-being and human development.

Here too the capabilities approach helps. As discussed, Sen's approach focuses on 'substantive freedoms – the capabilities – to choose a life one has reason to value' (1985, 1993, 1999: 74) and the freedoms that an individual can enjoy. Here the focus is not on the quantity of the bundles of entitlements but instead on the principle of equality and capability to do and to be. When translated to water, such an approach would mean that a basic amount of water is required for basic human functioning (drinking, washing and to be free of disease) and one could also argue that this minimum requirement for human functioning should also capture livelihood and subsistence purposes (see Mehta, 2014). This strongly resonates with work by Jepson et al. (2020) who argue that water and access to it should be understood as a hydrosocial and cultural process to 'include water flows, water quality and water services in all of their socio-natural complexity' (Jepson et al., 2020).

The capabilities approach refrains from outlining what exactly this minimum threshold should be. Evidence from the water sector in setting up standards around what constitutes a 'basic water requirement' highlights the variations by country and by institution. Basic water requirements have been suggested by various donor agencies, regardless of culture, climate or technology. The WHO definition – seemingly blind to contexts such as culture, climate, livelihoods, urban and rural – prescribes between 20 and 100 litres a day (WHO, 2003), but this amount excludes water for productive or survival activities such as growing food. Clearly, this low end provision takes a very narrow view of the water needs of the poor, not considering livelihood and subsistence needs. By contrast, there is rarely any discussion of what the 'maximum amount of water' people can consume per day, for example, in the United States it is about 300 litres.[2]

People ultimately need different basic amounts of water to enjoy the same standard in terms of capability. Take the case of South Africa which was one of the first countries that explicitly recognized the RTW. Its Free Basic Water policy provided a minimum of 25 litres per capita per day based on a household size of eight people initially free to all citizens (see McDonald and Ruiters, 2005). But implementing the RTW in South Africa was fraught with difficulties and there have been huge debates regarding whether the right has had a significant impact on improving the well-being of poor South African citizens (ibid., see also Flynn and Chirwa, 2005). There are further heated debates about whether the RTW is compatible or not with parallel trends of water privatization or if this created new forms of poverty and ill-being reinforcing racialized and class inequalities (see McDonald and Ruiters, 2005; Bond, 2020). Last but not least, it has also raised questions concerning the sufficiency of 25 litres per day per person, especially if the household number is large. All these issues further emphasize how a human development perspective building on CA supports the case for having a broader conceptualization of the RTW to also encapsulate wider productive uses of water.

Conclusions

This chapter argued for the need to advance a human development approach to water. It demonstrated how the EA allows us to move away from aggregate views of water scarcity and water crises to focus on the structural and institutional arrangements (including market-based mechanisms) that exclude the poor and intensify water-related inequalities. The chapter also demonstrated how merely having access to water is not enough. Instead, a person needs a certain kind of access in order to derive certain freedoms or functionings (i.e., capabilities) which in turn depend on a host of factors and capabilities. Linked to these issues are people's ability (or not) to participate in decision-making processes around water resources management and development. Policy makers therefore need to focus on the multi-dimensional aspects of water and their links with human well-being and how individuals and communities can play an active role in shaping their water futures. There needs to be far more focus on creating participative freedoms for local women and men to relate to the multiple and fluid ontologies and properties of water and make water-related decisions in a culturally appropriate way. Providing for the freedom and capabilities of people in relating to water and its uses must also necessarily entail questioning dominant discourses and relations of power that shape issues concerning water scarcity, water commodification, dam-related displacement and so on.

The CA helps strengthen the case for the human RTW and governments need to prioritize providing poor people with access to water that is safe, affordable and allows them to flourish. Both the EA and CA help to break down the false distinction between water for domestic and productive purposes to be more true to how water is understood and embedded in the daily lives of local women and men around the world.

Thus, the human development approach provides strong analytical tools to understand why many people in reality cannot access and/or realize their freedoms vis-à-vis water and how poor and disenfranchised people's entitlements and capabilities to water can be strengthened. This is important because even though water policy rhetoric may be about rights and equity, in practice utilitarian, efficiency and sufficiency notions of value persist (cf. Schmidt, 2020) which may not always have the interests of the marginalized upfront. Thus, a human development approach to water that focuses on equality, the 'good' and human freedoms of all is highly necessary.

Acknowledgements

This is an updated and summarised version of Mehta, 2014 which was based on a background paper I did for the 2006 Human Development Report. This article was published in World Development, vol 59, issue C, Mehta, L. Water and Human Development. pp 59–69. Copyright Elsevier (2014)." I thank Sofie Hellberg for her useful comments & patience.

Notes

1 Seehttps://www.unicef.org/drcongo/en/what-we-do/water-sanitation-and-hygiene#:~:text=The%20Democratic%20Republic%20of%20Congo,lack%20access%20to%20quality%20water, accessed January 2023.

2 https://www.epa.gov/watersense/statistics-and-facts, accessed March 2022.

References

Anand, N. (2017). *Hydraulic City. Water and the Infrastructures of Citizenship in Mumbai*. North Carolina: Duke University Press.

Anand, P. B. (2007). *Scarcity, Entitlements and the Economics of Water in Developing Countries*. Cheltenham: Edward Elgar.

Anand, S. & Sen, A. (2000). Human development and economic sustainability. *World Development, 28*(12), 2029–2049.

Boelens, R., Perreault, T. & Vos, J. (Eds.) (2018). *Water Justice*. Cambridge: Cambridge University Press.

Bond, P. (2020). Class, Race, Space and the "Right to Sanitation". In F. Sultana & A. Loftus (Eds.) *Water Politics*. London: Routledge.

CA (Comprehensive Assessment of Water Management in Agriculture). (2007). *Water for Food, Water for Life: A Comprehensive Assessment of Water Management for Agriculture*. London Earthscan & Colombo: International Water Management Institute.

Cumming, O. & Slaymaker, T. (Eds.) (2018). *Equality in Water and Sanitation Services*. London: Routledge.

Falkenmark, M. & Widstrand, C. (1992). Population and water resources: A delicate balance. *Population Bulletin, 47*(3), 1–36.

Flynn, S. & Chirwa, D. M. (2005). The Constitutional Implications of Commercialising Water in South Africa. In D. McDonald & G. Ruiters (Eds.) *The Age of Commodity: Water Privatization in Southern Africa*. London: Earthscan, pp. 59–77.
Gasper, D. (2006). What Is the Capability Approach? Its core, Rationale, Partners and Dangers. *Institute of Social Studies Working Papers*, 428. The Hague: Institute of Social Studies.
Hellum, A., Kameri-Mbote, P. & van Koppen, B. (2015). The Human Right to Water and Sanitation in a Legal Pluralist Landscape: Perspectives of Southern and Eastern African Women. In A. Hellum, P. Kameri-Mbote & B. van Koppen (Eds.) *Water Is Life: Women's Human Rights in National and Local Water Governance in Southern and Eastern Africa*. Harare, Zimbabwe: Weaver Press, pp. 1–31.
HLPE. (2015). *Water for Food Security and Nutrition*. Report by the High Level Panel of Experts on Food Security and Nutrition of the Committee on World Food Security. Report No. 9, Rome: FAO.
Jepson, W., Wutich, A. & Harris, L. M. (2020). Water-Security Capabilities and the Human Right to Water. In *Water Politics*. London: Routledge, pp. 84–98.
McDonald, D. & Ruiters, G. (Eds.) (2005). *The Age of Commodity: Water Privatization in Southern Africa*. London: Earthscan.
Mehta, L. (2005). *The Politics and Poetics of Water: Naturalising Scarcity in Western India*. New Delhi: Orient Longman.
Mehta, L. (2014). Water and Human Development. *World Development*, *59*(C), 59–69.
Mehta, L., Oweis, T., Ringler, C., Schreiner, B., Varghese, S. (2020). *Water for Food Security, Nutrition and Social Justice*. London: Routledge.
Mehta, L. & Punja, P. (2006). Water and Well-Being: Explaining the Gap in Understandings of Water. In A. Baviskar (Ed.) *Waterscapes: The Cultural Politics of a Natural Resource*. Delhi: Permanent Black, pp. 188–210.
Mitra, A. & Rao, N. (2019). Gender, water and nutrition in India: An intersectional analysis. *Water Alternatives*, *12*(1), 169–191.
Movik, S. (2012). *Fluid Rights: South Africa's Water Allocation Reform*. Cape Town: Human Sciences Research Council Press.
Nussbaum, M. C. (1987). *Nature, Function, and Capability: Aristotle on Political Distribution*. Helsinki: World Institute for Development Economics Research.
Nussbaum, M. C. (2003). Capabilities as fundamental entitlements: Sen and social justice. *Feminist Economics*, *9*(2–3), 33–59.
Razavi, S. (1999). Gendered poverty and well-being: Introduction. *Development and Change*, *30*(3), 409–433.
Robeyns, I. (2005). The capability approach: A theoretical survey. *Journal of Human Development*, *6*(1), 93–114.
Schmidt, J. (2020). Valuing Water: Rights, Resilience and the UN High-Level Panel on Water. In F. Sultana and A. Loftus, op. cit. pp. 15–28.
Sen, A. (1981). *Poverty and Famines: An Essay on Entitlement and Deprivation*. Oxford: Clarendon.
Sen, A. (1983). Development: Which way now? *The Economic Journal*, *93*(372), 745–762.
Sen, A. (1985). *Commodities and Capabilities*. Amsterdam: North-Holland Press.
Sen, A. (1999). *Development as Freedom*. Delhi: Oxford University Press.
Sultana, F. & loftus, A. (Eds.) (2020). *Water Politics. Governance, Justice and the Right to Water*. Oxford: Routledge.
Toye, J. (2005). What Did Economists Do With Scarcity. In *Presentation at the Conference 'Scarcity and the Politics of Allocation'*. Brighton: IDS.
UNDP (United Nations Development Programme). (2006). *Beyond Scarcity: Power, Poverty and the Global Water Crisis*. Basingstoke: Palgrave.
United Nations Environment Programme (UNEP). (2011). *Water issues in the Democratic Republic of the Congo: Challenges and Opportunities*. Nairobi: Kenya.
UNSCN (United Nations Standing Committee on Nutrition). (2020). *'Water and Nutrition. Harmonizing Actions for the United Nations Decade of Action on Nutrition and the United Nations Water Action Decade*. Rome: UNSCN.
Van Koppen, B., Hellum, A., Mehta, L., Derman, W. & Schreiner, B. (2017). Rights-Based Freshwater Governance for the Twenty-First Century: Beyond an Exclusionary Focus on Domestic Water Uses. In E. Karar (Ed) *Freshwater Governance for the 21st Century*. London, UK: Springer, pp. 129–143.
WHO. (2003). *The Right to Water. Health and Human Rights* (3). Geneva: WHO.
Wilson, N. J. & Inkster, J. (2018). Respecting water: Indigenous water governance, ontologies, and the politics of kinship on the ground. *Environment and Planning E: Nature and Space*, *1*(4), 516–538.
World Commission on Dams (WCD). (2000). *Dams and Development: A New Framework for Decision-Making*. London and Sterling, VA: Earthscan.
World Health Organisation (WHO) and the United Nations Children's Fund (UNICEF). (2017). *Progress on Drinking Water, Sanitation, and Hygiene: 2017 Update and SDG Baselines*. Geneva and New York: WHO and UNICEF.

5

CRITICAL AND POST-STRUCTURAL APPROACHES TO WATER AND DEVELOPMENT

Sofie Hellberg

Introduction

This chapter provides an overview of critical and post-structural approaches to water and development. Critical approaches, in this context, are approaches that draw on Marxian and neo-Gramscian theories, while post-structural approaches mainly depart from Foucauldian understandings.

Notwithstanding that these theoretical approaches are different, ontologically, and epistemologically, these strands have communalities in terms of their research focus. What is at the heart of this chapter is not so much the differences between these strands in relation to the theory of science but rather what they have contributed to the understanding of the relationship between water and development.

One communality between these diverse bodies of literature is that they have questioned taken-for-granted aspects of water and have queried into what water *is* and *does* (Hellberg, 2018; Linton, 2010; Linton & Budds, 2014; Yates, Harris & Wilson, 2017). The two strands approach questions of the allocation, distribution and management of water as issues of power and rule, and, furthermore, as related to the creation of particular societies and subjects. From these perspectives, "society and water can be understood to make each other" (Linton, 2010). In this regard, water and its governance are central for understanding "development" in the sense that water is a core component of the making of the societies in which we live.

Overall, what we learn from this body of work, sometimes labelled under the umbrellas of political ecology or hydrosocial research, is that it is an inherently political act to identify the "nature of water" (Linton, 2010: 63). Furthermore, the meanings and roles ascribed to water have effects in terms of "who gets access to it and on what terms" (Linton, 2010: 69; see also Johnston, 2003; Strang, 2004). We also learn about inequality, justice, inclusion and exclusion in relation to water access and how these themes and processes are related to neoliberal forms of governing, understood through different theoretical lenses. Another important contribution of this literature is to demonstrate how interactions with water and water technologies are co-constitutive of identities, (gendered[1]) subjectivities and connected to specific notions of citizenship.

In a time when the critique of critical perspectives and post-structuralism abound and water challenges are on the increase, not least because of climate change, it is important to reflect on the role of our perspectives as well our positionality. To deconstruct the dominant narratives of a

DOI: 10.4324/9781003095545-6

global water crisis does not, by any means, imply that critical and post-structural water and development scholars disregard the water problems of the world. On the contrary, this chapter argues, it is a characteristic of both critical and post-structural perspectives on water to take seriously the material aspects of water through its biological function, fluid properties and close connection to technology and infrastructure, and to the devastating effects of climate change. Understanding the latter, which in essence will centre around situations in which there will be either too much or too little water, requires analyses that place the issues of power and inequality in focus as it will hit in uneven and unequal ways.

The chapter begins by introducing how critical and post-structural approaches address the power and politics of water which is followed by a presentation of key concepts in this literature. The chapter then proceeds to give an overview of the contribution of these perspectives in relation to three areas that are central to water and development: (1) The deconstruction the global water crisis (2) neoliberalization of water and (3) the relationships between water, modernity and development. The chapter is then concluded.

The Power and Politics of Water

Critical and post-structural approaches, especially in critical geography and in political ecology, have drawn attention to how power relations affect the regulation, distribution and access to, or exclusion from, water (Johnston, 2003; Loftus, 2009; Swyngedouw, 1999; 2004). From these perspectives, water is not just a chemical substance – H_2O – but a deeply political matter (Linton, 2010).

Neo-Marxian and Gramscian approaches to analysing this dialectic relationship between water and society remain dominant in this literature. This means that social struggles around water have been placed at the centre of academic debates. A Gramscian conceptualization of hegemony has been instructive for much of the critical water literature. This concept is useful for explaining dominance over subordinate groups by the simultaneous use of consent and control and how these social relations are reproduced in relations between the state and civil society (Ekers & Loftus, 2008).

The concept of hegemony has been applied in studies on transboundary issues, and more specifically to frameworks of hydro-hegemony in order to capture complex power dynamics between states and other actors (Menga, 2016; Zeitoun & Warner, 2006). These issues are closely connected not only to security but also to questions of development (Öjendal, Hansson & Hellberg, 2011).

Hegemony as a concept has also been applied in studies on the right to water and to water service delivery more broadly. In this context, studies departing from a neo-Gramscian perspective have placed emphasis on the relationships between ideology, water infrastructure and lived practise in particular places (Ekers & Loftus, 2008). Such scholarly work has also problematized the role of the scholar in terms of how the relationship between theory and practise is conceived of. The core argument here is that critical scholars actively need to learn from water struggles in order to be able to "transform socio-natural relations" (Loftus, 2015).

Some of the work discussed in this chapter draws explicitly on Foucauldian power theory and the notions of (green) governmentality and biopolitics in order to address power relations and modes of governing in water politics and management (see, e.g., Birkenholz, 2009; Boelens, Hoogesteger & Baud, 2015; Gandy, 2006; Hellberg, 2018; Kooy & Bakker, 2008a; 2008b). Some contributions are (careful) dialogues between Gramscian and Foucauldian approaches (Birkenholz, 2009; Ekers & Loftus, 2008). In this vein, Ekers and Loftus (2008) show both the differences between Gramscian and Foucauldian approaches but also makes clear the similarities in their efficacy in understanding the way that ideas and knowledge are infused into the materiality and practices of water, water infrastructures and everyday lives.

In this context, Foucauldian approaches place particular focus on how specific technologies of government are employed. This body of literature has explored the specific shape that water governance takes, in particular places and contexts, and how neoliberal techniques of governing reshape the relations between state, market and society (Boelens, Hoogesteger & Baud, 2015). These studies have explored how hydropolitical relations produce subjects (both policy makers and water users) that take an active part in their own governing (Birkenholz, 2009; Hansson, 2014; Hellberg, 2014), the role of identities in relation to water infrastructure, access to water, and the uncertainty of rule and of resistance (Birkenholz, 2009; Boelens, Hoogesteger & Baud, 2015; Kooy & Bakker, 2008a; 2008b). Some of this work specifically uses the concept of "hydromentality" a concept which will be presented further on in this chapter.

Such a research endeavour, to employ Foucauldian theories on water, has also included seeing water issues and water governance through the lens of biopolitics (Bakker, 2012; Hellberg, 2018). Biopolitics as it is understood here, refers to the regulation of the conditions of life at the general level of the population (Dean, 1999; Foucault, 1998: 139; 2003). Karen Bakker recognizes water as biopolitical in how "modern governments seek to optimize both water resources and our individual water-use practices in order to secure the health and productivity of the population" (Bakker, 2012: 619). Other studies, which have explored water as biopolitical, have explained the biopoliticality of water through the ways, in which it works so as to produce and make appear natural, hierarchizations of different forms of lives, both human and non-human (Hellberg, 2020). Connected to these hierarchizations is also that scarcity gains the function of a regulatory device through which some users are regarded as incapable of becoming sustainable and responsible (Ibid.). Again, such an argument around scarcity as a governing tool does however not mean that problems of water and water access are disregarded. The point here is rather that a biopolitical *dispositif* affects the survival, lifestyles and the possibilities of living a good life of different populations (human and non-human) in different and uneven ways. We will come back to the role of scarcity, as both a material and constructed phenomenon, below.

It also needs to be stated here that this critique, that dominant understandings of water and water governance is productive of hierarchizations of different forms of life, is connected to post-decolonial, feminist and indigenous understandings of water. Contemporary water governance may not only involve a continuation of colonial relationship, discrimination and marginalization, but also a neglect of other ways of knowing water and the relationality towards the "more than human" world (see, e.g., Davis, Arriagada & Harris, this volume; Macgregor, Sritharan & Whitaker, this volume; Wilson & Inkster, 2018; Zwarteveen, this volume). Hence, while this chapter focuses mainly on Marxian, neo-Gramscian and Foucauldian approaches, there are many ties to other, equally important and insightful bodies of work which address issues in relation to the power and politics of water, but which will be discussed elsewhere in this volume.

Key Concepts

When entering into the conceptual world of critical and post-structural scholarship on water, the concept of the "hydrosocial" cycle is important because of its usefulness in "theorizing and analysing water-society relations" (Linton & Budds, 2014: 170). The hydrosocial cycle – in contrast to the hydrological cycle on which the concept is based – attends to the social aspects and political nature of water (Linton & Budds, 2014; Swyngedouw, 2009). The hydrosocial cycle can be conceived of as a "socio-natural process by which water and society make and remake each other over space and time" (Linton & Budds, 2014: 170).

While the hydrosocial cycle is a concept that is useful for understanding water–society relations in general terms, the concept of the *waterscape* is useful for understanding hydrosocial formations and articulations in particular contexts and landscapes, both in the global north and the global south. In Eric Swyngedouw's influential study of the Spanish waterscape (1999), he described it as a "hybrid" "a thing like appearance [...] that is part natural and part social, and that embodies a multiplicity of historical-geographical relations and processes" (Ibid.: 445). Studies of the waterscape have politized water issues through making power relations and political economy central to the analysis of water (Karpouzoglou & Vij, 2017).

A related strand of research has also analysed, and called for, *water justice* (Boelens, Perreault & Vos, 2018; Sultana, 2018; Zwarteveen & Boelens, 2014). According to Sultana, working for water justice involves recognizing that water problems are not purely technical but that they are ecological, political and social issues too. The concept is furthermore based on principles of fairness equality and participation and is characteristically context sensitive (Joy et al., 2014). Such an understanding of water justice makes it impossible to once and for all define what water justice is. Rather, studying water justice becomes an endeavour to "analyze, in all their diversity, how living people experience injustice, facing political oppression, cultural discrimination and economic marginalization" and to relate them both to local perceptions and hegemonic discourses (Boelens, Perreault & Vos, 2018).

Another related concept is "Hydrosocial territories." Hydrosocial territories have been defined as "as spatial configurations of people, institutions, water flows, hydraulic technology and the biophysical environment that revolve around the control of water" (Boelens et al., 2016; see also Duarte-Abadía & Boelens, 2019). The differences between the two concepts have been systematically studied by Flaminio, Rouillé-Kielo and Le Visage (2022). These authors argue that even though both concepts include spatial relations through their focus on landscape and territory respectively, they tend to focus on different issues. While studies of waterscape cover a wider range of topics, often with a focus on water is related to the (re)production of socio-spatial inequalities, studies of hydrosocial territories tend to focus on large-scale projects and the relationship between different stakeholders as well as how water struggles reconfigure power relations between stakeholders (Ibid.: 17–18).

Yet another relevant concept for this chapter is "(neoliberal) hydromentality" (Hellberg, 2014; 2018; 2020; see also the Lankford, 2013, conceptualization with an emphasis on infrastructure). Hydromentality draws on the Foucauldian concept of governmentality and has been defined as "assemblages of governing rationalities, techniques of rule and ways of thinking about, and defining, water" (Hellberg, 2018: 33). Neoliberal governmentality, which is based on the understanding of neoliberalism in a Foucauldian sense (something to which we return below) is in turn, understood as "a hydromentality that includes a focus on market-based solutions to water problems but at the same time recognizes the need for interventions, such as guaranteeing a human right to water and water for the environment, which in the end will make market rationalities function" (Hellberg, 2018: 103–106).

Deconstruction of the Global Water Crisis

One of the important contributions of critical and post-structural studies has been to make visible the political aspects of water and water governance and to deconstruct notions of water and water scarcity. In this regard, critical and post-structural research has politicized and deconstructed the so-called global water crisis, which became manifested as a dominant narrative around water and water governance in the 1990s. Globally, not all are affected by such a crisis and certainly not in the same ways. Linton (2010) argues that gathering water problems "under the rubric of global water crisis diverts attention from the political and social circumstances that produce such problems and frames their solution in predominantly technical and hydrological terms" (2010: 7). To deconstruct the global water crisis in this way also involves deconstructing the notion of scarcity as it is central to the global water crisis narrative. In this narrative, scarcity is described as a physical phenomenon, while from the point of view of critical water scholarship water scarcities are understood as localized and co-constructed by social, political, cultural, geographical and climatological factors, and which has already been indicated above. Therefore, in contrast to mainstream policy discourse, which often portrays scarcity as the physical absence of water and thereby naturalizes (and, thus de-politicizes) water insecurity (Loftus, 2015), hydrosocial research argues that water scarcities are constructed within prevailing political economy (Swyngedouw, 2009).

Lyla Mehta's work is central in this endeavour as she has unpacked different notions of water scarcity. She argues, in a recent publication together with colleagues (Mehta, Huff & Allouche, 2019) that "Scarcity is a dangerous idea and has long been a totalizing discourse in resource politics and mainstream economics." In order to problematize the notion of scarcity Mehta has put forward the concept of "manufactured" (or "constructed") scarcity (2010). This kind of scarcity becomes manufactured through socio-political processes and is used by powerful actors to legitimize certain solutions to water access (Mehta, 2010). Such solutions have historically been large dam constructions. More recently, desalination plants have increasingly become portrayed as a solution to scarcity challenges. Oftentimes, Mehta has argued, such large scale and expensive technological solutions work as obstacles for a more equal distribution of water (Mehta, 2010: 7). From such a viewpoint, analysing scarcity, but without losing sight of its very material aspects and the consequences for people, which Mehta has termed lived/experienced scarcity, the task for the critical water scholar becomes to ask questions such as: *How and by whom has scarcity been created? Who is affected? Who benefits? What do conventional understandings of water scarcity obscure? It also begs for new visions of how the "scare of scarcity" can be challenged* (Mehta, 2010).

Neoliberalization of Water

Another common theme in critical and post-structural research has been to analyse the rationales, and effects, of neoliberalizations of water. Marxist and neo-Gramscian inspired work has taken on to study the relationship between capitalist development and the production of inequalities in access and usage of water. Such studies have analysed neoliberal trends in water governance, including the effects of cost recovery policies, water pricing systems and processes of commodification/privatization (Bakker, 2005; 2010; Loftus, 2006; McDonald & Ruiters, 2005; Swyngedow, 2005). This literature has carefully investigated the processes of making water a site for capital accumulation (see also Budds & Loftus, this volume), including a focus on the involvement of the private sector in water services provision as well as on economic rationalities and marketization in/of both public and private water service delivery utilities.

When understanding neoliberal aspects of water governance there is a notable difference between Marxian/post-Gramscian approaches and post-structural research that depart from a Foucauldian understanding. While the term liberalism in the former is used with reference to a political ideology, it takes on a very different meaning in Foucault's conceptualization. From a Foucauldian perspective, liberalism is understood as a practical art of governing which governs through using the subjects' agency and freedom in order to secure the ends of government (Dean, 1999: 15).

Hence, when analysing the neoliberalization of water governance from a Foucauldian perspective, emphasis is placed on the employment of neoliberal governing "techniques" such as responsibilization and the steering of people's behaviour around water. Another important aspect here is that such a perspective highlights the need for constant intervention and regulations in neoliberal governmentality in order to make market rationality function. Such a perspective questions the conventional idea of (neo)liberalism as an ideology (or form of government) that is first and foremost concerned with allowing things (people, goods and markets) to be "free."

Water, Modernity and Development

Drawing on the concept of the hydrosocial cycle, presented above, Jamie Linton has problematized what he calls "modern water" (Linton, 2010). Modern water, Linton proposes, can be understood as: "the dominant, or natural way of knowing and relating to water, originating in western Europe and North America, and operating on a global scale by the larger parts of the twentieth century" (2010: 14). Not only has this dominant discourse ruled out local knowledge, through making water known simply as "an abstract, isomorphic measurable quantity," (Ibid.: 14; see also Strang, 2004) but it has also been productive of the construction of the (idea of) modern societies with "civilized" citizens as well as too particular understandings of nature.

Most distinct in terms of the work that the notion of modern water does is to reduce water to a "single substance" and detach it from its local, social and cultural, religious and ecological contexts. In this process, water is rendered measurable by the same standards everywhere, a necessary step in order to be able to construct the hydrological cycle (Linton & Budds, 2014). Water thereby becomes identified as a resource that can be managed controlled and utilized, which has been a foundation for the growth of the modern state, manifested in large infrastructural projects, most notably dams. This is also something that has been highlighted by water scholars who have studied water governance though the concept of the hydraulic mission (Molle, Mollinga & Wester, 2009) and the so-called state hydraulic paradigm (Bakker, 2012; see also Öjendal & Hellberg, this volume).

In this line of research, critical geographers Maria Kaika's (2005) and Matthew Gandy (2004; 2006; 2014) has shown how water has been central to the relationships between modernity, nature and urban life. Kaika has explored how, in the late 19th and 20th centuries, water in the cities was "tamed and controlled" through the construction of water supply and sewage networks. These new practices of treating and purifying, and thereby promoting the health of urban populations, became central to the construction of the identity of the West and Western cities, representing the good and orderly urban life (Kaika, 2005: 86).

In this process, the private house played a significant role as it became the place for "good" water, in contrast to water in the public domain, understood to be associated with threat and danger and hence "bad" water. These processes were of course marked by both gender and class differences. Not everyone could enjoy the luxury of a private bathroom, for example (Ibid.: 55), and the work around water was divided between men's work, taming the bad water through infrastructure projects, and women's work with good water in the domesticated sphere (Ibid.: 58–59). Importantly also, in these processes both nature and society were "othered" from the (Western and middle-class) private home as it became an independent sphere, separated for example from dirt and from the homeless (Ibid.: 7).

In the problematizations between water, society and modernity, one aspect that is central is the role of technology. Critical and post-structural scholarship has taken on the task to study how water technologies mediate water – society relationships (see, e.g., Bakker, 2012; Gibbs, 2009; Hellberg, 2014; 2018; Kaika, 2005; O'Reilly, 2006; Swyngedouw, 2004; 2015). Such studies explored how technology networks can be seen as "carriers of modernist promise" and symbols of "advancement and progress," "freedom and emancipation" (Kaika, 2005: 33), as well as productive of notions of hygiene, cleanliness and self-presentation (see Bakker, 2012; Gandy, 2004; Shove, 2003; Sofoulis, 2005). They can also however, and especially with regards to large dams, be described as "'spectacular concretisations' of colonial and/or dependency relations" (Sofoulis, 2005: 452). This body of work has also explored how interactions with water and water technologies such as urban water supply networks, water metres and boreholes are co-constitutive of identities, (gendered) subjectivities and notions of citizenship (Birkenholz, 2009; Ekers & Loftus, 2008; Kooy & Bakker, 2008a; 2008b; O'Reilly, 2006; Rodina & Harris, 2016; Strang, 2004; Sultana, 2009; 2011; 2013; von Schnitzler, 2008).

The above is intrinsically related to the question of development since this literature engages with the making of human societies and subjects as well as with ideas of advancement progress and modernity. Some critical scholars have however *explicitly* addressed the relationship between water and development (and technology) in the global South (Kooy, 2014; Sultana, 2013). While the mainstream story of the relationship between water and development is that of an extended provision of safe drinking water through development interventions (Sultana, 2013), critical scholars have complicated such development narratives around the provision of water access. In this body of work the delivery of safe drinking water through instalment of water technologies has been productive of uneven waterscapes as well as public health crises (Sultana, 2013). In a similar way to how water becomes understood in the modern city (Kaika, 2005), narratives of development become muddled with notions of "good" and "bad" water (Sultana, 2013). The Bengal Delta is a case in point, where a development intervention involving the instalment of tubewell technology, with the aim of providing alternatives to people consuming unhealthy surface water, altered the story of success when it became know that the water from tubewells poisoned people because of the naturally occurring arsenic (Ibid.).

Another theme that has shone light on the relationship between water and development has been work focusing on the role of the informal sector in water services provision (Kooy, 2014; Ranganathan, 2014). This work provides important perspectives on the relationship between state and society, public and private as well as the formal and informal (Ranganathan, 2014) and on the relationship between particular modes of water governance and development. The existence of the informal sector, Kooy contends, should not be seen as a failure of development, but rather as an outcome of particular politics of development processes (2014).

Hence, from these perspectives, the search for "rational," "efficient" and "neutral" water management – or development governance – are futile, as they are legitimated through regimes of power/knowledge and intrinsically tied up with the political economy as well as part of constituting agents and their interests (cf. Lemke, 2007).

Conclusion

This chapter has given an overview of critical and post-structural approaches to water and development. This literature opens the seemingly self-evident question of what water *is* and *does* and it deconstructs notions of the grand narrative of the global water crisis as well as the notions of scarcity.

It is challenging to summarize such rich and diverse bodies of literature and there is not always clear-cut boundaries between Marxian and post-Gramscian inspired work in water scholarship. The chapter has aimed outlining both communalities and differences between Marxian and post-Gramscian inspired work. It has also proposed a few concepts that are at the centre of these approaches to water and development. These concepts – the hydrosocial cycle, waterscapes, water justice, hydrosocial territories and hydromentality – are tools that productively can be used to explore the political aspects and the power dimensions to issues of water and development. Such explorations complicate common understandings and assumptions around the relations between water and development and thereby make an important contribution to this field.

Another core message of the chapter is that the fact that these approaches, regardless of being Marxian or Foucauldian, take seriously the materiality of water make these perspective important and relevant for addressing contemporary water and development challenges. This is because of their efficacy to understand the complexity of water issues (and development) in their social, political and ecological contexts.

Note

1 Feminist perspectives and Gender will be covered in two chapters in this handbook and these perspective and research field will therefore not be expanded on in this chapter.

References

Bakker, K. (2005). "Neoliberalizing Nature? Market Environmentalism in Water Supply in England and Wales." *Annals of the Association of American Geographers*. 95(3), 542–565.

Bakker, K. (2010). *Privatizing Water: Governance Failure and the World's Urban Water Crisis*. Ithaca, NY: Cornell University Press.

Bakker, K. (2012). "Water: Political, Biopolitical, Material." *Social Studies of Science*. 42(4), 616–623.

Birkenholz, T. (2009). "Groundwater Governmentality: Hegemony and Technologies of Resistance in Rajasthan's (India) Groundwater Governance." *The Geographical Journal*. 175(3), 208–220.

Boelens, R., Hoogesteger, J. and Baud, M. (2015). "Water Reform Governmentality in Ecuador: Neoliberalism, Centralization, and the Restraining of Polycentric Authority and Community Rule-Making." *Geoforum*. 64, 281–291.

Boelens, R., Hoogesteger, J., Swyngedouw, E., Vos, J. and Wester, P. (2016). "Hydrosocial Territories: A Political Ecology Perspective." *Water International*. 41, 1–4.

Boelens, R., Perreault, T. and Vos, J. (Eds.). (2018). *Water Justice*. Cambridge, UK: Cambridge University Press.

Dean, M. (1999). *Governmentality: Power and Rule in Modern Society*. Los Angeles, CA: Sage.

Duarte-Abadía, B. and Boelens, R. (2019). "Colonizing Rural Waters: The Politics of Hydro-Territorial Transformation in the Guadalhorce Valley, Málaga, Spain." *Water International*. 44(2), 148–168.

Ekers, M. and Loftus, A. (2008). "The Power of Water: Developing Dialogues between Foucault and Gramsci." *Environment and Planning D: Society and Space*. 26(4), 698–718.

Flaminio, S., Rouillé-Kielo, G. and Le Visage, S. (2022). "Waterscapes and Hydrosocial Territories: Thinking Space in Political Ecologies of Water." *Progress in Environmental Geography*. 1(1–4), 33–57.

Foucault, M. (1998) [1976]. *The Will to Knowledge: History of Sexuality I*. Harmondsworth, UK: Penguin books.

Foucault, M. (2003). *"Society Must Be Defended", Lectures at the Collège De France 1975–1976*. New York, NY: Picador.

Gandy, M. (2004). "Rethinking Urban Metabolism: Water, Space and the Modern City." *City*. 8(3), 363–379.

Gandy, M. (2006). "Zones of Indistinction: Some Thoughts on the Bio-Politics of Urban Space." *Cultural Geographies*. 13(4), 1–20.

Gandy, M. (2014). *The Fabric of Space: Water, Modernity, and the Urban Imagination*. Cambridge, MA: The MIT Press.

Gibbs, L. M. (2009). "Water Places: Cultural, Social and More-than-Human Geographies of Nature." *Scottish Geographical Journal*. 125(3–4), 361–369.

Hansson, S. (2014). "Dependence and Autonomy: Conditions of Possibility for Nigerien Ownership in the Water Sector." *Journal of Intervention and Statebuilding*. 9(2), 256–277.

Hellberg, S. (2014). "Water, Life and Politics: Exploring the Contested Case of Ethekwini Municipality Through a Governmentality Lens." *Geoforum*. 56, 226–326.

Hellberg, S. (2018). *The Biopolitics of Water: Governance, Scarcity and Populations*. London, UK: Routledge/Earthscan.

Hellberg, S. (2020). "Scarcity as a Means of Governing. Challenging Neoliberal Hydromentality in the Context of the South African Drought." *Environment and Planning E: Nature and Space*. 3(1), 186–206.

Johnston, B. R. (2003). "The Political Ecology of Water: An Introduction." *Capitalism, Nature, Socialism*. 14(3), 73–90.

Joy, K. J., Kulkarni, S., Roth, D. and Zwarteveen, M. (2014). "Re-Politicising Water Governance: Exploring Water Re-Allocations in Terms of Justice." *Local Environment*. 19(9), 954–973.

Kaika, M. (2005). *City of Flows: Modernity, Nature and the City*. London, UK: Routledge.

Karpouzoglou, T. and Vij, S. (2017). "Waterscape: A Perspective for Understanding the Contested Geography of Water." *WIREs Water*. 4(3), 1–5.

Kooy, M. (2014). "Developing Informality: The Production of Jakarta's Urban Waterscape." *Water Altern*. 7, 35–48.

Kooy, M. and Bakker, K. (2008a). "Splintered Networks: Urban Water Governance in Jakarta." *Geoforum*. 39(6), 1843–1858.

Kooy, M. and Bakker, K. (2008b). "Technologies of Government: Constituting Subjectivities, Spaces, and Infrastructures in Colonial and Contemporary Jakarta." *International Journal of Urban and Regional Research*. 32(2), 375–391.

Lankford, B. (2013). "Infrastructure hydromentalities: Water sharing, water control and water (in)security." In B. Lankford, K. Bakker, M, Zeiton, D. Conway (eds.). *Water Security: Principles, Perspectives and Practices*. New York, NY and London, UK: Earthscan/Routledge. pp. 256–272.

Lemke, T. (2007). "An Indigestible Meal? Foucault, Governmentality and State Theory." *Distinktion: Journal of Social Theory*. 8(2), 43–64.

Linton, J. (2010). *What Is Water? The History of a Modern Abstraction*. Vancouver, BC: UBS Press.

Linton, J. and Budds, J. (2014). "The Hydrosocial Cycle: Defining and Mobilizing a Relational-Dialectical Approach to Water." *Geoforum*. 57, 170–180.

Loftus, A. (2006). "The metabolic processes of capital accumulation in Durban's waterscape." In N. Heynen, M. Kaika, E. Swyngedouw (eds.). *In the Nature of Cities: Urban Political Ecology and the Politics of Urban Metabolism*. London, UK: Routledge.

Loftus, A. (2009). "Rethinking Political Ecologies of Water." *Third World Quarterly*. 30(5), 953–968.

Loftus, A. (2015). "Water (in)Security: Securing the Right to Water." *The Geographical Journal*. 181(4), 350–356.

McDonald, D. A. and Ruiters, G. (Eds.). (2005). *The Age of Commodity: Water Privatization in Southern Africa*. London, UK: Earthscan.

Mehta, L. (2010). *The Limits to Scarcity*. London, UK and Washington, DC: Earthscan.

Mehta, L., Huff, A. and Allouche, J. (2019). "The New Politics and Geographies of Scarcity." *Geoforum*. 101, 222–230.

Menga, F. (2016). "Reconceptualizing Hegemony: The Circle of Hydro-Hegemony." *Water Policy*. 18, 401–418.

Molle, F., Mollinga, P. P. and Wester, P. (2009). "Hydraulic Bureaucracies and the Hydraulic Mission: Flows of Water, Flows of Power." *Water Alternatives*. 2(3), 328–349.

O'Reilly, K. (2006). "Traditional" Women, "Modern" Water: Linking Gender and Commodification in Rajasthan, India." *Geoforum*. 37(6), 958–972.

Öjendal, J., Hansson, S. and Hellberg, S. (2011). *Politics and Development in a Transboundary Watershed: The Case of the Lower Mekong Basin*. Dordrecht: Springer.

Ranganathan, M. (2014). "Mafias' in the Waterscape: Urban Informality and Everyday Public Authority in Bangalore." *Water Alternatives*. 7(1), 89–105.

Rodina, L. and Harris, L. (2016). "Water Services, Lived Citizenship, and Notions of the State in Marginalised Urban Spaces: The Case of Khayelitsha, South Africa." *Water Alternatives*. 9(2), 336–355.

Shove, E. (2003). *Comfort, Cleanliness and Convenience: The Social Organization of Normality*. Oxford, UK and New York, NY: Berg Publishers.

Sofoulis, Z. (2005). "Big Water, Everyday Water: A Sociotechnical Perspective." *Continuum: Journal of Media & Cultural Studies*. 19(4), 445–463.

Strang, V. (2004). *The Meaning of Water*. Oxford, UK: Berg Publishers.

Sultana, F. (2009). "Fluid Lives: Subjectivities, Gender and Water in Rural Bangladesh." *Gender, Place and Culture*. 16(4), 427–444.

Sultana, F. (2011). “Suffering for Water, Suffering from Water: Emotional Geographies of Resource Access, Control and Conflict.” *Geoforum*. 42(2), 163–172.

Sultana, F. (2013). “Water, Technology, and Development: Transformations of Development Technonatures in Changing Waterscapes.” *Environment and Planning D: Society and Space*. 31, 337–353.

Sultana, F. (2018). “Water Justice: Why It Matters and How to Achieve It.” *Water International*. 43(4), 483–493.

Swyngedouw, E. (1999). “Modernity and Hybridity: Nature, Regeneracionismo, and the Production of the Spanish Waterscape, 1890–1930.” *Annals of the Association of American Geographers*. 89(3), 443–465.

Swyngedouw, E. (2004). *Social Power and the Urbanization of Water: Flows of Power*. Oxford, UK: Oxford University Press.

Swyngedouw, E. (2009). “The Political Economy and Political Ecology of the Hydro-Social Cycle.” *Journal of Contemporary Water Research & Education*. 142(1), 56–60.

Swyngedouw, E. (2015). *Liquid Power: Contested Hydro-Modernities in Twentieth-Century Spain*. Cambridge, MA and London, UK: MIT Press.

von Schnitzler, A. (2008). “Citizenship Prepaid: Water, Calculability, and Techno-Politics in South Africa.” *Journal of Southern Africa Studies*. 34(4), 899–917.

Wilson, N. J. and Inkster, J. (2018). “Respecting Water: Indigenous Water, Governance Ontologies and the Politics of Kinship on the Ground.” *Environment and Planning E: Nature and Space*. 1(4), 516–538.

Yates, J., Harris, L. and Wilson, N. (2017). “Multiple Ontologies of Water: Politics, Conflict and Implications for Governance.” *Environment and Planning D: Society and Space*. 35(5), 797–815.

Zeitoun, M. and Warner, J. (2006). “Hydro-Hegemony – A Framework for Analysis of Trans-Boundary Water Conflicts.” *Water Policy*. 8(5), 435–460.

Zwarteveen, M. Z. and Boelens, R. (2014). “Defining, Researching and Struggling for Water Justice: Some Conceptual Building Blocks for Research and Action.” *Water International*. 39, 143–158.

6

FEMINIST CONTRIBUTIONS TO WATER AND DEVELOPMENT SCHOLARSHIP

Margreet Zwarteveen

Introduction

In the not-so-distant past, choosing a career as a water expert was among the obvious choices when one had the aspiration to become a hero. Water engineering education granted its bearer – provided he was white and male, but that went without saying – with the scientific credentials as well as the moral superiority to re-design and re-shape large areas of land and re-direct flows of water in the name of progress, development and modernity. When working in developing countries, the work of water engineers was seen as part of "the white man's burden" as expressed in the eponymous Rudyard Kipling poem of 1899. It posited white men as having the moral obligation to rule over and encourage the development of people from other ethnic and cultural backgrounds until they could take their place in the world by fully adopting Western ways. Many historical water engineers indeed coupled a Samaritan need to do good with a belief in their own moral and scientific superiority. Do such great water heroes still exist today; those strong and intelligent men who have the audacity to turn their dreams about making this world into a better place into reality through scientific engineering designs? My suspicion is that they do not, and perhaps – as I will argue – that is a good thing.

What I want to suggest in this chapter is that compared to the past, the terms of the existence of water heroism have drastically changed. Or, to say it differently, what it takes today to become a successful water professional – and to develop effective water knowledge – is unlike what it took a few decades ago. This is so because today's water problems are radically different from those of the past. Solving them, therefore, requires different types of knowledge and new types of water experts. In the rest of this chapter, I'll first explain in more detail how current water problems are different from those of yesterday. I then go on to argue that insights developed by feminist scholars provide useful ingredients for developing the new types of knowledge that are necessary for solving these problems.

Changes in Water as an Object of Knowledge and Intervention

Today's water problems differ from those of yesterday, in at least three ways. The first two are related to the two meanings of the term "development" when used in relation to water: development as the more general term for progress, and development as more narrowly referring to the construction of

DOI: 10.4324/9781003095545-7

new hydraulic projects to divert water and make it available for societal purposes – something that required engineering types of knowledge. These two developments used to closely overlap. After all, in the past, the narrow water development of hydraulic projects was considered to lead to, or be an almost intrinsic part of, more generic development. Water resources development by the state was part of a "hydraulic mission," a political strategy for controlling space, water and people as well as an economic strategy to spur development (see Molle et al., 2009). Today, the main premise of the hydraulic mission – the idea that all available waters should be harnessed to benefit humankind – has become untenable. This is importantly associated with the realization that water availabilities are limited, something that for instance becomes manifest in the closure of river basins (Molden, 2007; Molle, 2003). Hence, the first way in which today's water problems are different is that questions of conservation, protection and equitable distribution assume much greater importance. Water problems also increasingly are about controlling flows to protect people and other beings from floods and pollution.

A second way in which water problems are (seen as) different from the golden ages of engineering is related to the first, more generic, meaning of development. The belief that ever-better engineering or hydraulic knowledge and more advanced technologies will lead to ever-increasing levels of progress and modernity – a belief that provided an important justification for and sense of mission to engineers – has seriously waned. It is no longer so self-evident what progress, change (modernity and development) are and what they should be about. Doubts are based on the realization that the Western World is not and cannot be taken as the standard and model for the rest of the world, and on the awareness that Western development was and is achieved at the cost of mal-and underdevelopment elsewhere, that wealth and poverty are connected by unequal terms of trade that are maintained through economic, political and military powers. For water, these doubts importantly stem from accounts of the high social and environmental costs of large water development projects, such as the Darmodar Valley (Lahiri-Dutt, 2012) or Sardar Sarovar projects in India (Mehta, 2001) or the Three Gorges Dam in China (Crow-Miller et al., 2017). The increased prominence of environmental concerns on policy and political agendas, further triggered by the haunting spectre of the water-related disasters climate change may provoke (floods, droughts and pollution) have generated troubling new questions about the role of water investments and technological infrastructures in processes of change or development. These questions trespass administrative boundaries or national territories. Although water conflicts play out at local and basin levels, they are – and maybe increasingly so – subject to the dynamics of the global political economy and geopolitics (Allan, 2011). The 4,800 litres of water that it takes to produce a hamburger is just one illustration of such global water connections; the financing of large dams for the production of hydropower in the Himalayas by Western countries under the Global Environmental Facility as a mechanism to off-set their own carbon dioxide emissions another (Ahlers et al., 2015).

The third way in which the meaning of water has changed is that water development and management no longer are (seen as) the exclusive realm of the state. Water knowledge and decision-making powers used to be concentrated in often powerful public agencies. Partly because of privatization and other neo-liberal reforms, and partly in response to the uncertainty and complexity that characterizes water problems, new actors have entered the water management scene, including private companies and civil society organizations. Scales and territories of water sovereignty are also changing, with de facto control over how water is used and allocated increasingly coming to lie with international companies, who often appear to be more accountable to their overseas headquarters and the final clients of their products than to governments or people in the regions where they operate (see Zwarteveen et al., 2017, also see Allen, 2006; Ballabh, 2008). Alongside and partly because of this diversification of actors and scales, water is no longer just managed, but it is governed (Gleick, 2002). A change in terminology that some associate with a shift to a *polycentricity* (Andersson & Ostrom, 2008; Neef, 2009) and that others have more generically labelled as a deliberative turn in natural resource management (Neef, 2009).

These changes are prompting a serious reconsideration of the kinds of knowledge that are needed to understand and help solve water problems, a reconsideration that reflects wider debate about the role of expertise in dealing with and solving societal problems. This debate hinges on at least two phenomena which also characterize water resources: (1) the recognition that water problems are wicked or complex problems and that understanding these problems requires relaxing or revisiting assumptions of knowability, causality and predictability (Funtowicz & Ravetz, 2003; Kurz & Snowden, 2003; Moench et al., 2003) and (2) the realization that the interfaces between science, policy and society are radically different from what they used to be in the past, with expert knowledge increasingly being interrogated and questioned from diverse directions, unsettling public trust in scientific expertise. This debate is generating proposals for new forms of knowledge – often called order-2, discursive or post-normal knowledge (Funtowicz & Ravetz, 2003; Norgaard et al., 2008; Whatmore, 2002) – that are associated with new integrative or adaptive ways of managing and governing water.

My contribution to this debate hinges on the suggestion that new water knowledges need to be much more explicit about the fact that water management, at heart, is a deeply political affair (see Boelens et al., 2010; Menga, 2017; Molle et al., 2008, 2009; Roth et al., 2005). Water is about allocation and distribution – about the distribution of water, about the distribution of technologies to access water, as well as the distribution of public investments in water. It is about distributing the costs involved in accessing water and about distributing risks to floods, but also about the distribution of voice and authority and about the organization of water decision making, and about how to arrange transparency, accountability and democracy. Answers to all these distributional questions necessarily entail difficult to disentangle discussions of what is true with opinions about what is right (see Joy et al., 2014; Zwarteveen et al., 2017).

My proposition therefore is that new approaches to understanding and solving water problems need to much more seriously acknowledge and deal with the deeply contested and political nature of water. I would like to suggest that insights developed by feminist scholars are particularly useful for doing this as exposing – and dealing with – the political nature of scientific and technological knowledge has been and still is at the heart of much feminist scholarship. First, feminist concerns about the equity of interventions and policies have produced sharp awareness of how distributional choices may be screened off from democratic deliberation because of how they are "rendered technical" (see Li, 2007), often by prioritizing questions of efficiency over those of allocation and justice. The effect is that far-reaching re-distributions and re-allocations of water remain implicit or appear as inevitable and scientifically rational (cf. Donahue & Johnston, 1998). A second and related insight which feminist scholars have importantly helped developing is that knowledge and politics are intertwined in complex ways. This is also an important insight for water. Current water knowledge continues to be dominated by engineering and natural sciences, disciplines that often continue to hold on to the desirability and possibility of separating knowledge from politics (or from society and culture more generally). In what follows in this chapter, I further elaborate how feminist proposals to deal with these two insights can improve water knowledge.

Questioning the Division Between Nature and Society

If there is one basic insight that informs feminist scholarship, it is that the distinction between men and women is not (just) natural, but importantly also social, cultural and political. Simone de Beauvoir's phrase:" On ne naît pas femme, on le devient" is a famous way to express this. That women are made, not born is an acknowledgment of empirical and historical diversity, but also helps see that inequalities and power differences between men and women are not naturally ordained and therefore self-evident but the result of more or less conscious human choices. To theorize this, the concept of "gender" was coined, a concept that served to distinguish between those differences between women and men (sex) that could be attributed to nature and those that are the result of nurture – that are social or cultural. In this way, the concept allowed to shift questions of inequality and difference from the domain of the undisputable to the domain of the disputed. What it did, in other words, was that it allowed questioning what it means to be a woman or a man.

Yet, the concept of gender turned out to be itself problematic as it remained based on the premise of a fixed and therefore knowable divide between nature and culture. This is unsatisfactory, as people are not "Cartesian minds that happen to be located in biological matter in motion," as Sandra Harding observed, but they are "embodied creatures" (Harding, 1986). In other words, feminist scholars helped point out that the dichotomous distinction between nature and culture is a human-made and often ideological one. The implication is that the definition of what is natural (or what natural differences consist of) itself forms part of what needs to be challenged and questioned.

> Rather than marking a categorically determined pole, "nature" or "woman's body" too easily means the saving core of reality distinguishable from the social impositions of patriarchy, imperialism, capitalism, racism, history, language. That repression of the *construction* of the category "nature" can be and has been both used by and used against feminist efforts to theorize women's agency and status as social subjects.
>
> *(Haraway, 1991: 135)*

These reflections about gender produced the important insight that categories (definitions) of "nature" and "society" (or "culture") do not straightforwardly mirror pre-existing realities, but are constituted through specific practices that form part of larger knowledge traditions and cultures. Donna Haraway was among the first to further re-theorize human-environment relations in this way. With concepts such as "cyborg" (Haraway, 1991) and "natureculture" (Haraway, 2008) she articulated how "the social" and "the natural" (or the "technical") continuously influence each other in processes of co-evolution: there is no "essential" difference between the two prior to their interaction with each other. In this way, the concepts form part of an attempt to re-theorize nature-society relations as continuously shaping and indeed constituting each other. This re-theorization helps acknowledge that the conceptual divide between society and nature does not originate from an observable reality that is distinct from the observer, but is the product of very particular "modern" approaches to seeing, using and managing nature (for water, see Linton, 2010). Haraway also shows that denominating something as either natural or social is always itself a political act, as it determines what can be opened up for questioning by separating the contingent from the essential. Positing something as truly or authentically natural, dangerously designates as an *origin* and *cause* those categories that are in fact the *effects* of institutions, practices and discourses with multiple and diffuse points of origin.

The idea that the boundaries between nature, technology and society (or humans) are not pre-given, but are themselves the products of human minds and social conventions that reflect the hard construction work of the actors involved is one that is extremely useful for water, as the act of relegating phenomena to the realms of nature is a well-known and much used strategy to de-politicize water questions, placing contentious questions of distribution outside of the domain of public deliberation and debate. To avoid raising critical questions, for instance, water scarcity is often referred to as a *natural* problem caused by climate change and changing weather conditions, rather than as a problem of distribution or of social relations of power. Bakker (2004) has famously shown this with her analysis of the Yorkshire drought, which took place after one of the wettest winters of the century. The drought occurred, she argues, because it was the least cost option for water industry. Instead of investing in overcoming leakage, improving storage capacity and monitoring climate change, the company sluiced profits to its parent company, outsourced many of its tasks surrendering overview, and trucked in water when crisis became unbearable, making record profits that year (Bakker, 2004). In a similar vein, Ahlers (2005) shows how "drought" was born in Mexico as a result of increased demand for water by commercial agriculture and growing textile industry. Although water supplies showed regular climatic trends, the drought was "felt" and became an issue because of the removal of state subsidies to farmers which used to guarantee them a fairly stable income during periods of climatic stress. Without such subsidies, farmers were left to fend for themselves (Ahlers, 2005: 24). Notions of water scarcity obviously may have "absolute" subsistence and survival properties but always, as in all relations of demand and supply, they are deeply mediated by humans and determined by power relationships that construct "scarcity" far beyond just the wickedness of nature.

The ambition to see and conceive nature and society as co-constituted can draw upon a rapidly growing body of scholarship within the environmental social sciences, much of which takes inspiration from the early work of feminist scholars such as Haraway. This body of work retheorizes human-environment relations in terms of the mutual constitution of society and nature (see Bakker, 2010; Castree, 2002; Dingler, 2005; Goldman & Schurman, 2000; Harvey, 1996; Latour, 1993; Robbins, 2004; Swyngedouw, 1999; Whatmore, 2002). In addition to the already mentioned concepts which are proposed by Haraway, words to express this "co-production thesis" include "hybrid" (e.g., Whatmore, 2002), "assemblage" (e.g., Whatmore, 2002), as well "quasi-object" (Latour, 1993). More specifically for water, terms proposed are "waterscapes" (Baviskar, 2007; Swyngedouw, 1999) and hydrosocial networks (Bakker, 2010) or cycles (Linton & Budds, 2014). This work sees human activity and the bio-physical as mutually interacting to shape landscapes that are dynamic and continuously contested in a process that is constituted by, and simultaneously constitutes, the political economy of access and control over resources (Harvey, 1996; Peluso & Watts, 2001; Swyngedouw, 1997, 2003). Infrastructural and institutional water developments develop "part natural part social," as material dynamic reflections of historic and never ending socio-political-geographical struggles (Swyngedouw, 2003: 95). These re-theorizations of society-water relations help recognize that growing pressures on water resources, as well as the resulting scarcities, are not natural processes but the outcome of specific histories and practices of water resources exploitation or indeed development (in the two meanings of the term).

Redefining Objectivity: Situatedness, Positionality and Identity

As noted, the prevailing epistemological norm in conventional water science is – or used to be – positivist. It is based on the belief that reality exists independent of scientific knowledge of it, and that it can best be discovered by looking for regularities revealed in normal circumstances (cf. Langton, 2000: 135). This discovery can and should be done by interchangeable knowers whose specificities of embodiment and subjective location disappear in the process. Adherence to this epistemological norm, which is of course linked to a belief in the possibility of transcendence, makes much water knowledge speak "as if" from nowhere. As if from a value-free and god-like position, from the perspective of someone without interests and background, someone who benevolently represents the universal good – or some version of this as when speaking on behalf of national interests – and the universal truth.

The co-production thesis, or the idea that nature and society mutually constitute each other, importantly challenges this epistemological norm. Partly by insisting that meanings and definitions are part of, rather than external to, water (Dingler, 2005; Norgaard et al., 2008). Linton (2010) has for instance shown how the concept of the hydrological cycle embodies a particular representation of nature that has a specific history and politics, having been created by particular people at certain moments for specific ends, and promoted with the aim of achieving dominance. Similarly, concepts of river basin and watershed, which are often invoked as representing natural units for water management, reflect instead particular epistemic and management traditions and political choices. As pointed out by Blomquist and Schlager, "the definition of a watershed and the selection of boundaries are matters of *choice*. As soon as the matter of choice is present, there is a role for politics – which among other things is about who decided and how and with what effects" (Blomquist & Schlager, 2005: 104–110; also see Warner et al., 2014).

The recognition that knowledge and power are intrinsically entangled requires replacing the old idea of science as distinguishable or even in opposition to power or politics with a new one. Feminist scholars have been particularly influential in proposing how to do this. In their efforts to understand and unravel the causes and effects of the dominance of white men in the production of scientific facts, they have shown in different ways how produced knowledge always bears the imprints of the social and cultural values of communities of knowers (see, for instance, Harding, 1986; Hartsock, 1998; Longino, 1990), and have demystified the wish to provide truth statements in the strict positivist tradition as impossible. According to them, any effort to describe reality in one consistent all-encompassing narrative or discourse is politically dangerous, because it privileges and totalizes one perspective and source of knowing, thereby excluding all others. It is also academically suspect because it hides the knower and her identity and affiliations in cloaks of objectivity (Flax, 1986, cited in Harding, 1986: 193). In water, the old positivist norm of objectivity which called for knowledge producers to separate their identity and specific location from the knowledge they generate, did in fact serve to maintain a specific power-knowledge hierarchy which used to be dominated by white men. Water experts often continue to use this norm of objectivity to positively distinguish their own scientific knowledge from indigenous, layman's or traditional beliefs and knowledges, thus asserting their superiority and justify their powers of intervention and authority (see Sehring et al., 2022; Zwarteveen, 2008, 2011).

Haraway asserts that knowledge is never universal, but emerges from specific places – it is a situated, rather than a place-less, activity – and comes from specific people, with identities and particular values and interests (Haraway, 1991). Making such knowledge travel, or making it generic or universal, requires concerted efforts on the parts of involved actors. Part of this work is cultural and performative, and has to do with packaging knowledge into words and languages, shapes and forms, that are recognized, accepted and attractive. In this way, knowledge about water realities not just represents realities "in the field," but also represents prevailing epistemic communities and expert identities in water (with their own configurations of power, status, authority and funds). Epistemic norms always form part of a larger range of cultural resources through which water professionals represent and identify themselves, and that contribute to legitimizing their professional activities and choices.

Gender and ethnicity often are important symbolic markers in this construction of experts and expertise. Ruth Oldenziel in a book called "Making Technology Masculine" provides a fascinating illustration of this by showing how engineering was *made* masculine (and white, for that matter). Through conscious efforts to positively distinguish themselves and gain status and respect as a professional group, mechanical and civil engineers during the late nineteenth century in the United States succeeded in delimiting the definition of technology as consisting of those activities they engaged in. In doing so, they reduced the significance of existing technologies; the artefacts and forms of knowledge associated with women and black people. This resulted in the rise of engineers as an elite with exclusive rights to technical expertise. And, crucially, it involved the simultaneous creation of a male professional identity, based on educational qualifications and the promise of managerial positions, sharply distinguished from shop-floor engineering and blue collar workers (Oldenziel, 1999). In my own research (Sehring et al., 2022; Zwarteveen, 2011, 2008) I have shown that similar processes are and have been at work in water, with definitions of what constitutes a good bureaucracy or a good engineer (or a good manager, or good science) being deeply shaped by – and forged around – prevailing ethnic and gendered identities, hierarchies and power relations.

The realization of the situatedness of knowledge, and of its cultural and performative dimensions, invites critical reflection on how knowledge is constructed and by whom, and on how the identity and social situatedness of knowledge producers impacts not just the type of truth claims they make, but also the legitimacy of these claims. This is precisely the plea of critical and feminist scholars (as well as of decolonial scholars, for that matter). They have argued that there is merit in trying to be much more explicit about how facts are fabricated and by whom, showing the production process and the identity and perspectives of producers. The point is that rather than speaking from a non-identifiable position and admitting to desires of detachedness and universality, knowledge becomes more credible when admitting and accounting for differences between knowers. This maybe most difficult for those who normally embody scientific authority, those who have never had to explain themselves because they stood for normalcy – white men – but comes almost spontaneously to many others, who – precisely because of their gender, class or ethnicity – are already marked as different from the norm.

Conclusions: Towards Reflexivity and Plurality

How then, to use these insights for developing new and better water knowledge – or for better solving water problems? There are at least three important lessons to be learnt from feminist scholarship. The first is the importance of awareness of the situatedness and partiality of *all* water knowledges, including scientific ones. Water experts should become much more explicit about the ways in which "facts" or "truths" are constructed or made to travel, by whom, and for what purpose. This starts by rejecting the subject- object binary in the production of knowledge, and by acknowledging "the critical and interpretative core of all knowledge" (Haraway, 1991: 191). It includes serious efforts to be much more explicit about concepts, delimitations and theoretical assumptions used, and about the effects of these for ordering water realities, for discussing questions of distribution and indeed distributing water. As Nicholson noted: "Any discursive move which attempts to place itself beyond question automatically invokes suspicion" (Nicholson, 1995: 5). And it implies developing critical sensitivity to the cultural and performative aspects of scientific authority, developing a much sharper alertness to how it authority attaches itself much more easily to some bodies and identities than to others.

Second, recognition of the situatedness of knowledge counters dreams of universality and generalization, and relaxes the search for one most accurate and reliable account of water problems and realities. This opens the door to accepting other types of knowledge about processes of change that are related to water, diversifying the publics with whom experts can and should collaborate, and redistributing environmental water expertise (cf. Whatmore, 2009: 10). Here, it becomes important to accept that diverse knowledges cannot always be easily combined, which is why it may sometimes be better to stay with the differences and tensions between them. The one-world (universe) quest for commensuration or coherence as the condition for collaboration across epistemic communities needs to be replaced with a more modest search for the *partial connections* – between different worlds, places, disciplines (Strathern, 2004; also see Cadena & Blaser, 2018). This allows shifting the attention to creating the translations (Callon, 1984) or co-ordinations (Mol, 2002) needed to make insights travel (e.g., see Dominguez-Guzman et al., 2017). Rather than the single best, optimal or most accurate representation, the debate becomes a more pragmatic and indeed more political one: one about which enactments (worlds, realities, imaginaries and futures) are plausible and desirable for a specific situation, place or community.

And thirdly, feminist insights trigger much more attention to the process dimensions of producing knowledge. Many water scholars agree to this, yet they tend to remain rather optimistic about the possibility of arriving at consensual solutions and shared analyses. The lesson of feminist scholarship is that it is important to remain vigilant about questions of inclusion and exclusion, about whose knowledge and expertise count and why, as well as about the politics of integration and translation. After all:

> […] it matters what we use to think other matters with; it matters what stories we tell to tell other stories with; it matters what knots knot knots, what thoughts think thoughts, what descriptions describe descriptions, what ties tie ties. It matters what stories make worlds, what worlds make stories.
>
> *(Haraway, 2016)*

To return to where I started this chapter: I think it is better for new water knowers to not gain the same heroic status as the grandfathers of water engineering. The all-knowing, well-intending experts who travel the world to disseminate their wisdoms and technologies are no longer needed, if at all they ever were. Instead, solving today's water problems requires the burgeoning of many different kinds of knowledge and many different knowing identities and cultures. And it entails accepting that these knowledges and knowers will never be easily reconciled. Rather than masterful forms of knowing that come from no-where, feminist-inspired modes of engaging with and intervening in water are more modestly anchored in always specific attachments and connections.

References

Ahlers, R. (2005). *Fixing water to increase its mobility: The neoliberal transformation of a Mexican irrigation district.* PhD diss., Cornell University.

Ahlers, R., J. Budds, D. Joshi, V. Merme and M. Zwarteveen. (2015). "Framing hydropower as green energy: Assessing drivers, risks and tensions in the Eastern Himalayas." *Earth System Dynamics* 6, 195–204.

Allan, J. A. (2011). *Virtual water. Tackling the threat to our world's most precious Resource.* London: I.B. Tauris & Co.

Allen, T. (2006). "IWRM: The new sanctioned discourse?" In: P. Mollinga, A. Dixit and K. Athukorala (eds.), *Integrated water resources management: Global theory, emerging practice and local needs.* New Delhi: Sage, 38–63.

Andersson, K. and E. Ostrom (2008). "Analyzing decentralized resource regimes from a polycentric perspective." *Policy Sciences* 41(1), 71–93.

Bakker, K. (ed.). (2004). *An uncooperative commodity: Privatizing water in England and Wales.* Oxford, UK: Oxford University Press.

Bakker, K. (2010). "The limits of 'neo-liberal natures': Debating green neoliberalism." *Progress in Human Geography* 34(6), 715–735.

Ballabh, V. (ed.). (2008). *Governance of water: Institutional alternatives and political economy.* New Delhi: Sage Publications.

Baviskar, A. (ed.). (2007). *Waterscapes. The cultural politics of a natural resource.* New Delhi: Orient Longman.

Blomquist, W. and E. Schlager (2005). "Political pitfalls of integrated watershed management." *Society and Natural Resources* 18(2), 101–117.

Boelens, R., D. Getches and A. Guevara-Gil (eds.). (2010). *Out of the mainstream. Water rights, politics and identity*. Washington, DC: Earthscan.

Cadena, M. de la and M. Blaser (eds.). (2018). *A world of many worlds*. Durham and London: Duke University Press.

Callon, M. (1984). "Some elements of a sociology of translation: Domestication of the scallops and the fishermen of St Brieuc Bay." *The Sociological Review* 32(1_suppl), 196–233. DOI: 10.1111/j.1467-954X.1984.tb00113.x

Castree, N. (2002). "False antitheses? Marxism, nature and actor-networks." *Antipode* 34, 111–146.

Crow-Miller, B., M. Webber and S. Rogers (2017). "The techno-politics of big infrastructure and the Chinese water machine." *Water Alternatives* 10(2), 233–249.

Dingler, J. (2005). "The discursive nature of nature: Towards a post-modern concept of nature." *Journal of Environmental Policy and Planning* 7(3), 209–225.

Dominguez-Guzman, C., A. Verzijl and M. Zwarteveen. (2017). "Water footprints and 'Pozas': Conversations about practices and knowledges of water efficiency." *Water* 9(1), 0016. DOI: 10.3390/w9010016

Donahue, J. M. and B. R. Johnston (eds.). (1998). *Water, culture, and power. Local struggles in a global context*. Washington, DC: Island Press.

Funtowicz, S. and J. Ravetz (2003). *Post-normal Science. International Society of Ecological Economics*; Internet Encyclopaedia of Ecological Economics. [Accessed 12 November 2012]. Available at http://leopold.asu.edu/sustainability/sites/default/files/Norton,%20Post%20Normal%20Science,%20Funtowicz_1.pdf

Gleick, P. (2002). "Water management: Soft water paths." *Nature* 418, 373 (25 July 2002). DOI: 10.1038/418373a

Goldman, M. and R. Schurman (2000). "Closing the 'great divide': New social theory on society and nature." *Annual Review of Sociology* 26, 563–584.

Haraway, D. (ed.). (1991). *Simians, cyborgs and women. The reinvention of nature*. London, UK: Free Association Books.

Haraway, D. (2008). *When species meet*. Minneapolis, MN/London, UK: University of Minnesota Press.

Haraway, D. (2016). *Staying with the trouble: Making kin the Chthulucene*. Duke University Press.

Harding, S. (ed.). (1986). *The science question in feminism*. Ithaca, NY and London, UK: Cornell University Press.

Hartsock, N. C. M. (ed.). (1998). *The feminist standpoint revisited & other essays*. Boulder, CO: Westview Press.

Harvey, D. (1996). *Justice, nature and the geography of difference*. Oxford, UK: Blackwell.

Joy, K. J., S. Kulkarni, D. Roth and M. Zwarteveen (2014). "Re-politicising water governance: Exploring water re-allocations in terms of justice." *Local Environment* 19(9), 954–973.

Kurz, C. F. and D. J. Snowden (2003). "The new dynamics of strategy: Sense-making in a complex and complicated world." *IBM Systems Journal* 42(3), 462–483.

Lahiri-Dutt, K. (2012). "Large dams and changes in an agrarian society: Gendering the impacts of Damodar Valley Corporation in Eastern India." *Water Alternatives* 5(2), 529–542.

Langton, R. (2000). "Feminism in epistemology. Exclusion and objectification." In: M. Fricker and J. Hornsby (eds.), *The Cambridge companion to feminism in philosophy*. Cambridge, UK: Cambridge University Press, 127–145.

Latour, B. (ed.). (1993). *We have never been modern* (translation C. Porter). London, UK: Longman.

Li, T. M. (ed.). (2007). *The will to improve. Governmentality, development, and the practice of politics*. Durham, NC: Duke University Press.

Linton, J. (ed.). (2010). *What is water? The history of a modern abstraction*. Vancouver, BC: University of British Columbia Press.

Linton, J. and J. Budds (2014). "The hydrosocial cycle: Defining and mobilizing a relational-dialectical approach to water." *Geoforum* 57, 170–180.

Longino, H. (ed.). (1990). *Science as social knowledge: Values and objectivity in scientific inquiry*. Princeton, NJ: Princeton University Press.

Mehta, L. (2001). "The manufacture of popular perceptions of scarcity: Dams and water-related narratives in Gujarat, India." *World Development* 29(12), 2012–2041.

Menga, F. (2017). "Hydropolis: Reinterpreting the polis in water politics." *Political Geography* 60, 100–109.

Moench, M., A. Dixit and S. Janakarajan (2003). *The fluid mosaic. Water governance in the context of variability, uncertainty and change: A synthesis paper*. Kathmandu: Water Conservation Foundation.

Mol, A. (2002). *The body multiple: ontology in medical practice*. Durham and London: Duke University Press.

Molden, D. (ed.). (2007). *Water for food, water for life. A comprehensive assessment of water management in agriculture*. London, UK: IWMI/Earthscan.

Molle, F. (ed.). (2003). *Development trajectories of river basins. A conceptual approach. IWMI Research Report No. 72*. Colombo, Sri Lanka: IWMI.

Molle, F., P. P. Mollinga and R. Meinzen-Dick (2008). "Water, politics and development: Introducing water alternatives." *Water Alternatives* 1(1), 1–6.

Molle, F., P. P. Mollinga and P. Wester (2009). "Hydraulic bureaucracies and the hydraulic mission: Flows of water, flows of power." *Water Alternatives* 2(3), 328–349.

Neef, A. (2009). "Transforming rural water governance: Towards deliberative and polycentric models." *Water Alternatives* 2(1), 53–60.

Nicholson, L. (1995). "Introduction." In: S. Benhahib, J. Butler, D. Cornell and N. Fraser (eds.), *Feminist contentions. A philosophical exchange*. New York, NY and London, UK: Routledge, 1–16.

Norgaard, R. B., G. Kallis and M. Kiparsky (2008). "Collectively engaging complex socio-ecological systems: Re-envisioning science, governance, and the California delta." *Environmental Science and Policy* 12(2009), 644–652.

Oldenziel, R. (ed.). (1999). *Making technology masculine. Men, women and modern machines in America 1870–1945*. Amsterdam: Amsterdam University Press.

Peluso, N. L. and M. Watts (ed.). (2001). *Violent environments*. Ithaca, NY: Cornell University Press.

Robbins, P. (2004). *Political ecology: A critical introduction*. Oxford, UK: Blackwell.

Roth, D., R. Boelens and M. Zwarteveen (eds.). (2005). *Liquid relations. Contested water rights and legal complexity*. New Brunswick: Rutgers University Press.

Sehring, J., R. Ter Horst and M. Zwarteveen (2022). *Gender dynamics in transboundary water governance. Feminist perspectives on water conflict and cooperation*. Oxon and New York: Routledge.

Strathern, M. (2004). *Partial connections. Updated edition*. Walnut Creek, Lanham, New York, Toronto and Oxford: Altamira Press.

Swyngedouw, E. (1999). "Modernity and hybridity: Nature, *Regeneracionismo*, and the production of the Spanish Waterscape, 1890–1930." *Annals of the Association of American Geographers* 89(3), 443–465.

Swyngedouw, E. (2003). "Modernity and the production of the Spanish waterscape 1890–1930." *Annals of the Association of American Geographers* 89(3), 443–465.

Warner, J. F., P. Wester and J. Hoogesteger (2014). "Struggling with scales: Revisiting the boundaries of River basin management." *Wiley Interdisciplinary Reviews; Water* 1, 469–481.

Whatmore, S. (ed.). (2002). *Hybrid geographies: Natures, cultures and spaces*. London, UK: Sage.

Whatmore, S. (2009). "Mapping knowledge controversies: Science, democracy and the redistribution of expertise." *Progress in Human Geography*, 1–12.

Zwarteveen, M. (2008). "Men, masculinities and water powers in irrigation: A plea for making men visible in irrigation." *Water Alternatives* 1(1), 111–130.

Zwarteveen, M. (2011). "Questioning masculinities in water." *Economic and Political* Weekly XLVI(18), 40–48.

Zwarteveen, M., J. S. Kemerink-Seyoum, M. Kooy, J. Evers, T. A. Guerrero, B. Batubara, A. Biza, A. Boakye-Ansah, S. Faber, A. Cabrera Flamini, G. Cuadrado-Quesada, E. Fantini, J. Gupta, S. Hasan, R. ter Horst, H. Jamali, F. Jaspers, P. Obani, K. Schwartz, Z. Shubber, H. Smit, P. Torio, M. Tutusaus and A. Wesselink. (2017). "Engaging with the politics of water governance." *WIREs Water* 4, e1245.

7

INDIGENOUS PEOPLES, SUSTAINABLE DEVELOPMENT, AND ONTOLOGIES OF WATER

Deborah McGregor, Mahisha Sritharan, and Steven Whitaker

Introduction

What does it mean to understand water and sustainable development and the relationship between them from the perspective and experience of Indigenous peoples? How do Indigenous ontologies of water differ from the "water-as-resource/commodity" perspective currently held by global, regional, and local government authorities? How do different water ontologies contribute to or help to redress the global water crisis? In exploring these questions, this contribution will examine Indigenous perspectives of water, in part through the various international water declarations and reports that highlight key ways in which Indigenous relationships to water differ from non-Indigenous views, particularly in a sustainable development context. These questions will privilege the views, knowledges, and experiences of Indigenous peoples through their own voices, as expressed in various international fora for decades. These international Indigenous environmental and water declarations offer anti-colonial critiques and insights, thus creating the space for alternative sustainable water futures that embrace Indigenous ontologies. The quotes below highlight the value placed on water as essential to human survival, yet also introduce glimpses into the contrasting ontologies of water:

> Unsustainable development pathways and governance failures have affected the quality and availability of water resources, compromising their capacity to generate social and economic benefits. Demand for freshwater is growing. Unless the balance between demand and finite supplies is restored, the world will face an increasingly severe global water deficit.
>
> *(United Nations, 2015: 3)*

> We recognize, honor and respect water as sacred and [sustaining of] all life. Our traditional knowledge, laws and ways of life teach us to be responsible in caring for this sacred gift that connects all life.
>
> *(Indigenous Peoples Kyoto Water Declaration, 2003)*

> The value we place on water varies, depending upon who is using it, and why. Value can be a guide to what our goals should be, what actions are needed, and where we should invest.

DOI: 10.4324/9781003095545-8

> Many of our problems arise because we don't value water highly enough; all too often water is not valued at all.
>
> *(United Nations, 2021: vii)*

> The spiritual connection to water that indigenous societies maintain as an integral aspect of their culture is a basis for countless water conflicts with outside, predominantly Western, forces for development. While Western cultural values do give some attention to a spiritual dimension of water, it is very much a minority view. The dominant value system determining how water is utilized in Western culture is basically an economic one.
>
> *(Groenfeldt, 2006: 108)*

Based on these ontologies of water, we suggest that there is a need to decolonize internationally adopted conceptions of sustainable development expressed more recently through the post-2015 United Nations sustainable development agenda and the World Water reports (2014–2022). Indigenous water insecurity is very much an outcome of unsustainable and detrimental "development," as well as gross violations of human and Indigenous rights as pointed out by Indigenous peoples globally for decades. Indigenous peoples have formulated alternative forms of "sustainable development" based in part on anti-colonial critiques of the concept, and have asserted the right to their own self-determined sustainable future since the Earth Summit in 1992 (McGregor, 2021). There has been much critique levelled at sustainable development, yet the discussion regarding Indigenous ontologies of water remains largely unexplored.

The United Nations World Water Development Reports (2014–2019) (United Nations,2014; 2015; 2016; 2019; 2020; 2021; 2022) all reveal serious challenges ahead, globally, regionally, and locally, in relation to climate change impacts on water security. By 2030, the world is projected to face a 40% global water deficit under the business-as-usual (BAU) scenario (United Nations, 2015) and it is expected that climate change will exacerbate these conditions. In the absence of genuine efforts at supporting a sustainable future, the planet and its inhabitants (all life) are now in a multi-facted ecological, economic, and climatic crisis, including a crisis of values. The continued deterioration of the planet's health clearly demonstrates that current efforts at sustainability are inadequate, as evidenced by the recent IPCC reports (IPCC, 2021; 2022).

We will also briefly examine how water as a human right may be an area where Indigenous and state perspectives of water converge. Critical international declarations, including the *United Nations Declaration on the Rights of Indigenous Peoples* (UNDRIP); UN Resolution 64/292 *The Human Right to Water and Sanitation*; and other instruments have been developed internationally to protect Indigenous and other vulnerable and marginalized peoples. These declarations establish vital links among human rights, water rights and self-determination in an Indigenous context, thus expanding the role of water beyond that of a resource for human consumption.

Indigenous Water Governance and the Settler State

Many settler-colonial systems embody dominant Western ontologies of water in their approach to governance (Hartwig et al., 2022; Wilson et al., 2021). This has in many cases led to a lack of Indigenous voices/perspectives in water governance discourse, including the frequent exclusion of such views from decision-making in relation to water (OHCHR, 2022). Indigenous people's distinct water perspectives and experiences remain conspicuously absent. As Parsons, Fisher, and Crease state, "Indigenous responsibilities for water encapsulate the maintenance and protection of water, including its quality and quantity, to enable and enhance the health and wellbeing of human and more-than-human beings (including biological and metaphysical entities)" (2021: 286). For many Indigenous communities, water ontologies transcend the primacy of access for humans, and instead require that water be shared equitably among all living things.

One of the greatest challenges for Indigenous communities in terms of water governance is that dominant Western approaches/discourse around water governance greatly differ from Indigenous approaches (McGregor, 2014; Moggridge & Thompson, 2021; Wilson & Inkster, 2018). Indigenous knowledge, perspectives, and voices are not reflected in the dominant Western approaches to water governance globally (Root, 2021). In New Zealand, the Māori legal system "encapsulates a certain way of life that depends on the relationships between all things, including between: people and gods; different groups of people; and people and everything in the surrounding world" (Ruru, 2018: 216). The Māori people fought for over 160 years to have the Whanganui River obtain legal personhood status and protection, which was finally granted in 2017 (Evans, 2020). In Australia, Indigenous peoples are also subject to colonial ontologies of water and water governance as well as being subjected to culturally inappropriate measures that fail to take into account the communities' views and needs in terms of water governance (Hartwig et al., 2022; Maclean, 2015). Many Indigenous communities around the world struggle to have their voices heard in water governance and this is primarily due to the lack of recognition and inclusion of Indigenous knowledge and approaches to water governance based on responsibilities (OHCHR, 2022). As will be pointed out in this contribution, dominant Western discourse of viewing water as a commodity or resource offers a limited view of water and thus limits future visions of water security and sustainability.

Taylor, Longboat, and Grafton highlight that "[s]ettler states have tended to claim authority over water, excluding and obscuring Indigenous peoples' systems of water governance" (2019: 4). For many Indigenous communities, the differences in how water is experienced, conceptualized, and understood are not the only challenge. A significant barrier facing Indigenous involvement in water governance is a systematic failure to recognize some groups of Indigenous peoples as self-governing entities. "Colonial understandings of sovereignty and jurisdiction assume that the States alone exercise the final authority over a given territory" (Wilson et al., 2021: 3). For example, Indigenous people/tribes in the United States that are not recognized federally face additional hurdles as they struggle to influence water governance (Emanuel & Wilkins, 2020). However, "in order for water governance to be equitable and effective, it should be collaborative, and should include affected populations – specifically Indigenous peoples – in shared decision-making processes" regardless of whether or not they are recognized by the state as a distinct group or self-governing entity (Simms et al., 2016: 6). The contrasting views of water and dominance of one over the other can be characterized as a form of "ontological and material violence" that "permeate[s] water law, infrastructures, and assessments of water insecurity and risk, to the exclusion of Indigenous cultural, spiritual and physical health as well as the health of water itself" (Wilson et al., 2021: 3).

This scenario puts Indigenous peoples at heightened risk for water insecurity in years to come (OHCHR, 2022).

Indigenous and Non-Indigenous Ontologies of Water

Indigenous ontologies of water were made evident in the 12th Annual Indigenous Environmental Network's *Protecting Mother Earth Conference*, held in British Columbia, Canada, in 2001. Indigenous delegates at this conference prepared a water declaration stating that:

> As Indigenous Peoples, we recognize, honor and respect Water as a sacred and powerful gift from the Creator. Water, the first living spirit on this earth, gives life to all creation. Water,

> powerful and pristine, is the lifeblood that sustains life for all peoples, lands and creation. We know that by listening to the songs of the Water, all creation will continue to breathe. Our knowledge, laws and ways of life teach us to be responsible at all times in caring for this sacred gift that connects all life.
>
> *(Indigenous Declaration on Water, 2001)*

International Indigenous water declarations such as this are often generated in response to dissatisfaction with the views held by state governments and the United Nations as part of World Water Forums (UNESCO, 2006). Indigenous conceptions of and relationships with water often contrast sharply with the underlying ontologies of water espoused by state organizations, such as the one evident at the International Conference on Water and the Environment in Dublin in 1992. The resulting "Dublin Statement on Water and Sustainable Development" (NGOCE, n.d.) set out guiding principles aimed at strengthening global water conservation efforts, as listed below:

Principle 1: Fresh water is a finite and vulnerable resource, essential to sustain life, development, and the environment.
Principle 2: Water development and management should be based on a participatory approach, involving users, planners, and policy-makers at all levels.
Principle 3: Women play a central part in the provision, management, and safeguarding of water.
Principle 4: Water is a public good and has a social and economic value in all its competing uses.

These principles, while aimed at better conserving water, are based on a problematic set of assumptions along the following lines:

- Water is a resource.
- Water is separate from people and the environment.
- Water has economic value and is therefore a public good.

Despite their increasingly recognized flaws, such assumptions have endured and remain evident in the UN World Water Development context nearly 30 years later. Water in this narrative is understood as a commodity, a resource or property to be bought and sold. This approach in turn means that, as is the case with any commodity (even ones that humans depend upon for life), access is necessarily unevenly and often unfairly distributed.

Indigenous ontologies of water expand beyond as well as contradict the dominant narrative, fundamentally relating to water not only as life-giving and essential for the well-being for all Creation, but also *living* in and of itself. Water is a gift, bound to people through a set of obligations and responsibilities. The Kyoto Declaration describes this relationship as follows:

> Our relationship with our lands, territories and water is the fundamental physical, cultural and spiritual basis for our existence. This relationship to our Mother Earth requires us to conserve our freshwaters and oceans for the survival of present and future generations. We assert our role as caretakers with rights and responsibilities to defend and ensure the protection, availability and purity of water. We stand united to follow and implement our knowledge and traditional laws and exercise our right of self-determination to preserve water, and to preserve life.
>
> *(Indigenous Peoples Kyoto Water Declaration, 2003)*

Not only do Indigenous people have their own conception of water that guides such relationships, but we have developed knowledge and legal systems to support the responsibility to care for water, as one might support and care for a relative.

The Nature of Water: Beyond H_2O

How a culture defines water provides insights into how water is conceptualized in that culture's water discourse. Often, what we (as humans generally) refer to as water remains unquestioned. It just is. Common dictionary definitions (which tend to focus on water's chemical makeup and properties) illustrate how water is understood and thus related to in dominant-western worldview. In these definitions, water is clearly a non-living thing, and the underlying assumption is that this is the only truth about water. For example, according to the online UK dictionary, water is:

> A colourless, transparent, odourless liquid that forms the seas, lakes, rivers, and rain and is the basis of the fluids of living organisms.
>
> Water is a compound of oxygen and hydrogen (chemical formula: H_2O) with highly distinctive physical and chemical properties: it is able to dissolve many other substances; its solid form (ice) is less dense than the liquid form; its boiling point, viscosity, and surface tension are unusually high for its molecular weight, and it is partially dissociated into hydrogen and hydroxyl ions.
>
> *(Lexico Dictionaries, 2021)*

Other dictionaries offer similar perspectives of water. The result of such conceptions of water is expressed in the World Water Development Report 2003 as follows:

> We have made good progress in understanding the nature of water in its interaction with the biotic and abiotic environment. We have better estimates of climate change impacts on water resources. Over the years, our understanding of hydrological processes has enabled us to harvest water resources for our needs, reducing the risk of extreme situations.
>
> *(UNESCO World Water Assessment Programme, 2003: 10)*

Without exception, every world water report refers to water as a *resource*. While over time, various other aspects of water have been highlighted, the underlying foundational understanding and purpose of water has remained the same. The World Water Reports have called for transformation of how water is governed, and have widely analysed and critiqued the "business as usual approach" to water development efforts, yet have not deconstructed their understanding of the very subject of the reports itself: water. To truly embrace transformation and not simply tweak the BAU approach to global water governance, we need to learn from Indigenous peoples' concepts of water.

Author David Groenfeldt, in his analysis of different ontologies of water, describes Western concepts of water below:

> Water, in Western cultural theory, is a resource. It is not alive, it is inert, and it can be fully defined in terms of its physical properties. It has no consciousness, and it has no life. It is neither plant nor animal; it is a type of mineral, a liquid one (usually) but very much a mineral. It has no value in itself, but has great potential value in being applied to some

> productive purpose. There is no benefit from water's existence other than the extent to which humans can benefit, directly or indirectly, from the water itself, or the environments that water supports (e.g., stocks of food fish that depend on the viability of a lake ecosystem). Water is a resource, much like coal or oil or phosphate or gold. It is not only culturally permitted, but actually culturally preferred to make use of the resource by 'mining' it, or recovering it in whatever way is technologically feasible. Choosing not to recover the resource, electing not to utilize the potential benefits of the resource, is considered to be wasteful, and in this sense, even sinful.
>
> *(Groenfeldt, 2006: 109)*

In contrast, Groenfeldt states, Indigenous participants at earlier World Water Forums explained that:

> Water is not only an aspect of Indigenous spirituality, but a very major component of that spiritual world. Water, whether as a substance, or in the form of water bodies (rivers, lakes) and meteorological phenomena (rain, snow, fog, clouds), is seen through a spiritual – not an economic – lens.
>
> *(Groenfeldt, 2006: 112)*

In UNESCO's report, *Water and Indigenous Peoples,* based on views expressed by Indigenous peoples at the 2nd and 3rd World Water forums, Indigenous concerns around how water is viewed are highlighted as follows:

> Water [is seen] as an *economic good* and a human need, not a human right. [A] criticism against the Forum reports is the reduction of water to an economic good, and the strong push for its privatisation and full cost recovery. The right to water, which includes access and control over water resources, is one of the most basic demands of indigenous peoples. Water is an integral part of our ancestral territories and resources, and when we claim our right to our ancestral territories, this includes our right to the water found there.
>
> *(UNESCO, 2006: 29, emphasis mine)*

Indigenous water declarations are further instructive in providing insight into the deeply rooted understanding of water based on Indigenous worldviews. The Declaration of the Indigenous Peoples Parallel Forum of the 4th World Water Forum (2006), or the *Tlatokan Atlahuak Declaration,* for example, asserts that we must, "[...] honor and respect water as a *sacred being that sustains all life.* Our traditional knowledge, laws and forms of life teach us to be responsible and caring for this sacred gift that connects all life" (emphasis mine).

The Garma International Indigenous Water Declaration (2008), an outcome document of the International Indigenous Water Experts Exchange, sought to facilitate discussions around how best to achieve appropriate protection of Indigenous interests in water and explored how Indigenous knowledge can be recognized as expert knowledge and used alongside Western scientific knowledge in water management systems. The *Garma Water Declaration* (2008) asserts that, "[...] *water has a right to be recognized as an ecological entity*, a being with a spirit [that] must be treated accordingly. For the Indigenous Peoples water is essential to creation; Ancestral beings are created by and dwell within water" (emphasis mine). Water in this sense is a *being with agency.* Indigenous peoples don't view their relationship with water as simply having a spiritual component; rather, the relationship is understood as one shared with another living entity.

Writing from an Anishinaabek context, Indigenous legal scholar, Aimee Craft, explains that:

> [...] *nishinaabe nibi inaakonigewin* (water law) tells us that water is life: *nibi onje biimaadiziiwin*. We are born of water, and we are primarily composed of water. Not only does it give and take life, it is also a living being in and of itself that relies on a larger web of relationships to be well and to bring wellness to other beings.
>
> *(Craft, 2018: 56)*

There are thus striking differences between Western and Indigenous foundational understandings of what water is. The binary, materialistic water-as-resource/commodity/property way of conceptualizing water has influenced how humans relate to water to such a negative extent that it has greatly contributed to the global water crisis and the water injustices we see today.

The Water Crisis and Sustainable Development

Among the numerous global threats to human existence, the state of water continues to be of central concern. Indigenous peoples and water development agencies share the common view that a water crisis is indeed underway and that resulting pressures are increasing over time. Most analyses attribute the causes to failures of water governance and management regimes. As noted earlier, the World Water Development Reports have brought attention to the water crisis in every release since the first one in 2003:

> At the beginning of the twenty-first century, the Earth, with its diverse and abundant life forms, including over six billion humans, is facing a serious water crisis. All the signs suggest that it is getting worse and will continue to do so, unless corrective action is taken. This crisis is one of water governance, essentially caused by the ways in which we mismanage water. Yet of all the social and natural resource crises we humans face, the water crisis is the one that lies at the heart of our survival and that of our planet Earth.
>
> *(UNESCO World Water Assessment Programme, 2003: 4)*

An Indigenous statement on the matter, the Tlatokan Atlahuak Declaration (2006), again highlights the contrast between what is seen as a problem of improper application of economic resource management, versus a problem caused by trying to apply a Western economic approach to water management in the first place:

> Our lands, territories and natural resources, particularly our water (rivers, springs, wells, lakes, groundwater) continue to be stolen or ruined with extreme pollution. The water multinationals, with the support of the international finance agencies like the World Bank and the Interamerican Development Bank are accomplices in the privatization of our territories and our water. This creates a scarcity of water, raising its economic value and furthering the view of water as an object of commerce.
>
> *(2006)*

These ontological differences in relation to water result in conflict. Water development projects within Indigenous lands/territories are often imposed by external forces of government and industry. Indigenous peoples have stated that understanding the root of the water crisis requires a deep look at "[...] the significant social, political, economic, ecological, cultural and *even spiritual roots of the crisis*. It is not enough just to look into the technical, managerial and economic aspects, and not analyse the frameworks used by the think-tanks and donor agencies" (UNESCO, 2006: 28, emphasis mine). The world water forums in themselves are often seen as problematic, as when they are "[…] mainly designed and run by the water industry and its supporters, like the World Bank and the IMF as well as so-called water specialists and experts, there is a clear disconnect with the assertion of the UN Human Rights bodies, Indigenous peoples and civil society that water is a basic human right" (UNESCO, 2006: 31). The root of the crisis may in fact reflect a *spiritual crisis*.

Water Colonialism and Indigenous Peoples

Water colonialism is when water is disrespected and exploited with disproportional impacts on Indigenous peoples, their livelihoods and territories. Capitalist and development agendas have contributed to water colonialism, more often than not, along with the specific colonization of Indigenous peoples.

As the Kyoto Water Declaration explains:

> We see our waters increasingly governed by imposed economic, foreign and colonial domination, as well as trade agreements and commercial practices that disconnect us as peoples from the ecosystem. Water is being treated as a commodity and as a property interest that can be bought, sold and traded in global and domestic market-based systems. These imposed and inhumane practices do not respect that all life is sacred, that water is sacred.
>
> *(2003)*

According to the Garma Water Declaration:

> Nation-States, in asserting competing sovereignty over the lands and waters, have introduced and enforced unlawful and unjust mechanisms resulting in trespass of the legal entitlements of Indigenous Peoples to the ownership, use, management and benefit of the lands and the waters, without consultation, consent or just compensation.
>
> Furthermore, Nation-States have grossly mismanaged the lands and waters of Indigenous Peoples, causing ecosystem collapse, human induced climate change, severe water quality degradation, extreme stress upon ecologies and species extinction at a scale and rate which is unprecedented.
>
> *(2008)*

Indigenous water declarations thus reveal that historical and ongoing colonialism also contribute to the root causes of the water crisis. The recognition of the human right to water and sanitation is of critical importance because many peoples, including Indigenous peoples, do not have sufficient access to water.

Water as a Human Right

Water is, of course, essential for the survival of humanity, a basic necessity for us to live, and as such has been recognized as a human right since 2010 by the United Nations. The UN adopted Resolution 64/292: *Human Right to Water and Sanitation* which "recognizes the right to safe and clean drinking water and sanitation as a human right that is essential for the full enjoyment of life and all human rights" (UNGA, 2010). Despite this recognition of the right to water for all, there remain millions of people around the world who lack access to safe drinking water or sanitation. Roughly "40 percent of the global population lacks access to clean, safe drinking water" (IPBES, 2019: 29). Indigenous peoples as a group tend to face a greater burden than others in relation to such water insecurity (World Bank, 2017).

According to the UN, "at the core of sustainable development" lies water (United Nations, n.d.). "Water and sanitation for all" is one of the United Nations Sustainable Development Goals. Despite such efforts to highlight the critical role water plays, to safeguard this essential resource, and to ensure it is used sustainably, water is increasingly under pressure and demand for it only continues to grow (United Nations, 2015; 2021). Historical and current practices of colonialism, water governance and policies, and the complete disregard for Indigenous rights, continue to disenfranchise Indigenous peoples and impact their access to safe and clean drinking water (Galway, 2016; Jiménez et al., 2014). As one example, Canada, a country known for its abundance of fresh water, has for decades perpetuated a drinking water crisis impacting primarily First Nations peoples (Galway, 2016). As of April 2022, there remained 33 long-term boil water advisories impacting 28 First Nations communities in the country (Government of Canada, 2022). A recent United Nations Report of the Special Rapporteur on Human Rights, based on an official visit to Canada, found that there exists in this country a:

> [...] jurisdictional quagmire faced by indigenous peoples, whose reserves often fall between the cracks of federal and provincial jurisdiction, posing a risk of unregulated toxic exposure. For example, throughout Canada, the provincial drinking water quality standards are not applicable on reserves and the federal standards, which have yet to be set, are not legally binding. Jurisdictional separation is not an excuse for shortcomings by the Government in taking prompt action to address toxic exposure.
>
> *(UNGA, 2020: 3)*

Water and Indigenous Self-Determination

Self-determination is a critical aspect of realizing water security for Indigenous peoples. As the 2003 Kyoto Declaration states, "Self-determination includes the practice of our cultural and spiritual relationships with water, and the exercise of authority to govern, use, manage, regulate, recover, conserve, enhance and renew our water sources, without interference." According to UNDRIP, "Indigenous peoples have the right to maintain and strengthen their distinctive spiritual relationship with their traditionally owned or otherwise occupied and used lands, territories, waters and coastal seas and other resources and to uphold their responsibilities to future generations in this regard" (UNGA, 2007, Article 25). The challenges with water insecurity and Indigenous peoples relate to a colonial legacy of destructive policies and practices that have allowed for the mismanagement of Indigenous peoples' lands and waters along with the complete disregard for their inherent rights. An important component of the antidote to this situation is the full realization of self-determination and rights of Indigenous peoples as expressed in UNDRIP and Indigenous water declarations.

Conclusion: Beyond Material Values of Water

> We challenge the dominant paradigm, policies, and programs on water development, which includes among others; government ownership of water, construction of large water infrastructures; corporatization; the privatization and commodification of water; the use of water

> as a tradeable commodity; and the liberalization of trade in water services, which do not recognize the rights of Indigenous Peoples to water.
>
> *(Indigenous Peoples Kyoto Water Declaration, 2003)*

Western-based approaches to water security generate disproportionate negative impacts on Indigenous peoples that threaten their very existence. The principle of treating water-as-resource/commodity has led to the exploitation and misuse of water around the world (Chiblow, 2019; Jiménez et al., 2014; McGregor, 2014; Whyte, 2017). This issue applies not only to water, but to how the land, animals, and all of the Earth are seen as resources to be exploited by human activity (Wilson & Inkster, 2018).

The imbalance of power between Western and Indigenous perspectives is crucial to understanding the discourse that dominates water governance, as:

> Those who control how water is valued control how it is used. Values are a central aspect of power and equity in water resources governance. The failure to fully value water in all its different uses is considered a root cause[...] of the political neglect of water and its mismanagement.
>
> *(United Nations, 2021: 1)*

Western discourse around water governance and management has clearly fallen short as every World Water Development Report reveals. There has not been the space for other views of water. Applying an Indigenous perspective can highlight how inadequate this framing of water-as-commodity is.

Indigenous relational ontologies of water are built on "principles of respect, relationality, responsibility, and reciprocity" (Wilson & Inkster, 2018: 526). Indigenous world views are rooted in seeing water as part of Mother Earth, nature, spirit, and/or Creation (McGregor, 2012; Ruru, 2018; Wilson & Inkster, 2018). For many Indigenous peoples, water is not viewed as a commodity, but as a gift, a living part of Creation (Chiblow, 2019; Craft, 2018; McGregor, 2012; Taylor et al., 2019). Just as the Earth is seen as a living being, water is also considered to be a living entity (McGregor, 2012; Zinga & Styres, 2011).

It is through such understandings, where water is viewed by all of us with the kind of reverence it not only deserves but which would also render unthinkable the devastation of water in which we currently engage, that sustainable human-water relations may eventually be (re-)established. Indigenous peoples will therefore continue to challenge the dominant water development discourse and in so doing generate possibilities for humanity to reassume its role in supporting all life.

References

Chiblow, S. (2019). "Anishinabek women's Nibi Giikendaaswin (water knowledge)," *Water*, 11(2), 209. https://doi.org/10.3390/w11020209

Craft, A. (2018). "Navigating Our Ongoing Sacred Legal Relationship With Nibi (Water)." In *UNDRIP Implementation More Reflections on the Braiding of International, Domestic and Indigenous Laws*. Waterloo, ON: Centre for International Governance Innovation, pp. 53–61.

Declaration of the Indigenous Peoples Parallel Forum of the 4th World Water Forum: Tlatokan Atlahuak Declaration. (2006). *Australian Indigenous Law Reporter*, 10(2), 99–100.

Emanuel, R.E. and Wilkins, D.E. (2020). "Breaching barriers: The fight for indigenous participation in water governance," *Water*, 12, 2113.

Evans, K. (2020). "The New Zealand river that became a legal person." *BBC*. [Accessed 10 February 2023] Available at https://www.bbc.com/travel/article/20200319-the-new-zealand-river-that-became-a-legal-person

Galway, L.P. (2016). "Boiling over: A descriptive analysis of drinking water advisories in first nations communities in Ontario, Canada," *International Journal of Environmental Research and Public Health*, 13(5), 505.
"Garma Water Declaration." (2008). [Accessed 10 February 2023] Available at: https://www.afn.ca/uploads/files/env/garma-international.pdf
Government of Canada. (2022). "Ending long-term drinking water advisories." [Accessed 10 February 2023] Available at https://www.sac-isc.gc.ca/eng/1506514143353/1533317130660
Groenfeldt, D. (2006). "Water Development and Spiritual Values in Western and Indigenous Societies." In: Boelens, R., Chiba, M. and Nakashima, D. (eds.), *Water and Indigenous Peoples*. Paris: UNESCO, pp. 108–115.
Hartwig, L.D., Jackson, S., Markham, F. and Osborne, N. (2022). "Water colonialism and indigenous water justice in South-Eastern Australia," *International Journal of Water Resources Development*, 38, 30–63. https://doi.org/10.1080/07900627.2020.1868980
"Indigenous Declaration on Water." (2001). *Freshwater Future*. [Accessed 10 February 2023] Available at https://freshwaterfuture.org/community-resources/publications/freshwater-voices-newsletter-archive/volume-10-number-4-•-fall-2002/indigenous-declaration-on-water/
"Indigenous Peoples Kyoto Water Declaration." (2003). *Third World Water forum*, Kyoto Japan. Available at: http://www.waterculture.org/uploads/IPKyotoWaterDeclarationFINAL.pdf
Intergovernmental Panel on Climate Change (IPCC). (2021). "Climate change 2021: The physical science basis." In: Masson-Delmotte, V., P. Zhai, A. Pirani, S.L. Connors, C. Péan, S. Berger, N. Caud, Y. Chen, L. Goldfarb, M.I. Gomis, M. Huang, K. Leitzell, E. Lonnoy, J.B.R. Matthews, T.K. Maycock, T. Waterfield, O. Yelekçi, R. Yu, and B. Zhou (eds.). *Contribution of Working Group I to the Sixth Assessment Report of the Intergovernmental Panel on Climate Change*. Cambridge, United Kingdom and New York: Cambridge University Press, p. 2391. https://doi.org/10.1017/9781009157896
Intergovernmental Panel on Climate Chang (IPCC). (2022). "Climate change 2022: Impacts, adaptation and vulnerability." In: H.-O. Pörtner, D.C. Roberts, M. Tignor, E.S. Poloczanska, K. Mintenbeck, A. Alegría, M. Craig, S. Langsdorf, S. Löschke, V. Möller, A. Okem, B. Rama (eds.). *Contribution of Working Group II to the Sixth Assessment Report of the Intergovernmental Panel on Climate Change*. Cambridge, UK and New York: Cambridge University Press, p. 3056. https://doi.org/10.1017/9781009325844
IPBES. (2019). *Summary for Policymakers of the Global Assessment Report on Biodiversity and Ecosystem Services of the Intergovernmental Science-Policy Platform on Biodiversity and Ecosystem Services*. Bonn, Germany: IPBES secretariat, p. 60. https://doi.org/10.5281/zenodo.3553579.
Jiménez, A., Cortobius, M. and Kjellén, M. (2014). "Water, sanitation and hygiene and indigenous peoples: A review of the literature," *Water International*, 39(3), 277–293.
Lexico Dictionaries. (2021). *Meaning of Water, English*. [Accessed 10 February 2023] Available at: https://www.lexico.com/definition/water
Maclean, K. (2015). "Crossing cultural boundaries: Integrating indigenous water knowledge into water governance through co-research in the Queensland wet tropics, Australia," *Geoforum*, 59, 142–152.
McGregor, D. (2012). "Traditional knowledge: Considerations for protecting water in Ontario," *International Indigenous Policy Journal*, 3(3), 1–21.
McGregor, D. (2014). "Traditional knowledge and water governance: The ethic of responsibility," *AlterNATIVE: An International Journal of Indigenous Peoples*, 10(5), 493–507.
McGregor, D. (2021). "First Nations, Traditional Knowledge, and Water Ethics." In: Stefanovic, I.L., Adeel, Z. (eds.), *Ethical Water Stewardship, Water Security in a New World*. Cham: Springer International Publishing, pp. 147–163. https://doi.org/10.1007/978-3-030-49540-4_8
Merriam Webster. (2021). *Definition of Water*. [Accessed 10 February 2023] Available at: https://www.merriam-webster.com/dictionary/water
Moggridge, B. J. and Thompson, R. M. (2021). "Cultural value of water and western water management: An Australian Indigenous perspective." *Australasian Journal of Water Resources*, 25(1), 4–14. https://doi.org/10.1080/13241583.2021.1897926
NGO Committee on Education (NGOCE). (n.d.) "The Dublin Statement on Water and Sustainable Development." [Accessed 10 February 2023] Available at http://www.un-documents.net/h2o-dub.htm
Office of the High Commissioner for Human Rights (OHCHR). (2022). "Special thematic report on climate change and the human rights to water and sanitation." *Special Rapporteur on the human rights to safe drinking water and sanitation*, Pedro Arrojo Agud. [Accessed 10 February 2023] Available at https://www.ohchr.org/en/documents/thematic-reports/special-thematic-report-climate-change-and-human-rights-water-and

Parsons, M., Fisher, K. and Crease, R.P. (2021). "Transforming River Governance: The Co-Governance Arrangements in the Waikato and Waipa¯ Rivers." In Parsons, M., Fisher, K., Crease, R.P. (eds.), *Decolonising Blue Spaces in the Anthropocene: Freshwater Management in Aotearoa New Zealand*, Palgrave Studies in Natural Resource Management. Cham: Springer International Publishing, pp. 283–323.

Root, R.L. (2021). "Why an Indigenous approach to water is needed. Devex." [Accessed 10 February 2023] Available at: https://www.devex.com/news/sponsored/why-an-indigenous-approach-to-water-is-needed-100657

Ruru, J. (2018). "Listening to Papatūānuku: A call to reform water law," *Journal of the Royal Society of New Zealand*, 48(2–3), 215–224.

Simms, R., Harris, L., Joe, N. and Bakker, K. (2016). "Navigating the tensions in collaborative watershed governance: Water governance and indigenous communities in British Columbia, Canada," *Geoforum*, 73, 6–16.

Taylor, K.S., Longboat, S. and Grafton, R.Q. (2019). "Whose rules? A water justice critique of the OECD's 12 principles on water governance," *Water*, 11(4), 809.

UNESCO World Water Assessment Programme. (2003). "Water for people, water for life: The United Nations world water development report: A joint report by the twenty-three UN agencies concerned with freshwater." [Accessed 10 February 2023] Available at: https://unesdoc.unesco.org/ark:/48223/pf0000129726.page=576

UNESCO World Water Assessment Programme. (2006). "Water: A shared responsibility; The United Nations world water development report 2." [Accessed 10 February 2023] Available at: https://unesdoc.unesco.org/ark:/48223/pf0000144409?posInSet=2&queryId=f7d9322b-19c4-49fb-ae10-d848f1f07ef1

United Nations. (n.d.). *Water and sanitation, Sustainable Development Goals – Knowledge Platform*. [Accessed 10 February 2023] Available at: https://sustainabledevelopment.un.org/topics/waterandsanitation

United Nations. (2014). *The United Nations World Water Development Report 2014: Water and Energy*. Paris: UNESCO. [Accessed 10 February 2023] Available at: https://www.unwater.org/publications/world-water-development-report-2014-water-energy/

United Nations. (2015). *The United Nations World Water Development Report 2015: Water for a Sustainable World*. Paris: UNESCO. [Accessed 10 February 2023] Available at: https://www.unwater.org/publications/world-water-development-report-2015/

United Nations. (2016). *The United Nations World Water Development Report 2016: Water and Jobs*. Paris: UNESCO. [Accessed 10 February 2023] Available at: https://www.unwater.org/publications/world-water-development-report-2016/

United Nations. (2019). *The United Nations World Water Development Report 2019: Leaving No One Behind*. Paris: UNESCO. [Accessed 10 February 2023] Available at: https://www.unwater.org/publications/world-water-development-report-2019/

United Nations. (2020). *The United Nations World Water Development Report 2020: Water and Climate Change*. Paris: UNESCO. [Accessed 10 February 2023] Available at: https://www.unwater.org/publications/world-water-development-report-2020/

United Nations. (2021). *The United Nations World Water Development Report 2021: Valuing Water*. Paris: UNESCO. [Accessed 10 February 2023] Available at: https://unesdoc.unesco.org/ark:/48223/pf0000375724

United Nations. (2022). *The United Nations World Water Development Report 2022: Groundwater: Making the invisible visible*. Paris: UNESCO. [Accessed 10 February 2023] Available at: https://unesdoc.unesco.org/ark:/48223/pf0000380721

United Nations Educational, Scientific and Cultural Organization (UNESCO). (2006). *Water and Indigenous Peoples*. Boelens, M. Chiba, and D. Nakashima (eds.). Paris: UNESCO. [Accessed 10 February 2023] Available at: https://unesdoc.unesco.org/in/documentViewer.xhtml?v=2.1.196&id=p::usmarcdef_0000145353&file=/in/rest/annotationSVC/DownloadWatermarkedAttachment/attach_import_e7e52280-39fc-427e-8381-ff2f091de235%3F_%3D145353eng.pdf&locale=en&multi=true&ark=/ark:/48223/pf0000145353/PDF/145353eng.pdf#%5B%7B%22num%22%3A436%2C%22gen%22%3A0%7D%2C%7B%22name%22%3A%22XYZ%22%7D%2Cnull%2Cnull%2C0%5D

United Nations General Assembly (UNGA). (2007). *United Nations Declaration on the Rights of Indigenous Peoples*. Geneva: Office of the Commission for Human Rights. [Accessed 10 February 2023] Available at https://www.un.org/development/desa/indigenouspeoples/wp-content/uploads/sites/19/2018/11/UNDRIP_E_web.pdf

United Nations General Assembly (UNGA). (2010). *Resolution adopted by the General Assembly 64/292: The human right to water and sanitation*. [Accessed 10 February 2023] Available at: https://digitallibrary.un.org/record/687002?ln=en

United Nations General Assembly (UNGA). (2020). "Visit to Canada" – *Report of the Special Rapporteur on the implications for human rights of the environmentally sound management and disposal of hazardous substances and wastes*.

Whyte, K. (2017). "The Dakota access pipeline, environmental injustice, and U.S. Colonialism," *RED INK: An International Journal of Indigenous Literature, Arts, & Humanities*, 19(1), 154–169.

Wilson, N.J. and Inkster, J. (2018). "Respecting water: Indigenous water governance, ontologies, and the politics of kinship on the ground," *Environment and Planning E: Nature and Space*, 1(4), 516. https://doi.org/10.1177/2514848618789378

Wilson, N.J., Montoya, T., Arseneault, R. and Curley, A. (2021). "Governing water insecurity: Navigating indigenous water rights and regulatory politics in settler colonial States," *Water International*, 46(6), 783–801.

World Bank. (2017). *Reducing Inequalities in Water Supply, Sanitation, and Hygiene in the Era of the Sustainable Development Goals: Synthesis Report of the WASH Poverty Diagnostic Initiative*. Washington, DC.

Zinga, D. and Styres, S. (2011). "Pedagogy of the land: Tensions, challenges, and contradictions," *First Nations Perspectives*, 4(1), 59–83.

PART II

Case Studies on Water and Development

8
CAMBODIA

Joakim Öjendal

Introduction[1]

As is argued elsewhere in this volume, water has been at the core of the birth of civilizations, empires, as well as modern state-building. How it has been captured, used, and distributed have been vital for the fate of many societies. While this has an archaic historical foundation (Wittfogel, 1957; cf. Caponera, 1992), the brutal significance of water utilization is coming back in the contemporary era where classic modernity is turning towards unsustainability. Global water scarcity is spreading and an accelerating global climate change will – albeit for different reasons – again forefront water as a human survival issue.

The availability and usage of water have considerably shaped the history of Cambodia. Positioned in mainland Southeast Asia, transgressed by the Mekong River, residing almost entirely within its basin, large swathes of the country are subject to severe flooding in the monsoon season, as well as lengthy dry spells outside the rains. Livelihood in Cambodia has always drawn heavily on natural resources, fishing, and rice growing, hence water is both physically and culturally a key resource, in historical as well as in contemporary Cambodia (Öjendal et al., 2012). It is commonly mentioned as a "hydraulic society," implying that the origin of the state and the base for its sustainability is to be found in its water regime (Groslier, 1979). Whether this is, a historical fact is controversial (cf. van Liere, 1980), but it is certainly the case that a particular water regime was central for the Khmer Kingdom's early success in the Angkor era from the 9th to the 15th century. With this glorious past – being the pre-eminent power in Southeast Asia for six centuries – the colonial era and its subsequent decolonization process were all the more problematic. Even worse, when being caught in the midst of the cold war, Cambodia experienced devastating infrastructural destruction – civil war, genocide and renewed civil war – from 1970 up to the early 1990s. In the post–Cold War era, a period of reconstruction commenced in a relatively orderly fashion and with major involvement of the international community and a certain focus on rehabilitating the water sector (cf. Halcrow, 1994). In each of these phases, water played a major role in the (re-)creation of Cambodia (Öjendal, 2000).

Cambodia is historically formed, as we shall argue below, by its particular water regime, and its fluctuating eras of progress has to a large extent to do with how water has been used. The governing of the water resources in Cambodia is a fundamental political concern with deep implications;

DOI: 10.4324/9781003095545-10

the ones controlling the water resource harness power, and those using this source of power strategically gains political legitimacy and financial strength. As we shall see below, this is not a trivial endeavour. Already in the early Angkor Wat era, water harnessing, dam-building, and irrigation practices were central for generating a food surplus. This is still a central ambition in the country. And in doing so, Cambodian water governance is perpetually struggling to balance bottom-up and top-down practices (Öjendal, 2000).

With a deep historical perspective, we will review the centrality of water in the foundation and subsequent formation of Cambodia, as well as its role in modern Cambodia. We will conclude with a reflection on challenges in contemporary Cambodia as regards water and development.

Water Management During the Angkor Era (9th–15th Century)

Cambodia or rather the Khmers (Mabbett & Chandler, 1995) achieved historical prominence in the era of the Angkor Wat. It was a period stretching from the 9th to the 15th century. At its height its sphere of influence covered large parts of present day Vietnam, Laos, Thailand, and Cambodia. It dominated the region, produced the temples at Angkor Wat[2] and was a well-organized society with sophisticated cultural traits, a tight social organization and an advanced political administration.[3]

The Angkor Empire and its lustre was, it is said, based on the clever and centralized management of the large-scale water resources. "Thanks to the hydraulic policy of its kings, a policy that stretched over many centuries, Cambodia was able to enjoy great prosperity" (Nguyen, 1999: 16). From the vantage point of the north-western tip of the Ton le Sap lake, the Angkor Empire based, according to some authors, its welfare on superior and centrally controlled, large-scale water management, which facilitated three to four harvests a year (Nguyen, 1999: 14; cf. Fletcher et al., 2015). Indeed, it has been claimed that the water resources were physically controlled from the palace area by the king (Fletcher et al., 2015). Consequently, the Angkor state has frequently been called a "Hydraulic Civilization" (Caponera, 1978: 92f; cf. Groslier, 1966; 1979; cf. Nguyen, 1999: 15), implying not only that the water was strictly and centrally managed, but also that political power and the actual state formation was based on this central control. This, in turn, it has been argued, promoted a hierarchical and power-based culture with utter dependence on those at the centre of power (cf. Wittfogel, 1957).

Cambodia fits the pattern of the hydraulic civilization: hierarchical in its cosmology; a great early civilization; heavily dependent on water management; and some physical evidence for the water distributive role of the palace (Groslier, 1979; cf. Fletcher et al., 2015). The perception that the great Khmer Empire was based on large-scale irrigation is still cultivated. FAO writes:

> The golden age of irrigation in Cambodia is considered by many to have been the Angkorian period. The Angkor Wat irrigation system was based on four reservoirs built between the tenth and thirteenth centuries, and storing some 100 to 150 million m3 of water diverted from adjacent rivers.
>
> *(FAO, 1994: 10)*

A joint UNDP/government report writes about "The Angkorian hydraulic City of the Khmer Empire" such that:

> Evidently, the distribution of the water and the maintenance of the reservoirs would not be effective unless the entire irrigation system was controlled by the strong central power. [...] It is clear that the succeeding kings [to Indravarman I, 877-889] followed up to the hydraulic

development by creating new reservoirs or artificial lakes which ensured the prosperity of an agrarian society.

(UNDP-CNMC, 1996: 14)

The foremost advocate of the great "hydraulic" Angkor civilization, Bernard Groslier, writes:

> The entire region of Angkor was meticulously divided into square rice fields [...] that were irrigated by the always visible network [...] of collectors and sub-collectors that brought water from rivers and barays to the tiniest plots. Thanks to this organization, the Angkor region was abundantly wealthy.
>
> *(Groslier, taken from Nguyen, 1999: 14)*

Although the hands-on involvement of the central level may have been over-emphasized and the parallel with Wittfogel's "hydraulic civilisation" is imperfect (Vickery, 1986), water still permeated all parts of society from rituals to food production, and from the court to the peasants/fishermen. Even though central power may not have directed water flows, society functioned under a generally accepted "hydraulic order" (Stott, 1992; van Liere, 1980). The physical geography is conducive to centrally controlled water management around the area of the former palace, however, nowhere near the size for the up-keep of the Angkor empire as a whole. The hypothesis on the hydraulic civilization must then, in its pure form, be questioned for Cambodia. Rather:

> [the] Angkor-era agriculture was based instead on bunded-field transplanted wet-rice cultivation that allowed the planting of approximately fifty million fields. [...] A network of dams and bunds diverted and retained the receding floodwaters of the Great lake after the rainy season.
>
> *(Hal, 1992: 230)*

Even if we reject the somewhat exoticized views on amazing hydraulic civilizations, the importance of water and the "order" under which it was utilized remained a key asset for agricultural surplus production and the power of the regime:

> [The early highly developed] civilizations [in Southeast Asia] had their economic base in agricultural productivity, which was achieved through water management. The millions of small plots, neatly leveled and bunded to keep the rainwater on the land, and the thousands of small devices to manipulate water in the wet season at village and farm level reshaped the lowlands of the lower Mekong Basin[...].
>
> *(van Liere, 1980: 265)*

Perhaps Nguyen provides the most balanced historical description of water management in ancient Cambodia when indicating that *central policies* were crucial although not necessarily *central management* and that individual interventions were limited in scale:

> The striking fact concerning the water control works of Angkor or of the Mun or Chi cultures of the Northeast Thailand is that they dealt with tributaries, small-scale rivers, and streams but none directly harnessed great, powerful rivers like the Mekong or the Chao Praya.
>
> *(Nguyen, 1999: 16)*

It can be argued that the review above is of little relevance for the contemporary state-building and development. However, the opposite is more accurate. The Angkor empire and its water management is the primary contender for Khmer greatness and a cornerstone in the (re-) building of the nation-state in Cambodia. Commonly, the grand story of the centrally controlled hydraulic regime is embraced with an orientalized attraction. However, instead the ideal of a decentralized, small-scale, community based approach appears as the most viable, credible, efficient and significant historical lesson (Öjendal, 2000).

The reasons for the collapse of the Khmer Empire in the 15th century are highly debated among historians and others. A number of theories have emerged where one of the most elaborated one is that the deteriorating water management led to increased salinization due to unsustainable irrigation, and decreasing soil fertility lowering agricultural production, weakening the material base for the empire (Groslier, 1966; Nguyen, 1999: 16). As such, water was as central to the empire's demise as it was to its rise. This is, as we shall see below, casting long shadows into the modern era.

Water Management during the Colonial Era 1863–1953/1954

After the fall of the Angkor empire, the remnants of the empire were for a long time under severe pressure from its neighbours and it could be claimed that France stabilized Cambodia's territory – being gradually squeezed by its neighbours for centuries – through the colonization in the late 19th century (cf. Osborne, 1969). With the treaty of 1863 Cambodia became a French protectorate and faced French "development" activities. While French colonization domestically and internationally consolidated the territory of Cambodia, it did little to enhance infrastructure for water (or anything else) in the country. The French made a vain attempt to establish a deep presence, but Cambodia remained a relative "backwater" in the French colonial endeavour. "Colonial efforts were concentrated in the Mekong Delta" (largely in present-day Vietnam), Nguyen writes (1999: 34).

In terms of water management, two different approaches were introduced: the first, and most significant, was the "Colmatage"-system where annually inundated areas were controlled by temporary reservoirs. Canals "were dug through the natural levees of the rivers in order to allow silt bearing flood waters to penetrate behind the levees to agricultural land or areas where farmers wished to create new arable land" (Dennis, 1990: 214). This system is still in use in many parts of the country. The other major "French" invention was large-scale and "modern" (Kong, 1993: 49) irrigation schemes normally supplied by a major dam. These were constructed with a conscripted work force, so called *corvée* labour, at high human toll, which still, in spite of genocide and civil wars, stands out and sends chills down the spine of the older part of the local population.[4] These constructions are now either in disrepair or inefficiently used. The best known ones are perhaps Kampong Sne in Prey Veng and Bovel in Battambang. While not very successful in and out of themselves, the water infrastructure projects were among the deepest physical interventions the French colonial rule produced.

Water Management during the Independence Era – 1953/1954–1970[5]

With the hasty formal independence 1953/1954, the country was poorly prepared for the difficult existence as an independent country in the conflictive Southeast Asia. Domestically, Prince Sihanouk (who abdicated in favour of his father in 1955 in order to become more politically active as a Prime Minister) led a conservative, repressive regime, while he internationally tried to stay neutral in the cold war, or hesitantly leaning towards China. During this *Sangkum Reastr Niyum* period (1954–1970), the nation entered into an ambitious development programme, so also in the water sector. "The people must have enough water to drink and for their fields and rice paddies, even during the dry season" Prince Sihanouk said in 1958 (taken from Than, 1982: 24). The (re-) construction of the water sector in fact assumed a leading role and a special water policy was outlined:

> The only way to have sufficient water for cultivation of rice and other secondary crops in the dry season is to trap and store the water from the monsoon rain and rivers during the floods by building dams and dikes in the appropriate places nearby the Mekong River and its tributaries where the floods occur yearly.
>
> *("Sihanouk Water Policy," taken from Than, 1982: 41)*

To distance himself from colonialism and the attempts by the French to continue to dominate the country, the "Sihanouk Water Policy" attempted to counter the heavy top-down "French" approach that was evident in the work of the colonial technical bureaus of "La Societée des Grand Barages," "Genie Rural," and the "Bureau Infrastructure."[6] Its working principle was "To explain and to persuade the rural population to voluntarily provide manual labour in carrying out the task of digging these water works facilities" (Than, 1982: 43). This policy was partly initiated because the large-scale projects were expected to take decades to carry out and development was urgently needed.

As a result of the Sihanouk Water Policy, also small and medium scale projects were initiated during this period. A fair number of projects were completed and reported to work satisfactorily. Most interestingly, it has been reported that:

> [...]during the period of capital shortage that handicapped the repair work [in the mid 50s], the local inhabitants, without compensation, continued the work by their own means to complete some sections before the next monsoon began [...].
>
> *(Than, 1982: 32)*

However, the Sihanouk regime had also its fair share and own style of hierarchical and top-down approach, with little grassroots participation.

The big investments were planned together with the "Joint Khmero-American Policy" which was a result of an initiative by the International Cooperation Agency (ICA) and the US Operation Mission (USOM, later USAID). The reason for the US engagement in the Mekong Committee was that the US had an interest in stabilizing the young state and "developing it away" from any communist sympathies. Its strategy was based, with a few exceptions on irrigation and dam-building projects ranging from large to gigantic. In the case of development planning for the entire Mekong Basin, the US experience in water management (cf. Chapter 2, this volume) was seemingly quite uncritically transferred to Cambodia (as it was to many other developing countries).

Although the Sihanouk era has a questionable reputation for its achievements in rural development, due to, *inter alia,* a weak presence in the local contexts and a corrupted civil administration, a fair number of hydraulic works were carried out during this period, and they seem to have been reasonably successful (Himel, 1996a; cf. Than, 1982). Compared to what was to come later, this period is often from within described as a "golden era" and is often associated with romanticized memories by the Cambodian rural population.

Water Management during the Khmer Rouge – 1975–1978

> Democratic Kampuchea must [...] as the first priority[...] attain rice yields more than those of Japan [...] who can attain 7 tons per hectare.
>
> *(Pol Pot, 1978, taken from Martin, 1983: 1)*

The Khmer Rouge – or formally "Democratic Kampuchea" – took power in a revolution in 1975 and up to 1978 Cambodia was ruled by one of the most brutal and repressive regimes known in the modern world. It was subjected to a social experiment resulting in (auto-)genocide where, at least, some 15%–25% of the population were killed or died from mismanagement. One of the Khmer Rogue's more distinctly articulated ambitions was to create an irrigation- and rice-based, self-sufficient version of the great Angkor. As often is the case in rushed nation-building attempts, the history was glorified and brought forward in order to create and resuscitate the nation. It caused a major disaster – humanly as well as agriculturally – through their ignorance and incapacity. The ambition was to construct a criss-cross pattern of nationwide irrigation structures – amounting to 7,000 km of waterways – which would facilitate several rice-crops per year. The forced and back-breaking work on the construction of these structures by bare hand and spades, took perhaps the highest human toll of all the Khmer Rouge abuses (Tyner et al., 2018).

Everything of value would, in the Khmer Rogue's "Super Great Leap" (Chandler, 1991), come from the land, and irrigated rice cultivation was the absolute priority (as it was in the mythologized version of Angkor). The major increase in output was expected to come from irrigating the fields and muster several harvests a year. This "National Reclamation Plan" aimed more precisely at: (i) at least two crops of rice should be achieved; (ii) other crops than rice were curtailed; (iii) high yielding variety rice should be used; and (iv) land should be reclaimed from the tropical forest.

The first and the fourth goal were the ones that proved most disastrous; the second was largely in line with Khmer cultural traits and the third did little harm. The fourth point killed scores of people who were forced into areas affected by malaria to carry out inhumanly hard work. To fulfil the first, the entire country was divided up into neat plots of approximately one hectare, often replacing already existing ones. Irrigation structures were dug across the entire country in order to create a grid system of irrigation canals. Today, from aerial photos, or from an ordinary plane, the "Pol Pot canals" are still clearly visible all across the country.

> The country was covered in dams thrown across watercourses, reservoirs, canals and dikes. Everywhere where there was existing rice paddy, everywhere where it was decided to create new paddy, an irrigation system was [built].
>
> *(1983: 15)*

The system was centralized and rigid. Structures were built with limited references to hydrology and topography, and even less to quality of soil, salinization, siltation, potential for maintenance, and/or communication between various interventions (cf. Tyner et al., 2018). There was no technical capacity for carrying out the plan and people who wanted to apply "abstract" techniques such as mathematics and hydrology were generally discarded, or even punished. No instruments were available to undertake such studies and no central body was capable of coordinating it. The people were mobilized in a military fashion, including the children, and forced to work incredibly long hours. Often the local cadre was responsible for achieving a predetermined, often impossible, harvest (based on the theoretical output from double or triple cropping), when they failed, the level of violence tended to rise in a vain attempt to force "development" to occur. Much of the work, and probably the highest death toll, was experienced in mobile "work brigades" consisting of approximately 200 persons (usually men) and moving from site to site and in which very few survived.

In short, an impossible (water) strategy was attempted with violence and coercion as its primary driving forces and ignorance as its hallmark. Its large-scale, centralized, mechanical, ignorant, and rigid approach were basic flaws. Neglect of respect for human needs and human rights made it worse. Although it might have been successful in a narrow meaning, in terms of the number of reservoirs constructed and kilometres of canals dug, it was a disaster in a developmental sense. People died in large numbers from overwork, disease, and lack of access to medical care. The dams were often ill considered and many of the structures were entirely wrongly constructed given the topography and the geography.

It is still a lingering problem, both for ecological and social reasons. The forced labour, the endless digging and the high death toll has made the Cambodian farmers wary of digging irrigation and drainage canals, which now is a very low status job. Or put differently, the only ones digging canals are the people who have no other alternative. Doing that implies that you recognize that you are destitute, lack resources and powerful connection, and that you "belong" at the bottom of society. In addition, the high failure rate during the Khmer Rouge may have made farmers wary, on a more general level, of working for an abstract common good in a distant future. Moreover, a seemingly reasonable idea is that a fair share of the canals dug during the Khmer Rouge could be renovated, improved and re-used in a more technically educated fashion. This has been attempted during the 1980s and again in the post-conflict era (cf. Halcrow, 1994).

The situation is of course not the same everywhere in the country. In some places the schemes might have succeeded and one possible positive effect of the Khmer Rouge experiment is that the farmer's awareness of the potential of irrigation has risen (Himel, 1996b: 21). Also some recent research (Tyner et al., 2018) cautiously opens up for the interpretation that in places where local initiative "took over" and adapted the grand irrigation plan, success rate might have been considerably higher. They even indicate that more structures than anticipated have, with time been renovated and put to use (ibid.).

Water Management during 1979–1991[7]

In 1978/1979, the Khmer Rouge was overthrown by Vietnamese forces and a new Cambodian regime was installed. Although civil war raged on for another decade, it was dominated by the KPRP (Khmer People´s Revolutionary Party) that was in close alliance with the Communist Party in Vietnam. A *Krom Samaki* system ("Solidarity Groups") of collective work, largely based on a Vietnamese model, was introduced in the rural areas. Even the political system on the national level was highly influenced, and initially dominated, by the Vietnamese one. The regime was, however, plagued by an international economic embargo, meagre resources, lack of educated people, and it was facing an externally fed civil war. At the same time people was starving and agricultural output needed urgently to be enhanced. A lot of reconstruction was achieved during this period, but few irrigation projects were constructed although some of the failed Khmer Rouge ones were reconstructed.

In terms of water management, the development strategy at the time (socialism with central planning) directed the state administration to "deliver" irrigation structures, and as for water management it concentrated on the backbreaking work of rehabilitating the Khmer Rouge structures. This led to a situation where the farmers were waiting for the state to deliver development, and the state normally did not have the resources to do that. It had, moreover, an authoritarian approach to development trying to "command" farmers and development. While this is a strategy which might be successful for keeping political control it is rarely so for instigating development (and even less so after the Khmer Rouge experience). The production collectives soon broke down and was replaced with nothing but the core family and relatives. The local authorities enjoyed only limited legitimacy and commanded few resources. A development stalemate was facing the rural areas.

With the beginning in 1985/1986 the system was gradually liberalized, and from 1989, farmers disposed of the harvest themselves. While this provided a (small) boost for agriculture, it rather resulted in a negative outcome for water management and other activities requiring collective action.

The state was further impoverished, engaging even less in the reconstruction of water management infrastructure. The liberalization was carried out with little legal or other preparedness, resulting in a "smash-and-grab" capitalism reinforced by the declining state, a waning political authority, and collapsing moral values. Even though the SOC regime still had a very tangible political presence on grassroots level, it had a limited development ambition.

Water Management in the Era of Post-Conflict Reconstruction: 1993–2000

With the Paris Peace Agreement, concluded in 1991, a formula was reached under which peace would be restored to the country. The associated election was recognized by all major parties and a constitution was subsequently drafted in September 1993. From here on, the scene was open for major involvement of the international aid community and water was a sector gathering interest. This was channelled both through various NGOs (OXFAM, CIDSE; AFSC, CWS), through the UN-system (UNICEF, UNDP, ILO, etc.) as well as increasingly through bilateral aid. A stalemate was replaced with a frenzy under the banner of development with a sense of urgency. The Prime Minister, Hun Sen, seems to have the same fears in mind when in a statement was quoted as saying:

> It is my conviction that Cambodia has no other choice than to implement scrupulously and rigorously the reform programmes or we will remain bogged down in poverty, underdevelopment and instability.
>
> *(Hun Sen, taken from Reuters, 14 June 1999)*

The state budget for development purposes was minimal, as well as its capacity to initiate improvement in the water sector. Instead, IOs and NGOs were rehabilitating irrigation structures in various parts of the country, but with little or no state involvement. Often the state administration was further drained of educated and capable people by the fact that various organizations recruited state employees and competed with prospects of far higher salaries. The complex water sector with its troubled history proved too much for many. The combination of the withdrawing state and the rapid arrival of many "new" organizations have created a situation which is not directed by policies and plans, or not even by laws or the general will of the Government of Cambodia. Himel states that:

> [...] the current situation is very much one of a series of isolated and uncoordinated programs being run by different organizations [...] in different areas with different philosophies, methods, goals and timeframes.
>
> *(1996a)*

After this immediate and chaotic post-conflict period, things slowly normalized, and under an overall liberal paradigm, market functions started to increasingly direct water investments.

Contemporary Water Management: 2000 and Ongoing

Cambodia is an a phase of intense modernization and growth, putting pressure on the water sector in many different ways. Asian Development Bank wrote in their country sector review in 2012 that:

> Cambodia's water resources are relatively underexploited and its aquatic ecosystems are healthy in many locations, although they are vulnerable to damage through the impedance of

drainage flows by roads and the effects of irrigation. The development of a market economy, agriculture, industry, and fisheries is likely to bring increased pressure on water resources and ecosystems.

(ADB, 2012: 4)

Combining the contextual facts that Cambodia is subjected to the annual monsoon, possesses a low lying territory and where the mainstay of the population rely on the traditional water regime and its natural resources, make it vulnerable to interventions. Yet increased irrigation, more intensive fisheries, and an aggressive marketization of agricultural produce is under way. On top of this, ADB choose not to mention the energy sector and its proposed dam-building plans. in Cambodia and further up in the Mekong system (which is a major source of water for the entire system in Cambodia).

In 2003, a national sector review for hydropower was produced by the government (CNMC, 2003). This report identified a vast number of dam sites. Initially there were 60 possible sites for hydropower development in Cambodia and it was estimated that the country's total generation potential were at 10,000 MW. Of this possibly half is on the mainstream of the Mekong which is highly controversial and plans on this has not moved forward since. Although controversial and criticized, the hydropower energy production has grown explosively, 50 times in the period from 2011 to 2016 for instance.[8] The mega-dams on the mainstream of the Mekong River – the real game changer – remain so far elusive and far away.

Fuelled by the conducive topography, the historical experience (or the myth of them), the presence of half-finished project by the Khmer Rouge-regime, abundantly available donor funding, and, again, the underlying urgency for increased agricultural production, there were many and massive attempts at jumpstarting irrigation. This was also publicly acknowledged in the Government's "Rectangular strategy" (RGC, 2013), declaring a focus on large-scale water infrastructure development (especially for irrigation). Hence, irrigation has, again, been aiming for large-scale, industrial-sized and highly visionary development. However, most of these attempts missed the target or right out failed. Of the officially recognized irrigation schemes, 1,574 do not function at all; 802 function partly and only 149 schemes function well (Sithirith, 2017: 7). Interestingly the historical pattern (and the reality beneath mythologization) repeats itself, where the grand plans do not work, but instead the medium and small scale approaches have considerably higher success (Sithirith, 2017).

Conclusion – Historical Patterns of Water and Development in Cambodia

As should be obvious from the above, water and its usage has considerably shaped the history of Cambodia, and still does. As often is the case in the water sector, the grand narratives of large-scale water management have time after time taken precedency and lured various regimes into creating/saving the nation with large-scale plans harnessing the water for the creation of a "development leap." These ambitions have often been founded in unrealistic visions or in more or less bizarre political ideologies. Interestingly, irrespective of whether they are under French colonial rule, revolutionary communism, or liberal modernism, the same mega-projects are brought forward, with the same mindset of "nirvana-solutions" (cf. Molle, 2008) to development issues. Throughout, there is an awkward relation between "water" and sustainable "development" and the ambition to better people's lives. Instead, water seems to trigger the "seeing like state"-rationale (cf. Scott, 1999), becoming clinical, politicized and subject to technical mega-projects/visions. Instead of grand "once-for-all" solving-the-problem-at-stake, these "nirvana-solutions" tend to marginalize people they would supposedly assist; falter through its overwhelming complexity; emphasizing the problem to be solved; and mostly serve the elite in their political strife.

The contemporary era may be slightly different with a more fine-tuned approach and some historical insights of the challenges involved (cf. ADB, 2022; RGC, 2018). There is however the risk of falling into the same trap of again of harbouring an excessive belief in large-scale, nation-saving mega projects. There are several major irrigation structures and huge dam-building plans in the pipeline. The classic modernist schemes and the "Seeing like a state" are still present still there. With water scarcity spreading, global energy crisis prevailing, and overall climate change accelerating, mistakes in water management will be costlier than ever.

Notes

1 The author is grateful to Ella Öjendal for an eminent work with an early version of this text.

2 Angkor Wat is one of the world's greatest archeological remains. It lies just outside present day Siem Reap at the northern tip of the "Great Lake" (Tonle Sap Lake), in north western Cambodia.

3 We have basically three kinds of sources for this period: inscriptions on temples (Vickery, 1977), one eyewitness account from the 13th century (Chou, 1967), and various archeological traces (Vickery, 1977). Chandler (1983) is the most well-known historian bringing this era into contemporary knowledge. Increasingly, aerial photos and satellite screenings have become increasingly sophisticated and is able to trace historical and physical patterns with high degree of accuracy (cf. Fletcher et al., 2015).

4 Based on my own interviews with the older residents of a village in Prey Veng. This village supplied labor force for the nearby Kampong Sne Dam, which was built in the 1930s.

5 This section draws on Than (1982). For a general historic overview over this period in Cambodia, see, e.g., Chandler (1991).

6 "The National Organization for Big Dams," "Department of Rural Engineering," and the "Infrastructure Bureau," respectively.

7 The section below is a revised version of the authors previous review (cf. Öjendal, 2000).

8 https://opendevelopmentcambodia.net/topics/hydropower-dams/, updated 2019.

References

ADB. (2012). *Cambodia Water Supply and Sanitation Sector Assessment, Strategy, and Road Map*, ADB, Manila.

ADB. (2022). *Cambodia: Water Resources Management Sector Development Program*, ADB, Manila.

Caponera, D. (1978) "Water Laws in Hydraulic Civilisations," in Ulmen, G.L. (ed.) *Society and History*, De Gruyter Mouton.

Caponera, D. (1992). *Principles of Water Law and Administration – National and International*, A. A. Balkema, Amsterdam, Rotterdam.

Chandler, D. (1983). *A History of Cambodia*, Westview, Boulder, CO.

Chandler, D. (1991). *The Tragedy of Cambodian* History, Yale University Press, New Haven, CT and London, UK.

Chou, T.-K. (1967). *Notes on the Customs of Cambodia*, Social Science Association Press, Bangkok.

CNMC (Cambodia National Mekong Committee). (2003). "National Sector Review 2003: Hydropower." *CNMC, Phnom Penh*. pp. 5, 8–9.

Dennis, J. (1990). "Kampuchea's Ecology and Resource Base: Natural Limitations on Food Production Strategies," in David, A. and Hood, M. (eds.) *The Cambodian Agony*, M.E. Sharpe, London, UK.

FAO. (1994). "Agricultural Development Options Review (Phase 1). Annex 4: Water Resources and Irrigation, (Draft)," *in Report No: 42/94 UNDP* – CMB 1 SR., Rome.

Fletcher, R., Penny, D., Evans, D., Pottier, C., Barbetti, M., Kummu, M. & Lustig, T. and APSARA. (2015). *The Water Management Network of Angkor*, Cambridge University Press, Cambodia [published online].

Groslier, B. P. (1966). *Indochina*, Archeologia Mundi Series, F, Muller, London, UK.

Groslier, B. P. (1979). "La Cité hydraulique Angkorienne: exploitation ou surexploitation du sol." *Bulletin, l'Ecole Francaise d'Extreme-Orient*, 66, 161–202.

Halcrow. (1994). *Irrigation Rehabilitation Study in Cambodia – Inventory and Analyses of Existing Systems. Volume 1, Main Report*, Mekong Secretariat, Bangkok.
Hall, K. R. (1992). "Economic History of Early Southeast Asia," in Tarling, N. (ed.) *Cambridge History of Southeast Asia*, Cambridge University Press, Cambridge, UK.
Himel, J. (1996a). *Some Ideas About Food-For-Work Irrigation Projects*, Mimeo, Phnom Penh.
Himel, J. (1996b). *CWS Kompong Thom Reintegration Project – Small Scale Irrigation Consultancy*, CWS, Mimeo, Phnom Penh.
Kong, S. (1993). *Water Resources Management in Cambodia: A Case Study on Development and Sustainability*, Khon Kaen University (M.Sc.).
Mabbett, I. & Chandler, D. (1995). *The Khmers*, Wiley-Blackwell, London, UK.
Martin, M. A. (1983). "Rice Culture and Water Control in Democratic Kampuchea," in *Rural Studies Magazine*, July–September 1983.
Molle, F. (2008). "Nirvana Concepts, Narratives and Policy Models: Insights from the Water Sector." *Water Alternatives, 1*.
Nguyen, T. D. (1999). *The Mekong River and the Struggle for Indochina – Water War and Peace*, Praeger, London, UK.
Öjendal, J. (2000) *Sharing the Good – Modes of Managing Water Resources in the Lower Mekong River Basin*, PhD thesis, Göteborg University.
Öjendal, J., S. Hansson & S. Hellberg (eds.). (2012). *Politics and Development in a Transboundary Watershed – The Case of the Lower Mekong Region*, Springer, London, UK.
Osborne, M. (1969). *The French Presence in Cochinchina and Cambodia*, Cornell, Ithaca, NY.
RGC (Royal Government of Cambodia). (2013). "Rectangular Strategies Phase III," *The Council for Development of Cambodia*, Phnom Penh, Cambodia.
RGC (Royal Government of Cambodia). (2018). *Rectangular Strategy for Growth, Employment, Equity and Efficiency*, Phase IV, RGC, Phnom Penh.
Scott, J. C. (1999). *Seeing Like a State – How Certain Schemes to Improve the Human Condition Have Failed*, Yale University Press, New Haven, CT and London, UK.
Sithirith, M. (2017). "Water Governance in Cambodia: From Centralized Water Governance to Farmer Water User Community," *Resources*, 6(3), 44.
Stott, P. (1992). "Angkor: Shifting the Hydraulic Paradigm," in Rigg, J. (ed.) 1992, *The Gift of Water – Water Management, Cosmology and the State in Southeast Asia*, SOAS, London, UK.
Than, S. (1982). *Alternative Strategies of Irrigation in Development in Cambodia*, Cornell University, Ithaca, NY.
Tyner, J. A., Munro-Stasiuk, M., Coakley, C., Kimsroy, S. & Rice, S. (2018). "Khmer Rouge Irrigation Schemes During the Cambodian Genocide," *Genocide Studies International*, 12(1), (Spring 2018), 103–119.
UNDP-CNMC. (1996). *Rural Development – Water Policy*, Mimeo, UNDP, Phnom Penh.
van Liere, W. J. (1980). "Traditional Water Management in the Lower Mekong Basin," *World Archeology*, 11(3), pp. 265–280.
Vickery, M. (1977) *The Chronicular Evidence for the Fourteenth to Sixteenth Centuries*, Ann Arbor, Michigan (Ph.D. dissertation).
Vickery, M. (1986). *Kampuchea – Economics, Politics and Society*, Pinter/Rienner, London, UK.
Wittfogel, K. A. (1957). *Oriental Despotism. A Comparative of Total Power*, Yale Univ. Press, New Haven, CT.

9

SOUTH AFRICA

Richard Meissner, Stephen Rule, Karen Nortje, and Inga Jacobs-Mata

Introduction

The development of South African society and its state apparatus is closely entwined with the development and exploitation of the country's water resources. Although a perception may exist that the development trajectory of the country was for the greater good of the entire population, and not just certain race groups, like white people,[1] historical research of the water sector suggests otherwise. Development before and during apartheid was to further the interests of the minority population at the expense of the majority. During the colonial era and at the height of apartheid, a large portion of the country's population was excluded from state-led socio-economic development activities and practices. This was followed, after 1994, with the new dispensation after the first democratic elections, with a wholesale reform of society and the water sector. Since then, water allocation was also under successive African National Congress (ANC)-led governments that catered for the needs of all and not a small section of the population. Later, during the so-called missing ten years (from 2007 to 2017) of former President Jacob Zuma's presidency, another exclusionary dynamic emerged. This time around the ruling elite, that, through state capture, was responsible for the looting of billions of South African rand that not only placed water sector reforms on hold but eroded some of the progress made to supply water for all.

In this chapter, we will, firstly, report on the role of water management in South Africa's colonial era and the diverse influences it has had on various population groups. The second part of the chapter investigates the role of water during apartheid as a means of control and how it facilitated the building of the apartheid state. The third section deals with the contemporary era wherein successive post-apartheid governments are not only facing a growing demand for effectively functioning water supply and sanitation services, but also water scarcity in the face of systemic corruption, the erosion of the state, climate change, *and* climate variability. We end with a conclusion.

The Colonial Era

It was during the 19th century that plans to develop and exploit the country's water resources started to surface. It was particularly in the Cape Colony, after successive droughts and floods that plans emerged to "conserve the water flowing into the sea" and put it to good use for agricultural

DOI: 10.4324/9781003095545-11

purposes. Since 1860, various reports, published by the government of the Cape Colony promoted the establishment of irrigation works for the development of agriculture in that part of South Africa. During this time, and before, farmers relied on rain-fed agriculture to produce commodities. Even so, many farmers still relied on dry-land farming and in a sense looked down on irrigation and no irrigation projects were established. The situation changed towards the end of the 19th century, where irrigation would only be implemented on a large scale in the south-western parts of the Colony (Kanthack, 1909) and the Orange River (Funke et al., 2007; Turton et al., 2004).

The discourse around the development of South Africa's water resources is usually told through a state-centric lens, with successive local and national governments playing the leading role in the unfolding drama of dam building and irrigation scheme construction. However, in the latter part of the 19th century on the banks of the Orange River one of South Africa's first irrigation schemes was developed by an individual farmer and later by the Dutch Reformed Church in 1885. According to Legassick (1996: 371–372 cited in Meissner, 2014: 4), the Dutch Reformed Church "[i]n organising [the Upington canal's construction for irrigation purposes], [Reverend Christian] Schroeder [...] took [his] lead from Abraham September, who had first led water from the Orange River." This indicates that an individual and non-state institution were some of the first change-enabling agents (Meissner, 2014) to develop infrastructure to exploit water resources for agricultural purposes.

Prior to this, John G. Gamble, hydraulic engineer appointed by the Colonial government in 1875/1876 until 1885, reported in 1877 that the colony had no "previous successes or failures" (Cape of Good Hope, 1877: 9) to follow regarding the construction of irrigation works. He therefore recommended that South Africa should learn from other countries, most notably India, Italy, Spain, Egypt, Germany, Belgium, Chile, and Victoria (Australia). One of Gamble's most significant recommendations regarding the establishment of irrigation works was the question of the role of the state. He asked whether government must be responsible to "construct or to aid?" indicating that the government of the colony was not clear of its role at the time. Gamble suggested that should the government decide to construct irrigation schemes, which were not to be located on "crownland" or government-owned land, it should acquire such land before work commences and lease the land. "Otherwise, the nation does not benefit directly from the increased value of the land, but only the price given for the water" (Cape of Good Hope, 1877: 9 cited in Turton et al., 2004: 115). The state should therefore "claim sovereignty to the land" (Cape of Good Hope, 1877: 9) for the purpose of serving the interests of the nation, with the land playing a more important role than the river's water. Yet, these two elements were still seen as interrelated (Turton et al., 2004: 115). Considering the construction of the irrigation works themselves, Gamble suggested that the government should support irrigation projects and not directly construct them. He also suggested the establishment of "associations of proprietors on any particular river from its source to the sea for the promotion of new works," which should be encouraged by government with the purpose of placing power in the hands of the associations so that "the majority can compel the minority either to take the water or to sell the farms" (Cape of Good Hope, 1877: 10 cited in Turton et al., 2004: 115). Gamble, therefore, suggested the formation of what is today know as water user associations (WUAs), some of which would, after 1994, become catchment management agencies (CMAs). That said, South Africa has an interesting history of modern irrigation (Tempelhoff, 2011; Van Vuuren, 2012) showing that during the colonial era the Cape Colony government pushed for agricultural sustainability and increasing its food security prospects.

This historical description should not detract from the important role that South Africa's vast mineral resources played in generating South Africa's wealth (Swatuk, 2010). With the discovery of first, diamonds, and then, gold, water resource exploitation through infrastructural development took on a new meaning. Kimberley and Johannesburg, the towns and later cities that sprang up around the diamond and goldfields, respectively, had to have reliable water supplies. At first local rivers and streams where able to cater for the demand. But as populations expanded and commercial and industrial activities took root, water had to be sourced from further away. For both towns, the Vaal River became a reliable source, and later for the entire Witwatersrand area, the highlands of Lesotho where the Orange/Senqu River has its source (Meissner and Turton, 2003; Turton et al., 2004). From a critical theory perspective, Swatuk argues that "Perhaps less well known is the significant contribution of (colonial, settler, independent state, and primarily white) agriculture not only to national wealth but to the shape and texture of South Africa's current socio-ecological problems" (2010: 523–524). Swatuk (2010) goes further to say that: "The expansion of colonial settlement was made possible by a combination of factors over time, in particular, the use of state power in the alienation of land 'owned' by indigenous Africans, creating a legal structure in support of farming efforts (e.g. private property rights, government subsidies for crop-production and water resources development), and ensuring social order and control" (523).

In 1910, with the establishment of the Union of South Africa translating into South Africa becoming a self-governing state entity, South African society consisted of "English mining and Afrikaner agricultural capital as predominant social forces in South Africa" (Swatuk, 2010: 524). This dichotomous social arrangement would also manifest itself in the water sector, where these interests, and particularly Afrikaner interests would play a role in the allocation of water resources, to agriculture, domestic, and industrial sectors. According to Funke et al., "While the government went to great lengths to supply the mineral-rich, financially lucrative areas of South Africa with water, vast numbers of rural South Africans were neglected" (2007: 13). Local governments, on the other hand, were apprehensive of providing sufficient services and infrastructure in the townships, knowing that the majority of inhabitants would be unable to pay for them. This meant that service delivery of drinkable water, sewerage, and household refuse management was either non-existent – or woefully poor – posing a serious health risk (Tempelhoff, 2017). Water services were more reliably allocated first to white residential areas, industry and then agriculture, controlled by these racially characterised interests and geographical areas.

Apartheid Era

Since the 1940s, the government has used mining and industrialisation to achieve economic growth and socio-economic development. The water sector emerged as one of the sectors that would provide the most basic input variables for achieving such goals (Tempelhoff, 2017). It should be noted that racial segregation had been a characteristic of both colonial and the Union governments' policies from around the late 19th century. The National Party (NP), driven by a strong independent Afrikaner identity, came to power in 1948. Soon after the party assumed power, it introduced apartheid as the government's raison d'être (Funke et al., 2007; Turton et al., 2004). The hydraulic mission, which had started in the previous century and accelerated in the early 1930s, received new impetus as a series of prime ministers promoted an increase in the size and number of large dam building and irrigation construction initiatives to stimulate economic development (Funke et al., 2007; Turton et al., 2004).

In 1956, the government promulgated the Water Act of 1956, and the Department of Water Affairs (then known as the Department of Irrigation) was established under the Act (Funke et al., 2007; Tempelhoff, 2017; U of SA, 1956). The predominant task of the Act and the Department was to provide and allocate water for predominantly agricultural development. In this sector, a large part of the NP's support base was situated (Funke et al., 2007). This meant that from 1948, and as part of the government's apartheid strategy, there was considerable energy poured into the improvement of farming, particularly farming practices, in what became known as the homelands.[2] The government viewed agriculture as the mainstay of local food production, and for this to work, water was a vital input factor (Tempelhoff, 2017; Turton et al., 2004). Earlier strategies, particularly before, during and after the turn of the 19th century, irrigation was used to root out the so-called poor white problem, and by so doing benefiting a small portion of South Africa's population (i.e. white people). The government realised that these projects were quite successful and set out to replicate them after the Second World War (Tempelhoff, 2017; Turton et al., 2004). Regarding the African population's access to water and related services, the government reacted in a command-and-control manner (i.e., authoritarianism). Fearing the fragmentation of management following the government's declaration of rivers flowing through the homeland areas as international drainage basins governed by the 1966 Helsinki Convention, the Department regulated water in the homelands by establishing Government Water Control Areas (Tempelhoff, 2017).

The Tomlinson Commission, investigating the socio-economic problems of the "native reserves" with a view of increasing their sustainability to support an increasing human population (Davenport and Saunders, 2000; Turton et al., 2004) recommended that homeland farmers be settled on plots of land of between 1.3 and 1.6 hectares. This would be sufficient, the Commission argued, for the provision of household food security (Perret, 2002; Tapela, 2008; Tempelhoff, 2017). Despite the government's drive to establish irrigation farming in the homelands, the agricultural production of African farmers was below (Tempelhoff, 2017) that which white farmers produced. For instance, the 1951 census indicated that, while white farmers were able to produce 11.85 bags of maize per 0.85 hectares, African farmers could only manage 2.25 bags on a similar sized area. Officials from the Department of Native Affairs, the government department that governed the homeland system, believed white farmers should receive preferential treatment over African farmers since the former held a "superior knowledge" of irrigation farming. African farmers were seen to be adept at cultivating tomatoes, which was, in any case government subsidised (Tempelhoff, 2017; U of SA, 1954). Regarding this perception, it should also be noted that under apartheid, the homelands had no local government and were governed by traditional authorities. The Department of Water Affairs also focused on the supply of bulk water to agriculture and livestock, while water for human consumption was under the competence of the homeland governments and Regional Service Councils with a focus on tapping groundwater. Typically, boreholes were sunk with little consideration of the locality and with little or no consultation with local communities. These boreholes also received little maintenance, and as a result, they fell into ruin (Galvin and Habib, 2003).

All-in-all, the apartheid government invested in irrigation beginning in the 1960s to improve local food security and to "avoid "black inundation" of white areas" (van Koppen and Schreiner, 2014: 547). The result of the apartheid government's "hydraulic mission" was to skew the distribution of water. Van Koppen and Schreiner state that "[i]n rural areas, 1.2% of the people use 95% of the water. The other 98.8%, most of whom depend on agriculture-based livelihoods, only access 5% of the water" (2014: 547). According to Tempelhoff:

> Annual reports by the Department of Irrigation/Water Affairs for the period 1948–1960 suggest that the department kept a firm focus on planning within the paradigm of rapid development and economic growth[…] [T]he movement towards the postcolonial phase of African development saw South Africa increasingly swimming against the proverbial tide of trends in the rest of Africa. Despite the country's growing isolation, the government was shored up by the support of the majority of eligible (white) voters[…] [The government]

> was determined to preserve white minority rule, a salient feature of the apartheid policy. In 1955, a year before the passing of the Water Act of 1956, the African opposition and its supporters, made a profound statement in the Freedom Charter.
>
> *2017: 207*

These opposition forces demanded that all should have an equal share in resources, including water. In political and legal terms, this aspiration would only be realised in the 1990s (Tempelhoff, 2017), starting with former State President F.W. de Klerk's opening of Parliament speech on 2 February 1990, removing obstacles to constitutional negotiations by announcing the release of Nelson Mandela and other political prisoners from prison and the unbanning of the ANC, South African Communist Party (SACP), and the Pan Africanist Congress (PAC) as well as a number of other organisations. He also lifted the State of Emergency Regulations in force since the mid-1980s that affected the media and education (Klerk, 1998). This moment in South African history ushered in the political transformation that culminated in the country's first democratic election in 1994.

The Contemporary Era

Since 1994, the year that heralded South Africa's transition to multi-racial democracy, several socio-economic and political transformations have had an influence on water infrastructure (Tempelhoff, 2009) development and water sector reforms (Backeberg, 2005). Although these reforms are commendable, questions remain about addressing the inequalities in practice. Kane-Berman argues that "[...] despite South Africa's own pioneering history in water, the country now faces a water crisis in the form of crumbling infrastructure, looming shortages, and serious contamination" (2020: 2). He goes on to note that while the ANC has made delivering water to people who previously did not have it, a top priority, one of the consequences has been a failure to maintain, update, and create infrastructure to the volume that is required. Using Statistics South Africa (Stats SA) data, he notes that in 2017 around 1.84 million households did not have access to piped water. He goes on to say that, according to the DWS, in 2018 more than 3 million people have not received a basic water supply service and 14.1 million people do not have access to safe sanitation services (Kane-Berman, 2020).

Starting in the late 19th century, the state took a centralised and active responsibility for the development and exploitation of the country's water resources. Towards the turn of the 20th century, South African society decided, through democratic processes to reject apartheid policies which ushered in a new political dispensation in 1994. Successive ANC-led governments promulgated new water acts (i.e., Water Services Act, No. 108 of 1997 and the National Water Act, No. 36 of 1998) (RSA, 1997, 1998) and vowed to redress the inequalities of the past regarding water resources allocation and utilisation. In other words, the democratic government has, since 1994, "sought to redress the inequalities of apartheid's legacy by setting revolutionary water reforms in motion" (Funke et al., 2007:12), which is often interpreted as actions to benefit the poor and marginalised. Whereas the development of water resources was, during apartheid, based on racial discriminatory policies, government initiated new policies and programmes to provide water for all based on equal human rights. Regarding this, Hellberg argued that water scarcity "[...] in the post-apartheid period has mainly been used as a means of governing the poor[...]" (2020: 1), meaning that scarcity has been used to "govern the poor" while other users were able to use water as much as they wanted given that they pay for the water used. However, the success of these aspirations is patchy. Although access to water for the poor and marginalised had been expanded, service delivery protests have highlighted the lingering inequalities in access to water and sanitation services.

Since the transition to a multiracial democracy in 1994, South Africa's socio-economic development progress showed significant progress (Tempelhoff, 2011) until around the 2008 global financial crisis, since which economic growth has been stagnant or negative.

Since 1994, several issues have had an influence on the development of the water sector. These are decentralisation as part of water sector reform and, in the last decade, state capture.

Considering decentralisation as a political process (Galvin and Habib, 2003) a programme of water allocation and management reforms had been carried out since 1994 with institutional changes in the policy, legal and organisational dimensions. These influenced all water sub-sectors including allocation of water to the environment culminating in the new National Water Act, a new national water policy and a national water strategy. With this in mind, the focus shifted from centralised water management to a more decentralised dispensation, user participation and the allocation of water through a licensed-based system (Backeberg, 2005). CMAs give effect to decentralisation and user participation. However, since 1998, only two of the nine envisioned CMAs had been established: the Breede-Gouritz and the Inkomati-Usuthu (Meissner et al., 2016).

Despite the slow progress, analysts and commentors have hailed the reforms as a shining example to the rest of the world. The National Water Act (No. 36 of 1998) (RSA, 1998), for instance, is described as one of the most progressive water acts in existence (Schreiner, 2013). In so far as criticism is concerned, many analysts draw the same conclusion; there is nothing wrong with the purpose; it is the implementation that is problematic (Schreiner, 2013; Schreiner and Hasan, 2011). One of the most inhibiting factors in this regard, is systematic corruption manifesting as state capture and accompanied cadre deployment.

State capture is an extreme and specific form of systemic corruption, where private interests are used and abused by state decision-makers for their own benefit or the benefit of a particular group (Duvenhage, 2020). Since 2018, the South African Judicial Commission of Inquiry into Allegations of State Capture has been hearing testimony. At the centre of the Commission's inquiry are members of the ruling ANC including party members with business links to the ANC, CEOs of state-owned enterprises, directors general, previous ministers and former, and current presidents. Economically, the estimate of looted financial resources varies between R500 billion and R1 trillion (USD 340.5 million to USD 690 million). This has had a debilitating impact on the government and the Department of Water and Sanitation (DWS) in supplying water services to the public. According to Merten,

> The cost of State Capture hovers at around R1.5-trillion over the second term of the Jacob Zuma administration. That's just short of the R1.8-trillion Budget for 2019. Put differently: State Capture has wiped out a third of South Africa's R4.9-trillion gross domestic product, or effectively annihilated four months of labour and productivity of all South Africans.
>
> *2019*

The ANC, since the time of former presidents Thabo Mbeki and Jacob Zuma, had substantially conflated and ultimately fused the ANC with the state to such an extent that it is frequently inextricable and indistinguishable from the state (Booysen, 2015). When scholars and practitioners highlight politics as a hindrance, they refer to the DWS and specifically problems at ministerial and director-general levels. Incompetence and internal politics are advanced as *the* central problems, and specifically the frequent appointments and suspension of directors general and ministers (Schreiner, 2013: 242). Internal politics is at the heart of the problem, but analysts, often ANC and former Departmental insiders of the reform process, stop short of going into detail (Merrey, 2011).

At the centre of these leadership challenges was not so much the internal politics of the department, but the ANC ideology as manifested in former President Zuma's propensity for regular cabinet reshuffles. As stated earlier, the result led to high turnover rates of directors general and ministers. An analysis indicated that from May 2009 to August 2013 there had been 114 directors general of the 33 national government departments (Booysen, 2015; Van Onselen, 2013) as well as 24 new ministers, 30 new deputy ministers, and 81 new directors general. This destabilised departmental operations, consequently detracting from accountability in policy implementation and financial oversight (Booysen, 2015). In DWS, between 2004 and 2013 three ministers held the office during four terms, with a debilitating effect on implementing the Act (Schreiner, 2013) and, by default, water sector reforms such as the implementation of the CMAs, not to mention water services to the population. Recently it came to light that "cadre deployment" also has a debilitating influence on lower echelons within government departments, where according to the Minister of Public Service and Administration indicated around 35% or 3,301 of the 9,477 senior managers in government service do not have the necessary qualification (Prince, 2021). According to Professor Erwin Schwella, this is the effect of cadre deployment and its negative impact on service delivery (Prince, 2021).

In 2016, Mike Muller, a former water affairs director general, noted that a corrupt elite captured the state through the guise of radical economic transformation initiatives. Systemic corruption has had a debilitating impact on the country's water and sanitation service to such a measure that the number of people without a reliable water supply increased by 2 million between 2011 and 2015 (Galvin and Roux, 2019; Muller, 2016). Also, The Vaal River System water supply scheme, was in 2016 more than five years late (Muller, 2016) and telemetry within the Vaal River system no longer functions properly because of finances that had been diverted elsewhere, such as to fund election campaigns. Due to cadre deployment, the country is rapidly losing its ability to manage the country's water resources. Other aspects in the South African context that are beyond the control of decision-makers, are the biophysical environment and climate change.

Climate variability has, since South Africa's hydraulic mission started in the late 19th century, been one of the driving forces to kick-start the mission. Successive governments reacted to cyclical droughts and floods through the hydraulic mission. Even so, with climate change also having an influence on climate variability, the hydraulic mission is not always able to cope with the demand of water from a growing population. This was highlighted during the drought that began in 2015 and ended in many parts of the country in 2019/2020. One of the most dramatic manifestations of the drought played itself out in Cape Town between 2015 and 2018.

In the Western Cape, and other parts of the country, an extreme drought driven by a strong El Niño (Meissner and Jacobs-Mata, 2016) occurred when water levels fell to 28% of capacity in August 2017 (Rule, Parker, Majikijela, and Lunga, 2020). According to the media and experts, in 2017, the third consecutive "very dry year" (Sousa et al., 2018), Cape Town was expected to run out of water. "Day Zero", predicted as the point at which the city's 3.7 million residents would have no water as taps would literally run dry (Wolski, 2018), was postponed as the situation unfolded and rains fell intermittently (Warner and Meissner, 2021). Experts argued that the drought was only a mild one, and that the actual cause of the situation was poor planning, mismanagement, and the wastage of water by residents (Wolski, 2018).

Comprehensive, and sometimes controversial interventions were put in place by the municipality and provincial and national governments. These included uncoordinated tapping of ground water resources (Rule, Parker, Lunga, and Majikijela, 2020). Households were motivated by various measures to reduce water consumption from 540 to 280 litres per household per day over the 36 months between January 2015 and January 2018 (Booysen et al., 2019). Since human behaviour is one of the main forces in the utilisation of natural resources (Meissner et al., 2018), it was imperative for local, provincial, and national government to focus on water supply management as well as the demand side, where the consumer holds the key to water conservation. What the Day Zero situation indicated to society is that reliance on government driven measures to secure more water could, under circumstances such as climate change and climate variability, be inadequate. This is particularly the case in a country where the average suburban family of four consumes around 300 litres per person per day (Jacobs-Mata et al., 2018) in a society where inequalities are the rule and not the exception (Hellberg, 2020).

A vast water infrastructure network, from large dams to purification plants and reticulation systems, is necessary to deliver for such a demand. The irony of water infrastructure development is that, as more water becomes readily available, there is also a tendency on the part of consumers to waste more of the resource. The key is not only to management supply, but also demand. What is necessary is that urban water infrastructure, technologies, water use practices and cultural demands require social and political reform. This could start with the questioning of taken for granted practices and norms (Jacobs-Mata et al., 2018; Hellberg, 2020). According to the DWS's National Water and Sanitation Master Plan,

> South Africa is facing a water crisis caused by insufficient water infrastructure maintenance and investment, recurrent droughts driven by climatic variation, inequities in access to water and sanitation, deteriorating water quality, and a lack of skilled water engineers. This crisis is already having significant impacts on economic growth and on the well-being of everyone in South Africa.
>
> *2018: 2*

This, almost 30 years after the country's first democratically elected government took control of the state and its governing apparatus with the aim of transforming society from the discriminatory allocation of water resources to a society where there is "some, for all, forever" (e.g., WRC, 2021).

Conclusion

Throughout the history of water resources for development in South Africa, successive governments since the 19th century have played a dominant and often domineering role. They either developed water resources for only a portion of the population based on racial prejudices, discrimination, or through systemic corruption denied access to a vital resource for large parts of the population. However, it is not only the socio-political situation that has a determining influence on how, for whom and what South African society has developed its water resources. We should also remember that the reality emanating from the natural environment is that South Africa is a semi-arid to arid country, with climate variability having had an influence on how the country manages water resources. How future successive governments will manage water for development will not only depend on the biophysical environment, but also how various stakeholders in society perceive their role and influence within the water sector.

Notes

1 During apartheid South Africa's peoples were categorised according to race, which is still entrenched in society and used during demographic surveys such as the censuses and academic research (Posel, 2001).

2 Established by the apartheid government, these were areas where the majority of the black population was moved to.

References

Backeberg, G. R. 2005. "Water institutional reforms in South Africa." *Water Policy*. 7, 107–123.

Booysen, S. 2015. *Dominance and decline: The ANC in the time of Zuma*. Johannesburg: Wits University Press.

Booysen, M. J., Visser, M. and Burger, R. 2019. "Temporal case study of household behavioural response to Cape Town's "Day Zero" using smart meter data." *Water Research*. 149, 414–420.

Cape of Good Hope. 1877. *Report of the hydraulic engineer*. Office of the Commissioner of Crown Lands and Public Works, 27th February 1877. *Report No. G.10-'77Report No. G.10-'77.*

Davenport, T. R. H. and Saunders, C. 2000. *South Africa: A modern history*. New York, NY: Houndmills.

Department of Water and Sanitation (DWS). 2018. *National water and sanitation master plan, volume 1: Call to action: Ready for the future and ahead of the curve*. Pretoria: Department of Water and Sanitation.

Duvenhage, A. 2020. "Suid-Afrika in 'n smeltkroes van verandering: eietyds en toekomsperspektiewe." *Journal of Christian Scholarship*. 56(3&4), 137–162.

Funke, N., Nortje, K., Findlater, K. M., Burns, M., Turton, A., Weaver, A. and Hattingh, H. 2007. "Redressing inequality: South Africa's ne water policy." *Environment: Science and Policy for Sustainable Development*. 49(3), 10–23.

Galvin, M. and Habib, A. 2003. "The politics of decentralisation and donor funding in South Africa's rural water sector." *Journal of Southern African Studies*. 29(4), 867–884.

Galvin, M. and Roux, S. 2019. "Dam state capture: Its cascading effect on the department of water and sanitation." *Transformation: Critical Perspectives on Southern Africa*. 100, 153–178.

Hellberg, S. 2020. "Scarcity as a means of governing: Challenging neoliberal hydromentality in the context of the South African drought." *Environment and Planning E: Nature and Space*. 3(1), 186–206.

Jacobs Mata, I., Funke, N. S. and Banoo, I. 2018. *A focus on water use attitudes and behaviours in South Africa's metropolitan areas: A people-centric approach to urban water conservation and demand management*. Pretoria: Council for Scientific and Industrial Research.

Kane-Berman, J. 2020. *Thirsty land: South Africa's water crisis and how to overcome it*. Johannesburg: Institute for Race Relations.

Kanthack, F. E. 1909. *Irrigation development in the Cape Colony: Earlier, now and in the future* (Dutch). Proceedings of the First South African Irrigation Congress, held at Robertson, Cape Colony, on 18 May 1909. Cape Town: Cape Times Limited. Report No. 39A-1909.

Klerk, D. 1998. *The last trek: A new beginning*. London: Macmillan.

Legassick, M. 1996. "The will of Abraham and Elizabeth September: The struggle for land in Gordonia, 1898–1995." *Journal of African History*. 37, 371–418.

Meissner, R., N. Funke, and K. Nortje. 2016. The politics of establishing catchment management agencies in South Africa: the case of the Breede-Overberg Catchment Management Agency. Ecology and Society21(3):26. http://dx.doi.org/10.5751/ES-08417-210326

Meissner, R. 2014. "Who wants to be an agent? A framework to analyse water politics and governance." *Water SA*. 40(1), 1–10.

Meissner, R. and Jacobs-Mata, I. 2016. "South Africa's drought preparedness in the water sector: Too little too late?" *SAIIA Policy Briefing 155*, November 2016.

Meissner, R. Jacobs-Mata, I. and Murambadoro, M. 2018. "Busting myths around urban household water use in South Africa." *Water Wheel. 17*(3), 41–43.

Meissner, R. and Turton, A. R. 2003. "The hydrosocial contract theory and the Lesotho highlands water project." *Water Policy*. 5(2), 115–126.

Merrey, D. J. 2011. "Time to reform South African water reforms." Book review of Schreiner and Hassan (eds.), *Transforming water management in South Africa: Designing and implementing a new policy framework*. Springer. *Water Alternatives*, 4(2), 252–255.

Merten, M. 2019. "State capture wipes out third of SA's R4.9-trillion GDP – Never mind lost trust, confidence, opportunity." *Daily Maverick*, 1 March.

Muller, M. 2016. "South Africa's water sector: A case study in state capture." *The Conversation*, December 1.

Muller, M. 2020. 'Money Down the Drain – Corruption in South Africa's Water Sector'. Johannesburg and Berlin, Corruption Watch and the Water Integrity Network.

Perret, S. 2002. "Water policies and smallholding irrigation schemes in South Africa: A history and new institutional challenges." In Department of agricultural economics, *extension and rural development*, Working Paper 2002–19, University of Pretoria, Pretoria.

Posel, D. 2001. "What's in a name? Racial categorisations under apartheid and their afterlife." *Transformation*. 47, 50–74.
Prince, L. 2021." Meer as derde van senior staatslui se kwalifikasies gebrekkig." *Beeld*, 13 April.
Republic of South Africa (RSA). 1997. *Water Services Act, No. 108 of 1997*. Pretoria: Government Printer.
Republic of South Africa (RSA). 1998. *The national water act, no. 36 of 1998*. Pretoria: Government Printer.
Rule, S., Parker, S., Lunga, W. and Majikijela, Y. A. 2020. *Drought disaster management: Lessons from the Western Cape*. Pretoria: Human Sciences Research Council.
Rule, S., Parker, S., Majikijela, Y. and Lunga, W. 2020. "Neighbourhood responses to drought in the Western Cape." *GeoJournal*. 86, 2267–2278.
Schreiner, B. 2013. "Why has the South African national water act been so difficult to implement?." *Water Alternatives*. 6(2), 239–245.
Schreiner, B. and Hassan, R. 2011. "Lessons and conclusions." In: Schreiner, B. and Hasan, R. (eds.), *Transforming water management in South Africa: Designing and implementing a new policy framework*. Dordrecht: Springer.
Sousa, P. M., Blamey, R. C., Reason, C. J. C., Ramos, A. M. and Trigo, R. M. 2018. "The 'Day Zero' Cape Town drought and the poleward migration of moisture corridors." *Environmental Research Letters*. 13, 124025.
Swatuk, L. A. 2010. "The state and water resources development through the lens of history: A South African case study." *Water Alternatives*. 3(3), 521–536.
Tapela, B. N. 2008. "Livelihoods in the wake of agricultural commercialisation in South Africa's poverty nodes: Insights from small-scale irrigation schemes in Limpopo province." *Development Southern Africa*. 25(2), 181–198.
Tempelhoff, J. W. N. 2009. "Civil society and sanitation hydropolitics: A case study of South Africa's Vaal River barrage." *Physics and Chemistry of the Earth, Parts A/B/C*. 34(3), 164–175.
Tempelhoff, J. W. N. 2011. "Water history and transdisciplinarity: A South African perspective." *TD The Journal for Transdisciplinary Research in Southern Africa*. 7(2), 316–336.
Tempelhoff, J. W. N. 2017. "The *Water Act*, no. 54 of 1956 and the first phase of apartheid in South Africa (1948–1960)." *Water History*. 9, 189–213.
Turton, A. R., Meissner, R., Mampane, P. M. and Seremo, O. 2004. *A hydropolitical history of South Africa's international river basins*. Pretoria: Water Research Commission. WRC Report No. 1220/1/04.
Union of South Africa (U of SA). 1954. *Report of the native affairs commission for the period 1 January 1948, to 31 December 1952*. UG36. Pretoria: Government Printer.
Union of South Africa (U of SA). 1956. *Water act, no. 54 of 1956*. Cape Town: Cape Times Ltd.
Van Koppen, B. and Schreiner, B. 2014. "Moving beyond integrated water resource management: developmental water management in South Africa." *International Journal of Water Resources Development*. 30(3) 543–558.
Van Onselen, G. 2013. "The short life of a DG in the regime of Jacob Zuma." *Sunday Times*, 1 September.
Van Vuuren, L. 2012. *In the footsteps of giants – Exploring the history of South Africa's large dams*. Pretoria: Water Research Commission.
Warner, J. F. and Meissner, R. 2021. "Cape Town's "Day zero" water crisis as a manufactured media event?" *Journal of Disaster Risk Reduction*. 64, 102481.
Water Research Commission (WRC). 2021. *Some, for all, forever – Water ecosystems and people*. Pretoria: Water Research Commission. [Accessed on: 4 October 2021] Available at: http://www.wrc.org.za/mdocs-posts/tt-176-complete/.
Wolski, P. 2018. "How severe is cape Town's "Day zero" drought?" *Significance*. 15(2), 24–27.

10
PERU

Patricia Urteaga-Crovetto

Introduction

Peru is in the top eight countries with the most fresh-water resources, far above the worldwide average. Yet access to this water is unevenly distributed both geographically and socially. The majority of freshwater resources are in the Amazon region, while the Andean and the coastal regions are characterized by water scarcity and fierce competition between water users. Of the annual water volume of 2,484,078 Hm^3/sec, only 1.8% is located on the Pacific slope where 65% of the population and most agroindustry is concentrated (ANA et al., 2015). According to official data from the ANA (2012), almost 88% of water resources are used for agriculture, 10% is for domestic uses and 2% is for mining and industry.

Historically, water management in Peru has been closely related to state power dynamics. Despite some attempts to democratize the agricultural sector by means of agrarian reforms, the concentration of land and water has long characterized the Peruvian agrarian reality.

The organization of the agricultural and water sectors in Peru is closely linked to the process of nation-state formation in the 19th century. After Independence in 1821, both mining and agriculture were considered the poles of development in Peru, for which water management was a key component. In the midst of political turmoil, policies in the 19th and early 20th centuries were aimed to centralize water management to promote development projects. Following this trend, during the mid-20th century, high-tech, state megaprojects were constructed in an attempt to expand agricultural production. With the support of International Financial Institutions (IFIs), another reform wave came during the first decade of the 2000s to foster institutional changes that would boost private investment. The resulting projects driven by neoliberal policies contributed to the multiplication of water conflicts in Peru. In 2020 and 2021, 190 conflicts were recorded, 65% of which are related to extractive industries and infrastructural mega projects (Defensoría del Pueblo, 2021). One of these is the Olmos Tinajones project (OTP). Conceived in the early 20th century, the Olmos Tinajones project for irrigation and hydroenergy is of analytical interest here because it epitomizes almost one century of the history of water management in Peru showing how water governance is eventually captured by powerful economic interests that cement power imbalances.

This chapter is organized as follows. First, following from a historical perspective I touch on the institutional transformation of the water sector and its effect on local users. Second, I focus on

DOI: 10.4324/9781003095545-12

the implementation of the National System of Water Resources and the 2009 Law of Water Resources. Third, the Olmos Tinajones project is depicted to demonstrate how development projects reinforce unequal water access. Finally, some conclusions are drawn to the significance of water for development in Peru.

The Nation-State and the Modernization of Water Management in Peru

Modernization of the water sector has taken place in Peru steadily since the 19th century following the dire necessities of large landowners. The organization of the incipient nation-state focused on relevant economic sectors, such as agriculture and water. However, centralized policies to promote agriculture had to grapple with decentralized water management as, according to the 1834 Organic Act of Municipalities, the Municipal Boards of the villages and cities were in charge of water allocation (Guevara-Gil, 2013). This trend continued with the Law of Municipalities of 1856, which did not affect water grabbing by powerful local water users. During the late 19th century, land and water accumulation occurred in the northern coastal regions to produce sugar, cotton, grapes, cacao, and wheat on a large scale (Gutierrez Rivas 2004; Orrego, 2000). The centralization of water management started with hydraulic projects that replaced irrigation ditches to the benefit of large landholders and plantation owners known as *hacendados* (Aldana & Diez, 1994; Núñez, 2004; Ploeg, 2006). The international economic crisis caused by the Great Depression of 1873–1896, natural disasters, the war with Chile in 1879 and population growth coalesced to increase state intervention in water management. The nation-state formation process reinforced institutional water governance consolidating the control of land and water by powerful elites. In the midst of political turmoil, a representative of the Aristocratic Republic (*República Aristocrática*[1]), President Nicolas de Pierola guaranteed landowners a leading role in the modernization of agriculture, which was often used for their own advantage (Orrego, 2000; Urteaga-Crovetto, 2016).

During the early 20th century, foreign investment particularly from the United Kingdom and the United States contributed to integrate the country into the world economy prompting the modernization of agriculture (Klaren, 1976). Projects were designed to extend the agricultural frontier, identify main watersheds for infrastructure works, organize the water sector and create scientific institutions that would rationalize water management (Aldana & Diez, 1994; Orrego, 2000; Ploeg, 2006). As in early 20th-century Egypt, water was transferred to the hand of experts (Mitchell, 2002). The Water Code of 1902 contemplated a mixed water management regime according to which some waters were under public ownership while others belonged to the owner of the land. Gradually, the increased centralization of water management occurred by means of specialized State agencies, such as the Water and Agriculture Directorate and the Technical Commissions created in 1911 (Oré, 1989). These agencies disputed the control of water by local municipalities and landowners, for which they were criticized by the elites. Irrigation communities, including representatives of municipalities and indigenous communities, were established in each agricultural district. The taxes on irrigation water that the Municipalities had introduced for the purposes of water conservation, administration and distribution were eliminated. Interestingly, while the ordinances of the irrigation communities were approved, the local water uses and customs were respected as long as they did not contravene the Water Code (Guevara-Gil, 2013).

Centralized water management was the target of criticism. Engineers and technicians disapproved of the 1902 Water Code because, among other things, it was a copy of the Spanish Water Code. They also complained that Code was inefficient at challenging the unrestrained power of the landowners over the water. In addition, the rules for the operation of the new irrigation unions were unclear. On the other hand, landowners complained that the new water management had not overcome various obstacles that prevented the development of large agriculture as the elites desired. Nonetheless, the Water Code contained a provision according to which water was distributed by day shifts for the landowners and at night for the indigenous people, as was established during the 17th century. Eventually, indigenous and small farmers led a large movement in many parts of the country that forced the government to reform the Water Code during the first decade of the 20th century, and create a body of state officials to monitor irrigation and prevent abuses in water distribution (Oré, 1989).

Peasant and indigenous responses to state-centralized water management ranged between resistance and adaptation. The history of the Laraos community during the transition of the 19th to the 20th century, reflects the local effects of the transformations brought by the institutional and legal regimes for water management. A group of peasants who politically controlled the community proposed to change the communal land regime. A regime of private land and water management was established in Laraos which replaced the traditional collective administration. This implied a significant institutional and normative change. The government sent trustees to the community to verify if the Water Code and regulations were being complied with, and in 1905, it appointed water allocators to work in the community. After more than a decade under the private water management regime, in 1920 the communal leaders decided to go back to the collective water management as they realized the former had harmed most families (Mayer, 2004).

Modernization of water management went on until the mid-20th century without really altering the *status quo*. Following the military coup of 1968, the government of General Velasco Alvarado (1968–1975) identified the most productive valleys with the purpose of conducting a long-awaited reform that limit the agrarian estates and distribute the land among the poorest. The centralization of water management was reinforced as the military government named the Ministry of Agriculture as the National Water Authority. In 1969, the General Water Act (17756) replaced the Water Code of 1902 to modernize the water sector and facilitate big hydraulic works to improve water supply (Peña Pozo, 2004). However, the results were flawed with an increased silting of reservoirs and more than 200,000 irrigated hectares that became salinized (Zegarra et al., 2007). After the military dictatorship (1968–1980), liberal governments that took office gradually reversed all the reforms concerning land. Nonetheless, their water policies continued to be based on a supply-driven approach to facilitate private investment. Particularly during Alberto Fujimori's regimes (1990–2000) several attempts to privatize water were unsuccessfully made, though the seeds for a neoliberal water reform were already in place.

The National System of Water Resources and the 2009 Law of Water Resources

During the early 2000s, in order to ensure that the principles of a free-market economy were not obstructed, the modernization of the water sector included not only changes in legislation, but also a thorough transformation of the institutional structure for water management echoing the models in Mexico and Brazil (Andersen 2019; Hendriks, 2006; Oré, et al., 2009; Oré and del Castillo, 2006). The National System of Water Resources was conceptualized as an intersectoral platform where all public sector institutions and users with competences and functions related to water management participate. In addition to the creation of the National System of Water resources, an important piece of the modernization process for water management was the approval in 2004 of the National Strategy for Water Resources and in 2009 of the Law of Water Resources (29338), which was financed by the World Bank and the Interamerican Development Bank (Urteaga, 2010). The process was not straightforward, with some political parties arguing in favour of local water rights while others supporting blatant open market policies (Roa et al., 2013). This polarization is reflected in the law, as some articles acknowledge local water rights and protect the environment, and other openly follow neoliberal premises. And even though the Law of Water Resources recognizes peasant and indigenous water rights, the regulation of the law mandates individualizing collective water rights.

In both the National Strategy for Water Resources and the Law of Water Resources, technology was used to obscure socioeconomic differences in access to water that the law and strategy both reproduce rewarding the behaviour of more affluent users. Accordingly, water users obtain prioritized water allocation if three requirements are met: the highest efficient use, important incomes produced, and the minimum (or none) environmental impact. "Through this definition of the public interest, the law puts less affluent farmers at a clear disadvantage since they have less capacity to invest in water efficiency improvements" (Roa et al., 2013: 6). The essence of the neoliberal character of the law is that all users are assumed to have the same capacity to improve water management, and under this idea competition and incentive mechanisms seem natural and fair.

The local response to the implementation of the National System of Water Resources Management and the Law of Water Resources (29338) has been rather disparate. As in the 20th century, some communities heralded the new system; while others wholeheartedly rejected it on mostly economic and cultural grounds (Paerregaard, 2013). Such communities refused to pay the water tariff because either the state did not provide them with hydraulic infrastructure, or they assured that local deities and not the state are the main water suppliers. In the Andes and the Amazon basin, most peasant and indigenous communities cultivate a nuanced, and ever-changing relationship with water (Boelens, 2015; Boelens et al., 2010; Gerbrandy & Hoogendam, 1998; Murra, 1984; Sherbondy, 1998; Spalding, 1984), which is regarded as a living sentient entity. In the southern Andes, local water management is part of a tapestry of relationships including humans and other-than-human beings. Inasmuch as water management is dynamic and culturally specific, these relations are constantly activated and renovated in mutual dependence with non-humans, local space, and climate change conditions (Paerregaard, 2019; Paerregaard & Andersen, 2019; Stensrud, 2019). Usually, water management in the Andes comprises not only a masterful understanding of the physical complexity of water but also a cultural language and specific relational network that anthropomorphizes water, thus transcending the material.

Undoubtedly, development interventions alter the cultural substance of local water management in unpredictable ways. Paerregaard (2019) historical account notes unexpected results of a planned development in water management in the rural Andes of Arequipa. In Cabanaconde, where the state built agricultural infrastructure to modernize water management, the community still pays tribute to the water that flows through the cemented water sluices and ditches. Similar practices have been registered among peasants in Santa Rosa, a district in the central Andes, where *despachos*[2] were paid to the water circulating on the infrastructure inaugurated in 2005 (Guevara-Gil, 2013). Meanwhile, in Tapay, users renovated its links to the spirit of the mountain to guarantee an annual water supply (Paerregaard, 2019).

Driven by the necessity to modernize water management, the Law of Water Resources barely acknowledges the meaning of water for Andean people. The economic rationality that underlies development projects typically simplifies a more complex reality. Altruist justifications of development projects serve the state to locally reinforce ideas of progress, capturing water for corporate purposes, while simultaneously transforming it into "modern water" (Linton, 2010). Prevailing mining and agroindustrial development have exacerbated social inequalities in water access that ended up transforming the whole waterscape (Urteaga, 2011). In mining regions, water availability has been significantly reduced due to impacts on quality and volume (Budds & Hinojosa, 2012; Li, 2016). By reducing water to its economic value, "water extractivism" is conducted (Stensrud, 2019), as is evident in the Olmos Tinajones project.

Irrigation and Water Transfer Megaprojects: The Olmos Tinajones Special Project

The Olmos Tinajones Project (OTP) for hydroenergy and irrigation is of interest here because it shows how the centralization of water management has not necessarily implied the ending of privileges enjoyed by powerful users. Designed during the early 1920s,[3] the OTP was initially conceived to irrigate more than 40,000 hectares of desert land on the Pacific coast by diverting water from the Huancabamba River in the Andes (Del Castillo, 2013; Zegarra et al., 2007). For the young nation-state, hydraulic projects such as the OTP were meant to reinforce its legitimacy by guaranteeing local welfare. The general goal was to generate economic development and better living conditions for local people, for which it planned the construction of the Limon reservoir dam to store 1,000,000 m^3 of water for irrigation and hydroenergy. With a rational discourse that emphasized technology and efficiency in water management, the promoters of the project gradually resumed the intention to further the interests of large transnational corporations and powerful Peruvian economic groups with serious evidence of corruption. Although corruption is a general problem in many countries, the water sector is particularly known to be the target of corruption practices.[4] The OTP is a pertinent example of corruption in hydro projects.

Due to adverse climatic conditions and economic constraints, the project did not see the light of day until the 1990s, during the administration of then-president Alberto Fujimori (1990–2000). The government sought the participation of the national and transnational private sector (the Brazilian Odebrecht Group), which unfolded a sophisticated scheme of corruption. The initial infrastructure works began in 2001. In 2004, *Concesionaria Trasvase Olmos S.A.* (Odebrecht Group) was granted a 20 years concession to construct the trans-Andean tunnel to transfer water from the Eastern Andes to irrigate desertic lands in the Pacific coast. The costs for constructing the tunnel and the Limon reservoir dam were initially calculated as being USD$112 million. However, by the end of the construction phase, the cost reached USD$580 million (AmericaTV.com, 2017; Escobedo, 2017).

Regarding the irrigation component of the project, the first stage contemplated the irrigation of 43,500 ha, of which 38,000 ha were assigned to 16 big agricultural companies by the regional government, 2,500 ha were for local companies, and 3,000 ha were given to the Santo Domingo de Olmos peasant community as a kind of compensation. The Santo Domingo de Olmos community was forced to cede part of its lands for the OTP. As the members of the community refused, in 1999 Fujimori's government confiscated 15% of the community's land with the future promise of providing the community irrigated land and affordable water. This promise was far from reality. The agricultural model predicated on the OTP considered irrigation as solely for large-scale export agriculture. The design of the irrigation infrastructure made by the H2Olmos company (Odebrecht group) disguised social inequalities. While the water infrastructure reached the plots farmed by the aforementioned 16 agricultural companies, smaller scale farmers had to construct canals along with a hydrant to direct water from the major hydraulic infrastructure to their own plots. In addition, the cost of water for all users was US$0.06125 per cubic meter, which was unaffordable for small farmers. In many ways, this pattern echoes the asymmetric condition that peasants faced vis-à-vis the landowners in the northern coast during the 19th and early 20th centuries (Del Castillo, 2013; Zegarra et al., 2007).

The second stage of the project included diverting water from Tabaconas and Manchara rivers in the Atlantic Basin and using groundwater to irrigate 100,000 hectares of land in the Pacific Basin. The initial goal to benefit poor farmers was abandoned and redesigned as the powerful Gloria Group business conglomerate acquired 15,600 hectares. In June 2011, the regional government of Lambayeque granted a 25-years concession to the H2Olmos company (Odebrecht group) to manage the irrigation project. Although the contract clauses prohibited Odebrecht to acquire land, the rules were changed to allow the Brazilian company to acquire 11,000 hectares (AmericaTV.com, 2017). Medium and small-scale farmers were unable to compete. In the years that followed, the companies involved in the project developed several schemes of corruption that are still under investigation by the Office of the Controller General of the Republic (Contraloría General de la República del Perú, 2020). The initial goal of the project to improve living conditions of local farmers finally changed. By 2019, more than 25,000 ha owned by export companies were producing avocados, asparagus, grapes, cranberry, and sugar cane to supply export markets worth US$206 million (Proyecto Especial Olmos Tinajones, 2020).

After the inauguration of the OTP in 2014, earlier altruist discourses surrounding the common good were abandoned to advance particular economic interests. The legal framework allowed forced displacements, resettlements, and socio-environmental impacts, which caused profound transformations in the local milieu (Del Castillo, 2013; Escobedo, 2017; Santos Arias, 2013; Zegarra et al., 2007). The hydroenergy component generated conflicts with other irrigation projects competing for water. The district of San Felipe, where displaced people from Olmos were forced to resettle, was regarded as one of the worst in the 2007 Human Development rankings with very few natural resources to count on. The living conditions and cultural identity of resettled farmers were undeniably damaged in the name of progress.

The OTP is an example of a development project that has contributed to cementing the asymmetrical foundations of the Peruvian nation-state formation process. Initially proposed as an economic pole that would generate incomes for the poor, it ended up dispossessing local communities of land and water, illegally benefitting the companies involved and reinforcing corruption and socioeconomic imbalances. The OTP represents the epitome of the history of water management in Peru, as it reproduced anew the discriminatory character of the modernization processes of water management in Peru.

Conclusion

Throughout this historical account I have shown how water governance in Peru has been key to the formation of the nation-state as linked to powerful economic interests. During the late 19th century, water management was modernized and concentrated in the hands of local elites to guarantee large-scale production. During the 20th century, the centralization of water management was part of the nation-state formation process to guarantee modernization. However, it failed in levelling the historical unequal water access. During the 21st century, institutional transformations and new rules based on the participation of the private sector and the liberal idea of equality before the law aimed at homogenizing local water management, leaving middle and small water users unprotected. The legal and institutional framework fostered by IFIs smoothed the construction of large hydro projects such as the Olmos Tinajones Project for irrigation and hydroenergy, whose discursive appropriation of the common good aimed to create local consent on progress, development and technology. Actually, it served primarily corporate agriculture for export. Thus, the state reinforced its economic interests by engaging in corporate corruption schemes that guaranteed the companies' control of land and water, with detrimental consequences for the poorest.

Notes

1 The Aristocratic Republic refers to a period of the history of Peru (1895–1919), whereby the oligarchy consolidated its political and economic power with the support of the United Kingdom and the United States.
2 A "despacho" is a symbolic gift that Andean peasants offer to a deity, be it a lake, the land, or a mountain. Their content differs, as it may have "coca" leaves, flowers, food, alcohol, etc. It is an important piece of a ritual of gratitude to nature and propitiation.
3 It was proposed in 1924 by engineers Manuel Mesones Muro and Charles Sutton. For more on the OTP see Escobedo (2017) and Del Castillo (2013).
4 In fact, 10% of investment in the water sector was lost to corruption (Transparency International, 2008).

References

Aldana, S. R. and Diez, A. H. (1994). *Balsillas, Piajenos y Algodon. Procesos Historicos en Piura y Tumbes.* Lima: Cipca Tarea.

América Tv.com. (2017). *En el contrato inicial se contemplaba un costo de 242 millones de dólares.* [Accessed 2 October 2021] Available from: https://www.americatv.com.pe/noticias/actualidad/odebrecht-proyecto-olmos-termino-costando-580-millones-dolares-n265994

Andersen, A. O. (2019). "Assembling commons and commodities: The Peruvian water law between ideology and materialisation." *Water Alternatives*. 12(2), 470–487.

Autoridad Nacional del Agua. (2012). *Recursos Hídricos en el Perú.* 2nd Edition. Lima, Perú: Metrocolor S.A.

Autoridad Nacional del Agua, Ministerio de Agricultura y Riego, Banco Interamericano de Desarrollo. (2015). *Plan Nacional de Recursos Hídricos.* Resumen Ejecutivo, Lima, Mss.

Boelens, R. (2015). *Water, Power and Identity. The Cultural Politics of Water in the Andes.* London, New York, NY: Routledge.

Boelens, R., Getches, D. and Guevara Gil, A., (ed). (2010). *Out of the Mainstream. Water Rights, Politics and Identity.* London, Washington, DC: Earthscan.

Budds, J. and Hinojosa, L. (2012). "Restructuring and rescaling water governance in mining contexts: The co-production of waterscapes in Peru." *Water Alternatives*. 5(1), 119–137.

Contraloría General de la República del Perú. (2020). *Report N° 875-2020-CG-GCOC: Contraloría detecta favorecimiento a empresa vinculada a Odebrecht en venta de lotes agrícolas.* [Accessed 02 October 2021] Available from https://www.contraloria.gob.pe/wps/wcm/connect/cgrnew/as_contraloria/prensa/notas_de_prensa/2020/lambayeque/np_875-2020-cg-gcoc

Defensoría del Pueblo. (2021). *Reporte Mensual de Conflictos Sociales* N° 207. May. [Accessed 02 May 2021]. Available from https://www.gob.pe/institucion/defensoria-del-pueblo/informes-publicaciones/1973375-reporte-de-conflictos-sociales-n-207-mayo-2021 https://www.gob.pe/institucion/defensoria-del-pueblo/informes-publicaciones/1973375-reporte-de-conflictos-sociales-n-207-mayo-2021

Del Castillo, L. (2013). "La comunidad y la irrigación de Olmos: una relación nada justa." In: R. Boelens and A. Arroyo (eds.). *Aguas Robadas. Despojo hídrico y movilización social.* Lima: CBC, Cusco, IEP, 83–101.

Escobedo, J. (2017). "Olmos: Negocio Redondo." *La Revista Agraria*, No. 184, May. CEPES, 19–21.

Gerbrandy, G. and Hoogendam, P. (1998). *Aguas y Acequias. Los derechos al agua y la gestión campesina de riego en los Andes bolivianos.* Cochabamba: PEIRAV, Plural editores/CID, Bolivia.

Guevara-Gil, J. A. (2013). *El Derecho y la gestión local de agua en Santa Rosa de Ocopa, Junín – Perú.* Lima: Universidad de Ámsterdam and IPROGA.

Gutierrez Rivas, J. (2004). "Cap. VI. La Independencia." In: J. A. del Busto and Rosales (eds). *Historia de Piura.* Piura: Universidad de Piura, Departamento de Humanidades, Instituto de Investigaciones Humanísticas, 321–388.

Hendriks, J. (2006). "Legislación de aguas y gestión de sistemas hídricos en los países andinos." *Derechos Colectivos y Políticas Hídricas en la Región Andina.* In: P. Urteaga and R. Boelens (eds.). Lima: WALIR, Instituto de Estudios Peruanos, pp. 47–111.

Klaren, P. F. (1976). *Formación de las Haciendas Azucareras y Orígenes del APRA.* Lima: IEP.

Li, F. (2016). "In defense of water: Modern mining, grassroots movements, and corporate strategies in Peru." *The Journal of Latin American and Caribbean Anthropology*. 21(1), 109–129.

Linton, J. (2010). *What Is Water? The History of a Modern Abstraction*. Vancouver: UBC Press.
Mayer, E. (2004). *Casa, Chacra y Dinero: Economías Domésticas en los Andes*. Lima: IEP.
Mitchell, T. (2002). *Rule of Experts: Egypt, Techno-Politics, Modernity*. Berkeley, CA: University of California Press.
Murra, J. (1984). "Andean societies", *Annual Review of Anthropology*. 13, 119–141.
Núñez, D. (2004). "Capítulo VIII: La Guerra con Chile y su secuela." In: J. A. Del Busto and J. Rosales (eds.). *Historia de Piura*, Piura: Universidad de Piura, Departamento de Humanidades, Instituto de Investigaciones Humanísticas, 483–515.
Oré, M. T. (1989). *Riego y organización: evolución histórica y experiencias actuales en el Perú*. Lima: IT.
Oré, M. T. and del Castillo, L. (2006). *La legislación de Aguas en el Perú*. Lima: Perú. Mss.
Oré, M. T., Del Castillo, L., Orsel, S. V. and Vos, J. (2009). *El agua, ante nuevos desafíos: actores e iniciativas en Ecuador, Perú y Bolivia*. Lima: Oxfam International.
Orrego, J. L. (2000). La República Oligárquica 1850–1950. In: Lexus (ed.). *Historia del Perú*. Perú: Lexus Editores.
Paerregaard, K. (2013). "Governing water in the Andean community of Cabanaconde, Peru: From resistance to opposition and to cooperation (and back again?)." *Mountain Research and Development*, 33(3), 207–214.
Paerregaard, K. (2019). "Liquid accountability: Water as a common, public and private good in the Peruvian Andes." *Water Alternatives* 12(2): 488–502.
Paerregaard, K. and Andersen A. O. (2019). "Moving beyond the commons/commodity dichotomy: The socio-political complexity of Peru's water crisis." *Water Alternatives*. 12(2), 459–469.
Peña Pozo, R. E. (2004). "Capítulo X. La Hacienda reciente: 1950–2000." In: J.A. del Busto and J. Rosales (eds.). *Historia de Piura*. Piura: Universidad de Piura, Departamento de Humanidades, Instituto de Investigaciones Humanísticas, 575–641.
Ploeg, J. D. (2006). *El futuro robado. Tierra, agua y lucha campesina*. Lima: IEP, WALIR.
Proyecto Especial Olmos Tinajones. (2020). *Beneficios del Proyecto Olmos al año 2020*. [Accessed 29 08 2021] Available from BENEFICIOS DEL PROYECTO OLMOS AL AÑO 2020 – YouTube.
Roa, C., Urteaga, P. and Bustamante, R. (2013). "Water laws in the Andes: A promising precedent for challenging Neoliberalism." *Geoforum*. [Accessed 30 08 2021]. Available from http://www.sciencedirect.com/science/article/pii/S001671851300256X
Santos Arias, L. (2013). *Impactos socioambientales de reasentamiento por Proyectos de Desarrollo. Caso: Caserío de Huabal – Cajamarca por el Proyecto Especial Olmos Tinajones (PEOT)*. Tesis de Maestría en Desarrollo Ambiental. Pontificia Universidad Católica del Perú.
Sherbondy, J. (1998). "Andean Irrigation in History." In: R. Boelens and G. Davila (eds.). *Searching for equity: Conceptions of justice and equity in peasant irrigation*. Assen: Van Gorcumh, 210–214.
Spalding, K. (1984). *Huarochirí: An Andean Society under Inca and Spanish Rule*. Stanford: Stanford University Press.
Stensrud, A. B. (2019). "The formalization of water use and conditional ownership in Colca Valley, Peru." *Water Alternatives*. 12(2), 521–537.
Transparency International. (2008). *Informe Global de la Corrupción 2008. Corrupción en el sector del agua*. Transparency International, Water Integrity Network.
Urteaga, P. (2010). "Ingeniería legal, acumulación por desposesión, y derechos colectivos en la gestión del agua." In: Bustamante, R. (ed.). *Lo Colectivo y el Agua*. Lima: IEP, Programa Concertación, 51–74.
Urteaga, P. (ed.). (2011). *Agua e Industrias Extractivas: Cambios y continuidades en los Andes*. Lima: Concertación, IEP.
Urteaga-Crovetto, P. (2016). "Between water abundance and scarcity: Discourses, biofuels and power in Piura, Peru." *Antipode, A Radical Journal of Geography*. 48(4), 1059–1079.
Zegarra, E., Oré, M. T. and Glave, M. (2007). "El proyecto Olmos en un territorio árido de la costa peruana." In: J. Bengoa (ed.). *Territorios Rurales. Movimientos sociales y desarrollo territorial rural en América Latina*. Santiago de Chile: Catalonia, 514–550.

11
JORDAN

Neda Zawahri

Introduction

Jordan is one of the most water-scarce nations in the world. The combination of climate change, rapid population growth, the inflow of refugees from conflict-affected neighbouring states, and the mismanagement of existing supplies will continue to aggravate the nation's already severe water shortage into the near future. Despite the significant water scarcity, over 50% of Jordan's domestic water resources are consumed by the agricultural sector, which in turn contributes marginally to the gross domestic product. To appreciate the agricultural sector's continued over-consumption of a scarce natural resource that is essential to human security and welfare along with political and economic security, it is essential to consider the nature of the state-society relationship that arose during the state-building process and the role of water in nation building. As this chapter will demonstrate, since the Hashemite Kingdom's state-building process, agriculture and water have been used as tools for constructing regime legitimacy and state security. The byproduct of the policies, negotiations, and decisions that were made during the state-building process has been the creation of powerful elite segments of society with a substantial interest in maintaining their existing over-consumption of cheap water within the agricultural sector. In other words, the consequence has been the rise of a powerful elite that over-consumes water and, in the process, succeeds in limiting the state's autonomy in regulating its domestic water resources.

States confronting severe water scarcities generally have two options available to meet their domestic demands, they can either augment supplies or use existing resources more efficiently. The latter option involves embracing and implementing policies that cut subsidies, increase water prices, punish theft of water, or impose more efficient use of water on the agricultural sector. Policymakers in Jordan and international donors might discuss the inefficient use of domestic water resources and the need to improve water demand management, but government institutions lack the power, interest, and will to oversee the implementation of these policies. The difficulty of implementing policies designed to improve the domestic management of water resources exists because of strong opposition by powerful agricultural elites with an interest in maintaining their levels of water consumption. Moreover, upsetting existing consumption patterns within the agricultural sector is generally perceived by the ruling monarchy as challenging state-society relations and hence regime stability. As a result, instead of implementing effective water demand

DOI: 10.4324/9781003095545-13

management policies, the regime has opted to focus on the less politically sensitive policies that increase prices on marginal sectors of society, such as tourists, and work towards the construction of hydrological infrastructure with the goal of augmenting existing water supplies.

To demonstrate how Jordan's historical development process contributed to the evolution of weak governing institutions that lack the capacity, power, and interest to impose more efficient water allocation and utilization on powerful actors in society, the chapter first presents an overview of Jordan's water crisis, which is followed by an evaluation of the nation's state-building process and the bargaining that took place between the young state and its society. This is followed by an evaluation of how the agricultural elite is able to protect its existing consumption of water resources.

Jordan's Water Supplies

In the relatively arid Middle East and North Africa, Jordan is one of the poorest states in terms of available fresh water resources. The majority of its water supply comes from both renewable and non-renewable groundwater resources. Renewable aquifers are being over-extracted by 55% beyond their safe yield, and non-renewable sources are being mined to meet increasing demands (Humpal et al., 2012).

Other sources of water available to the Kingdom are transboundary. Jordan is a downstream riparian of the transboundary multilateral Jordan River system, which it shares with Israel, Lebanon, Palestine, and Syria. While Lebanon and Syria have alternative river systems, the Jordan basin is the only perennial river for Israel, Jordan, and Palestine. Historically the Kingdom has had a relatively unstable supply of water from the basin because of increased upstream pumping by Israel and Syria which decreased the water flowing into the Yarmouk tributary and the lower Jordan River, portions of the basin it relies upon (Zawahri, 2008). After years of unstable cooperation over the Yarmouk tributary and the lower Jordan River, in 1994, Israel and Jordan signed a peace treaty that formalized the management of their shared portion of the basin.

Jordan is predominantly landlocked, with a small coastline located in the south of the country along the Red Sea. The majority of its population, however, is located in the far north of the country. Given the substantial cost it will incur in building, maintain, delivering, and operating a large desalinization plant, economically poor Jordan faces many challenges in unilaterally developing its own desalination capacity. Consequently, for many years, it has pushed for the construction of the Red Sea and Dead Sea Canal (RSDSC) project, and in 2013, a trilateral understanding was signed between Israel, Palestine, and Jordan to undertake a feasibility study to construct phase one of the RSDSC, which was designed to provide Jordan and the south of Israel with water, save the declining Dead Sea, and encourage peace along with prosperity in the region (Zawahri & Weinthal, 2014). A combination of high costs and concern over the project's environmental impact on the Dead Sea has decreased Israel's interest in the project. After the completion of the feasibility study, the project remained stalled.

The stalled implementation of the RSDSC project along with a general deterioration in overall bilateral relations between Israel and Jordan led to a "cold peace" between the riparians that threatened their otherwise cooperation over water (Weinthal & Zawahri, 2020). When Prime Minister Benjamin Netanyahu was ousted and replaced by Neftali Bennett as prime minister in 2021, bilateral relations improved. Soon after taking office, Prime Minister Bennett set out to repair bilateral relations with Jordan, which included salvaging their water cooperation.

As part of the improvement in bilateral relations between the riparian states, in November 2021, Israel, Jordan, and the United Arab Emirates (UAE) signed a Letter of Intent (LOI) that combined the sale of solar power for desalinated water. According to the LOI, Jordan is to sell Israel

600 megawatts of solar energy it will generate from solar power plants to be constructed within its territory. Israel will sell 200 million cubic meters of desalinated water produced by a plant it will construct. The UAE has committed to provide funding and technology for the solar power plants through its government-owned company, Masdar. A feasibility study is set to begin in 2022 and it remains uncertain when construction begins. This exchange of solar power for water was made possible not only by the rise to power of a new Israeli prime minister, but also the 2020 Abraham Accords that formalized relations between the UAE and Israel.

As for water demand, the agricultural sector consumes over 50% of the national water budget, while the municipal and industrial sectors consume 45% and 3%, respectively (Yoon et al., 2021). Jordan's annual water deficit is projected to increase, not only because of the increased upstream consumption of transboundary water resources but also because of population growth and an influx of refugees from regional conflicts (Humpal et al., 2012). Since 1948, the country has experienced many waves of refugees. After several waves of Palestinian refugees, the 2003 war in Iraq brought about 800,000 refugees (Hummel, 2008), while the current turmoil in Syria has increased the number of refugees in Jordan by 660,000 (Mercy Corps, 2023). With population growth and the various waves of refugees, Jordan's population is expected to double by 2047, placing additional stress on its limited supply of water (Mercy Corps, 2023).

Climate change is another important force impacting Jordan's current and future water security. Studies indicate that climate change is expected to decrease the domestic water resources available to Jordan in the near term (Waha et al., 2017). While it is difficult to predict the exact impact of climate change on the water sector, the majority of the models show certain trends such as an increase in temperature, a decrease in precipitation, an increase in evaporation, and an increase in weather extremes. These combine to decrease the overall supply of fresh water (World Bank, 2014). Models indicate that a 2°C increase in temperature could decrease water resource availability in the Middle East region by 15%, while a 4°C increase in temperature could decrease water availability by 45% (Waha et al., 2017). The water supply in Jordan will be directly impacted by recurrent droughts. By 2100, models expect a doubling in the occurrence of droughts, which are also expected to be more severe and prolonged (Rajsekhar & Gorelick, 2017).

Studies indicate that Jordan is already experiencing the impact of climate change. Analysis of data from the past 45 years reveals a decrease in annual precipitation between 5% and 20% in the majority of the country, a decrease in the number of rainy days, and an overall warming trend (Assi & Ajjour, 2009). Studies also indicate that droughts increased in occurrence and severity in Jordan between 1981 and 2010 (Rajsekhar & Gorelick, 2017).

The combination of the impacts of climate change, population growth, and an influx of refugees is expected to impact Jordan's ability to meet the Kingdom's national and individual food security needs. Models indicate that a 1.5–2°C increase in temperature and the associated decrease in precipitation and water supplies could result in about a 30% decline in crop yields in Jordan (World Bank, 2014). The reduction in crop yield will challenge individual food security for poorer farmers. Jordan is heavily dependent on food imports, buying over 70% of its stable food needs from abroad. Any further reduction in domestic food supplies brought on by climate change will leave Jordan vulnerable to the capricious international market in order to meet national food needs in wheat, barley, and rice (Fathallah & Robertson, 2021). The Kingdom witnessed this vulnerability during the Covid pandemic due to export restrictions imposed by states from which it traditionally imports grains (Fathallah & Robertson, 2021).

Even if all planned small and mega-projects (RSDSC, increase in consumption of treated wastewater, and implementation of the LOI to exchange desalinated water for solar power) are implemented, Jordan is still projected to continue to operate on a water budget deficit (Yoon et al., 2021). Given the increasing shortages and ever-dwindling supplies, Jordan must improve its management of existing water resources and increase its efficiency in using its water. These policy changes will be critical to meet domestic needs. However, the nature of the power structure that arose during the state-building process continues to complicate the possibility of addressing the inefficient and ineffective management of water resources.

Nation-State Building

In the spring of 2021, Jordan celebrated its centennial. The COVID-19 pandemic imposed subdued celebrations marking the birth of what was then referred to as "Transjordan," which was established as a result of a political bargain between Great Britain and Amir Abdullah. The borders of the Protectorate of Transjordan, whose governance was overseen by Great Britain until it gained full independence in 1946, extended from the East Bank of the Jordan River, the Yarmouk tributary in the north, and west through the *Badia* (rangeland). Jordan's state building process and historical development was associated with the need to establish and secure the regime's power domestically within society and throughout the new borders. In the 1920s, this relatively arid land was inhabited by armed nomadic Bedouin tribes along with predominantly settled agriculturalist tribes. To form the new state and stabilize the borders, Great Britain and King Abdullah needed to settle and disarm the nomadic Bedouin tribes, establish a source of indigenous revenue through taxation, and incorporate the population in rural and urban lands into a socially constructed national identity (Fischbach, 2000).

Settling nomadic Bedouin tribes required providing them with jobs for their livelihood and then creating a system to govern over the community. To achieve this objective, Great Britain and King Abdullah worked to stabilize land tenure and the cultivation of farmland. From 1927 until 1952, they undertook what is known as the Land Program, which sought to survey and register lands throughout the country along with demarcating the boundaries between villages and cities. During this period, the great majority of the settled population was concentrated in the agricultural sector and one of the many goals of the demarcation of land via the Land Program was to increase the efficiency of the farming sector. International organizations also promoted the development of irrigated agriculture in attempt to settle nomadic and semi-nomadic Bedouin tribes, which was considered essential for border security, economic integration, and modernization (Van Aken et al., 2009).

While the great majority of the land within the new border was predominately desert, the Jordan Valley, Northern Highlands, and portions of the Badia had the potential for agricultural expansion. Approximately 85% of Jordan's fertile land is located in the Jordan Valley, which is also a unique land in its ability to permit year-round cultivation that produce fresh crops months ahead of schedule in European markets (Aresvik, 1976). To stabilize the predominantly rain-fed and capricious agricultural sector that existed during the state-building process, construction of water development projects, such as canals, wells, and dams, were undertaken. This hydrological infrastructure also enabled the irrigation of farmland, the settlement of nomadic pastoral groups, and it provided the Bedouins with a livelihood. It also enabled the young state to spread its population throughout the border, and in the process, secure its control of the land (Van Aken et al., 2009).

Soon after independence, King Abdullah also confronted the need to settle waves of Palestinian refugees from the Arab-Israeli conflict. In the Jordan Valley, approximately 774,000 arrived in 1948, which added to the need to develop the region's irrigation potential in attempt to settle people (Van Aken et al., 2009). Again, agriculture and water development projects funded by the

United States became part of not only tribal but also Palestinian refugee settlement in the Jordan Valley and the national incorporation of powerful tribes in rural and arid areas (Khouri, 1981; Peters & Moore, 2009).

In the 1950s, in order to stabilize the agricultural sector and expand the frontier of production through irrigation, the government provided soft loans and licenses permitting the construction of wells in the Northern Highlands and the Badia (Van Aken et al., 2009). Due to the completion of these hydrological infrastructure, Jordan increased the productivity of the agricultural sector leading to the export of produce (Fischbach, 2000). The Land Program also enabled tribal leaders to secure large amounts of land, which permitted them to increase their economic and political influence (Alon, 2009). While the Land Program established a cooperative environment and relationship between the farmers and the monarchy, it also led to the accumulation of agricultural land by moneylenders and merchant families, which will ultimately rise to become powerful forces influencing the regime's ability to govern its domestic water supplies.

Large landholding elites further increased their powers in the 1960s, when King Hussein incorporated them into the state by rewarding them with government and military jobs, such as heading ministries and the parliament along with key positions in the military. Having penetrated into the government and military structure, agricultural elites were able to then secure their continued access to water (Yorke, 2016). Thus, land tenure and access to water became state instruments used to gain the loyalty and support of tribal leaders, former nomadic tribes, and Palestinian refugees. While the ultimate goal of the Land Program and decisions during the state building process was to achieve political stability, it resulted in the creation of a social contract that is a patron-client relationship between the state and its society.

From the period of state formation until now, Jordan has been a resource-poor, underdeveloped country whose leadership has maintained a *rentier* state by initially using land tenure and then foreign assistance to finance a patronage system and substantial subsidies that served to purchase legitimacy along with elite and popular acquiescence, especially among the tribes (Muasher, 2011). The Kingdom has used agriculture and access to cheap water as part of its patronage system or social contract with key constituencies in order to gain their support. That is, from its establishment the Hashemite kingdom has provided its elite farmers and tribal elders with privileged access to cheap, highly subsidized, and excessive use of water for irrigation in return for their unquestioned allegiance to the regime, which has helped to build political stability and reinforce the regime's legitimacy. It is important to understand that not all tribal members receive this patronage. Rather, within individual large tribes there are members that are economically or politically influential that have direct contact to those in authority and are therefore able to resist, change, or shape water policies. Power is therefore not bestowed on an entire tribe, rather it is bestowed onto individuals within the tribe and the Sheikhs.

Governing Domestic Water Supply

As a hereditary monarchy, all power and authority in Jordan ultimately resides with the king, who assigns and dismisses ministers at will. Officially, Jordan's domestic water resources are governed by the Minister of Water and Irrigation who oversees and coordinates the activities of the Ministry of Water and Irrigation (MWI) along with two autonomous authorities, the Water Authority in Jordan (WAJ) and the Jordan Valley Authority (JVA), that are directly linked to the MWI. While the JVA is responsible for water management in the Jordan Valley, an important agricultural area, the WAJ is responsible for municipal and industrial water supply and wastewater treatment in the rest of the country. The Minister of Water and Irrigation chairs the WAJ's Board of Directors.

MWI is the overarching institution responsible for formulating the water sector's strategies, planning, prioritization, and policies, and developing the national master plan. While officially the MWI governs the nation's domestic water supply, in reality there is a "shadow state" that is also heavily involved in water policy. The shadow state can include the aforementioned powerful agricultural or political elites, or even commissions that the king establishes to overcome bureaucratic deadlock. In the case of water governance, the Royal Commission for Water, headed by Prince Faisal Ibn al Hussein, King Abdullah II's brother, is one of the most powerful bodies governing domestic water resources. Another organization is the Higher Agricultural Council, which is a parliamentary body chaired by the Prime Minister and has a great amount of influence over water policy for the agricultural sector (Zeitoun et al., 2012).

As it governs its limited domestic water supplies and attempts to meet its ever-increasing demand, the MWI can either focus on using existing supplies more efficiently or focus on augmenting supplies of water through the construction of hydrological infrastructure. The Ministry has often preferred the latter approach by promoting the construction of large projects. In fact, the National Water Strategy for 2016–2025 highlights the critical importance of improving and building new hydrological infrastructure to augment existing supplies (National Water Strategy, 2016). Addressing water demand management is generally perceived by all governing institutions as a political liability (Zeitoun et al., 2012). Due to the nature of the rentier state and the regime's use of agriculture along with water as patronage, about 60% of the water used by the agricultural sector is unaccounted for water (Barham, 2012). Attempts to restrain this heavy consumption have largely failed.

Given that the King's priority is political stability and regime survival, the fear of upsetting a powerful elite constituency, such as elite farmers, has meant that the governing institutions lack the ability needed to implement policies that can enforce more efficient use of water resources. While at the state-building process, agricultural elite were predominantly prominent members of tribes or tribal leaders, with the passage of time this group expanded. Powerful elites in the agricultural sector today can be prominent figures from the public sector, military, intelligence services, ministries, sheiks, tribal leaders, and influential tribal members (Hussein, 2018). Agricultural elites in the Jordan Valley and the Highlands have been successful in blocking, circumventing, delaying, and preventing the needed reforms that would increase the efficiency by which water is used through implementation of policies. It is important to note that not all farmers in Jordan are agricultural elites. Rather, in the Jordan Valley and the Northern Highlands, there are two types of farmers, poor farmers and powerful elites. It is only the latter group that has the political power to augment its water supply at cheap prices. Due to their lack of political connections, poor farmers lack the means to augment their water supplies and secure cheap water. To demonstrate the political power of the agricultural elite, the following draws on the regime's attempt to reign in the construction of illegal wells in the Northern Highlands and increase the price of water in the Jordan Valley with the attempt to limit the excessive use of water by powerful elites.

In the Northern Highlands, there is a practice known as prestige farming, whereby elites undertake farming as a leisurely activity. Instead of farming for financial returns, prestige farming in many cases is done by absentee farmers, who also plant water-intensive crops and build vacation homes with swimming pools. Approximately 30% of the irrigated water used by these farmers is used in low-value olive groves (Van Aken et al., 2009). These prestige farming elites are major tribal figures, high ranking state employees, and military officers. They have the direct ability to prevent the implementation of policies that can challenge their access to excessive and unlimited quantities of water. Farming in this area is done predominately through groundwater extraction directly by farmers. These farmers consume a total of 304 million cubic meters per year, which represents a substantial quantity of water for Jordan (Barham, 2012). Much of this consumption of water is done through illegal and unmetered wells, which has contributed to the extensive mining of the region's groundwater. While the government has attempted to reign in the over-extraction of this groundwater through the 2002 Groundwater By-Law No. 85, politically powerful elites have prevented the implementation of the policy which could regulate their extraction.

The 2002 Groundwater By-Law No. 85 sought to address the over-abstraction of groundwater, especially in the Northern Highlands through block tariffs, the closing of illegal wells, and penalizing over-extraction from licensed wells. Between 2010 and 2012, the MWI and the German Agency for International Cooperation attempted to negotiate with owners of these wells about the over-extraction of water and tariffs, but their efforts were met with little success (Barham, 2012). After pressure from the U.S. Agency for International Development, in 2013, Jordan undertook a campaign to implement the 2002 Groundwater By-Law No. 85 to close all illegal wells that have been over-extracting and damaging the country's groundwater supplies and resulting in environmental damage. Examples of the destruction of groundwater include the Jafr and Azraq basins. The campaign sought to close 50 illegal wells in an attempt to improve the country's water management policy. The majority of these illegal wells are located in the Northern Highlands and are owned by powerful elite farmers. As government officials attempted to implement and enforce the law, powerful farmers succeeded in evading this campaign while poor, marginalized, or unconnected farmers were the only ones to be impacted by this campaign (Hussein, 2018). That is, powerful elites in the Northern Highlands along with other regions of the country managed to secure continued privileged access to water through evading the law or whenever possible, receiving exemptions to the law that attempts to close illegal wells that permit excessive use of water (Van Aken et al., 2009). This constitutes a perfect example of the power that can be wielded by large farmers and tribal leaders of the region.

Leisure farming among politically powerful tribes, influential Bedouin tribes (such as the Adwani tribe), and elites also occurs in the Jordan Valley, where they tend to grow water intensive citrus fruits and banana farms, which remain profitable because of their political connections that secure for them higher water quotas (Van Aken et al., 2009). For example, the Ghzawi tribe's political power allowed it to evade laws regarding crop selection, and in 2004, the tribe was able to have the JVA legalize its previously illegally planted citrus orchards and then provide it with higher water allocations, which went against government policy at the time (Van Aken et al., 2009). In the Jordan Valley, two tribes that own the majority of large farms have succeeded in securing their production of citrus and bananas through water pricing policies and tariffs protecting domestic production. Growing these crops goes against government policies (Zeitoun et al., 2012). Some large citrus farms in the Valley are owned by absentee farmers who are interested in leisure and prestige farming regardless of the economic revenue generated (Molle et al., 2008).

In the Jordan Valley, the government has subsidized water for farmers who tend to pay 10% of the water costs, which is much less than the delivery cost (Barham, 2012; World Bank, 2018). In fact, from 1988 through 1992, revenue from water charges in the Jordan Valley covered only one sixth of the operation and maintenance costs (Molle et al., 2008). The combination of low prices for water and tariffs on imports complicate the incentive to increase on the farm conservation or to switch to higher-value and less water-intensive crops (Yorke, 2016). Due to the nature of state-society relations and the patronage system that was established during the state-building process, attempts to increase water prices or reduce quotas have the potential to result in protests and powerful farmers are likely to use their connections to weaken the implementation of the laws. In fact, Jordan has a long history of such protests against cuts in subsidies (Zawahri, 2012). For example, in 1997, the World Bank pushed Jordan to increase water tariffs to farmers in the Jordan Valley. Attempts to implement and enforce this policy resulted in farmers occupying the floor of the Jordanian

Parliament and requiring the King's intervention to defuse the crisis. Due to this political pressure, attempts to increase the water tariffs were halted (Pitman, 2004). As government officials attempted to collect unpaid water and land fees, they have been known to encounter violent conflicts and unrest as farmers protest and the military has been called to intervene to stop the unrest (Molle et al., 2008; Zawahri, 2012).

Conclusion

Confronted with a water deficit, states have several policy options, which include augmenting supplies or improving water demand management. Due to the nature of state-society relations established during the state-building process, agriculture and water became a tool of patronage provided by the state to powerful and important constituents whose allegiance is needed for regime stability and legitimacy. Consequently, attempts by international organizations, the donor community, and international banks to rely on technical solutions to Jordan's water problem by pushing for increased efficiency in domestic water resource management through pricing policies, containing the quantity of unaccounted for water, or the domestic legal system to attempt to change consumer behaviour have all proven to be ineffective thus far. In fact, one may argue that the social contract negotiated during the state-building process has survived to constrain the regime's current autonomy in formulating technical policy to address water demand and governing water resources more efficiently.

As this chapter has demonstrated through several examples, such as attempts to reign in illegal wells in the Northern Highlands or increase prices in the Jordan Valley, implementing policy to manage water demand has failed because powerful farmers managed to secure their interests regardless of attempts by government agencies to contain their consumption. Simply put, existing governing institutions lack the ability and power to enforce existing policy. The governing institutions that seek to promote improved domestic water management, such as the MWI, the Ministry of Environment, and the Ministry of Finance, lack the ability and power to challenge existing consumption patterns (Zeitoun et al., 2012).

Given the combined impacts of climate change and increasing demand, the current consumption pattern in the Kingdom is unsustainable in the near or long term. The Monarchy needs to modify its social contract with elite members of society in attempt to limit their water consumption. As it modifies this social contract, the monarchy needs to consider alternative means for building legitimacy and connecting with its populace. Leisure farming needs to be reined in, farmers need to be persuaded to avoid high water consuming crops, and water saving technology can be distributed to farmers at minimum costs or free. By impressing on the political elite the state's critical shortage of domestic water and the need to meet water security to protect the future stability of the regime, the political elite may be persuaded to voluntarily limit its consumption. Such critical changes in consumption can help the Kingdom save precious water as it searches to achieve water security.

References

Alon, Y. (2009). *The Making of Jordan: Tribes, Colonialism and the Modern State*. London: IB Tauris.

Aresvik, O. (1976). *The Agricultural Development of Jordan*. New York, NY, Washington, DC, and London: Praeger Publishers.

Assi, R. and Ajjour, R. (2009). "Jordan's Second National Communication to the United Nations Framework Convention on Climate Change." In *The Hashemite Kingdom of Jordan*. Ministry of Environment of Jordan and the United Nations Development Programme. Amman: Central Press.

Barham, N. (2012). *Is Good Water Governance Possible in a Rentier State? The Case of Jordan*. Odense: Analysis Center for Mellemøststudier, Syddansk Universitet..

Fathallah, H. and Robertson, T. (2021). *The Cost of Food Security in Jordan*. Carnegie Endowment for International Peace, Sada, Middle East Analysis, April 28. [Accessed on 5 February 2023]. Available at https://carnegieendowment.org/sada/84424

Fischbach, M. R. (2000). *State, Society, and Land in Jordan* (Vol. 75). Leiden: Brill.

Hummel, D. (2008). "The interaction of population dynamics and transformations in water supply systems in the Jordan River Basin." In F. Zereini and H. Hotzl (eds.) *Climatic Changes and Water Resources in the Middle East and North Africa*. Berlin, Heidelberg: Springer. pp. 497–518.

Humpal, D., El-Naser, H., Irani, K., Sitton, J., Renshaw, K. and Gleitsmann, B. (2012). *A Review of Water Policies in Jordan and Recommendations for Strategic Priorities*. Amman: United States Agency for International Development (USAID).

Hussein, H. (2018). "Tomatoes, tribes, bananas, and businessmen: An analysis of the shadow state and of the politics of water in Jordan." *Environmental Science & Policy*. 84, 170–176.

Khouri, R. G. (1981). The Jordan *Valley*: Life and *Society Below Sea Level*. London and New York: Longman.

Mercy Corps. (2023). The Facts: What you need to know about the Syrian Crisis. [Accessed 5 February 2023]. Available at https://www.mercycorps.org/blog/quick-facts-syria-crisis

Molle, F., Venot, J. P. and Hassan, Y. (2008). "Irrigation in the Jordan Valley: Are water pricing policies overly optimistic?" *Agricultural Water Management*. 95(4), 427–438.

Muasher, M. (2011). "A decade of struggling reform efforts in Jordan." The Resilience of the Rentier System. Washington, DC: The Carnegie Papers.

National Water Strategy 2016–2025. (2016). Ministry of Water and Irrigation. Amman. [Accessed 5 February 2023]. Available at https://faolex.fao.org/docs/pdf/jor156264E.pdf

Peters, A. M. and Moore, P. W. (2009). "Beyond boom and bust: External rents, durable authoritarianism, and institutional adaptation in the Hashemite Kingdom of Jordan." *Studies in Comparative International Development*. 44(3), 256–285.

Pitman, G. T. K. (2004). "An evaluation of bank assistance for eater development and management." *A Country Assistance Evaluation*. Washington, DC: World Bank.

Rajsekhar, D. and Gorelick, S. M. (2017). "Increasing drought in Jordan: Climate change and cascading Syrian land-use impacts on reducing transboundary flow." *Science Advances*. 3(8), e1700581.

Van Aken, M., Molle, F. and Venot, J. P. (2009). "Squeezed dry: The historical trajectory of the Lower Jordan River Basin." *River Basin Trajectories: Societies, Environments and Development*. Cham, Heidelberg: Springer. pp. 20–46.

Waha, K., Krummenauer, L., Adams, S., Aich, V., Baarsch, F., Coumou, D., Fader, M., Hoff, H., Jobbins, G., Marcus, R. and Mengel, M. (2017). "Climate change impacts in the Middle East and Northern Africa (MENA) region and their implications for vulnerable population groups." *Regional Environmental Change*. 17(6), 1623–1638.

Weinthal, E. and Zawahri, N. (2020). "Don't politicize water," *Foreign Policy*, September 17.

World Bank. (2014). *Turn Down the Heat: Confronting the New Climate Normal*. Washington, DC: World Bank. [Accessed 5 February 2023]. Available at https://openknowledge.worldbank.org/handle/10986/20595

World Bank. (2018). "The role of food and agriculture for job creation and poverty reduction in Jordan and Lebanon." *Agricultural Sector Note, Technical Note*. Washington, D.C.: World Bank Group.

Yorke, V. (2016). "Jordan's shadow state and water management: Prospects for water security will depend on politics and regional cooperation." In R. Huttl, O. Bens, C. Bismuth, and S. Hoechstetter (eds.) *Society-Water-Technology*. Heidelberg: Springer. pp. 227–251.

Yoon, J., Klassert, C., Selby, P., and Gorelick, S. 2021. "A coupled human-natural system analysis of freshwater security under climate and population change." Proceedings of the National Academy of Science. 118(4). E2020431118.

Zawahri, N. (2008). "Designing river commissions to implement treaties and manage water disputes: The story of The Joint Water Committee and Permanent Indus Commission." *Water International*. 33(4), 464–474.

Zawahri, N. (2012). "Popular protests and the governance of scarce fresh water in Jordan." *The Arab World Geographer*. 15(4), 267–301.

Zawahri, N. and Weinthal, E. (2014). "The World Bank and negotiating the Red Sea and Dead Sea water conveyance project." *Global Environmental Politics*. 14(4), 55–74.

Zeitoun, M., Allan, T., Al Aulaqi, N., Jabarin, A. and Laamrani, H. (2012). "Water demand management in Yemen and Jordan: Addressing power and interests." *The Geographical Journal*. 178(1), 54–66.

12
THE NETHERLANDS

Erik Mostert

Introduction

The Netherlands is a small country with a surface area of some 40,000 km^2 and a population of 17.5 million, located on the shores of the North Sea, in the delta of the Rhine, the Meuse and the Scheldt rivers. A thousand years ago, the area consisted of sparsely populated marshes and bogs in the north and the west, and poor sandy soils in the south and the east (Van de Ven, 2004). Since then, many works of water management have been undertaken, and the country presently is one of the most highly engineered, densely populated and richest countries in the world, taking eighth place on the Human Development Index (UNDP, 2020).

In this chapter, the different roles that water played in the development of the Netherlands will be discussed, using mostly historical literature. While the older literature focused on institutional development, in particular the development of the regional water boards (e.g. De Vries Azn., 1876; Beekman, 1905–1907; Fockema Andreae, 1934; 1952), the more recent literature pays much attention to the ecological, economic, political, and cultural context (e.g. Van Dam, 2004; Jensen, 2018; Van Tielhof, 2021). In this chapter, these two perspectives will be combined since the institutions and their context influenced each other profoundly.

Despite the small size of the Netherlands, regional and local conditions differ a lot (Van Bavel, 2010). Usually, most attention is paid to the low-lying western part of the country, in particular to Holland, the economic and political heartland of the country.[1] In this chapter, a more marginal area that developed later will be discussed as well: the Dommel basin in the south. Both areas will be used to show that water can play very different roles in development and that development is always multi-sectoral and multi-scalar.

From Swamps to Riches

According to an old saying, God created the world, but the Dutch created the Netherlands (Niemeijer, 2016). Yet, while they did shape their own environment, they did not create it out of nothing and sometimes they destroyed it as well.

After the last glacial period, around 10,000 BCE, the level of the North Sea rose and the Netherlands were separated from England. When the sea level rise slowed down, around 3000 BCE,

DOI: 10.4324/9781003095545-14

dunes developed and behind the dunes peat soils formed, consisting of half-decomposed plant remains. In some areas these soils grew well above the surface water level and acidic raised bogs developed, as in the central part of Holland. Near the estuary of rivers, the peat often eroded and salt marshes developed (Van de Ven, 2004).

The peat areas could not support a large population, but from the 9th century CE onwards many were drained to make agriculture possible. Starting from a baseline, for instance, a riverbank, the area was cleared and drainage ditches perpendicular to the baseline were dug. As water was removed from the soil, air could penetrate and the peat started to shrink and decompose, resulting in land subsidence. This caused new drainage problems. The initial response was to drain new areas further inland, but this was possible only as long as undrained land was available (Besteman, 1997; Aten, 2007; Van Tielhof, 2007). In addition, dams, culverts and locks were constructed to prevent the inflow of water, and new drainage canals were dug to provide shorter evacuation routes for the water with a higher gradient. From the early 1400s onwards, artificial drainage was introduced, at first by means of windmills and from the late 18th Century onwards using steam pumps, diesel pumps and electric pumps. Furthermore, dykes were built or strengthened.

To meet fuel demands for domestic heating, brickmaking and brewing, much land was used for peat mining, which resulted in the creation of many lakes. Until salt began to be imported from southern Europe, around 1500, peat was also mined for salt production. Peat that had been flooded by seawater was burned and from the ashes salt was extracted. If mined close to dykes, this could weaken these dykes and result in flooding and loss of land, especially when the costs of dyke maintenance were high and agricultural prices were low. But as soon as land was lost to the waves, silt and clay would start to be deposited, and after several decades new dykes could be constructed and the area was turned into fertile agricultural land (e.g. Van Dam, 2003; Leenders, 2004).

It is not known how the first drainage works were organised, but we do know that in the late 9th century a dynasty of territorial lords settled in the area: the Counts of Holland. The Counts of Holland successfully claimed rights over all "wild" areas, including fens and bogs. They often transferred the right to exploit these areas to their vassals or to groups of free investors for services rendered or against payment. The upkeep of the necessary infrastructure, such as dykes and sluices, was the responsibility of the individual landowners and users (e.g. Henderikx, 2001). They were controlled by the local government, consisting of the representative of the count, the "schout", and the aldermen or "schepenen", who represented the local community but were usually appointed by the schout.

As the land continued to subside and the scale of the drainage and flooding problems increased, local water management no longer sufficed and from the 11th century onwards, regional water boards were established to oversee works of regional interest. The initiative for their establishment came either from the count, who granted them rights and a charter, or from the interested communities or local elites (Van Tielhof, 2021: 33–38). From the 16th century onwards, the regional water boards increasingly executed the works themselves and charged the landowners and users for this. Also at the local level contributions in-kind were increasingly replaced by charges.

Land drainage and the subsequent agricultural development stimulated the growth of market towns and cities, often located at strategic places on waterways, for instance at a dam in a river (e.g. Amsterdam and Rotterdam). As employment possibilities in agriculture were limited, many people turned to fishing or commercial shipping (Borger, 1997; De Neve & Van Heezik, 2007). Commercial shipping was stimulated greatly by the geographical location of Holland in the delta of the Rhine and Meuse rivers and by the early development of commercial markets in Holland, and from the 15th century onwards Holland was dominant in North Sea shipping (Van Bavel,

2010). From 1600 onwards, shipping outside Europe developed as well. In America, Africa and Asia, several trading posts and colonies were established, and in Holland, several waterways were improved and new canals were dug to facilitate passenger transport by barges and promote trade and industry (De Neve & Van Heezik, 2007). The profits of commerce and trade were often invested in land reclamation projects. In the first half of the 17th century alone, some 270 km^2 of lakes were reclaimed (Van de Ven, 2004; Aten, 2007).

Water and in particular transport over water also played an important role in the industrialisation of the Netherlands in the late 19th century. They continue to be important to this day. The port of Rotterdam is currently the biggest port outside of Asia. Moreover, there is still a serious risk of flooding in much of the country. Furthermore, water is essential for the production of drinking water and for agriculture, which has a production value of 27 billion Euros and uses 66% of the land surface (2015 data: CBS, 2021). In addition, there is much water recreation and the biggest nature areas are open waters, such as the Wadden Sea.

Water also has a cultural significance. According to a popular theory, the Dutch "struggle against the water" has shaped the Dutch national identity. It required initiative, ingenuity and perseverance, which laid the foundation for later national successes. In addition, it required cooperation. This would have resulted in an egalitarian society and democratic institutions, such as the water boards. In this context, the commercial and bourgeois ("burgerlijk") character of the country is often mentioned as well and this too is linked with water, especially with shipping and trade (Huizinga, 1968; Schama, 1987; Mostert, 2020a).

This theory, however, is an oversimplification. There have been many examples of cooperation in Dutch water management, but also many examples of conflict, for instance of polders[2] refusing to contribute to the upkeep of the dyke of another polder that protected them as well (Maas, 1963; Mostert, 2012). Moreover, the water boards were not democratic in the modern sense. Until the mid-19th century, board members were co-opted or selected from the same families that ruled the cities and were on the provincial councils. At the local level, a broader group of people could hold office, at least initially, but later involvement often was restricted to large landowners only (Van Tielhof, 2021).

For centuries water in the Netherlands was managed in the absence of a Dutch nation and a Dutch national identity. Politically, the Netherlands became a nation in the late 16th century, when the different territories in the Netherlands joined forces in their struggle against the Habsburg kings of Spain, who by means of marriage and inheritance had become the rulers of these territories. Culturally, the Netherlands became a nation only later as a result of increasing interaction between the territories and the development of a national press in the early 18th century (Van Sas, 2005).

Water was first advanced as part of the national identity in the late 18th century by the "Patriots", who called for a regeneration of the Dutch nation and demanded more political power (Schama, 1977). They presented the Dutch fight against the water as one of the major feats of the nation. Since then, water has been used many times by different social groups as a source of national pride and identity, especially in times of social change, such as the economic crisis of the 1930s and the Second World War. Since 1990, "water" is very popular again, possibly in reaction to the erosion of traditional bonds in the post-war era and neoliberal politics with its international outlook and emphasis on individualism (Mostert, 2020a).

The Dutch water identity has also been used as an argument in favour of large hydraulic works. For example, in the late 19th century the Zuiderzee works were discussed, involving the construction of a 32 km long closure dam (the Afsluitdijk) to improve protection against flooding and facilitate the construction of large new polders. The government and many MPs saw these works as too expensive. According to the Zuiderzeevereeniging, an association lobbying for these works, however, history had shown that the Dutch people "consider building dykes and land reclamation as their vocation; it is in their character". The Zuiderzee works would be a national interest on which all parties across the political spectrum could agree (Van der Houven van Oordt, 1898: 165). More recently, the Dutch relation with water has been used as an argument to promote the export of Dutch water expertise (e.g. Kuper, 2020).

The Dommel Basin[3]

In the preceding section the discussion has shifted almost imperceptibly from Holland to the Netherlands, but the Netherlands are larger than Holland. In this section a river basin from another part of the Netherlands is discussed: the Dommel basin. The Dommel has its source in Belgium and then flows northwards into the Netherlands until the city of Den Bosch, where it joins the Aa River and changes its name to the Dieze. Five kilometres downstream of Den Bosch it discharges into the Meuse. The basin area upstream of Den Bosch is circa 1,800 km^2, of which 400 km^2 are located in Belgium and 1,400 km^2 in the Netherlands (Bongaerts, 1919).

The Dutch part of the Dommel basin is presently part of the province of North-Brabant, but between 1648 and 1795 it was part of Staats-Brabant. Staats-Brabant was a kind of domestic colony of the Netherlands that was not represented at the national level. Most inhabitants were Roman Catholic, who until 1795 were not allowed to hold public offices (Van den Eerenbeemt, 1996). The agricultural conditions were bad. The upstream parts of the basin consisted mostly of heath on poor sandy soils, interspersed by narrow river valleys with swampy meadows. Around 1800 there were ideas to develop agriculture, but the soils were too poor and the costs of importing fertilisers, such as domestic refuse from Holland, were too high, given the poor transportation possibilities at the time. The economy in the basin was essentially a peasant economy, with most of the farm produce being consumed on the farm itself (Crijns & Kriellaars, 1987).

The valley of the Dommel flooded regularly, but until around 1900 only summer floods were seen as problematic as these resulted in a loss of the hay harvest. Winter floods were seen as increasing the fertility of the soils (Bongaerts, 1919; Deckers, 1927). In the 1840s flooding increased as a result of the construction of 4,000 ha of water meadows upstream in Belgium. These meadows were flooded in winter with Calcium-rich water from the Meuse, imported via the newly dug Kempens Canal, and in spring the surplus water was discharged onto the Dommel, which caused flooding downstream.

Even more important for flooding were the introduction of artificial fertilisers in the late 19th century and urbanisation. Artificial fertilisers made it possible to convert large tracts of heathland to agriculture (Crijns & Kriellaars, 1987; Crijns & Kriellaars, 1992; Bieleman, 2010). To improve field drainage, ditches were dug, resulting in higher runoff peaks and more flooding downstream. Urbanisation was the result of industrialisation, which was stimulated by the construction of shipping canals from 1825 onwards and railways from 1860 onwards (Van den Eerenbeemt, 1996). Between 1842 and 2020, the population of Eindhoven, the largest town in the basin, increased from 3,000 to 227,000, and the population of Tilburg, on the edge of the basin, increased from 14,000 to 194,000. The resulting increase in hard surfaces further increased runoff peaks.

To cope with the flooding problems, several river regulation works were executed. Between 1875 and 1893, many small river bends were cut off and obstacles removed; between 1907 and 1911, an extra river outlet was constructed, the Drongelens Canal; between 1931 and 1941 the Dommel further upstream was connected to the Wilhelmina Canal, a recently completed shipping canal; and between 1933 and 1936 a large river bend in the village of Boxtel was cut off (Bongaerts, 1919; Roeffen, 1963). Moreover, many land re-allotment projects were executed, especially in the 1960s and early 1970s, to reduce the number of land plots and increase efficiency in agriculture (e.g. Lohman, 1963). As part of these projects new drainage ditches were dug and the receiving brooks were regulated to cope with increased runoff peaks. This further increased discharges onto the Dommel and consequently plans were developed to channelise the main river.

To execute the first river regulation works and charge the beneficiaries, the Province of North-Brabant established a regional water board in 1863 (Roeffen, 1963). Initially, only the owners of the flood-prone land, some 4,300 ha. in total, were charged and could vote for the water board, but in 1921 the water board was extended to cover the whole basin and a few years later the owners of houses and other buildings were charged as well (Bongaerts, 1919; Deckers, 1927). Even so, the river regulation works projects would not have been possible without large subsidies from the State and the province. The first river regulation works (1875–1893) were subsidised for more than 50%, the construction of the Drongelens Canal was paid for completely by the State, and subsidies for post-war land re-allotment projects were typically in the order of 75%.

To cope with increasing water pollution problems, the water board got a new task in 1950: wastewater treatment. Representation of the municipalities on the board was increased to one-fifth of the seats and industry also got one-fifth of the seats. This increase in representation did not immediately lead to more attention for the "urban" interests of nature, landscape and recreation. Already before the Second World War there were protests against the river regulation works and land re-allotment projects, and sometimes plans were modified to accommodate the interests of nature and landscape. However, during the economic crisis of the 1930s, the Second World War and the early post-war years, agricultural interests prevailed. This started to change only around 1970. Plans to channelise the Dommel were effectively opposed and from 1990 onwards river restoration works were executed, nowadays as part of the implementation of the European Water Framework Directive (Didderen et al., 2009; Mostert, 2020b).

The change around 1970 can be explained by broader socio-economic developments in the post-war period in the Netherlands. Working hours were reduced and paid holidays were introduced, and the population of the ever-larger towns finally had time to visit the countryside and enjoy nature and the landscape (Maas, 1963). In addition, the general level of education had increased a lot and traditional authority was increasingly questioned (cf. Welzel et al., 2003).

Ironically, while there used to be a shortage of nutrients in the basin, there is currently an overabundance. Between 1950 and 2020, the number of pigs in the province of North-Brabant increased from 270,000 to 5.7 million and the number of chickens from 3.6 million to 26.5 million (Crijns, 1998; CBS, 2021). This resulted in large amounts of manure and severe surface and groundwater pollution (as well as deforestation, erosion and social problems in the countries where the fodder for the livestock was grown). The problems were first recognised in the late 1960s, but the agricultural sector in the Netherlands is well-organised and well-connected with politics, and therefore it took until 1984 before any regulatory action was taken (an act to limit the increase in livestock). Since then, regulation has become stricter, often in reaction to European Union directives, such as the Nitrates directive (91/676/EEC), and decisions of the European Court of Justice (e.g. Keessen et al., 2011; Squintani, 2019).

Unlike in Holland and the Netherlands more generally, water does not have a large cultural significance in the Dommel basin. The area is not generally known as the Dommel Basin, but as part of the "Meierij" or the "Kempen". In the 1920s and 1930s, a strong regional identity developed, but it was province-wide and not based on water, but on an image of traditional village life and Roman Catholicism. Despite urbanisation and secularisation, this image still persists (cf. Janssen, 2006).

Discussion

The two cases presented in this chapter show that water can play quite different roles in development, even in a small country such as the Netherlands. In Holland, land drainage and flood protection were essential for the transition to an agricultural economy, and its location on the North Sea coast in the delta of major rivers was essential for the development of shipping, trade and industry. In addition, water became part of the self-image of the country and this self-image was in turn used to promote large hydraulic works. In the Dommel basin, the introduction of artificial fertilisers was essential for agricultural development, which brought with it the need to drain land and caused flooding and water quality problems. These problems were exacerbated by industrialisation and urbanisation. Industrialisation and urbanisation were stimulated by the construction of shipping canals and later rail and road transport, but shipping and trade never played such a large role as they did in Holland, nor did water become part of a regional identity.

Simplifying the differences, we can state that in Holland water management triggered development, while in the Dommel basin development triggered water management. Yet, there are also many similarities. In both areas water management needs led to institutional changes to meet these needs, such as the establishment of water boards. Furthermore, in both areas development took place in a broader social, economic and political context and was influenced by factors such as the emergence of new markets, technical innovations and the establishment of the European Union. And finally, in both areas development was not exclusively national. Development started well before the Dutch nation came into being, especially in Holland, and also afterwards many relevant developments took place at other levels, ranging from the local level to the EU level and the global level.

Despite the wealth of information available on the importance of water for the development of the Netherlands, relatively little is known about the cultural dimension of this relation. For ten years now, the role of flood disasters in the development of a national culture has received a lot of attention (e.g. Bosch, 2012; Sundberg, 2015; Jensen, 2021), but less attention has been paid to drainage and land reclamation, even though they have played an important role in the self-image of the nation. Interesting questions include whether land drainage and land reclamation actually stimulated the development of a national culture or were used only strategically as arguments; whether in the process social and political differences were glossed over; what impact the national culture had on water management practices and national development; and which groups benefitted and which did not or less. Moreover, interesting international comparisons could be made, for instance with Spain (e.g. Swyngedouw, 1999).

Notes

1 "Holland" is the name of the historic County of Holland, which coincides more-or-less with the present provinces of North-Holland and South-Holland. It is sometimes used as a synonym for the whole of the Netherlands.
2 "Polder" can mean both an area and the board in charge of water management in that area. Polder areas are hydrologically isolated from their environment with an artificially controlled water level. They include peat areas drained by means of windmills or pumps, reclaimed lakes and coastal areas that have been enclosed by dykes.
3 This section reuses parts of Mostert (2018, section 4).

References

Aten, D. (2007). "Een afgerond geheel; Waterstaat en waterschappen ten noorden van het IJ." In: Beukers, E. (ed.) *Hollanders en het water: twintig eeuwen strijd en profijt*. Hilversum: Verloren. 1, 23–60.

Beekman, A. A. (1905–1907). Het dijk- en waterschapsrecht in Nederland voor 1795. 's-Gravenhage: Martinus Nijhoff.

Besteman, J. C. (1997). "Van Assendelft naar Amsterdam. Occupatie en ontginning van de Noordhollandse veengebieden in de middeleeuwen." In Boer, D. E. H., Cordfunke, E. H. P. and Sarfatij, H. (eds.) *Holland en het water in de middeleeuwen: strijd tegen het water en beheersing en gebruik van het water*. Hilversum: Verloren. pp. 21–39.

Bieleman, J. (2010). *Five centuries of farming: a short history of Dutch agriculture 1500–2000*. Wageningen: Wageningen Academic Publishers.

Bongaerts, M. (1919). *Verbetering der afwatering van de rivieren de Dommel en de Aa; Voorlopig rapport van den ingenieur Bongaerts omtrent de Dommel*. 's-Gravenhage: ingenieur Bongaerts.

Borger, G. J. (1997). "De strijd tegen en om het water: een samenvattende terugblik." In Boer, D. E. H., Cordfunke, E. H. P. and Sarfatij, H. (eds.) *Holland en het water in de middeleeuwen: strijd tegen het water en beheersing en gebruik van het water*. Hilversum: Verloren. pp. 141–147.

Bosch, T. (2012). "Natuur en cultuur; Modernisering van hulpverlening na catastrofale overstromingen in de Nederlandse Delta, 1740–1861." *Tijdschrift voor Waterstaatsgeschiedenis*. 21(1–2): 39–47.

CBS. (2021). *StatLine*. [Accessed 28 June 2021]. Available from https://opendata.cbs.nl

Crijns, A. H. (1998). *Van overgang naar omwenteling in de Brabantse land- en tuinbouw, 1950–1985: schaalvergroting en specialisatie*. Tilburg: Stichting Zuidelijk Historisch Contact Tilburg.

Crijns, A. H. and Kriellaars, F. W. J. (1987). *Het gemengde landbouwbedrijf op de zandgronden in Noord-Brabant 1800–1885*. Tilburg: Stichting Zuidelijk Historisch Contact.

Crijns, A. H. and Kriellaars, F. W. J. (1992). *Het gemengde landbouwbedrijf op de zandgronden in Noord-Brabant 1886–1930*. Tilburg: Stichting Zuidelijk Historisch Contact.

Deckers, J. H. F. (1927). *De waterstaatstoestanden in Noord-Brabant binnen het stroomgebied der Maas voorheen en thans: uit een economisch en landbouwkundig oogpunt beschouwd*. Tilburg: Bergmans.

De Neve, R. G. and Van Heezik, A. A. S. (2007). "Verbonden door het water; Binnenvaart en zeehavens in Holland." In Beukers, E. (ed.) *Hollanders en het water: twintig eeuwen strijd en profijt*. Hilversum: Verloren. 1, 168–273.

De Vries Azn, G. (1876). *Het dijks- en molenbestuur in Holland's Noorderkwartier onder de grafelijke regeering en gedurende de Republiek*. Amsterdam: Van der Post.

Didderen, K., Verdonschot, P., Knegtel, R. and Lototskaya, A. (2009). *Enquête beek(dal) herstelprojecten 2004–2008: evaluatie van beekherstel over de periode 1960–2008 en analyse van effecten van 9 voorbeeldprojecten*. Wageningen: Alterra.

Fockema Andreae, S. J. (1934). *Het Hoogheemraadschap van Rijnland. Zijn recht en zijn bestuur van den vroegsten tijd tot 1857*. Leiden: Eduard IJdo.

Fockema Andreae, S. J. (1952). Studien over waterschapsgeschiedenis. Dl. 8. *Overzicht van de Nederlandse Waterschapsgeschiedenis*. Leiden: Brill.

Henderikx, P. A. (2001). *Land, water en bewoning: Waterstaats- en nederzettingsgeschiedenis in de Zeeuwse en Hollandse delta in de Middeleeuwen*. Hilversum: Verloren.

Huizinga, J. (1968). *Dutch civilisation in the seventeenth century and other essays*. London: Collins.

Janssen, J. (2006). *Vooruit denken en verwijlen: de (re)constructie van het plattelandschap in Zuidoost-Brabant (1920–2000)*. Tilburg: Stichting Zuidelijk Historisch Contact.

Jensen, L. (ed.). (2021). *Crisis en catastrofe. De Nederlandse omgang met rampen in de lange negentiende eeuw*. Amsterdam: Amsterdam University Press.

Jensen, L. E. (2018). *Wij tegen het water. Een eeuwenoude strijd*. Nijmegen: Vantilt.

Keessen, A. M., Runhaar, H. A. C., Schoumans, O. F., Van Rijswick, H. F. M. W., Driessen, P. P. J., Oenema, O. and Zwart, K. B. (2011). "The need for flexibility and differentiation in the protection of vulnerable areas in EU environmental law: The implementation of the Nitrates Directive in the Netherlands." *Journal for European Environmental & Planning Law*. 8(2), 141–164.

Kuper, S. (2020). "Can the Dutch save the world from the danger of rising sea levels?; The Netherlands has learnt to manage flooding. Climate-threatened countries are taking note." *Financial Times Magazine* [online], 30 January 2020. [Accessed 11 August 2021]. Available from https://www.ft.com/content/44c2d2ee-422c-11ea-bdb5-169ba7be433d

Leenders, K. A. H. W. (2004). "De interactie tussen mens en natuur in de strijd om land en water in het zuiden van Holland, 1200–1650." *Holland; Regionaal-historisch tijdschrift*. 36(3), 142–161.

Lohman, H. F. A. (1963). "Ruilverkaveling." In Anon (ed.) *Het stroomgebied van de Dommel 1863–1963*. Boxtel: Waterschap 'Het stroomgebied van de Dommel'. pp. 329–345.

Maas, F. M. (1963). "De toekomst van de beken en riviertjes in het stroomgebied van de Dommel." In Anon (ed.) *Het stroomgebied van de Dommel 1863–1963*. Boxtel: Waterschap 'Het stroomgebied van de Dommel'. pp. 329–345.

Maas, W. J. M. (1963). De Dijkwet van 1810; een algemene bijdragenwet in de Franse tijd. Rotterdam/'s-Gravenhage: Nijgh & Van Ditmar.

Mostert, E. (2012). "Water management on the Island of IJsselmonde 1000–1953; Polycentric governance, adaptation and petrification." *Ecology and Society*. 17(3), article 12.

Mostert, E. (2018). "An alternative approach for socio-hydrology: case study research." *Hydrology and Earth System Sciences*. 22(1), 317–329.

Mostert, E. (2020a). "Water and national identity in the Netherlands; the history of an idea." *Water History*. 12(3), 311–329.

Mostert, E. (2020b). "Law and politics in river basin management; The Implementation of the Water Framework Directive in the Netherlands." *Water*. 12(2), 1–12.

Niemeijer, A. F. J. (2016). "Wie 'schiep' Nederland nu echt? Een reactie op een vrijzinnige gedachte." *Vitruvius*. (36), 6–13.

Roeffen, H. J. M. (1963). "De waterstaat in het stroomgebied sinds 1800 en flitsen uit de geschiedenis van het waterschap." In Anon (ed.) *Het stroomgebied van de Dommel 1863–1963*. Boxtel: Waterschap 'Het stroomgebied van de Dommel'. pp. 36–100.

Schama, S. (1977). *Patriots and liberators; revolution in the Netherlands, 1780-1813*. New York: Knopf.

Schama, S. (1987). *The embarrassment of riches: an interpretation of Dutch culture in the Golden Age*. New York: Knopf.

Squintani, L. (2019). "Balancing nature and economic interests in the European Union: On the concept of mitigation under the Habitats Directive." *Review of European, Comparative & International Environmental Law*. 29(1), 1–9.

Sundberg, A. D. (2015). *Floods, worms, and cattle plague: nature-induced disaster at the closing of the Dutch Golden Age, 1672–1764*. PhD thesis, University of Kansas. [Accessed 29 June 2021]. Available from https://core.ac.uk/download/pdf/213413027.pdf.

Swyngedouw, E. (1999). "Modernity and hybridity: nature, regeneracionismo, and the production of the Spanish waterscape," 1890–1930. *Annals of the Association of American Geographers*. 89, 443–465.

UNDP. (2020). *Human Development Report 2020: The next frontier. Human development and the Anthropocene*. New York: United Nations Development Programme.

Van Bavel, B. (2010). *Manors and markets: Economy and society in the Low Countries, 500–1600*. Oxford: Oxford University Press.

Van Dam, P. E. J. (2003). "Schijven en beuken balken; Een sociaal-ecologische transformatie in de Riederwaard." In B. Wouda (ed.) *Ingelanden als uitbaters; Sociaal-economische studies naar Oud- en Nieuw-Reijerwaard, een polder op een Zuid-Hollands eiland*. Hilversum: Verloren. pp. 11–43.

Van Dam, P. E. J. (2004). "De nieuwe waterstaatsgeschiedenis; De interactie tussen mens en natuur in Holland." *Holland; Regionaal-historisch Tijdschrift*. 36(3), 128–141.

Van den Eerenbeemt, H. F. J. M. (ed.). (1996). *Geschiedenis van Noord-Brabant*. Meppel: Boom.

Van der Houven van Oordt, H. C. (1898). *De economische beteekenis van de afsluiting en drooglegging der Zuiderzee*. Leiden: Brill.

Van de Ven, G. P. (2004). *Man-made lowlands history of water management and land reclamation in the Netherlands*. Utrecht: Matrijs.

Van Sas, N. C. F. (2005). *De metamorfose van Nederland; Van oude orde naar moderniteit, 1750–1900*. Amsterdam: Amsterdam University Press.

Van Tielhof, M. (2007). "Betrokkenen bij de waterstaat; boeren, burgers en overheden ten zuiden van het IJ." In: Beukers, E. (ed.). *Hollanders en het water: twintig eeuwen strijd en profijt*. Hilversum: Verloren. 1, 61–98.

Van Tielhof, M. (2021). *Consensus en conflict; Waterbeheer in de Nederlanden 1200–1800*. Hilversum: Verloren.

Welzel, C., Inglehart, R. and Klingemann, H. D. (2003). "Human development as a theory of social change: a cross-cultural perspective." *European Journal of Political Research*. 42(3), 341–379.

PART III

Governing Water and Development

13
GOVERNING WATER SERVICES

Klaas Schwartz and Mireia Tutusaus

Introduction

The concept of "governance" has become increasingly fashionable in the water services sector over the past decades. Although no common definition of governance exists and the concept can mean "anything and nothing" (Jessop, 1998: 29), "[t]he term governance […] marks a change in policy emphasis from infrastructure to the organizational, financial, and institutional arrangements needed to regulate and order flows of water" (Zwarteveen et al., 2017: 2). The concept of governance signals a greater emphasis on organizations, their mandates and, most importantly, the relationships established among different actors in different domains of the water provision chain (government to operators) and decisions about who gets water and how. Historically, the water services sector has been subject to technocratic solutions that seek to find the "right" combination of infrastructure and organizations to provide a desired level of water services. Over the past decades, governance discussions relating to actors involved in providing services and the conditions and terms under which services should be supplied have become more prominent.

In this chapter, we discuss the main governance discourses in the water services sector over the past decades. In doing so, we link these discourses to a framework of service modalities. Essentially, every provisioning of water services involves a certain service modality consisting of infrastructure through which water is provided; an organization that operates and maintains that infrastructure; and a financial arrangement that covers the costs of operating and maintaining that infrastructure. This combination of infrastructure, organization, and finance then leads to a level of water services provided to consumers. The service level consists of the quality and quantity of water provider, the price that the consumer has to pay for accessing this water and the convenience of accessing water. This service modality forms the visible manifestation of water services.

The choices for the infrastructure, organization, finance, and service level that constitute a service modality are strongly influenced by the policy choices forwarded by the institutional environment in which services are provided. These policy choices are, in turn, informed by the prevailing governance discourses. As such, these governance discourses constitute a less-visible set of ideals and principles that underlie the more-visible service modalities used for water service provision.[1]

Underlying the discussion of governance issues in this chapter, we identify two distinct discourses that strongly shape the context in which water providers operate. The first discourse

DOI: 10.4324/9781003095545-16

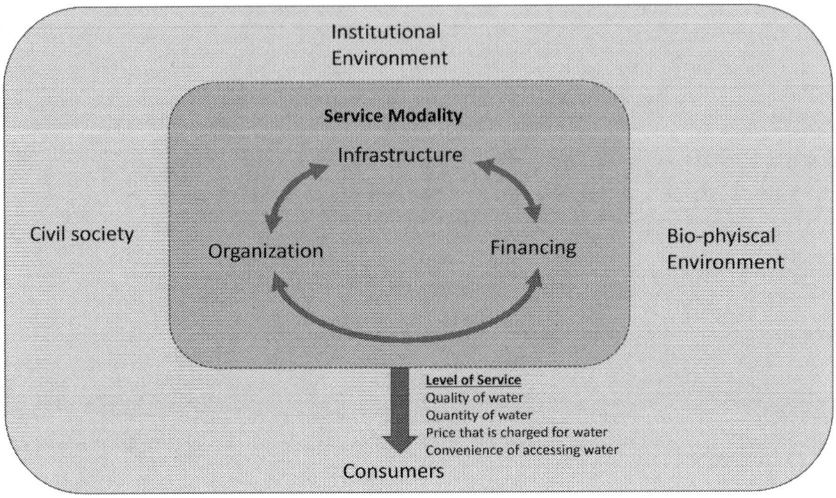

Figure 13.1 Water Service Modalities.

Source: Tutusaus (2019)

approaches the provision of water services as a commercial transaction. Water utilities are viewed as enterprises exchanging services for economic remuneration. In this discourse, water providers need to operate autonomously and on the basis of sound commercial principles if they are to be sustainable. In this discourse, the main unit of analysis concerns the water utility. Success in this discourse is portrayed in terms of operational and, particularly, financial performance indicators.

The second discourse highlights that water is a basic service, which is essential for human and economic development. Central to this perspective is an emphasis on the need to achieve universal access to services as stipulated in the human right to water and Sustainable Development Goal 6.1. For most national governments, public water utilities are the main instrument for achieving these targets. As such, most water utilities face considerable pressure from the government and politicians to extend services. In this perspective, the water provider is only an instrument, and the real focus is on the end-user (the consumer) rather than the water utility. Success in this approach is not measured by the commercial viability or operational performance of the water utility, but rather by the degree to which consumers can access safe and adequate water services.

The two discourses are informed by different problem definitions and result in distinct policy priorities. What characterizes the water services sector is that these two discourses co-exist simultaneously and as seemingly complementary approaches (Schwartz et al., 2017). We will illustrate how the combination of these two approaches translate into visible service provision realities that are characterized by the differentiation of services between consumer groups. This differentiation is visible in the infrastructure used to supply services, in the organizations operating this infrastructure, in the prices charged for water and ultimately in the level of service delivered to consumers.

In this chapter, we link governance discourses with service modalities. Although we try to show this link in general, the nature of this link is dependent on the existing water service realities in given locations. In locations where a high level of service coverage exists, discussions about the type of infrastructure, or facilitating access through different types of infrastructure, are less prominent. Under those conditions, discussions mainly concern organization and finance. In countries where service coverage is low, the question of the type of infrastructure used to provide services and the associated costs, becomes more prominent.

Infrastructure: From Standardized Technology to Appropriate Technologies

Until the International Drinking Water Decade (1980–1990) the Modern Infrastructural Ideal guided the water services sector. Central to this Ideal, was the principle that all consumers should be serviced through a centralized network allowing for in-house connections. In other words, every consumer was to receive the same level of service. The achievements of the International Drinking Water Decade turned out to be disappointing. Although approximately 1.5 billion people gained access to water services, the ambitious goal of universal service coverage was not achieved. As part of the reflection occurring during this time, the appropriateness and affordability of the technologies used for extending water services were increasingly questioned. An evaluation of the World Health Organization found that "[…] the Decade witnessed a move away from oversophisticated systems towards the use of technologies operated and maintained on a sustainable basis" (1991: 7). These technologies were frequently referred to as "appropriate technologies" and essentially constituted low-cost technologies such as standpipes and water kiosks. The underlying argument promoting appropriate technologies was that doing "more of the same" (Mara & Alabaster, 2008: 119) would not lead to universal service coverage and hence a new paradigm of service provision was required to ensure the extension of water supply access, particularly to low-income areas.

The promotion of low-cost technologies is largely based on two main arguments. The first relates to the characteristics of the consumers living in low-income areas. Often these consumers are employed in the informal sector and have low- and fluctuating-income levels, which makes it difficult to ensure monthly water bill payments. The challenges of paying monthly water bills have led some academics to conclude that "the poor have less consistent payment behaviour" (Berg & Mugisha, 2010: 592). Low-cost technologies are then offered as a solution as they allow a water utility to tailor service provision arrangements (and associated payments) to particular areas and customer groups. Technologies such as kiosks and standpipes lead to much lower consumption levels (and thus lower payment) and require payment at the moment of accessing water, relieving people of the financial burden associated with bulk monthly payments (Berg & Mugisha, 2010). In doing so, the technology is presented as better matching the needs and demands of low-income households (Heymans et al., 2014).

The second argument for low-cost technologies relates to the challenges, costs and risks that a water utility faces when expanding a conventional centralized water network to unserved areas. By servicing groups of households in these areas through low-cost technologies, rather than through individual household connections, a "huge reduction in costs" could be achieved (Mara & Alabaster, 2008: 120). The combination of lower costs of providing services with tailoring the service to the financial means of low-income households thus allows utilities to provide services to low-income areas with lower commercial risks. The use of low-cost technologies is thus seen as reflecting "a pragmatic move towards accommodating social and spatial disparities in a polarized city" (Jaglin, 2008: 1898).

The term "service differentiation" has been used to refer to the strategy of distinguishing different consumer groups and supplying them through different technologies. A fundamental feature of service differentiation is that it leads to different levels of service to which consumers have access. Proponents of service differentiation argue that the use of "appropriate technologies" allows the water utility to tailor the level of service to the particular consumer group (Heymans et al., 2014; Mara & Alabaster, 2008). Opponents argue that it leads to sustained inequalities in which "super-technologies of the rich" (Gerlach & Franceys, 2010) contrast starkly with the low-cost technologies servicing low-income households (Schwartz et al., 2017).

Organization

In terms of the organization for providing water services, the past decades have seen three significant developments, which are discussed below. These developments relate to the role of the private sector in providing services, the principles according to which water operators should operate and the incorporation of intermediaries in the provision of services.

Public versus private water operators

Although the private sector historically played an important role in the water services sector of many countries,[2] until the 1980s its role as a direct water provider was limited. In the 1980s, however, public water utilities in developing countries were frequently classified as poorly performing organizations. Spronk describes this dismal performance as "the three lows [...] low rates of cost-recovery, low productivity, and ultimately low service quality and coverage" (2010: 157). As a result, public utility reforms in the 1990s, saw donors and development banks advocating for more involvement of the private sector in providing water services. The establishment of a concession contract in Buenos Aires in 1993, the first of a series of large public-private partnership contracts, signalled the start of the "privatization decade," during which the involvement of the private sector in water service provision was strongly promoted by donors, lending agencies, and some academics. The privatization decade sparked off an intense debate on how to best organize water services provision, focusing on the incentive structure would make water operators perform well. The privatization decade was also a truly global phenomenon leading to heated debates in high-income countries (e.g., the Netherlands, Italy, and the United States) as well as low- and middle-income countries on how the water sector was to be organized.

During this decade private sector organizations were promoted as service providers, believing that these would realize efficiency gains (Donahue, 1989), increase access to technical and managerial expertise (Idelovitch & Ringskog, 1995), remove politics from the water services sector (Brown, 2002), and be a source of investment capital (Brown, 2002; Marin, 2009). The idea was that financial autonomy would accompany private sector participation as private water utilities were expected to operate on the basis of full-cost recovery. This, in turn, would make it possible to reduce or eliminate government subsidies for water services provisioning.[3]

Whereas in the 1990s academic and policy discussions in the water services sector, were strongly characterized by this public versus private dichotomy, the 2000s witnessed a change in momentum with respect to private sector participation. Large scale privatization projects collapsed or faced operational challenges. The most well-known of these collapses is the Cochabamba water concession, which is frequently referred to as the Cochabamba Water Wars. But also, in countries like Italy, citizens voted in a referendum against privatization of water services in 2010. Moreover, the increasingly available evidence on the perceived benefits of privatization, such as cost-savings and efficiency gains, was quite ambiguous at best (Bel et al., 2010).

Private organizations versus private principles

Whilst the end of the privatization decade[4] may have ended large-scale private sector involvement in the water services sector, the decade did not signal a return to business as usual. The privatization decade did not lead to the replacement of public utilities with private water operators, but it did have a strong impact on how public water utilities were to operate. Increasingly, public water utilities began to incorporate practices and principles that previously were associated with the private sector. Public water utilities were to operate at arms-length of the government, beyond the immediate control of politicians. Moreover, water utilities were to operate on the basis of cost-recovery, meaning they had to recover all service provision costs through tariffs. In other words, public water utilities needed to operate as if they were private water operators as it was argued that "[w]ell-run public utilities of the developing world have much in common with efficient private providers" (Marin, 2009: 147). What utilities needed was the introduction of private sector "discipline" to improve their performance (Furlong, 2015). This form of "commercialization" was considered more favourable and less controversial than models involving private sector organizations. Public water utilities, like the National Water and Sewerage Corporation in Uganda, increasingly started to profile themselves as "a successful example of a commercial public utility that combines public sector control with private sector efficiency" (Muhairwe, 2009).

The re-appraisal of alternative operators

As mentioned in the section on infrastructure, service differentiation has gained prominence in the past decades. This shift from the Modern Infrastructural Ideal to service differentiation has also led to organizational changes. Under the Modern Infrastructural Ideal, water services were to be provided by a single water utility that would be responsible for the entire water supply chain from the abstraction of water resources, to the production of potable water, to distribution to consumers. However, with the acceptance of service differentiation, also the principle of a single service provider fades. Increasingly, alternative providers are recognized as acceptable alternatives for the provision of services. Perhaps most illustrative of this development is the change in terminology when speaking of small-scale water providers that deliver water where the formal utility is unable to do so. Whereas previously these providers were often referred to as "informal water providers," increasingly they have been referred to as "small-scale water entrepreneurs" or "local water entrepreneurs" or providers whose "entrepreneurship" should be harnessed (Ahlers et al., 2013).

Finance

As highlighted above, water utilities are expected to operate as if they were private companies, meaning on the basis of commercial principles. For water utility finance this commercialization has two important consequences. First, most water utilities are to charge tariffs which allows for the recovery of the costs of service provision. Second, most water utilities try to minimize risks and costs of providing services. These two issues will be elaborated upon below.

Cost-recovery

The controversial Camdessus Report states that "the dream of pure water for all is within the reach of humanity" (2003: v) but then argues that the inability to implement cost-recovery is the main reason why water utilities have been unable to achieve universal service coverage (Rusca & Schwartz, 2018). The mantra of cost-recovery has become so prominent in the water services sector, that few dare to dispute the principle and many development actors now see subsidies as "an anathema" and an unsurmountable obstacle to financial sustainability of water utilities (Mitlin, 2008: 38). In arguing for cost-recovery, it is frequently stated that charging tariffs that fail to recover costs undermine the financial viability of the water utility. This in turn leads the utility to provide poor services, which ultimately lead to increasing inequalities (Kessides, 2004). The inability to set tariffs at levels of cost-recovery, often under the guise of achieving objectives of service expansion, may then actually constrain the water utility's ability to expand service to low income areas (Franceys & Gerlach, 2008). Moreover, cost recovering tariffs are viewed as a necessity for ensuring a utility's ability to repay loans for capital investments (Baietti et al., 2006). Without such cost-recovering tariffs the utility is not considered to be "bankable," meaning that it is unable to access loans from banks.

Although most water utilities are subject to regulations that dictate they must operate on the basis of cost-recovery, few, if any, water utilities actually are able to set their own tariffs (Hadzovich et al., 2021). In most cases, a government entity or regulator needs to approve tariffs. Most often, whilst referring to the affordability of water services for low-income households, these regulators do not allow the water utility to charge cost-recovering tariffs. The question of affordability is not something limited to developing countries. Teodoro and Saywitch (2020) have shown that low-income households in the United States, on average, spend close to 10% of their disposable income on water services.

Minimizing risks and costs

Without the ability to charge tariffs that recover the costs of service provision, water utilities need to engage in a strategy of reducing commercial risks and costs in order to achieve financial sustainability.

One strategy has already been referred to above and concerns the use of alternative providers to take over part of water supply provision.

For water utilities, extending water services to unserved, and often low-income, areas is a challenging undertaking, particularly given the need of most public water utilities to operate on the basis of commercial principles. Mangers in public water utilities that must operate on commercial principles tend to not view "low-income areas as a 'business opportunity', but rather a burden or a risk" (Heymans et al., 2014: 3). Although the challenge of servicing unserved areas often focuses on low-income countries, academics increasingly study unserved areas in high-income countries. Referring to the lack of water services as "plumbing poverty," Deitz and Meehan (2019) for example, highlight geographic and racial inequalities in access to water services in the United States.

In managing the commercial risks of servicing low-income areas, water utilities increasingly engage alternative providers to whom they transfer the commercial risks of operating in low-income areas. These organizations concern small-scale private operators, community-based organizations or landlords that undertake the last stage of the water supply provision chain by selling water to consumers. The public water utility limits its role to the supply of bulk water to these organizations.

A second strategy that a water utility can employ concerns reducing the costs associated with service provision. In this strategy the water utility decides to forego activities that would normally be considered to be part of the provision of water services, like delaying maintenance work or limiting treatment of water. Such measures to reduce treatment costs and economize on maintenance do not necessarily conflict with the concept of commercialization and could, in fact, be seen as increasing the efficiency of operations. However, it entails that cost recovery is achieved by reducing redundancies or trimming the fat from the organization (as the commercial principles of the original model would have it) and then charging tariffs to recover these costs. Cost-recovery is then also pursued by transferring or avoiding the normal, legitimate costs of any water provider (Tutusaus & Schwartz, 2020). This in turn can have significant consequences for the provision of water services. Delays in network maintenance, for example, may contribute to the running down of the infrastructure and to compromising the level of service offered in the medium to long-term.

The price of differentiation

The original idea behind service differentiation is that lower levels of service are to be charged lower tariffs (van Ryneveld et al., 2003). In doing so, the level of water services provided better match the demands and abilities of the consumer groups (Mara & Alabaster, 2008). In practice, however, water obtained through appropriate technologies is often of lesser quality, and comes at a price that is higher per cubic meter than water supplied through technologies enjoyed by middle- and high-income households (Hadzovich et al., 2021). What differentiation mainly appears to achieve is to keep water consumption levels of consumers low. Moreover, the proliferation of intermediaries, such as water user associations, landlords, or small-scale private providers, has made low-incomer consumers vulnerable to exploitation by these providers. For these intermediary providers, which are hardly regulated, the provision of water services represents a lucrative business. Based on research in three Kenyan cities, Boakye-Ansah et al. (2021) highlight that water sold by intermediaries is usually between three and seven times more expensive than water sold by the water utility. In extreme cases, the price of water charged by intermediaries is even increased 28-fold. As water is a basic necessity, consumers have little option but to pay the hefty prices charged by such intermediaries.

Conclusion: The Path to Differentiation

The service modalities through which water services are currently provided are largely informed by the dominant governance discourses to which water utilities must adhere. On the one hand, they are the main instrument of the government to achieve the human right to water and to ensure universal service coverage by 2030 as dictated by the SDG target 6.1. On the other hand, they are to operate on the basis of commercial principles, at arms-length from the government whilst recovering the cost of service provision. At the same time, the institutional context in which these utilities operate does not allow the water provider to charge tariffs which would allow for commercial viability nor does it provide adequate subsidies to extend services to all. This dual mandate of achieving commercial and social objectives, which stem from the two underlying discourses, are referred to as the mixed mandate of water utilities.

Confronted with the challenges of achieving this mixed mandate, water utilities have opted for differentiation, in which consumer groups are distinguished and supplied with distinct levels of service. This differentiation is visible in all the core elements of a water services modality. The "super-technologies of the rich" (Gerlach & Franceys, 2010) contrast with the "appropriate technologies" that supply low-income households. The water utility that supplies inhouse connection in middle- and high-income areas, contrasts with the alternative providers that operate kiosks and standpipes in slums or that ignores areas that are difficult to service. Moreover, the prices charged for these services differ depending on the technology and organization supplying the service. Contrary to the original idea of service differentiation, however, the differentiation seems to particularly benefit the middle and high-income households connected through inhouse connections serviced by the main water utility. They receive higher levels of service at lower prices.

The water utilities implementing differentiation find that it is the only workable way in which they can address the mixed mandate resulting from the prevailing governance discourses. In other words, the simultaneous co-existence of two distinct and seemingly conflicting discourses pushes water utilities to adopt a strategy of differentiation, visible in the service modality. In doing so, however, these utilities and the institutional environment in which they operate, also have to accept that inequalities in the levels of water service provision have become mainstream. These inequalities, be they "plumbing poverty" in the United States or the consumer exploitation by landlords in Kenya, are likely to be sustained for a considerable time. As long as the two discourses shaping the service modalities characterizing service provision remain co-existent and dominant in the water services sector, there is little opportunity for other approaches to develop.

Notes

1 The arrangements of infrastructure, organization and finance unfold in a specific context. Availability and quality of raw water resources, the topography, socio-demographic developments, legal and administrative frameworks and culture, and "civil-public" interactions all co-shape the service modalities for providing water (Mollinga 2008). Although we acknowledge the relevance of the context in shaping service modalities this chapter mainly focuses on the link between governance discourses and service modalities.
2 In many countries the initial development of the water services sector in the 19th century was a private sector initiative catering to high-income households that were willing to pay for better quality water.
3 During this decade, a large number of countries adopted new water policies and laws to allow for the involvement of the private sector in providing water services. These include Senegal (1995), Uganda (1998–2001), Mozambique (1999), Indonesia (2004), Mexico (1992), Namibia (2000), Zambia (1994–2001), Kenya (2002), Tanzania (2002), and South Africa (1998).
4 The end of the privatization decade does not mean a complete end to promoting private sector involvement. According to a senior staff member of the World Bank: "[t]he story on private sector participation is not over, it's simply being rewritten. When the emotions over this issue subside over the next few years, there will be a robust return to private sector participation that involves subtler forms of risk sharing" (Source Weekly, 2007, as cited in Ahlers et al., 2013: 176).

References

Ahlers R., Schwartz K. and Perez Guida, V. (2013). "The myth of 'healthy' competition in the water sector: The case of small scale water providers." *Habitat International*. 38, 175–182.

Baietti, A., Kingdom, W. and van Ginneken, M. (2006). *Characteristics of well-performing public water utilities*. Washington, DC: World Bank.

Bel, G., Fageda, X. and Warner, M. (2010). "Is private production of public services cheaper than public production? A meta-regression analysis of solid waste and water services." *Journal of Policy Analysis and Management*. 29(3), 553–577.

Berg, S. and Mugisha, S. (2010). "Pro-poor water service strategies in developing countries: Promoting justice in Uganda's urban project." *Water Policy*. 12(4), 589–601.

Boakye-Ansah, A., Schwartz, K. and Zwarteveen, M. (2021). Service differentiation as an improvement strategy for access to water in low-income areas: Evidence from Kenya." *International Journal of Water Resources Development*. 38(5), 766–782.

Brown, A. (2002). "Confusing means and ends: Framework of restructuring, not privatization, matters most." *International Journal of Regulation and Governance*. 1(2), 115–128.

Camdessus Report. (2003). *Financing water for all*. Marseille: World Water Council.

Deitz, S. and Meehan, K. (2019). "Plumbing poverty: Mapping hot spots of racial and geographic inequality in U.S. household water insecurity." *Annals of the American Association of Geographers*. 109(4), 1092–1109.

Donahue, J. (1989). *The privatization decision: Public ends, private means*. New York, NY: Basic Books.

Franceys, R. and Gerlach, E. (2008). *Regulating water and sanitation for the poor: Economic regulation for public and private partnerships*. London: Earthscan.

Furlong, K. (2015). "Water and the entrepreneurial city: The territorial expansion of public utility companies from Colombia and the Netherlands." *Geoforum*. 58, 195–207.

Gerlach, E. and Franceys, R. (2010). "Regulating water services for all in developing countries." *World Development*. 38(9), 1229–1240.

Hadzovich, L., Alda-Vidal, C., Rusca, M. and Schwartz, K. (2021). "Everyday practices and the production of uneven waterscapes: Water pricing regimes in Lilongwe, Malawi." *Environment and Planning C*. 39(2), 300–317.

Heymans, C., Eales, K. and Franceys, R. (2014). *The limits and possibilities of prepaid water in urban Africa: Lessons from the field*. [Accessed 14 May 2016]. Available from: http://wwwwds.worldbank.org/external/default/WDSContentServer/WDSP/IB/2014/08/26/000470435_20140826140705/Rendered/PDF/901590REPLACEM0Prepaid0Water0Africa.pdf

Idelovitch, E. and Ringskog, K. (1995). *Private sector participation in water supply and sanitation in Latin America*. Washington, DC: World Bank.

Jaglin, S. (2008). "Differentiating networked services in Cape Town: Echoes of splintering urbanism?" *Geoforum*. 39(6), 1897–1906.

Jessop, B. (1998). "The rise of governance and the risks of failure: The case of economic development." *International Social Science Journal*. 50(155), 29–45.

Kessides, I. (2004). *Reforming infrastructure: Privatization, regulation and competition*. Washington, DC: World Bank.

Mara, D. and Alabaster, G. (2008). "A new paradigm for low-cost urban water supplies and sanitation in developing countries." *Water Policy*. 10, 119–129.

Marin, P. (2009). *Public-private partnerships for urban water utilities: A review of experiences in developing countries*. Washington, DC: World Bank.

Mitlin, D. (2008). "With and beyond the State: Co-production as a route to political influence, power and transformation for grassroots organizations." *Environment and Urbanization*. 20(2), 339–360.

Mollinga, P. (2008). "Water, politics and development: Framing a political sociology of water resources management." *Water Alternatives* [Online]. 1(1), 7–23. [Accessed 2 February 2021]. Available from: https://www.water-alternatives.org

Muhairwe, W. (2009). *Making public enterprises work: From despair to promise: A turn around account*. London: International Water Association.

Rusca, M. and Schwartz, K. (2018). "The paradox of cost recovery in heterogeneous municipal water supply systems: Ensuring inclusiveness or exacerbating inequalities." *Habitat International*. 73, 101–108.

Schwartz, K., Tutusaus, M. and Savelli, E. (2017). "Water for the urban poor: Balancing financial and social objectives through service differentiation in the Kenyan water sector." *Utilities Policy*. 48, 22–31.

Spronk, S. (2010). "Water and sanitation utilities in the global South: Re-centering the debate on efficiency." *Review of Radical Political Economics*. 42(2), 156–174.

Teodoro, M. and Saywitch, R. (2020). "Water and sewer affordability in the United States: A 2019 update." *AWWA Water Science*. 2(2), 1–12.

Tutusaus, M. (2019). *Compliance or defiance?: Assessing the implementation of policy prescriptions for commercialization by water operators*. Leiden: CRC Press.

Tutusaus, M. and Schwartz, K. (2020). "Commercialisation as organised hypocrisy: The divergence of talk and action in water services in small towns in Uganda." *Water Alternatives* [Online]. 13(2), 248–265. [Accessed 16 March 2021]. Available from: https://www.water-alternatives.org

van Ryneveld, P., Parnell, S. and Muller, D. (2003). *Indigent policy: Including the poor in the city of Cape Town's income strategy*. Cape Town: CCT.

World Health Organization. (1991). *The International Drinking Water Supply and Sanitation Decade: End of decade review (as at December 1990)*. [Accessed 28 January 2021]. Available from: https://apps.who.int/iris/handle/10665/61775

Zwarteveen, M., Kemerink-Seyoum, J., Kooy, M., Evers, J., Acevedo Guerrero, T., Batubara, B., Biza, A., Boakye-Ansah, A., Faber, S., Cabrera Flamini, A., Cuadrado-Quesada, G., Fantini, E., Gupta, J., Hasan, S., ter Horst, R., Jamali, H., Jaspers, F., Obani, P., Schwartz, K., Shubber, Z., Smit, H., Torio, P., Tutusaus, M. and Wesselink, A. (2017). "Engaging with the politics of water governance." *WIREs Water* [Online]. 4(6), 1–9. [Accessed 8 March 2021]. Available from: https://pure.uva.nl/ws/files/29325943/Zwarteveen_et_al_2017_Wiley_Interdisciplinary_Reviews_3A_Water.pdf

14

WATER, NEOLIBERALISM, AND COMMODIFICATION

Jessica Budds and Alex Loftus

Introduction

From the 1980s, the involvement of the private sector and the use of market mechanisms were more strongly advocated as a means to increase efficiency and access in relation to water management (tools to measure and control water), and effective governance (decision-making structures and processes for water). While these approaches are often assumed to be relatively recent, they have been applied to water resources at least since the nineteenth century (Budds and McGranahan, 2003). What perhaps distinguished this new wave of "neoliberal" approaches, such as concession contracts for drinking water utilities and private property rights to raw water, was the rendering of water as a site of capital accumulation, facilitated by new institutional and governance arrangements (Harvey, 2003). This implies that water was no longer only considered as a necessary input into human and economic development, which could be managed by pricing among other approaches, but also as a site of investment in itself. Put more simply, while water has long flowed to capital, under neoliberal approaches, capital has also flowed to water. This, in turn, profoundly reshaped the principles of decision-making for water, as well as the type of actors who could participate in the process, resulting in a vigorous debate over the relationship between economic objectives and the public good. In this chapter, we explore this shift by examining how forms of neoliberalisation have come to sustain, and become sustained by, capital accumulation, and what this means for the social relations of control over water in places where these processes are underway. This entails considering governance shifts under neoliberalisation as not only processes to make decisions (a practice of social regulation), but also as means to pursue wider objectives through the reworking of social relations (a process of regulation-making) (Bridge and Perreault, 2006).

Following this introduction, we review the literature on the ways in which "neoliberal" principles have been applied to water, their underlying arguments and critiques, and the key outcomes in practice. In the following sections, we develop our focus on how water has been neoliberalised so as to secure capital accumulation and uphold dominant power relations through two specific forms: financialisation (a shift of core business from the production and sale of water towards strategies to profit from secondary (financial) markets) and marketisation (the transfer of water entitlements to promote efficient water use and to incentivise private investment in infrastructure

DOI: 10.4324/9781003095545-17

and industry), each supported by an empirical example. In our concluding section we argue that, together, these two processes of neoliberalising water demonstrate that what is at stake is no longer simply the application of economic and market principles to water directly as tools for effective and efficient management, as represented by the early literature. Rather, they have come to encompass broader strategies of the restructuring of the water sector that foster new opportunities for capital accumulation while reconfiguring water–society relations in ways that further consolidate the economic and political power that underpins these.

Neoliberalising Water

From the 1990s, a rich body of literature emerged that critically examined a variety of economic and market reforms in the water sector, from the engagement of private operators in water utilities to the growth of the bottled water industry, and from the application of property rights to water resources to the use of payments as incentives for ecosystem conservation. These related strategies that simultaneously increased the role of economic and market logic and reduced the role of the state in resource management were collectively defined as processes of neoliberalisation (Heynen and Robbins, 2005; McCarthy and Prudham, 2004). Heynen and Robbins (2005) define four key processes, comprising privatisation (the transfer of resources to private parties), enclosure (the capture of resources through the exclusion of customary users), valuation (the creation of commodities through pricing), and governance (the institutionalisation of these strategies in legal and policy frameworks). McCarthy and Prudham (2004) unpack the latter by identifying a series of institutional changes that reduce and/or delegate state responsibilities to non-state parties, including through privatisation, decentralisation, budget cuts, and the transition from binding regulation to self-regulation.

Bakker (2005, 2007) sets out three broad areas of neoliberal reform in the water sector, which she emphasises are not necessarily commensurate, concurrent, nor co-dependent: institutions, organisations, and governance. First, institutions comprise the "laws, policies, rules, norms and customs" that are used to govern water (Bakker, 2007: 434). These include instruments such as property rights and regulatory frameworks, as well as the institutional changes that introduce economic and market principles to guide water management (also referred to as commercialisation). Bakker (2007) defines commodification as the conversion of water from a previously non-commercial resource to an "economic good" that can be priced and sold through the market, and which comprises both material (e.g. enclosure) and symbolic (e.g. framing) practices. It is important to note that commodification does not require privatisation, since drinking water, in particular, has typically always been priced in some form. Yet, the uptake of neoliberal principles has often meant that pricing has been restructured to reflect market value, through full cost recovery (whereby users pay the full cost of provision) for example. However, commodification is also usually a prerequisite for privatisation, which is predicated upon commercial pricing (often in line with recovery).

Second, resource management organisations are the "collective social entities" through which water is governed (Bakker, 2007: 434). These include modes of asset management and organisational structure. This covers privatisation, which comprises a change of ownership of water resources (i.e. private property rights) or water infrastructure (e.g. asset sell-off), as well as corporatisation, which refers to the adoption of economic and market principles (commercialisation) but without a change in ownership. The model of corporatisation has been adopted by some public water utilities, which remain publicly owned and run, yet function in many respects on a commercial basis (Loftus and Budds, 2016).

Third, resource management governance is the "process by which organizations enact management institutions" to administer resource extraction, allocation and use (Bakker, 2007: 434). This includes the use of market mechanisms such as trading and auctions to regulate resource (re)allocation (marketisation), as well as forms of regulatory change such as deregulation, decentralisation, and the individualisation of responsibility for access to water.

Much early literature on the neoliberalisation of water focused on the privatisation of water supply (and sewerage) utilities, especially in the Global South, which had increased as a result of both the imposition of structural adjustment programmes during the 1980s and 1990s, as well as the search for new markets among private water companies from the Global North (Schulpen and Gibbon, 2002). This literature explored, both theoretically and empirically, the merit of the arguments that supported the engagement of the private sector, and in particular the use of corporate approaches to increase the coverage of infrastructure and quality of services. Here, some authors argued that water privatisation was simultaneously justified by the supposed effectiveness of the private sector on the one hand, and the presumed "failure" of the public sector on the other (Bakker, 2003a). These framings not only ignored examples of effective public water management, but also the economic inefficiency of private provision through higher costs, including to the public sector (Bakker, 2010). The links made between these assertions and the definition of water as a specific type of "good" in the language of institutional economics was also problematised, on the basis that water did not fit neatly into the categories – public good or private good – to which it was typically attributed. The wide range of institutional and contractual arrangements and responsibilities, especially commercialisation, further reinforced this misleading binary between "public" and "private" (Bakker, 2007).

Empirical studies emphasised the deep tension between commodifying the supply, circulation, and identity of water, and ensuring the (human) right to water, especially among those who had hitherto been excluded from arguably the most basic of human needs (Loftus and Budds, 2016). A number of studies thus focused on conditions of access to commercial water services, revealing the diverse ways in which private providers sought to protect their revenue by transferring costs, risks and responsibilities to the state and/or consumers (Budds and McGranahan, 2003). Practices documented included full cost recovery of new connections and the installation of pre-payment meters, delegating water distribution and billing in low-income settlements to community-based organisations, reneging on provision to informal settlements in contravention of contractual obligations, expectations of free household labour for installation and maintenance of infrastructure, and the transfer of responsibility for domestic water quality to households through apparatuses like filters (see e.g. McDonald and Pape, 2002).

In parallel, scholars also attended to the processes by which water sources managed as common pool resources were detached from local users and customary rights, and subjected to new institutional and contractual arrangements to enable their enclosure by the private sector (Harvey, 2003). A notable example is that of Cochabamba, Bolivia, where legislative reforms overrode existing customary rights to water sources that had been used for community drinking water networks in unserved urban neighbourhoods, as well as irrigation for small-scale agriculture in peri-urban and rural areas, in order to grant control over water resources to the future private concessionaire of the city's water utility (Assies, 2003). The dispossession of these water sources, alongside the deeply held cultural Andean notion that water should not be owned or traded, led to large-scale protests that eventually led to the cancellation of the private contract and the reform of the legislation. This illustrates Bakker's (2003b) argument that water's particular material properties and symbolic meanings rendered it resistant to processes of privatisation and marketisation, through her notion of an "uncooperative commodity."

It was in part due to these contradictions that some forms of neoliberalisation diminished, such as concession contracts for urban utilities like the one in Cochabamba, while others changed in both form and representation. Private sector involvement and market principles continued to be pursued, with ongoing emphasis on full cost recovery and the protection of capital investments and revenue streams, yet the terminology changed to concepts such as "partnership," and justifications emphasised social and environmental objectives, such as conservation (Loftus and Budds, 2016). This continued pursuit of neoliberal approaches in the face of economic, social, and environmental challenges suggests that they transcend the imperative to manage resources efficiently, effectively, and sustainably through evidence-based approaches. Rather, it signals that they form a means of leveraging new modes of capital accumulation, which are more detached from water as a resource. This, in turn, indicates a reciprocal dynamic, whereby water becomes neoliberalised in specific ways as a result of its exposure to capital, and that capital in turn becomes dependent upon the continuation of these specific forms of neoliberalisation. We will now outline two specific forms of this, illustrated by examples.

Financialisation

While we situate financialisation as one aspect of neoliberalisation, for Christophers (2015) "financialization" has become the buzzword of the 2010s. By a buzzword, Christophers points to the way in which a term is widely used even if its meaning is not always clear: indeed, the term means many different things to different people. Common definitions of financialisation refer to: the empowerment of the financial sector; the increasing importance of new actors such as sovereign wealth funds and institutional investors; and a shift in the locus of profit-making from the "real" economy to the "financial" economy. Lapavitsas (2013) captures the latter definition in his evocative description of financialisation as a process of "profiting without producing." In the case of water, financialisation by Lapavitsas' definition would mean profiting from secondary (financial) markets rather than the production and sale of water itself. In Christophers' view, the very range of these definitions – and the fact they often identify processes in existence well before the so-called dawn of an era of "financialisation" – diminish the term's utility.

A more helpful interpretation is given by Bracking (2012, 2016) who refers to "a technology of power which uses the money form to quantify human and physical contexts, privilege financial parameters in decision-making, and ultimately return decisions over resources in favour of money-holders in contexts of contested uses and meanings for the non-human world" (2012: 274). Bracking's interpretation includes direct reference to the non-human world and her definition of financialisation "as process" fits well with our own understanding of neoliberalisation. Bracking (2016) goes on to identify one particularly salient aspect to this new relationship – the creation of a fixed asset from which derivative income streams can be drawn.

Developing a Marxist approach to interpreting the growth of derivative income streams, Fine (2013) claims that financialisation refers to the growing importance of Interest Bearing Capital (IBC). IBC is money lent to generate future profits. Given that returns on IBC are utterly reliant on a future expansion in profits, there is an inherent riskiness and a fictitious element to the capital that circulates before future profits are realised. Any failure to generate expected returns will quickly identify the fiction behind the expansion in IBC. Part of the value in Fine's analysis is his ability to show how financialisation remains deeply implicated in the continued generation – or at least the expectation of a continued generation – of surplus value. Rather than "profiting without producing," financialisation is intimately tied to the generation of revenue streams through increasing surplus value. For Purcell et al. (2020), what both Fine and Bracking are describing – as "derivative income streams" or as the expansion of IBC – is the expansion of rents. Financialisation should therefore be understood as a process through which fictitious capital – circulating because of an increase in IBC – now actively seeks to generate new rents.

These seemingly abstract definitions of financialisation have a profound bearing on how one understands the relationship between water and neoliberalism. In very simple terms, the different definitions offer explanations as to why financial actors might invest in water and how they seek to generate returns on their investments. To shed further light, it is worth exploring the case of Thames Water, the utility providing water and sewerage services for London and much of the Thames river basin. Thames Water came into existence with the privatisation of water services in England and Wales in 1989. The UK government's sale of its golden share in each of the newly privatised water utilities in 1994 led to a reconfiguration of ownership structures. In the case of Thames Water, the utility was acquired by the German utility RWE in 2001 and then the Macquarie group – an Australian investment group – in 2006. Macquarie's acquisition of Thames Water is generally understood to have initiated a process of financialisation.

One of the Macquarie group's first moves was to change the corporate structure of the organisation into one that Allen and Pryke (2013: 423) suggest resembles a "wedding cake" structure. At the base of the structure is Thames Water Utilities Cayman Finance Ltd, which was able to issue bonds from the offshore tax haven of the Cayman Islands against future revenue streams generated within the utility itself (Thames Water Ltd). The issuance of bonds against future revenue streams can be defined as a process of securitisation whereby the income generated from the future payment of household bills was transformed into a financial commodity. The capture of these derivative income streams is the process identified in Bracking's definition of financialisation. As noted above, Purcell et al. (2020) view these derivative income streams as forms of rent and they distinguish between two different types of rent generated within Thames Water's corporate structure. The first they consider are referred to as monopoly rents. As noted above (see Bakker, 2005) water is often considered a natural monopoly – it is bulky and costly to transport, and it makes far more economic sense for one utility to distribute the resource within a given geographic space. Indeed, most consumers will only be able to buy piped water from one supplier. Such a status ensures that a private company can charge monopoly rents derived from the higher prices associated with water's material and geographic specificity. Such monopoly rents are mitigated somewhat through the involvement of a consumer regulator – OFWAT – that applies a system of price cap regulation to be reviewed every five years. Nevertheless, the entry of the Macquarie group resulted in the pursuit of new forms of rent – understood as "absolute rents" – and resulting from the erection of barriers to other investors, barriers that can only be circumvented through extracting a higher price.

The concrete outcomes of such a process have been several. First, as Allen and Pryke (2013) have shown, OFWAT is only able to regulate a very small part of the "wedding cake" structure of which Thames Water is comprised. The utility therefore evades accountability. Second, as Bayliss and Hall (2017) have shown, the securitisation of revenue streams has led to massive gearing. The payment of large dividends to institutional investors – the sovereign wealth funds of Abu Dhabi, China, and Kuwait, along with the Universities Superannuation Scheme and several others – have only exacerbated the indebtedness of the utility. Third, the price review system implemented by OFWAT has been skewed by the need for fictitious capital to pursue new forms of rent (Purcell et al.'s interpretation of financialisation). Price reviews now actively encourage an expansion of the terrain over which rents can be generated. The construction of the Thames Water Desalination Plant and the Thames Tideway Tunnel (even if the latter has been constructed using a Special Purpose Vehicle formally separate from Thames Water) enable new revenue streams to be generated from consumer bills (Loftus and March, 2016). These future revenue streams, when securitised, form the basis for new absolute rents.

Marketisation

The privatised water utilities described above also depend on secure sources of raw water, which have also increasingly been subjected to commercial forces, often as part of a package of neoliberal reforms as in Cochabamba. While bulk water resources have been historically managed by public authorities in most contexts, neoliberal principles have been applied to promote efficient management on the one hand, but also to foster new modes of capital accumulation on the other. One key way in which this has been implemented is through the creation of markets in water rights, which refer to the right of an individual or entity to extract a flow of water from a defined source under prescribed conditions.

The creation of a water market takes place when water rights are commodified (i.e. subject to pricing) and marketised (i.e. tradable between users). In some contexts, water rights have also been privatised (i.e. instituted as private property). While privatisation is in itself not a prerequisite for the creation of a water market, the water right must be clearly defined, so that what is being exchanged is clear. Water markets should be regulated, but, similarly, possible institutional arrangements include state or autonomous regulation, or self-regulation by users.

Water markets are primarily envisaged as tools of demand management, whereby the value attributed to water through commodification incentivises its allocation to the highest (economic) value use. In this way, water rights holders who may be using water for a low-value use (e.g. irrigating a crop for the local market), or who do not need water at all, are encouraged to transfer their water right to another user who is willing to pay for additional supply. Such transactions can either be temporary (i.e. leasing, or "spot market") or permanent (i.e. sale) where property rights are in place that enable this. Water markets are particularly applicable to high-volume water uses, especially irrigation, where it is not feasible to apply unit-based pricing and demand is instead managed by controlling access. Therefore, (re)allocation becomes determined by market forces, rather than (state) regulation. The direct transfers of water between users that markets enable are also often argued to be more efficient (i.e. less bureaucratic) than state allocation, although they can raise the need for regulation (e.g. where water could potentially be transferred from high-priority yet lower-value uses such as drinking water to lower-priority yet higher-value uses such as mining).

Chile is widely regarded to have established the most deregulated model of water markets worldwide. The 1981 Water Code was drawn up under the military dictatorship (1973–1990), by technocrats who were influenced by free market economics and political liberalism (Bauer, 1997). The model instituted water rights as private property, which were separated from land so that they could be freely traded between users with very few restrictions. The role of the state water authority was curbed to administering the system with little right of intervention, and regulation was delegated to water user organisations and the judiciary. The stated aims of the Water Code were to promote user investment in water infrastructure and economic activity through permanent property rights, to enable the transfer of water to the highest economic value uses through water trading, and to curb "politicised" decision-making by the state through the decentralisation of governance to water users. The Water Code thus transferred Chile's water to private parties on an indefinite basis, and with few obligations and no taxation, which was tremendously advantageous to the large-scale economic sectors – including commercial agriculture, mining, and hydroelectric power – that were promoted as the basis of Chile's economy during the dictatorship.

Authors examining the nature and outcomes of the Water Code have found that trading was limited for reasons including geography, culture, and access to information (Bauer, 1997). Budds (2004) asserted that this led to the over-exploitation of water sources as users turned to new sources – especially groundwater – rather than purchase water from others, which was often done illegally, an option facilitated by the absence of external regulation. Others argued that both the recognition and trading of water rights potentially fostered the dispossession of water from lower-income groups, especially indigenous communities (Prieto, 2015). Important structural limitations were also identified, such as the inability to institute integrated river basin management, and the confinement of all decisions about water management and use exclusively to water rights holders with almost no right of participation of public agencies, including the water authority, or other stakeholders (Bauer, 2015).

The water markets model can be seen as not merely a system of (re)allocating water between different uses and users, but also a means through which wider economic and political agendas are pursued. On the one hand, this is manifest through the vigorous defence of the Water Code against potential legislative reform by those who benefit from it. These actors include large-scale export-oriented natural resource industries and right-wing politicians, many of whom supported the neo-liberal model that fostered these industries and/or have stakes in them (Budds, 2013). These actors have lobbied to water down proposed reforms several times (1992–2018). On the other hand, it is demonstrated by the abandonment of the core principles of the model by those same actors when they find it convenient to do so. This included responses to ongoing water scarcity since 2010 (the "mega-drought"), which has been widely framed as an effect of climate change, but has been arguably reinforced by the over-exploitation of water fostered by the deregulated Water Code. Budds (2020) argued that the principal responses to a severe drought completely disregarded the institutionalised market mechanisms that were supposedly best-suited to manage water scarcity, and instead pursued supply led infrastructural solutions (e.g. reservoirs). These measures not only served to protect existing investments that were affected by drought, such as commercial fruit plantations, but also to stave off future reforms to the Water Code in the face of its potential failure. The economic and political power relations that were supported through the creation of the water markets model have thus in turn become dependent upon its protection.

These dynamics are further laid bare by proposals to build desalination plants – another supply led solution – as a means to protect water security against drought. Similar to London, the main business of Chile's regional water utilities (which closely resembled the UK's former regional water authorities, some of which were also privatised in a very similar way in the 1990s) has typically been assumed to comprise water supply. However, the introduction of desalination has enabled these utilities to scale up their engagement in water trading across sectors. In a region of northern Chile dominated by copper mining, one regional utility sought to expand its business from supplying urban consumers to brokering water rights between the drinking water and the mining sectors (Fragkou and Budds, 2020). Desalination formed the conduit that enabled it to do this, through a "water swap" arrangement whereby it diverted water extracted in the inland mountains, which had previously supplied the coastal cities, to the nearby mines, and replaced it with desalinated water extracted in situ from the sea for urban consumption. While at first sight this may appear to be an efficient and effective solution, the transfer of urban water supply from mountain water to desalinated water carries several significant implications: households may stop using tap water if they consider desalinated water unpalatable, desalination is expensive and may imply future domestic tariff rises, and desalination is a complex industrial process that could experience technical failures (Fragkou and Budds, 2020). In this way, switching the urban population to desalinated water to free up their customary water source for rent/sale to mining creates new capital accumulation opportunities for the utility and its beneficiaries, but largely by transferring the costs and risks to urban consumers.

Conclusion

Early analyses of the neoliberalisation of water focused on the logic and implications of the application of economic and market principles to water management and governance, in particular highlighting the tensions between treating water as a commodity and enabling access. More recent accounts have given greater attention to how varied forms of neoliberalisation have rendered water a site of capital accumulation, in ways that sustain constellations of economic and socio-political power. As we have shown through financialisation and (some modes of) marketisation, these processes entail institutional reforms that enable water sector or industry (financial) restructuring, and governance reforms that change how and by whom water is allocated, used, and regulated. These reforms are invariably configured in the interests of actors who control water, and at the expense of users. In short, these strategies of neoliberalisation are predicated upon reworking the social relations of control over water to extract new forms of rent.

In this way, capital accumulation is not only underpinned by neoliberal principles such as ownership and pricing, but has become increasingly dependent on processes that reconfigure these relations of control and fix them in place. One aim of our examples is to elucidate how this has unfolded in two different contexts, to reveal the deep-rooted changes that these dynamics have entailed to hydrosocial relations. For instance, Thames Water's "wedding cake" structure served as a means to bypass regulation, in particular the commercially inconvenient price cap mechanism – whose very function was to protect consumers from potential exploitation by companies like Thames Water, which enabled the cost of large infrastructure to generate new revenue streams from higher household water bills for years to come. Similarly, residents of northern Chile's coastal cities were effectively dispossessed of their customary mountain water source when the regional water company substituted this with desalinated water in order to be able to sell it to the expanding mining industry. This was not only done without consultation or consent, or even knowledge in many cases, but also reduced residents' (material or perceived) quality of tap water and exposed them to higher costs and risks over the long term. These modes of neoliberalisation thus enhance the tension between the imperatives of capital accumulation and the wider public interest, raising the crucial questions of how, by whom, and to what ends water should be managed (Bridge and Perreault, 2006). The answers to those questions, in diverse contexts around the globe, will provide a clearer picture of the economic and political power relations that both support and also uphold the ongoing neoliberalisation of water, and enable these to be challenged.

References

Allen, J. and Pryke, M. (2013). "Financialising household water: Thames water, MEIF, and 'ring-fenced' politics." *Cambridge Journal of Regions, Economy and Society*. 6(3), 419–439.

Assies, W. (2003). "David versus goliath in Cochabamba: Water rights, neoliberalism, and the revival of social protest in Bolivia." *Latin American Perspectives*. 30(3), 14–36.

Bakker, K. (2003a). "A political ecology of water privatization." *Studies in Political Economy*. 70, 35–58.

Bakker, K. (2003b). *An uncooperative commodity: Privatizing water in England and Wales*. Oxford: Oxford University Press.

Bakker, K. (2005). "Neoliberalizing nature? Market environmentalism in water supply in England and Wales." *Annals of the Association of American Geographers*. 95(3), 542–65.

Bakker, K. (2007). "The 'commons. versus the 'commodity': Alter-globalization, anti-privatization and the human right to water in the global South." *Antipode*. 39(3), 430–55.

Bakker, K. (2010). *Privatizing water: Governance failure and the World's Urban Water Crisis*. Ithaca: Cornell University Press.

Bauer, C. (1997). "Bringing water markets down to earth: The political economy of water rights in Chile, 1976–95." *World Development*. 25(5), 639–656.

Bauer, C. (2015). "Water conflicts and entrenched governance problems in Chile's market model." *Water Alternatives*. 8(2), 147–172.

Bayliss, K. and Hall, D. (2017). "Bringing water into public ownership: costs and benefits." [Accessed 10 February 2023]. Available at: https://gala.gre.ac.uk/id/eprint/17277/10/17277%20HALL_Bringing_Water_into_Public_Ownership_(Rev'd)_2017.pdf

Bracking, S. (2012). "How do investors value environmental harm/care? Private equity funds, development finance institutions and the partial financialization of nature-based industries." *Development and Change*. 43(1), 271–293.

Bracking, S. (2016). *The financialisation of power: How financiers rule Africa*. Abingdon: Routledge.

Bridge, G. and Perreault, T. (2006). "Environmental governance." In: N. Castree, D. Demeritt, D. Liverman, B. Rhoads (eds.). *A companion to environmental geography*. London: Wiley.

Budds, J. (2004). "Power, nature and neoliberalism: The political ecology of water in Chile." *Singapore Journal of Tropical Geography*. 25(3), 322–342.

Budds, J. (2013). "Water, power, and the production of neoliberalism in Chile, 1973–2005." *Environment and Planning D: Society and Space*. 31, 301–318.

Budds, J. (2020). "Securing the market: water security and the internal contradictions of Chile's water code." *Geoforum*. 113, 165–175.

Budds, J. and McGranahan, G. (2003). "Are the debates on water privatization missing the point? Experiences from Africa, Asia and Latin America." *Environment and Urbanization*. 15(2), 87–113.

Christophers, B. (2015). "The limits to financialization." *Dialogues in Human Geography*. 5(2), 183–200.

Fine, B. (2013). "Financialization from a Marxist perspective." *International Journal of Political Economy*. 42(4), 47–66.

Fragkou, M. C. and Budds, J. (2020). "Desalination and the disarticulation of water resources: Stabilising the neoliberal model in Chile." *Transactions of the Institute of British Geographers*. 45(2), 448–463.

Harvey, D. (2003). *The new imperialism*. Oxford: Blackwell.

Heynen, N. and Robbins, P. (2005). "The neoliberalization of nature: Governance, privatization, enclosure and valuation." *Capitalism, Nature, Socialism*. 16(1), 1–4.

Lapavitsas, C. (2013). *Profiting without producing: How finance exploits us all*. London: Verso.

Loftus, A. and Budds, J. (2016). "Neoliberalizing water." In: S. Springer, K. Birch, J. MacLeavy (eds.). *The handbook of neoliberalism*. London: Routledge.

Loftus, A. and March, H. (2016). "Financializing desalination: Rethinking the returns of big infrastructure." *International Journal of Urban and Regional Research*. 40(1), 46–61.

McCarthy, J. and Prudham, S. (2004). "Neoliberal nature and the nature of neoliberalism." *Geoforum*. 35(3), 275–284.

McDonald, D. and Pape, J. (2002). *Cost recovery and the crisis of service delivery in South Africa*. London: Zed.

Prieto, M. (2015). "Privatizing water in the Chilean Andes: The case of Las Vegas de Chui-Chiu." *Mountain Research and Development*. 35(3), 220–229.

Purcell, T. F., Loftus, A. and March, H. (2020). "Value–rent–finance." *Progress in Human Geography*. 44(3), 437–456.

Schulpen, L. and Gibbon, P. (2002). "Private sector development: Policies, practices and problems." *World Development*. 30(1), 1–15.

15
THE HUMAN RIGHT TO WATER

Peter H. Gleick

Introduction

The history of human rights extends back to the earliest civilizations and reflects the cultural and moral perceptions and beliefs of a given era. From ancient Mesopotamia to early Greece to post-Renaissance Western civilization, basic rights have been defined first for specific groups, typically privileged free men, and later extended to more groups and ultimately, for a class of rights, to all humans. In ancient Greece, free citizens were given the right to speak and vote. The Magna Carta, an English constitutional document written in 1215, set out rights against unlawful imprisonment and clarified disputed rights of English kings and their subjects. The English Bill of Rights (1689) proclaimed the right to be free from cruel and unusual punishment. The French Declaration on the Rights of Man and Citizen (1789) identified some rights to be universal, not just for French citizens, and the US Constitution and Bill of Rights (1791) identified a series of political and civil rights first available to free men and later to all citizens. Early writers and philosophers debated the concepts of rights, including John Locke (1632–1704) who promoted the idea that humans were all created equal and had "natural rights," including the right to "life, health, liberty, or possessions," and to religious freedom and expression, which superseded any rights provided by governments (Locke, 2010).

The concept of human rights took on a greater significance following World War II, when, in 1948, the UN General Assembly debated and passed the Universal Declaration of Human Rights (UDHR). The global community defines such human rights as those rights inherent to all, independent of nationality, religion, race, sex, ethnicity, or any other factor, without discrimination. Such rights originally focused on political and civil rights, but have increasingly been extended to social, economic, and cultural rights with the passage of the 1966 International Covenant on Civil and Political Rights (ICCPR) and the International Covenant on Economic, Social, and Cultural Rights (ICESCR). Rights include freedom from slavery and torture, the right to life and liberty, freedom of opinion, work and education, physical and mental well-being, an adequate standard of living, and others. Additional treaties have been developed in subsequent years that address specific rights for racial minorities, women, children, and people with disabilities. In addition to these treaties, the international community has created governmental and non-governmental organizations and mechanisms to protect and promote human rights, such as the International Criminal Court (ICC) with jurisdiction over the crime of genocide, war crimes, and crimes against

DOI: 10.4324/9781003095545-18

humanity, regional courts such as the European Court of Human Rights and the Inter-American Court of Human Rights, and numerous non-governmental organizations, such as Human Rights Watch, Amnesty International, Global Rights, and UN Rights Watch.

The Human Right to Water

Because of the initial focus on political rights, the early 20th-century human rights declarations said nothing explicit about a right to water or other basic necessities such as food, health, or education, but by the 1960s and 1970s, more social and economic rights were increasingly being acknowledged and protected (Gleick, 1998). Even these first broader rights agreements, however, said nothing explicit about a right to water or water services. This began to change in the mid-1970s as the global water community started to call attention to a wide range of water challenges, including the failure to provide safe water and sanitation for all.

The first efforts to include water as a formal human right were enunciated in 1977, at the Mar del Plata international water conference, where the water community issued a statement explicitly acknowledging the human right to adequate, safe water:

> [...] all peoples, whatever their stage of development and their social and economic conditions, have the right to have access to drinking water in quantities and of a quality equal to their basic needs.
>
> *(United Nations, 1977)*

Yet progress in formalizing this right was slow. At international water meetings through to the end of the 20th century, efforts to push for a comprehensive declaration of the human right to water were regularly rebuffed. The issue was debated at the 2nd World Water Conference in the Hague in 2000, the International Conference on Freshwater in Bonn in 2001, and the 3rd World Water Conference in Kyoto in 2003. Each time, however, influential government delegations either ignored or actively opposed the explicit inclusion of any language in final conference statements declaring a right to water.

International pressure on the issue was growing, however, and in 2002, the UN Committee on Economic, Social and Cultural Rights approved "General Comment 15" acknowledging that a human right to water was consistent with, and derived from, other previously acknowledged human rights:

> The right to water clearly falls within the category of guarantees essential for securing an adequate standard of living, particularly since it is one of the most fundamental conditions for survival... The right to water is also inextricably related to the right to the highest attainable standard of health and the rights to adequate housing and adequate food... States parties have to adopt effective measures to realize, without discrimination, the right to water, as set out in this general comment.
>
> *(UN Economic and Social Council, 2003)*

In 2003, the Vatican published a supportive message stating: "There is a growing movement to formally adopt a human right to water. The dignity of the human person mandates its acknowledgment [...]. The right to water is thus an inalienable right." In September 2003, the European Parliament declared, "access to drinking water is a basic human right." The European Council on Environmental Law passed a resolution in January 2004 stating that "access to drinking water and sanitation is a fundamental right of the individual. The implementation of this right shall be ensured by law[...]" (Langford et al., 2004).

Finally, these actions led, in July 2010, to the approval by the United Nations General Assembly of Resolution 64/292, which recognized the human right to water and sanitation and acknowledged that these are essential to the realization of all other human rights (United Nations General Assembly (UNGA), 2010a, b).

> *The General Assembly* [...] *Acknowledging* the importance of equitable access to safe and clean drinking water and sanitation as an integral component of the realization of all human rights [...] *Recognizes* the right to safe and clean drinking water and sanitation as a human right that is essential for the full enjoyment of life and all human rights.
>
> *(UNGA, 2010a,b)*

Part of the basis for this recognition is the right to life explicitly identified in the earlier ICCPR and the rights to an adequate standard of living and health acknowledged in the ICESCR. The UN Resolution explicitly notes that these acknowledged rights cannot be met without adequate, safe, affordable water and sanitation.

The UN resolution calls on nations and international organizations to provide money, technology, and other help to countries, particularly developing countries, seeking to provide safe, clean, accessible, and affordable water and sanitation for all. The July 2010 General Assembly resolution was shortly followed by the adoption of a comparable resolution of the UN Human Rights Council (UNHCR), which affirmed that:

> The human right to safe drinking water and sanitation is derived from the right to an adequate standard of living and inextricably related to the right to the highest attainable standard of physical and mental health, as well as the right to life and human dignity" and that "States have the primary responsibility to ensure the full realization of all human rights, and that the delegation of the delivery of safe drinking water and/or sanitation services to a third party does not exempt the State from its human rights obligations.
>
> *(UNHRC, 2010)*

The declaration by the United Nations in 2010, however, is not the end of the story as by itself it doesn't provide water and sanitation to those in need. These resolutions are political expressions by UN members and are non-binding. The UN formally recognized a human right to food in Articles 11.1 and 11.2 of the 1966 International Covenant on Economic, Social and Cultural Rights,[1] yet even today, hundreds of millions of people, mostly children, go to bed hungry every day. In 2019, the United Nations estimated that almost 690 million people, 8.9% of the global population, remain undernourished, and the number may grow to 840 million by 2030 if current trends continue (FAO, IFAD, UNICEF, WFP & WHO, 2020). Similarly, there should be no expectation that the official declaration of a human right to water, alone, will solve the long-standing failure of governments to provide safe water and sanitation for all. Indeed, more than a decade has gone by since the 2010 declaration and nearly 800 million people still lack access to basic safe and affordable water, and more than two billion still lack access to basic sanitation (WHO/UNICEF Joint Water Supply Sanitation Monitoring Programme, 2019).

Human Right versus Economic Commodity

There is confusion about the contradiction between the concept of the human right to water and the reality of water as a commodity, or water services as a product, to be provided at a cost by a utility. Declaring a human right does not require that water be provided free. It is reasonable to require that the costs of providing safe water and water services be recovered, to ensure that water services are sustainable and can be improved over time. Evidence also shows that even extremely poor communities are willing to pay for reliable, safe water (Acey et al., 2019; Chatterjee et al., 2017; Vásquez et al., 2009). The human right to water does, however, require that access to a basic water supply cannot be denied to those who are unable to pay for it and the declaration calls on all countries to move more aggressively to provide services to those who lack it. Further steps are needed by all, particularly since the enforcement of international human rights is the responsibility of individual nation states.

Implementing a Human Right to Water

Governments, non-governmental organizations, communities, and individuals are slowly stepping up to implement protections for the human right to water. Some countries are addressing this issue by acknowledging some form of the right to water in their national constitutions, including Niger, Egypt, the Republic of Congo, Bolivia, Ethiopia, Uruguay, Ecuador, and a growing number of others (Higuet, 2014). An early example is the South African Constitution drawn up in 1994 and adopted in 1996 after the fall of the apartheid government. Section 27(1)(b) of their Bill of Rights states: "Everyone has the right to have access to sufficient food and water" (South Africa, 1994). In 2003, the Supreme Court of India ruled "the right to access to clean drinking water is fundamental to life and there is a duty on the state under Article 21 [of the Indian Constitution] to provide clean drinking water to its citizens" (Mittal, 2015). The 2010 Kenyan Constitution states: "(1) Every person has the right [...] to reasonable standards of sanitation; and to clean and safe water in adequate quantities" (Kenya, 2010). In 2012, the State of California passed a bill recognizing that "every human being has the right to safe, clean, affordable, and accessible water adequate for human consumption, cooking, and sanitary purposes" (State of California, 2012).

The United States has never been proactive in supporting economic, social, and cultural rights, such as the right to food or water, and important problems with providing adequate water services to all remain. No federal law recognizes the human right to water, though laws in the states of California, Massachusetts, and Pennsylvania have some form of a right to water. As a result of underinvestment in water systems, systemic racism, and government failures, many low-income, minority, and indigenous communities still lack safe, affordable water and sanitation, despite sometimes paying high rates for water services. Examples can be found in the cities of Flint, Jackson, Baltimore, Boston, and communities in the Central Valley of California, Alabama, New Mexico, Puerto Rico, and on many Native American lands (Dig Deep & US Water Alliance, 2019). Drought and severe hurricanes have worsened the situation in Puerto Rico, where hundreds of thousands of residents lack reliable access to safe tap water. The contamination of groundwater with agricultural chemicals in the Central Valley of California, together with the overdrafting of groundwater by industrial agricultural operations is affecting hundreds of thousands of people in largely Hispanic, farmworker communities. Inadequate sanitation services plague poorer communities in Alabama, which suffer high rates of water-related diseases as a result (Santa Clara Law, 2015). All of these conditions are mirrored around the world for hundreds of millions of people.

Strategies for Implementing and Enforcing the Human Right to Water

In the years since the passage of the human right to water by the UN, the debate among water scholars and activists has shifted from whether such a declaration would be a positive step to how best to achieve this right in the context of the broader efforts for water security and justice. Sultana and Loftus (2019) recently brought together a wide array of voices to explore different perspectives, examples, and lessons that have been learned in recent years in efforts to advance the human right to water. The key actor in this effort has traditionally been considered the state. The UN declaration for example, is binding on states rather than individuals, local communities, or sub-national actors. Nation states however, have themselves often been responsible for perpetuating or at least failing to address inequitable water policies, while efforts to achieve universal access to safe and affordable water and sanitation have been left to non-governmental groups, local communities, private entities, or even supranational institutions like the European Union. This argues for a broader effort, integrated across both state and non-state actors, to achieve and protect a human right to water.

Another important evolution in the effort to provide a human right to water and sanitation is the realization that such a right is embedded in the much larger context of human security, including access to food, public health, reducing racism, sexism, or classism, or other critical objectives. Are there ways to produce multiple benefits across a broader set of objectives by linking them together? By doing so, the political attractiveness of addressing water issues may grow and new resources may become available (Sultana & Loftus, 2019).

While few governments have made formal commitments to acknowledge the right to water in their legal systems, or more importantly to actually provide that water, more and more are moving in this direction. The challenge now is how to implement this right to accelerate the provision of water and water services to the hundreds of millions who still lack it.

One mechanism for enforcing the international human right to water is the application and enforcement of international law. While there are few cases – and few venues where such cases can be heard – there are an increasing number of examples where the right to water has been acknowledged as a component of broader human rights protections. In the Inter-American Court of Human Rights, for example, several cases have been decided acknowledging the right to safe water and sanitation for indigenous populations and vulnerable communities, effectively recognizing that the human right to water derives from the human right to life (Murillo, 2014). Similarly, the African Commission on Human and Peoples' Rights has acknowledged that a right to water is core to the right to development, which is grounded in rights to health, a healthy environment, and dignity (Bulto, 2011).

There have been some important historical precedents in US law. One of the most important was an early Supreme Court case acknowledging water rights for Native American reservations. The right to access and use water in the United States varies with historical precedence and region, with a mix of traditional European riparian rights to parties along a water source in the eastern US, and "prior appropriation" rights for the first party to make use of a water source in the western US. A treaty signed in 1888 set aside reservation land for the A'aninin and Nakoda tribes, but made no explicit allowance for water, and water for the reservation was subsequently diverted to white settlers. A legal battle over these water rights led to the 1908 Supreme Court decision in *Winters v. United States* (207 U.S. 564), which explicitly acknowledged that the water rights of Native Americans were also guaranteed by any land agreements and were fundamentally related to their right of survival and self-sufficiency.

As with many such decisions, actual protection and provision of water to these communities have lagged behind the legal guarantee and many Native American communities still lack safe water. As of 2020, more than 50,000 American Indian and Alaska Native homes, or 12.5%, still lack adequate sanitation services and 6,600 lack access to safe water. Native American households are 19 times more likely to lack indoor plumbing than White households (Dig Deep & US Water Alliance, 2019; Indian Health Service, 2020). In mid-2023, a conservative majority on the US

Supreme Court refused to take any action to redress the failure to acknowledge or affirm Navajo Nation water rights. In federal courts in Australia a claim of the Yorta people to their right to Aboriginal lands and waters was rejected, with the court arguing the "tide of history" had swept away any traditional rights (High Court of Australia, 2002).

In an assessment of the water and sanitation situation in the United States, the UN Special Rapporteur on the Human Right to Safe Drinking Water and Sanitation concluded that all countries, including the United States must "do more to ensure that not only *de jure* but also *de facto* discrimination is eliminated regarding access to water and sanitation" (UNGA, 2011).

Strategies for doing this include the adoption of national human right to water laws, standards for water affordability, infrastructure investment policies focused on meeting basic water system needs, and mechanisms to enforce commitments to safe water and sanitation. Coalitions of interested parties can help. In 2015, 130 different advocacy groups, legal experts, and others in the US formed the National Coalition on the Human Rights to Water and Sanitation to strengthen and coordinate work on water and sanitation issues across the country. Where national efforts are unsuccessful, more local and regional efforts can still move forward.

Conclusion

The concept of a human right to water and sanitation in international law and practice is relatively recent. It was introduced as an idea in the latter part of the 20th century, long after other forms of human rights had been codified and accepted and was only formally acknowledged by the United Nations human rights organizations and the General Assembly after the turn of the millennium. Even today, the actual implementation and application of that right is in its early stages and there is little agreement about how the right should be applied and enforced. Human rights law, as originally implemented in modern times, initially focused on political and civil rights, but rapidly evolved to encompass social, economic, and cultural rights, including food, human health, education, and now water.

The drive to include water, and eventually sanitation, as a formal right grew in urgency as the consequences of failing to provide safe water and sanitation to all began to receive more attention, initially in the context of the growing concern about broad environmental issues and then as explicit objectives of the Millennium Development Goals (MDGs) and the subsequent Sustainable Development Goals (SDGs). It was also helped by the actions of South Africa and a few other countries who acknowledged early on the importance of meeting basic water needs for all. When the UN formally declared a human right to water and sanitation in 2010, it raised broader awareness of the failure to meet those basic needs and provided a push to governments to work more actively towards the progressive realization of those rights.

The formal declaration alone however, is not enough, and billions of people still lack access to these basic water services. More and more governments are slowly stepping up to implement protections for the human right to water in their laws, policies, and actions, but far more efforts are needed if the overall SDG target of 100% coverage by 2030 is to be met. As of writing this, it appears unlikely this goal will be achieved on time, but progress is being made.

New approaches are needed together with new financial investments and political commitments. One strategy is to embed the goal of safe water and sanitation for all in the broader context of human security, access to food and public health, and efforts to address racism, sexism, and classism. By combining these objectives, multiple benefits may be achieved. Another tool may be to expand the enforcement of national and international laws and the role of the human right to water in broader human rights protections. And overall, far greater economic investment in basic water infrastructure and improvements in water management and institutions will pay dividends in improvements of human rights, human health, and overall well-being.

Note

1 The 1966 International Covenant on Economic, Social and Cultural Rights acknowledges "the right of everyone to an adequate standard of living for himself and his family, including adequate food, clothing and housing, and to the continuous improvement of living conditions" and that urgent steps may be needed to ensure "the fundamental right to freedom from hunger and malnutrition" (United Nations, 1966).

References

Acey, C., Kisiangani, J., Ronoh, P., Delaire, C., Makena, E., Norman, G., Levine, D., Khush, R. and Peletz, R. (2019). "Cross-subsidies for improved sanitation in low income settlements: Assessing the willingness to pay of water utility customers in Kenyan cities," *World Development*, 115, 160–177. DOI: 10.1016/j.worlddev.2018.11.006.

Bulto, T. S. (2011). "The human right to water in the corpus and jurisprudence of the African human rights system," *African Human Rights Law Journal*, 11(2), 342–367.

Chatterjee, C., Triplett, R., Johnson, C. K. and Ahmed, P. (2017). "Willingness to pay for safe drinking water: A contingent valuation study in Jacksonville, FL," *Journal of Environmental Management*, 203, 413–421. DOI: 10.1016/j.jenvman.2017.08.008.

Dig Deep and US Water Alliance. (2019). "Closing the water access gap in the United States: A national action plan. DigDeep right to water project," *US Water Alliance*, p. 95. [Accessed 10 February 2023]. Available at: https://www.digdeep.org/close-the-water-gap

FAO, IFAD, UNICEF, WFP and WHO. (2020). "The State of Food Security and Nutrition in the World 2020." Rome: Food and Agricultural Organization of the United Nations. DOI: 10.4060/CA9692EN.

Gleick, P. H. (1998). "The human right to water," *Water Policy*, 1(5), 487–503. DOI: 10.1016/S1366-7017(99)00008-2

High Court of Australia. (2002). "Members of the Yorta Yorta Aboriginal Community v Victoria." [Accessed 18 March 2021]. Available at: https://www7.austlii.edu.au/cgi-bin/viewdoc/au/cases/cth/HCA/2002/58.html

Higuet, F. (2014). "States recognizing the right to water in their constitution, RAMPEDRE: From the Declaration to the Implementation of the Right to Water." [Accessed 26 January 2021]. Available at: http://www.rampedre.net/implementation/territories/national/world_table_constitution

Indian Health Service. (2020). "Safe Water and Waste Disposal Facilities | Fact Sheets, Fact Sheets: Safe Water and Waste Disposal Facilities." [Accessed 18 March 2021]. Available at: https://www.ihs.gov/newsroom/factsheets/safewater/

Kenya. (2010). "Constitution of Kenya 2010. National Council for Law Reporting (Kenya Law)." [Accessed: 27 January 2021]. Available at: http://kenyalaw.org:8181/exist/kenyalex/actview.xql?actid=Const2010

Langford, M., Jones, H., Khalfan, A. and Fairstein, C. (2004). "Legal Resources for the Right to Water: International and National Standards." *Geneva, Switzerland: Centre on Housing Rights and Evictions (COHRE)*. [Accessed: 3 February 2021]. Available at: https://www.google.com/search?q=ISBN+92-95004-11-6&rlz=1C1GYPO_enUS780US780&oq=ISBN+92-95004-11-6&aqs=chrome.69i57.1238j0j7&sourceid=chrome&ie=UTF-8

Locke, J. (2010). "The Second Treatise of Government (1690)." The Project Gutenberg. [Accessed: 2 February 2021]. Available at: https://www.gutenberg.org/files/7370/7370-h/7370-h.htm

Mittal, A. (2015). "Right to clean water," *Academike*, 22 October. [Accessed: 26 January 2021]. Available at: https://www.lawctopus.com/academike/right-to-clean-water/

Murillo, J. (2014). "The right to water in the case-law of the inter-American court of human rights," *Anuario Colombiano de derecho Internacional* – ACDI, 7, 39–68. DOI: 10.12804/acdi7.2014.02.

Santa Clara Law. (2015). "The Human Right to Water in the United States." *International Human Rights Clinic*. Santa Clara, CA: Santa Clara University School of Law, 30.

South Africa. (1994). "The South African Constitution: The Bill of Rights." [Accessed: 26 January 2021]. Available at: https://www.justice.gov.za/legislation/constitution/chp02.html

State of California. (2012). "State Water Policy, Assembly Bill No. 685." [Accessed 10 February 2023]. Available at: http://leginfo.legislature.ca.gov/faces/billNavClient.xhtml?bill_id=201120120AB685

Sultana, F. and Loftus, A. (eds). (2019). *Water Politics: Governance, Justice and the Right to Water*. Abingdon, Oxon: Taylor & Francis.

UN Economic and Social Council. (2003). "General Comment No. 15: The Right to Water (Arts. 11 and 12 of the Covenant)." E/C.12/2002/11. New York, NY: *UN Committee on Economic, Social and Cultural Rights* (CESCR). [Accessed: 26 January 2021] Available at: https://www.refworld.org/docid/4538838d11.html

United Nations. (1966). "International Covenant on Economic, Social and Cultural Rights. General Assembly Resolution 2200A (XXI) of 16 December 1966." *UN Office of the High Commissioner for Human Rights*. [Accessed 26 January 2021]. Available at: https://www.ohchr.org/EN/ProfessionalInterest/Pages/CESCR.aspx

United Nations. (1977). *Report of the United Nations Water Conference*, in. *United Nations Water Conference, March 14–25, 1997, Mar del Plata, Argentina*. New York: United Nations Publications.

United Nations General Assembly. (2010a). "Human Rights Obligations Related to Access to Safe Drinking Water and Sanitation, Note by the Secretary-General" A/65/254, 65th session. New York: UN General Assembly.

United Nations General Assembly (UNGA). (2010b). "The human right to water and sanitation. Resolution 64/292." New York, NY: United Nations. [Accessed 10 February 2023]. Available at https://digitallibrary.un.org/record/687002

United Nations General Assembly. (2011). "Report of the Special Rapporteur on the Human Right to Safe Drinking Water and Sanitation, Catarina de Albuquerque." A/HRC/18/33/Add.4. New York, NY: United Nations, p. 21. [Accessed: 18 March 2021]. Available at: http://www2.ohchr.org/english/bodies/hrcouncil/docs/18session/A-HRC-18-33-Add4_en.pdf

United Nations Human Rights Council (UNHRC). (2010). "The human right to safe drinking water and sanitation. Resolution A/HRC/RES/18/1." Geneva, Switzerland: United Nations. [Accessed 10 February 2023]. Available at https://www.right-docs.org/doc/a-hrc-res-18-1/

Vásquez, W. F., Mozumder, P., Hernández-Arce, J. and Berrens, R. P. (2009). "Willingness to pay for safe drinking water: Evidence from Parral, Mexico," *Journal of Environmental Management*, 90(11), 3391–3400.

WHO/UNICEF. (2019). "Joint Water Supply Sanitation Monitoring Programme. Progress on Household Drinking water, Sanitation and Hygiene 2000–2017." *Special Focus on Inequalities*. New York: United Nations Children's Fund (UNICEF) and World Health Organization (WHO).

16

WATER RESOURCES MANAGEMENT

The Missing Political Link

Kurt Mørck Jensen and Jens Christian Refsgaard

Introduction

The development of water resources management (WRM) concepts over time in developing (ICWE, 1992; GWP, 2000; WWAP, 2003; Molle, 2008; Hoff, 2011) and developed (Newig et al., 2005; Hering et al., 2010; Voulvoulis et al., 2017) country contexts have traditionally been described as separate processes as if belonging to two different worlds. Partly because the issues and challenges appeared to be quite different and partly because the author communities typically had professional experiences from only one of the two worlds.

The dominating WRM concepts, such as Integrated Water Resources Management (IWRM) (Jønch-Clausen, 2004) and Nexus (Hoff, 2011) have normative dimensions. They are frequently criticised for poor implementation efficiency when faced with the harsh reality of everyday water governance. It is therefore relevant to analyse the political process in WRM, and in particular the hardcore political advocacy from asymmetrical stakeholders (Mollinga, 2008; Savannan et al., 2008; Butterworth et al., 2010; Jensen et al., 2018).

Based on experiences from research and practical implementation of WRM concepts in both Denmark/EU and developing countries we have arrived at the conclusion that, in spite of obvious apparent differences, there are many fundamental similarities in the challenges as they unfold in developed and developing countries (Jensen et al., 2018).

The aims of this chapter are: (i) to describe the development over the past five decades of WRM concepts and their implementation challenges in developed and developing countries; (ii) to assess whether common experiences and lessons learned can be obtained; (iii) to assess the role of donors in developing WRM concepts; and (iv) to discuss perspectives for future developments.

WRM Concepts

Traditional concepts

Table 16.1 summarises some key characteristics of three dominating WRM concepts applied during the past 50 years: (i) top-down planned command and control water master planning (master planning); (ii) IWRM; and (iii) the Water, Energy and Food Security Nexus (Nexus). Although a number of other concepts, including variants of the three such as adaptive water management

DOI: 10.4324/9781003095545-19

Table 16.1 Characteristics of three key WRM concepts.

	Master planning	*IWRM*	*Nexus*
Origin	Always been here	Dublin/Rio UNCED – 1992	Bonn – 2011
Resource perspective	Unlimited availability – supply management (build more dams and drill more/deeper wells)	Finite resource – demand management – use water where it creates highest societal value	Finite resource – demand management – use water where it creates highest societal value
Aims and evaluation focus	Single bottom line: • Economic development	Triple bottom line: • Economic development • Environment • Social equity	Triple bottom line: • Water/energy/food security • Equitable and sustainable growth • Resilient productive environment
Focus	Technical product	Process, governance, institutional capacity	Sustainability and security issues in a more complex world
Level of public participation	Information (top-down) and some target group participation	Active involvement of stakeholders	Active involvement of stakeholders
Steering of planning and decision process	Sectoral ministries (limited coordination)	Cross-sectoral integration coordinated by Ministry of Water	Cross-sectoral integration coordinated by Ministry of Finance
Implementation and enforcement	Simple – top-down command and control	Complicated – enforcement not included	Complicated – enforcement not included
Reference	Loucks and van Beek (2005)	Jønch-Clausen (2004)	Hoff (2011)

(Pahl-Wostl, 2007) and human-right based approach to IWRM (WWAP, 2019), have been suggested, in this chapter we shall mainly be referring to these three major WRM concepts.

The political economy of water

The IWRM and Nexus concepts and their implementation approaches have normative dimensions. There has been a learning curve leading to concepts being more refined and holistic over time. The learning journey has taken on board the wider (sectoral) development context in which water is situated. It has also taken on board socio-economic approaches evolving in the general discourse of development such as stakeholder involvement and the economic value of water and the environment. This has added complexity to WRM making it a very ambitious proposition in terms of implementation and ownership by developing country governments and private sector stakeholders. Translating normative ideals into reality has become even more difficult and (probably) increased the gap between the (well-intended) global community of water experts (and their Nirvana approaches) and the realities on the ground in the developing (as well as developed) world thus exposing *politics as the missing link* (Molle, 2008; Mollinga, 2008; Swatuk, 2008; Cascão & Zeitun, 2010; Harris et al., 2011).

Although the water expert community was always aware of politics and economic interests working against the full implementation of the normative WRM approaches, it did not lead to a questioning of the relevance of these approaches. The thinking was that "over time they will understand" meaning that the logic and value of the WRM approaches would be the winner in the end as governments, politicians and all private sector stakeholders would come to their senses in the end and engage in harmonious decision-making on WRM.

As social and political scientists at the beginning of the 2000s started to equal and even outnumber the engineers (and their social engineering and normative approaches) in the water sector, so did the conceptualisation of WRM as being ultimately decided by political power and economic interests. This opened a platform for "the political economy of water" as an analytical framework or concept for understanding WRM in a larger development context. Although not an approach to WRM as such, the political economic challenged the normative approaches, particularly IWRM, and their social engineering solutions by emphasising that there are no model based smart solutions or quick fixes to WRM (Jensen & Lange, 2013).

The concept of the political economy of water considers not only water in a broader development context (like IWRM and the Nexus). As a fundamental distinguishing factor, the concept entails that political power is always asymmetrically and hierarchically distributed among water stakeholders – typically comprising of national governments (ministries, local governments), investors (private or public), civil society groups and local communities. The hierarchy of interests in a sectoral context is for example well illustrated by the higher importance of energy over water in the fast-growing development economies thus giving priority to hydropower dams on rivers at the expense of water for fisheries, agriculture and nature (ibid.).

WRM in Denmark and the EU

Before 1970. No acknowledged environmental problems

More than 99% of water use in Denmark for drinking water (as well as for industry and agriculture) has, during the past 150 years, originated from groundwater (Jørgensen et al., 2017). Denmark is not a water scarce country, so until water pollution occurred, and the sustainability of water dependent ecosystems became societal priorities, the WRM issues were mainly confined to identifying and ensuring suitable locations for groundwater abstraction. The first act on water supply, issued in 1926, established the doctrine that groundwater is a common good not belonging solely to the landowner. It also included regulations for the licensing of abstraction.

1970–2000. Building environmental management capacity in Denmark

As a consequence of rapid economic growth and industrialisation, the pollution of surface waters emerged as a recognised threat during the 1960s. In 1971, the Danish Ministry of the Environment was established, probably the first of its kind worldwide. This marked the initiation of three decades with a strong focus on environmental and water quality issues resulting in increasing national environmental legislation and capacity building.

This development took place in parallel with, and mutually inspired by, developments in other industrialised countries. In this respect the UN Conference on the Human Environment held in Stockholm in 1972 was the first major international conference on environmental issues and marked a turning point in the development of international environmental policies (UN, 1973). In line with the recommendations from the Stockholm conference, Danish policy in the subsequent years focussed on mitigating pollution, and ensuring and safeguarding both ecosystem quality and ecosystem dependence on both water quality and water quantity.

During the first couple of decades, following the conference, comprehensive capacity building took place both in the public and the private sectors and in academic institutions. The progress during the early years was significant with respect to mitigating water pollution from the industry and cities, with both sectors investing heavily in water treatment and better technology. Although some private companies complained about heavy water treatment costs, most of the industry managed the transition towards a cleaner production by introducing new technology that often turned out to become an advantage in international competition, and no major sectors were fundamentally threatened by the development. In this respect the development may be characterised as one of harvesting the low-hanging-fruits.

During the 1980s, the focus moved towards nutrients emitted to the terrestrial and aquatic environment from industrial and urban wastewater and from agricultural activities. The adoption of the first Action plan for the Aquatic Environment in 1987, aiming at reducing nutrient loads by 50%, represented a paradigm shift, because it involved regulation of the agricultural sector (Dalgaard et al., 2014). While the response from the municipalities and industry was a rapid improvement of the sewage treatment plants to contribute with their share of the nutrient load reduction, the agricultural sector was much more hesitant, because the heavy reduction targets imposed on the agricultural sector threatened the traditional way of farming as well as the economy of the sector. Denmark has traditionally been an agricultural country with more than 60% of the land area cultivated, and the agricultural sector was politically very influential. After some initial successes of reducing nutrient loads without severely damaging the profitability of the sector, the progress in reducing nutrient reduction more or less came to a standstill towards the end of the century.

The Danish WRM policy in these years included many elements characteristic of command and control water master planning (Table 16.1). The environmental administration, the monitoring programmes, as well as the research institutions, were largely organised sectorally, typically with groundwater, surface water and environment departments staffed by geologists, hydraulic engineers and biologists. The public participation was confined to information and consultation procedures. Strict enforcement mechanisms were built into the legislation to ensure compliance for the industry and the public sector, while the enforcement mechanisms for the agricultural sector were much weaker, because the political will to impose strong restrictions on agriculture did not exist.

After 2000. EU taking over

From the 1990s, EU policies on the environment increasingly influenced Danish legislation with EU directives on separate issues such as the EU Nitrate Directive (Directive 1991/676/EEC) and the EU Landfill Directive (Directive 1999/31/EC). With the adoption in 2000 of the EU Water Framework Directive (WFD) (Directive 2000/60/EC) the governance framework for all water management was now delegated from national to EU authority.

Many of the key elements in the WFD follow IWRM principles:

- Water shall be managed at a river basin scale, implying transboundary management in many cases.
- Water shall be managed in an integrated manner, both across water domains (groundwater/surface water, water quality/water quantity, etc.) and across sectors (agriculture, municipalities, industry, nature, etc.).
- Public participation at consultation level is mandatory, while stakeholder involvement is encouraged.

However, the WFD differed from IWRM in two important aspects:

- According to IWRM, assessments shall be made of economic, social and environmental aspects, without prescribing the political priorities among these three bottom lines. In the WFD, the aim

is predefined as achieving good ecological status of all EU water bodies by 2027, and the measures to achieve this aim shall be prioritised in a cost-effective manner leaving possibilities for derogation only in case of very severe economic costs and by use of troublesome procedures.
- Member states have delegated power (sovereignty) to EU, and the WFD includes strict legal implementation and enforcement mechanism.

The WFD implied a significant improvement and harmonisation of the governance framework across Europe. The progress on developing common methodologies and generating huge amounts of monitoring data has generally been good, but the implementation has been difficult, and the WFD timescale of goal achievements in 2027 is over-ambitious (Hering et al., 2010). A possible explanation of implementation problems may be that the intended policy shift from sectoral end-of-pipe solutions towards integrated management of land and water as a system (requiring an interdisciplinary, integrated and holistic approach) has not taken place (Voulvoulis et al., 2017).

The implementation problems can be illustrated by the lack of progress in reducing nutrient loads from agriculture to coastal water to the Baltic Sea (Box 16.1). After some initial progress (like in Denmark during the 1980s–1990s), progress has virtually stopped and the European

Box 16.1 Nutrient loads to the Baltic Sea (Reusch et al., 2018)

Nitrogen (N) and phosphorus (P) loads lead to eutrophication within the Baltic Sea (Gustafsson et al., 2012). Management to mitigate nutrient loads and the associated eutrophication started after the UN Conference on the Human Environment, Stockholm 1972. The first national initiatives responding to local and national problems were soon followed by international collaboration through the Helsinki Commission (HELCOM) and later EU. HELCOM, established in 1974 by all riparian countries, is a normative type of collaboration, where countries voluntarily agree to make certain actions without delegating enforcement power. EU, differing from HELCOM by its supranational legislation and enforcement mechanism delegated to EU institutions, has had an increasing importance, particularly after its enlargement with Poland, Estonia, Latvia and Lithuania in 2004, leaving Russia as the only HELCOM party not being EU member.

Nutrient loads to the Baltic Sea have been increasing between 1850 and 1980 and have decreased significantly since then. The agricultural sources, which today constitute most of the nutrient load (HELCOM, 2011), have been reduced slightly, partly due to collapse of the agricultural system in former communist states and partly due to improved agricultural practices in a few countries such as Denmark, where the nitrate loading from agriculture to the aquatic environment was reduced by around 40% (Dalgaard et al., 2014).

HELCOM has been successful in promoting systematic monitoring, data sharing, awareness raising and science-based analyses. While HELCOM has been successful in facilitating harvesting the low-hanging-fruits all can agree to, it has not had any substantial impacts on reducing the nutrient loads from agriculture, where a substantial reduction is required to achieve the Baltic Sea goals. Therefore, strong enforcement mechanisms, such as those connected to the EU directives, will be required. The EU directives are still in their initial implementation phases, and the lack of implementation in member states has already triggered court cases and political conflicts between EU and member states. A complication in this respect is that the EU environmental directives to some extents are undermined by the existing EU Common Agricultural Policy that subsidises intensive agriculture (Karlsson et al., 2016).

Commission has now opened cases against major agricultural countries such as Denmark, Poland, Germany and The Netherlands for non-compliance. The outcome of these processes is yet to be seen, so it is still difficult to assess the strengths of the EU's implementation muscles.

WRM in Danish Development Cooperation

Before 1992. Master planning

The first decades of Danish development cooperation in the water sector aimed at improving water supply and sanitation in rural areas. The first major Danish funded project was a water master planning and water supply implementation project for three regions in Tanzania initiated around 1980. This was soon followed up by a similar project in Sri Lanka, a major water supply project in Orissa, India and projects in many other countries.

These projects used traditional master planning approaches with focus on technical and economic feasibility and less priority given to environmental sustainability and the involvement of stakeholders in project design and implementation. The projects were typically carried out by Danish consultants and experts using relatively simple techniques such as piped gravity schemes for surface water supply and hand pumps for groundwater supply. Some of the projects, such as the Drinking Water Supply Project in Orissa, in addition tried to transfer Danish approaches of abstracting groundwater from deep wells. Many of the projects turned out to be unable to fully sustain their operations after project completion, partly due to poor maintenance and partly a lack of active support from local stakeholders.[1]

1992–2010s. IWRM

As part of the preparation for the UN Conference on Environment and Development in Rio 1992, the international water community devoted substantial efforts in rethinking water management concepts. One of the contributions was the Nordic Freshwater Initiative with Denmark as lead partner resulting in the Copenhagen statement (CIC, 1991) that provided significant inputs to the Dublin statement (ICWE, 1992). This resulted later in the formation of the Global Water Partnership, with Denmark as an active partner, and the crystallisation of the new concepts into IWRM (GWP, 2000).

Immediately after the Dublin conference, Denmark supported the development of a Water Action Plan for Uganda, where the new concepts were used and tested. In the subsequent years Denmark supported IWRM based programmes in other countries, such as Vietnam and Nicaragua.

The focus in these projects had shifted towards capacity development and institutional strengthening for an entire sector with priority now being given to environmental sustainability and a higher degree of stakeholder involvement and cross sectoral integration. These projects were typically carried out by a consortium of Danish or international consulting companies, but with a somewhat larger involvement of local experts than in previous (water master planning) projects.

2000–2010s. Transboundary water management

Around 2000 Denmark engaged in transboundary water management via support to the Mekong River Commission (MRC) (see Box 16.2), the Nile Basin Initiative (NBI) and the Zambezi Watercourse Commission (ZAMCOM). These transboundary water management initiatives were all backed by international treaties that recognised the IWRM principles. Over time they also

Box 16.2 The Mekong River Commission: Donors, politics and national interests

The Mekong River Commission (MRC) was established in 1995 based on the MRC Agreement between the four Lower Mekong Countries of Thailand, Laos, Cambodia and Vietnam. The Upper riparian countries of China and Myanmar were (and are) not part of the Agreement.

Denmark has supported WRM in the Mekong since 1991 under the then Mekong Committee. After the establishment of the MRC in 1995 and until 2015, Denmark was the single largest bilateral donor to the MRC providing approximately USD 90 million[5] support through grant funding and technical assistance to basin development planning, fisheries, environmental management and climate change adaptation.

The assumption behind Danish and other donor development aid to the MRC was that the knowledge and capacity developed through donor support would enable the four Lower Mekong Countries to jointly engage in WRM and agree on sustainable development interventions and thus prevent conflict of interest by riparian countries. However, as the MRC Agreement is a "soft law" agreement without "hard sanctions" based on the consensus driven "Mekong Spirit,"[6] it was not able to override national interests.

In response to the rapid economic growth of the region, China and by 2010 also Laos started the largescale construction of hydropower dams on the tributaries and mainstream of the Mekong. While Chinese hydropower dams supplies electricity to China's own economic needs, the hydropower projects built by Laos supply electricity primarily to Thailand. By implementing a cascade of hydropower projects on the Mekong, Laos has gained the title of "the battery of Southeast Asia."

Although the MRC through donor aid is equipped with the knowledge and tools to engage its member countries in sustainable WRM, it has been difficult to convince Laos to adjust its hydropower projects to a level that would retain water resources, fisheries and the environment at a sustainable level. At the same time, tension has risen between Laos and the downstream countries of Cambodia and Vietnam, where negative impacts of dam construction are now becoming a reality particularly during the dry season, when river flows are low. In the Mekong Delta this coincides with increased surface and ground water salinity caused by saline intrusion from the rising sea level as an effect of climate change. These ripple effects are potentially a perfect recipe for environmental and economic disaster. The good news from a political economy point of view is that from being a normative approach driven by donors, WRM for sustainable development is now driven by the downstream riparian countries at risk, Vietnam and Cambodia, either through the MRC governance institutions or in direct bilateral talks with Laos.

acknowledged the importance of cross-sectoral coordination and trade-offs as included in the Nexus approach.

The international support was mainly used to develop capacity at the MRC, NBI and ZAMCOM secretariats, which were staffed by a combination of international and riparian experts. Comprehensive efforts were used to develop monitoring network, quality-controlled data, publicly accessible data bases and modelling tools. In addition, environmental impact assessment procedures and protocols for water sharing were adopted. Finally, the data, modelling tools and assessment procedures were used in transboundary water resources planning processes where local and regional stakeholders were involved. The intention was that such commonly agreed data and tools and a transparent planning process together with regular meetings at cabinet ministerial level could prevent conflicts and secure a fair sharing of water resources among riparian countries.

After 2010s. New priorities – Less water

From being one of the lead water donors around 2000 when support to water constituted around 10% of Denmark's development budget, Denmark has now reduced its water sector engagement considerably as other political and humanitarian priorities have emerged. After 2015 Denmark no longer supports transboundary waters in the large river systems directly through bilateral cooperation but retains some support to WRM through the United Nations Environmental Programme (UNEP) and the World Bank. From having bilateral water cooperation in 11 countries[2] in Africa and Asia, Denmark now only supports water supply and WRM in three African countries.[3] However, with an increased focus on climate change adaptation, the water sector may gain increased importance in Danish development cooperation.[4]

Discussion – From Normative Ideals to Practical Implementation

Role of donors in developing WRM concepts

During the 1970s and 1980s donors were very much in control with supply driven master planning both for drinking water projects (wells and gravity) and building of dams. The focus was on poverty alleviation, and water along with health and education was considered as social infrastructure delivering social services otherwise out of reach for the poor. The priority to engineering and infrastructure at the time resonated well with recipient governments and their water authorities.

Donors were equally instrumental in the shift towards the more comprehensive IWRM and later the Nexus concepts and approaches during the 1990s. This shift reflected the learning that took place in donor organisations during the 1980s and 1990s, particularly on the sustainability of water infrastructure delivery. The narrow delivery of water infrastructure had severe problems in developing countries because of weak institutions, financial limitations and little tradition for stakeholder involvement. The donor learning resulted in a journey from an engineering only approach to an interdisciplinary and more contextual approach to aid delivery involving also socio-economic components meant to safeguard the sustainability of infrastructure. Adding to the new paradigm were the processes on water management linked to the Rio conference in 1992, which led to the creation of the IWRM concept (see *WRM concepts* section). Here donors along with water experts around the globe continued to be in the driver's seat rallying around the GWP with IWRM as its key normative approach to water management.

The IWRM approach was a good match with the donors' overall very idealistic and supply-driven approach to development. The logic behind donors supporting IWRM was that it was "the perfect solution" to all misgivings and challenges in the water sector. IWRM ideally provided an answer to almost everything, provided of course that all stakeholders in the IWRM system behaved according to the norms. One of the reasons why donors maintained the idea of the perfect solution was their accountability to their home constituencies, demonstrating that taxpayers' money was used in a good way. The same requirements were not necessarily added to reforms of environmental policies in their own developed countries, where strong stakeholder interest prevented ideal solutions. WRM in developed countries has always been layered with real life issues involving opposing stakeholder interests. Policy makers in developed countries were not so naïve that they thought they could change things dramatically through a new concept like IWRM. In hindsight, one wonders how the donors could be so idealistic and naïve to think that they could transform the realities in developing countries and insert IWRM as the solution, when this operation could not be accomplished in their home countries?

Over the last 10–15 years the IWRM and Nexus approaches have encountered problems in delivering ideal solutions because of, for instance, excessive dams and hydropower projects on rivers. However, donors appear not to have engaged in trying to identify realistic alternatives. As our analysis suggests, no such alternative exists without considering the political economy of water, which implies a shift away from the ideal solution towards political compromises and a second or third best solution, as we know it in developed countries. Donors seem not to have been willing to support such a political process and has left that with stakeholders in developing countries.

Although donors have had a large degree of shared interest and priorities towards the water sector including IWRM, they have all along differed in their political priorities, administrative procedures and criteria for quality assurance. Some donors put their weight towards policy and strategic direction, whereas others had a more hands-on approach focusing on identifying and measuring results on the ground. Often donors had very different requirements in terms of reporting and documentation. Some donors were easy, others more difficult to please. Thus, the notion of one "donor community" only applies in a very general sense.

Developed and developing countries – Similar lessons learned

In spite of obvious differences between developed and developing countries, our analysis suggests that many of the WRM experiences have fundamental similarities. We have seen the same development from centralistic, sector-based master planning without stakeholder involvement to normative IWRM/Nexus approaches and now a recognition that conflicts of interests and hard-core political advocacy of stakeholders play a dominant role in WRM.

The changes in WRM concepts have in many cases appeared stronger and more abrupt in developing countries, where donors have promoted these changes through support to policy formulation and reform. An obvious explanation for this difference is that the institutions and the civil societies are stronger in the developed countries, and hence it is more difficult and requires a longer process to incorporate new ideas in the entire water management system. Or in other words that developed countries have more resilience or institutional conservatism towards external inputs of new ideas.

Cross-sectoral integration

WRM based on cross-sectoral integration has proven extremely difficult. Cross-sectoral integration is built into the IWRM and Nexus concepts, and in the case of the WFD, it was designed to secure such integration. However, so far no attempts have fully succeeded. For instance, in EU the Common Agricultural Policy and the WFD are conflicting with competing stakeholder, both institutionally, culturally and economically. Similarly in the Mekong, the national and sectoral interests of member states have worked against cross-sectoral and multilateral integration in the MRC with some countries giving priority to hydropower, while fisheries or agriculture are more important for other countries. Although the Nexus approach to a much higher degree than IWRM acknowledges sectoral imbalances in terms of hierarchy of national priorities (e.g., energy over agriculture and fisheries) and thus implicitly the political economy of water, it still remains normative and is therefore also faced with implementation challenges.

A true cross-sectoral integration would require a transition towards a fully holistic approach that again would require a societal transformation in which conflicting stakeholder interests (by some miracle upper hand) are harmonised into a consensus about "a common good." By all practical circumstances, such consensus and transformation is probably unrealistic.

IWRM and Nexus normative concepts

Normative concepts have their limitations, especially if there is no delegation of power to an efficient enforcement mechanism. Although the MRC subscribes to IWRM and Nexus as normative paradigms, the MRC soft environmental mandate which is based on consensus, has no decision-making authority when it comes to fundamental disagreements between its member states over access to and use of its transboundary water resource. In the case of the Baltic Sea, the Helsinki Commission (HELCOM) is a normative arrangement without enforcement mechanisms. The experience from HELCOM shows, similarly to the MRC, that it is possible to achieve significant progress in terms of transparency in data, modelling tools and assessment procedures. These are valuable in advancing progress on issues where all countries have similar interests. Experiences from both the MRC and HELCOM, also show however, that progress stops when fundamental disagreements occur. The experience with the EU-WFD reveals that, although the built-in enforcement mechanism may significantly improve implementation, it also illustrates that when strong stakeholder interests are involved, it is not enough to have a strong enforcement mechanism. Rather, it also requires the political will to use such enforcement. Time will show to which extent the EU will use its muscles. Will the EU take Germany, The Netherlands, Poland and Denmark to court to ensure compliance, or will the member countries at the end make a political solution by renegotiating the WFD goals?

New challenges, Climate change and SDGs

Since the adoption of the UNs Sustainable Development Goals (SDGs) and COP15 in Paris, the climate change and broader sustainability issues have moved quickly towards the top of the agenda in most EU countries, including Denmark. While previous UN goals (e.g., Millennium Development Goals) were not part of the agenda in European institutions and companies, everybody, from schools to municipalities and private companies work with SDGs today, and in many cases incorporate it into their strategies.

Climate change will require major adaptation measures in the water sector. Climate mitigation measures will also affect water, in terms of more demand for hydropower, afforestation and land use change, for example. Climate change and the much broader SDG sustainability issues will be a major distortion for WRM. For instance, achieving all the many mutually dependent SDG goals will require a cross-sectoral integration far beyond what is considered in IWRM and Nexus and thus will require a fully holistic approach. As discussed above, we doubt that the existing WRM concepts will be able to cope with this.

Aid and political priorities

In addition to the climate agenda, an element of donor fatigue towards the water sector has resulted in reduced funding over the last decade. Aid budget allocations to water are now a far cry from its peak years in the 1980s and 1990s when, for example, 10% of the total Danish aid budget was spent on the water sector. Moreover, other political priorities in many donor organisations, particularly those operating within ministries of foreign affairs, have shifted humanitarian and development funding towards political hotspots such as Afghanistan and recently Ukraine leaving less funding for water and other traditional development aid sectors.

The missing political link

The hierarchy and imbalance of political power and influence invariably leads to conflict of interests over water resources, especially in situations and places with water scarcity and water quality problems. Although not fundamentally a water scarcity region, the Lower Mekong River basin is a case in point due to seasonal variations and the impact of climate change. Development financing (whether from international donors, development banks, developing county governments or private investors) adds fuel to the conflict between stakeholders, as we have seen with hydropower dams in Laos working against the interests of downstream countries in the Mekong. Moreover, such developments often further aggravate the asymmetrical distribution of political power between stakeholders and countries. According to a political economy of water perspective, WRM will always be political by nature as stakeholders compete for access to water resources. This makes the allocation, development and management of water resources a question of (political) governance.

Thus, the political economy of water focus on *conflicts of interest* challenges the harmonious social engineering approach of IWRM and the Nexus, where all stakeholders ideally get their share of the available water resources in a water regime. This means that WRM is a political process among stakeholders with winners and losers depending on their position in the stakeholder hierarchy.

We therefore conclude that the decision-making processes in the water resources sector cannot be fully understood from normative concepts such as IWRM and Nexus – it can only be understood by also considering the political economy of water. A logic consequence of this conclusion is that normative concepts are inadequate as policy paradigms. Hence, there is a need to develop new WRM paradigms that incorporate some of the idealistic elements of IWRM/Nexus as well as hardcore political analyses of asymmetrical stakeholder positions.

Notes

1 Poor operation and maintenance was repeatedly emphasised by reviews and evaluations of not only Danish financed but other donor funded drinking water projects.
2 Ghana, Burkina Faso, Niger, Kenya, Tanzania, Uganda, Afghanistan, India, Bangladesh, Bhutan, Vietnam.
3 Uganda, Burkina Faso and Niger.
4 At the time of writing Denmark is planning to again support transboundary waters in Africa through a fund mechanism under the EU.
5 Between 1995 and 2015 the MRC had received more than USD 300 million in development aid to support a wide array of projects in various sectors relevant for transboundary river management.
6 Mekong River Commission Agreement 1995.

References

Butterworth, J., Warner, J., Moriarty, P., Smits, S. and Bachelor, C. (2010). “Finding Practical Approaches to Integrated Water Resources Management.” *Water Alternatives*, 3(1), 68–81.

Cascão, A. E. and Zeitun, M. (2010). “Power, Hegemony and Critical Hydropolitics.” In A. Earle, A. Jägerskog and J. Öjendal, (eds.) *Transboundary Water Management. Principles and Practice*. London: Earthscan, pp. 27–42.

CIC. (1991). Copenhagen Report. *Implementation Mechanisms for Integrated Water Resources Development and Management. Report from the Copenhagen Informal Consultation*, 11–14 November 1991. Danida.

Dalgaard, T., Hansen, B., Hasler, B., Hertel, O., Hutchings, N. J., Jacobsen, B. H., Jensen, L. S., Kronvang, B., Olesen, J. E., Schjørring, J. K., Kristensen, I. S., Graversgaard, M., Termansen, M. and Vejre, H. (2014). “Policies for Agricultural Nitrogen Management—Trends, Challenges and Prospects for Improved Efficiency in Denmark.” *Environmental Research Letters*, 6, 115002.

Gustafsson, B. G., Schenk, F., Blenckner, T., Eilola, K., Meier, H. E. M., Müller-Karulis, B., Neumann, T., Ruoho-Airola, T., Savchuk, O. P. and Zorita, E. (2012). “Reconstructing the Development of the Baltic Sea Eutrophication.” *Ambio*, 41, 534–548.

GWP. (2000). "Integrated Water Resources Management. TAC Background Papers No. 4." Stockholm: Global Water Partnership, Technical Advisory Committee. 71 pp.

Harris, D., Kooy, M. and Jones, L. (2011). *Analysing the Governance and Political Economy of Water and Sanitation Delivery*. London: Overseas Development Institute.

HELCOM. (2011). "The Fifth Baltic Sea Load Compilation (PLC-5)." Baltic Sea Environment Proceedings" No. 128. Helsinki Commission.

Hering, D., Borja, A., Carstensen, C., Carvalho, L., Elliott, M., Feld, C. K., Heiskanen, A. S., Johnson, R. K., Moe, J., Pont, D., Solheim, A. L. and van de Bund, W. (2010). "The European Water Framework Directive at the Age of 10: A Critical Review of the Achievements with Recommendations for the Future." *Science of the Total Environment*, 408, 4007–4019.

Hoff, H. (2011). *Understanding the Nexus. Background Paper for the Bonn 2011 Conference: The Water, Energy and Food Security Nexus*. Stockholm: Stockholm Environment Institute.

ICWE. (1992). "The Dublin Statement and report of the conference. International Conference on Water and the Environment: Development issues for the 21st century. 26–31 January 1992," Dublin, Ireland.

Jensen, K. M. and Lange, R. B. (2013). "Transboundary Water Governance in a shifting development context. A comparative study of the Mekong and the Zambezi River Basins." *DIIS Report 2013*: 20, Copenhagen.

Jensen, K. M., Lange, R. B. and Refsgaard, J. C. (2018). "Pigs, Prawns and Power Houses: Politics in Water Resources Management." In L. A. Swatuk and C. Cash, (eds.) *Water, Energy, Food and People Across the Global South: Perspectives on the Nexus in an Era of Climate Change*. Palgrave, Macmillan, pp. 79–116.

Jønch-Clausen, T. (2004). "Integrated Water Resources Management (IWRM) and Water Efficiency Plans by 2005. Why, What and How"? *Global Water Partnership*. TEC Background Papers No. 10, Stockholm.

Jørgensen, L. F., Villholth, K. G. and Refsgaard, J. C. (2017). "Groundwater Management and Protection in Denmark – A Review of Pre-Conditions, Advances and Challenges." *International Journal of Water Resources Development*, 33(6), 868–889.

Karlsson, M., Gilek, M. and Lundberg, C. (2016). "Eutrophication and the Ecosystem Approach to Management: A Case Study of the Baltic Sea Environmental Governance." In M. Gilek et al., (eds.) *Environmental Governance of the Baltic Sea. MARE Publication Series 10*, Chapter 2, pp. 21–44.

Loucks, D. P. and van Beek, E. (2005). "Water Resources Systems Planning and Management. An Introduction of Methods, Models, and Applications." *Studies and Reports in Hydrology*. Paris: UNESCO.

Molle, F. (2008). "Nirvana Concepts, Narratives and Policy Models: Insights from the Water Sector." *Water Alternatives*, 1(1), 131–156.

Mollinga, P. P. (2008). "Water, Politics and Development: Framing a Political Sociology of Water Resources Management." *Water Alternatives*, 1(1), 7–23.

Newig, J., Pahl-Wostl, C. and Sigel, K. (2005). "The Role of Public Participation in Managing Uncertainty in the Implementation of the Water Framework Directive." *European Environment*, 15, 333–343.

Pahl-Wostl, C. (2007). "Transitions Towards Adaptive Management of Water Facing Climate and Global Change." *Water Resources Management*, 21(1), 49–62.

Reusch, T. B. H., Dierking, J., Andersson, H., Bonsdorff, E., Carstensen, J., Casini, M., Czajkowski, M., Hasler, B., Hinsby, K., Hyytiäinen, K., Johannesson, K., Jomaa, S., Jormalainen, V., Kuosa, H., Kurland, S., Laikre, L., MacKenzie, B. R., Margonski, P., Melzner, F., Oesterwind, D., Ojaveer, H., Refsgaard, J. C., Sandström, A., Schwarz, G., Tonderski, K., Winder, M. and Zandersen, M. (2018). "The Baltic Sea as a Time Machine for the Future Coastal Ocean." *Science Advances*, 4, eaar8195.

Savannan, V., McDonald, G. and Mollinga, P. (2008). "Critical Review of Integrated Water Resources Management; Moving Beyond Polarised Discourse." *ZEF Working Paper Series*, 29, 1–23.

Swatuk, L. A. (2008). "A Political Economy of Water in Southern Africa." *Water Alternatives*, 1(1), 24–47.

UN. (1973). "Report of the United Nations Conference on the Human Environment," *Stockholm*, 5–16 June 1972.

Voulvoulis, N., Arpon, K. D. and Giakoumis, T. (2017). "The EU Water Framework Directive: From Great Expectations to Problems With Implementation." *Science of the Total Environment*, 575, 358–366.

WWAP. (2003). "Water for people, water for life: The United Nations world water development report; A joint report by the twenty-three UN agencies concerned with freshwater." *UNESCO World Water Assessment Programme*. https://unesdoc.unesco.org/ark:/48223/pf0000129556

WWAP. (2019). "The United Nations World Water Development Report: Leaving No One Behind." *UNESCO World Water Assessment Programme*. https://www.unwater.org/publications/un-world-water-development-report-2019

17

WATER, PARTICIPATION AND DEVELOPMENT

Jeroen Warner and Richard Meissner

Introduction

From around the 1980s, participation has become popular in development projects to promote beneficiary "buy-in" and enhance legitimacy for interventions affecting local people's lives – ostensibly putting "people" central. Hitherto, projects had often failed to address the needs of the supposed beneficiaries and, at times, proved an ailment that was worse than the disease. Knowledge diffusion ("extension") tended to be one-way, assuming adoption by intended beneficiaries through exposure.

The present contribution explores the rise and promise of community participation in development, its setbacks, and the belated discovery by its practitioners in the water sector that power is fundamental to participatory processes.

We shall trace how the water sector belatedly caught up on participation in the late 1990s, despite the early wave of participation in development having been spearheaded by the apparent success of a participatory water project: Uphoff et al. (1985) observed how irrigator groups organized as Water Users Associations in Gal Oya, Sri Lanka, managed to significantly raise rice productivity in the face of water shortages and civil war. They were not only given a say on tertiary irrigation channels, but also on the irrigation system as a whole. Soon, participation became the next big thing in development and irrigation systems were part and parcel of this. Participation held out the prospect of more equitable and effective development outcomes, as well as more sustainable natural resource use.

Participation at all levels was one of the four internationally agreed guiding water management principles agreed at the International Conference on Water and the Environment (ICWE) in Dublin in 1992, requiring the involvement of water planners, managers, and users. Next, these principles were submitted to the UNCED (Sustainable Development) summit in Rio de Janeiro, the same year. The partnership model between nine "major groups" in society identified at that summit also proved a kind of model for this (Hemmati, 2012).[1] Multi-Stakeholder fora, Learning Alliances, innovation platforms, and more recently benefit-sharing dialogues mushroomed, seeking to represent a diversity of actors and interests, and harness their knowledge in a deliberative process to tackle common challenges. Participatory development also promised power sharing. Various ladders of participation were developed to measure the

DOI: 10.4324/9781003095545-20

degree of self-rule in such processes. The best-known ladder, developed by Arnstein (1969), has since been elaborated on by Bruns (2003) and others. Considering this emancipatory idea, it is surprising it took so long for power issues to gain serious standing in the participatory development literature.

The involvement of civil-society actors was expected to improve political support in implementing dialogue outcomes. Local involvement was hoped to tap local knowledge (enhancing the quality of decision-making) and mobilizing the support of the wider constituency outside of the participatory processes (Warner, 2005; Faysse, 2006; Carr et al., 2012). Bottom-up information-gathering techniques like Participatory Rural Appraisal (Chambers, 1994) promised a bottom-up and informed by the perspective and analytical capacities of the beneficiary community to gauge the development preferences and increase the support base in the implementation of interventions, trumping the usual tarmac-based "development tourism" of development project assessment.

As liberal states increasingly took a back seat, the state-led focus for development morphed into a diversity focus inviting a range of stakeholders to the table. Representing a diversity of stakeholders in dialogue carries the promise of continued support-generating capacity throughout the project cycle (Warner, 2016). The belief that no organization has the monopoly on knowledge, capacities and/or resources to solve independently complex social problems (Kooiman, 1993) promoted "multi-stakeholder platforms" as a governing system reflecting the complexity, diversity and dynamics of the governed system ("Ashby's Law") (Ashby & Goldstein, 2011).

The vogue for co-production and co-management in land-use planning (Aarts & Leeuwis, 2010) and the rise of network (Ansell & Gash, 2008) or collaborative governance (Margerum, 2011) is often contrasted with command-and-control regulation. State retreat, decentralization, subsidiarity, and public-private partnerships were the order of the day, for better and for worse. Transnational multi-stakeholder initiatives (MSIs) such as commodity roundtables aimed to promote good practices (Cheyns & Riisgaard, 2014).

By the turn of the century, participation was both hailed as "the new paradigm" in development in the 1990s (Chambers, 1997) and reviled as the new "tyranny." In the eyes of some, the assumption that participation brings better development showed fatal methodological flaws (Cooke & Kothari, 2001). A noticeable participation fatigue was prominent in practice as beneficiaries of development were "being participated" to distraction, mainly for the institutional needs of development bureaucracies. As both present authors can attest, community ranking exercises risked becoming a predictable piece of development theatre, easily dominated by local powerholders, and with depressingly little to show for them on the ground.

Participation in development is the process by which stakeholders can influence and share control over development initiatives and over the decisions and resources that affect them. More than "having a say" – it is held to be about taking responsibility for actions affecting other users and sectors. The corollary is the empowerment of economically and socially marginalized actors over their own lives (Guijt, 1998).

Participatory Water Management

At the turn of the millennium, the water sector experienced its own belated participatory turn, as a modality for (intersectoral) social learning (Steins & Edwards, 1999) and Alternative Dispute Resolution (Ramirez, 2001). Central to this thinking is social learning about complex resource management issues in platforms for rational dialogue as a conducive environment for social learning about a shared issue that will not go away, in so doing presenting a joint interest. Brouwer et al. (2016) moreover identify multiple loop learning as required to break through deadlocks.

Solving conflict through inclusive consensus-seeking rather than by majority decision is second nature to the Dutch "polder" model but promised a breath of fresh air in more adversarial cultures such as the United States or France. They seem particularly promising when problems are "wicked" and stakeholders are interdependent, which tend to be characteristics of contemporary water problems. Indeed, the water sector has become so smitten with the MSP concept that the Global Water Partnership made MSPs for improved water governance fundamental to its 2020–2030 strategy (MSPs – GWP).

Forerunners such as South Africa's progressive post-apartheid National Water Act (No. 36 of 1998) (RSA, 1998) and the European Water Framework Directive of 2000 both explicitly called for community and multi-stakeholder participation and a vogue for dialogues such as the Dialogue on Water, Food, and Climate was enthusiastically embraced by states, NGOs, and multilaterals alike. In South Africa, multi-stakeholder participation is facilitated by catchment management agencies acting as the promoters of the participation within the water management area under their jurisdiction (RSA, 1998; Meissner et al., 2016). Förster et al. (2016) "unmasked" the power asymmetries among actors during the establishment of a water user association in South Africa, resulting in a "flawed" establishment process from the mistaken assumption everyone will benefit equally.

Mandating stakeholder participation in legally binding documents would seem progress, but the nonvoluntary nature can also reduce motivation. Wesselink et al. (2011) found legally required ("legalistic") participation a prominent rationale for engaging in multi-stakeholder participation, which does not encourage opening the floor to more deliberative forms of decision-making. As a new "beasts" in the institutional ecology, MSPs created tensions with existing local and supralocal authorities (Warner, 2016; Mees et al., 2017).

The water sector was originally prone to taking a "not invented here" attitude to lessons learned in other domains where participatory processes had been tried before, such as in agrarian development projects in the 1970s. In community forestry, the power asymmetry issue has been addressed early on (Edmunds & Wollenberg, 2001, also Evans et al., 2021), well before the water sector came to terms with it.

The lack of lessons drawn from other sectors led to the repeated reinvention of the wheel and continued naiveté about power. While scholars of water and participation also echoed the earlier misgivings of the development sector (Faysse, 2006) and proposed reforms (Warner, 2006), this more critical work appeared to be largely ignored for the next decade. Most water MSP designs, such as integrated water resources management and adaptive water governance remained conspicuously "power-blind," wedded to sitting people round the table, creating the conditions for uninhibited deliberation, and may the best argument win. However, as the next section will argue, if power differences are ignored in the MSP design, its participants are set up for failure.

An underappreciated key weakness is the hydro-centricity of water governance. Many water problems do not originate in the water sector and cannot be solved in the water sector alone. The current vogue for the water-food-energy nexus at least brings food and energy actors to the table (Keskinen et al., 2016). In that sense, the current nexus approach (Hoff, 2011) is a step forward. The water-energy-food nexus considers the needs and interoperability of virtual water for food and water for energy, often at basin or landscape level. This nexus approach, however, tends to remain explicitly related to an ecosystem or common-pool resource. Such eco-centricity promises more sustainable water governance but ignores administrative, political, and cultural realities (Warner and Simpungwe, 2012), as people do not necessarily identify with an ecosystem. Except for small streams, the river basin may not be local enough for stakeholders to relate to it, so that sub-basin MSPs may be preferable (Mees et al., 2017).

Power-sensitive MSPs?

Despite their obvious limitations, expectations of transformative capacity remain high (Sarmiento Barletti et al., 2021). The mere event of convening a stakeholder forum is imbued with the expectation of magical emancipatory powers of balancing inequalities between stakeholders. The outcome the organizers have in mind, however, is not challenged, and as such the platforms reinforce the status quo. Cohen and Davidson (2011, fn. 3) point out that states may be ready to devolve responsibility, they are unwilling or unable to devolve authority. It is clear, that a "democracy first and socio-economic development will follow" mantra, has proven flawed time and again (Stel et al., 2012: 64). In a recent study of social networks and cooperation, Atwell and Nathan (2021) found that hierarchical positions in social networks "predict" collaboration. They found no evidence that communication increases coordination or cooperation, rather, they actively propagate social influence to tackle collective action problems.

The participation literature seems to increasingly wake up to the political. Since the 2010s, there has been greater recognition of the power aspect, as part of a broader power-sensitive wavein the governance literature (Brouwer et al., 2013). It is now dawning on the MSP community that it makes no sense to assume communities, or even households, have a common interest (Sarmiento Barletti et al., 2021) and that it is not enough to be "putting the right people in one room" (Hiemstra et al., 2012) or to assume that communication improves cooperation (Atwell & Nathan, 2021). The ideal of MSPs as an open space to freely express opinions, informed by the optimistic writings of Jürgen Habermas, gave way to a view of MSP as a space to express power relations (Hiemstra et al., 2012). Gaventa's (2019) power cube is a helpful tool to discuss different forms of power with an emancipatory objective. More powerful participants do not even have to use their privilege to intimidate other people; others will not dare to contradict them anyway. This moves the answer to the question: Why do people participate? away from instrumental (more effective and legitimate decisions, policy goals are fixed) and substantive (broader knowledge base, broadening the issue, policy goals may be rethought) to normative (can MSP mean countervailing power to elites?) (Wesselink et al., 2011).

Translation into local languages, tailored information, training, awareness workshops and experience-sharing activities (Bloomfield et al., 2019) can support stakeholders to address complex water planning issues – technology such as MapTables (interactive digital design tables) can assist here. Platforms can specifically support participant skill-building, learning to negotiate, or developing as a "partisan MSP" (Brouwer et al., 2013) or action platform as an alliance of opponents of an intervention (Warner, 2005). Separate pre-meetings (e.g., women-only or landless-only) supported in their participation by empowering facilitation can somewhat remedy this power-blindness.

Consensus or Compromise?

Consensus-seeking has often been regarded as the best possible way of ameliorating pressing environmental problems. Universities or environmental NGOs such as IUCN may function as "honest brokers" (Huntjens et al., 2017), acting in the best interest of the environment and those dependent on it. The assumption is that trust building can bring a dialogue overcoming political and institutional barriers. Barua and Vij (2018: 1038), for example refer to "open discussion among multiple stakeholders free from bureaucratic constraints" presumably under Chatham House rules of confidentiality. Such spaces can indeed be safe havens in a highly politicized environment (Santbergen, 2016).

This approach is also realistic (and power-sensitive) in that it accepts that states will jealously guard their sovereignty, loath to delegate responsibility to external multilateral authorities or responsibilities to domestic actors. As a result, however, state actors tend to take a Decide, Announce, Defend approach that brooks no modification.

It is, however, increasingly recognized that conflict between stakeholders is not always a bad thing. While earlier MSP analysts have tended to treat conflict as an aberration, it is now more readily recognized that various forms of (nonviolent) conflict can be necessary catalysts to propel systems towards betterment. After all, conflict reveals the state of play better than harmony and as Forester (2006) has noted, conflict is better than apathy. Compromise rather than consensus may well be the best that can be hoped for.

The naiveté of a focus on consensus within MSPs may be worse than the conflict that could ensue to thrash out issues where power asymmetry is a palpable reality. After all, mandated multi-stakeholder representation in an organization does not automatically translate into power and influence in such a structure; inclusivity is not only about rules written down on paper (e.g., South Africa's National Water Act) or being permitted in a committee or board, but about the agency of different stakeholders to influence decision-makers at various societal and government levels as well as institutional processes. Even if included in an organizational structure, emerging farmers will likely have less power than commercial farmers, due to the dearth of resources and communication channels at their disposal (Meissner & Funke, 2014). A South African example of institutionalized multi-stakeholder participation may illustrate the argument (Box 17.1).

Box 17.1 Participatory basin management: the experience of South Africa's CMAs

South Africa's water law sought to institutionalize multi-stakeholder participation in Catchment Management Agencies (CMAs). Since the National Water Act No. 36 of 1998 (RSA, 1998), the national Department of Water and Sanitation (formerly the Department of Water Affairs), initiated the idea of decentralized water resources management, with CMAs as its main vehicle for this. The initial plan was to create a CMA for every water management area, but after a promising start, this effort appears to have run out of steam (Meissner et al., 2016; Munnik, 2020). The intended number, 19 CMAs, was revised to nine, as the initial idea became a mammoth task for a fledgling democracy with limited resources and capabilities. Only two CMAs are presently operational: the Breede Gouritz and the Inkomati Usuthu. The remaining seven CMAs are at various stages of establishment: making them proto-CMAs enabling the spread of service provision (Meissner et al., 2016). There are also more formal and informal South African catchment fora, water user associations and irrigation boards, operational on smaller rivers like the Kat and Mthatha (Warner & Simpungwe, 2012). The decision to establish nine CMAs was based on financial considerations since 19 CMAs would be too costly to establish and operate.

The existence of CMAs places a central focus on stakeholder engagement and inclusivity since the Water Act makes provision for stakeholders to participate in a variety of institutional settings: catchment forums, catchment steering committees and as part of the CMA governing board. There is, therefore, a plethora of stakeholders involved in the CMA's organizational structures to represent as many sectors as possible including academics, environmental interest groups, local authorities and provincial government departments, business and industry, poor and rural settlements, agriculture, and NGOs (Meissner et al., 2016).

Power dynamics were an underestimated key factor inhibiting South African CMAs taking off. For instance, commercial farmers within a catchment management agency context, hold far better cards than emerging farmers due to the former's access to financial resources and negotiation capabilities. The commercial farmers, furthermore, have a vested interest for their views to be heard and translated into power dynamics (Meissner & Funke, 2014); a commercial farm in an irrigation scheme does not come cheap, loans have to be serviced, input costs must be recovered, and workers must receive wages. There is just too much at stake to involve everyone, let alone on an equal footing.

In the establishment of CMAs such as the Breede-Gouritz CMA, the scientific community and particularly individuals from engineering, hydrology and soil sciences exerted a strong influence on the authority's Catchment Management Strategy (CMS) formulation process. Technocracy had excelled at getting water to people bringing an impressive reduction of backlog access to water. The Nexus and like-minded concepts such as Green and Ecological Infrastructures has moved the goalposts in basin management. But many substantial research reports on the way in which CMAs should function and be managed regarding the Act, pro-poor governance approaches to rid the sector of inequality, and the notion of sustainability, were produced by former employees of DWA that became independent consultants (Meissner & Funke, 2014; Munnik, 2020).

Non-participation

Non-participation and exclusion may be considered failures of the democratic aspiration of multi-stakeholder processes (Flannery et al., 2018), when those who are not well-organized or have high opportunity costs decline attendance. The power asymmetries flagged up above extends to knowledge, negotiation skills and opportunity costs. Participants are often not trained to deal with complex, often highly technical issues with many uncertainties that preclude immediate feedback on the impact of the different alternatives brought forward. If the issues are complex, if language is jargony and if stakes and benefits (food on the table) are unclear or rendered technical (Vink, 2015), the opportunity cost of participation starts to sting especially for marginalized actors (Warner, 2005). Stakeholders may lose interest and or self-exclude, resorting to non-MSP alternatives such as direct access to power holders or mobilize an alternative arena (forum shopping). For some it might even be more advantageous to wait on the sidelines until things start to happen, which is the Best Alternative to a Negotiated Agreement (BATNA) option.

Focusing only on the actors represented at the table moreover diverts the focus away from the decision-makers behind the scenes, such as international investors, financiers, and multinational enterprises, who are unlikely to show up for a local or national multi-stakeholder consultation (De Groot & Warner, 2011). While unrepresented at the table, they may be the ones to make the real difference as veto actors, pulling strings, setting limits to action, and controlling the (non)agenda.

Powering and Puzzling

As noted, the tendency for organizers of participatory processes to promote consensus-seeking has led to a tendency to "puzzle" one's way together out of a shared problem situation, rendering it technical rather than political (Li, 2011). Where the platforms have a set, joint product – a map, a vision – this puzzling makes good sense. Drawing attention to the workings of power relations between dialogue participants predictably does not sit well with all involved. As power holders tend to resist discussing their power, Hiemstra et al. (2012) advocate using euphemisms like trust building.

But we may note that without a mandate and/or the mobilization of resources and power brokers, the link with the formal decision-making remained tenuous and the MSP ineffective. As a result, responsible management agencies may be unwilling to comply with the MSP's decisions, that could raise serious issues around accountability, raising questions such as: who is accountable to whom? Ignoring power asymmetries and hierarchy, however, is also a missed opportunity to facilitate "powering" (Warner & van Buuren, 2016). For this, a degree of inequality may even be beneficial. With hegemonic power comes responsibility, although this does not always translate into responsible leadership in practice (Zeitoun and Warner, 2006). More powerful players sympathetic to communities can however strengthen the latter's bargaining position (Köhne 2014). An authoritative chair/facilitator with the wherewithal to mobilize material and power resources can get much more done, though countervailing power/checks and balances will remain essential. Whether this set-up can work also appears to be dependent on the personality types of authority figures in a platform.

Limiting or Extending Participation?

While promising democratization of water governance, participatory platforms are not appropriate or even feasible in all contexts. The degree of problem structuring can make a water policy problem more or less suitable for multi-stakeholder participation – for example, too much conflict may forestall frank exchange, while too little conflict may make a task force more appropriate (Hurlbert & Gupta, 2015; Verhallen et al., 2016).

As a remedy to romanticized participation, Margot Hurlbert and Joyeeta Gupta developed a "split participation ladder" arguing participation is "not always necessary, not always useful, and may not always lead to consensus"; "different levels of engagement are likely appropriate in different contexts" (both quotes from Hurlbert & Gupta, 2015: 101). Indeed, water MSPs have run into similar problems as other MSP types. Benson et al. (2008) show that the normative desire underlying MSPs have outpaced the empirical shift to collaborative water partnerships on the ground. The contrast of top-down command-and-control regulation (from government to governance) is equally fictional in many contexts. Echoing Cooke and Kothari (2001), Rap et al. (2019) claim that if Participatory Irrigation MSPs are not strengthened, they may as well be abandoned. Indeed, public participation has been slammed as no more than a de-politicization strategy obscuring underlying power relations (Duncan & Claeys, 2018).

However, this criticism appears to refer to organized participation processes; unexpected or unintentional participants may manifest in- and outside the organized spaces (Turnhout et al., 2010). If we extend participation to social actors taking the arena, invited or uninvited, the picture looks far more dynamic and unpredictable.

And that is only to include the human participants. Of late, the eco-centric tendency in the water development sector has been boosted by a wider take on diversity incorporating the dialogue between multiple ontologies in the "pluriverse." Indigenous groups especially, often have very different cosmologies from Western cosmologies, experiencing rivers, rocks, and Mother Earth as spirits, so that a universalizing value system cannot be imposed (Escobar, 2018). Over the past few decades, we saw this playing out in the construction of large dams, where indigenous peoples have resisted the "Westernised" ideas of development through large infrastructure projects (Meissner & Warner, 2021). The development of assigning legal rights to non-human entities such as rivers in Australia, Canada, Ecuador, India, and New Zeeland may well promote sustainability, but could reinforce this depoliticizing tendency. Encouragingly, though, New Zealand has explicitly mandated a multi-stakeholder governing structure here (Kauffman & Martin, 2012; Gilissen et al., 2019).

Conclusion

Participation opens vistas of alternative, deliberative forms of decision-making, commanding greater legitimacy and social sustainability in development than majority rule. Notwithstanding its promise, participatory platforms however rarely have the funding and mandate at their disposal to solve real-life problems. The tools of participatory development are often "expropriated" by local elites creating a kind of "participatory exclusion" (Nadiruzzaman & Wrathall, 2015). It is still remarkably easy for decision-makers to ignore or dismiss the outcomes of participatory deliberation, particularly when the governing structure is populated by illiberal governors.

Participation, then, does not necessarily solve resource management problems or even bring decentralization (Sertas, 2018). One of the main reasons for the slow space in establishing South Africa's CMAs is the tendency of certain factions of the ruling African National Congress (ANC) to resist decentralization. Participatory platforms, however, have been shown to facilitate the development of social capital and bring different types of knowledge to the table (Tremblay-Lévesque et al., 2022).

Finally, it was noted in this contribution that the focus on consensus can make actors lose interest – in fact, without "healthy" conflict MSPs could wither away and achieve nothing. Also, if we look beyond formal participatory practice, we note that others may manifest in the arena as well outside the controlled "box" of multi-stakeholder dialogue. Getting to know and listening closely to such dissenters may increase the space to solve, or accommodate, the issues in hand.

Note

1 The chapter will not touch on higher-level multi-stakeholder policy processes such as the global processes described in Hemmati (2012).

References

Aarts, N. and Leeuwis, C. (2010). "Participation and power: reflections on the role of government in land use planning and rural development." *Journal of Agricultural Education and Extension*, *16*(2), 131–145.

Ansell, C. and Gash, A. (2008). "Collaborative governance in theory and practice." *Journal of Public Administration Research and Theory*, *18*(4), 543–571.

Arnstein, S.R. (1969). "A ladder of citizen participation." *Journal of the American Institute of Planners*, *35*(4), 216–224.

Ashby, W. R. and Goldstein, J. (2011). "Variety, constraint, and the law of requisite variety." *Emergence: Complexity and Organization*, *13*(1/2), 190.

Atwell, P. and Nathan, N. L. (2021). "Channels for influence or maps of behavior? A field experiment on social networks and cooperation." *American Journal of Political Science*, *65*(1), 1–18.

Bakker, K. and Morinville, C. (2013). "The governance dimensions of water security: a review." *Philosophical Transactions of the Royal Society A: Mathematical, Physical and Engineering Sciences*, *371*(2002), 20130116.

Barua, A. and Vij, S. (2018). "Treaties can be a non-starter: a multi-track and multilateral dialogue approach for Brahmaputra Basin." *Water Policy*, *20*(5), 1027–1041.

Benson, D., Jordan, A., Cook, H. and Smith, L. (2013). "Collaborative environmental governance: are watershed partnerships swimming or are they sinking?" *Land Use Policy*, *30*(1), 748–757.

Bloomfield, G., Meli, P., Brancalion, P. H. S., Guariguata, T. and Gare, M. R., E. (2019). "Strategic insights for capacity development on forest landscape restoration: implications for addressing global commitments." *Tropical Conservation Science*, 12, 1–11.

Brouwer, H., Hiemstra, W., van Vugt, S. and Walters, H. (2013). "Analysing stakeholder power dynamics in multi-stakeholder processes: insights of practice from Africa and Asia." *Knowledge Management for Development Journal*, *9*(3), 11–31.

Brouwer, H., Woodhill, J., Hemmati, M., Verhoosel, K. and van Vugt, S. (2016). *The MSP Guide: How to Design and Facilitate Multi-Stakeholder Partnerships*. Rugby, Warwickshire: Practical Action Publishing.

Bruns, B. (2003). "Water tenure reform: developing an extended ladder of participation." In: *Politics of the commons: Articulating development and strengthening local practices*, July 11–14, 2003, Chiang Mai, Thailand.

Carr, G., Blöschl, G. and Loucks, D. P. (2012). "Evaluating participation in water resource management: a review." *Water Resources Research*, 48(11), 1–17.

Chambers, R. (1994). "The origins and practice of participatory rural appraisal." *World Development*, *22*(7), 953–969.

Chambers, R. (1997). *Whose Reality Counts* (Vol. 25). London: Intermediate Technology Publications.

Cheyns, E. and Riisgaard, L. (2014). "Introduction to the symposium: the exercise of power through multi-stakeholder initiatives for sustainable agriculture and its inclusion and exclusion outcomes." *Agriculture and Human*, *31*(3), 409–423.

Cohen, A. and Davidson, S. (2011). "The watershed approach: challenges, antecedents, and the transition from technical tool to governance unit." *Water Alternatives*, *4*(1), 1–14.

Cooke, B. and Kothari, U. (eds.). (2001). *Participation: The New Tyranny?* London: Zed Books.

De Groot, W. T. and Warner, J. F. (2011). "From "All Stakeholders" to balanced process in participatory planning: A practical, stepwise ethics." In: Eliot, H. S. and Martin, L. E. (eds.). *River Ecosystems: Dynamics, Management and Conservation*. Pp. 179–202.

Duncan, J. and Claeys, P. (2018). "Politicizing food security governance through participation: opportunities and opposition." *Food Security*, *10*(6), 1411–1424.

Edmunds, D. and Wollenberg, E. (2001). "A strategic approach to multistakeholder negotiations." *Development and Change*, *32*(2), 231–253.

Escobar, A. (2018). *Designs for the Pluriverse*. Durham, NC: Duke University Press.

Evans, K., Monterroso, I., Ombogoh, D. B., Liswanti, N., Tamara, A., Mariño, H., Sarmiento Barletti, J. P. and Larson, A. M. (2021). *Getting It Right, a Guide to Improve Inclusion in Multistakeholder Forums*. Bogor, Indonesia: CIFOR Knowledge.

Faysse, N. (2006). "Troubles on the way: an analysis of the challenges faced by multi-stakeholder platforms." *Natural Resources Forum*, *30*(3), 219–229.

Flannery, W., Healy, N. and Luna, M. (2018). "Exclusion and non-participation in Marine Spatial Planning." *Marine Policy*, *88*, 32–40.

Forester, J. (2006). "Making participation work when interests conflict: moving from facilitating dialogue and moderating debate to mediating negotiations." *Journal of the American Planning Association*, *72*(4), 447–456.

Förster, J. J., Downsboroygh, L. and Chomba, M. J. (2016). "When policy hits practice: structure, agency, and power in South African water governance." *Society and Natural Resources*, *3*(30), 521–536.

Gaventa, J. (2019). "Applying Power Analysis: Using the 'Powercube' to Explore Forms, Levels and Spaces". In: McGee, R. and Pettit, J (eds.). *Power, Empowerment and Social Change*. London/New York: Routledge. pp. 117–138.

Gilissen, H. K., Suykens, C., Kleinhans, M., Van Rijswyk, M. and Van der Werf, K. (2019). "Towards a rights-based approach in EU international basin governance? Lessons from the Scheldt and EMS basins." *Water International*, 44, 6–7.

Guijt, I. (1998). *Participatory Monitoring and Impact Assessment of Sustainable Agriculture Initiatives: an Introduction to the Key Elements*. London: IIED.

Hemmati, M. (2012). *Multi-Stakeholder Processes for Governance and Sustainability – Beyond Deadlock and Conflict*. London: Earthscan/Routledge.

Hiemstra, W., Brouwer, J. H. and Vugt, S. M. (2012). *Power Dynamics in Multi-Stakeholder Processes: A Balancing Act*. Wageningen: Centre for Development and Innovation.

Hoff, H. (2011). *Understanding the Nexus*. Background paper for the Bonn 2011 Nexus conference: The Water, Energy and Food Security Nexus. Stockholm: Stockholm Environment Institute.

Huntjens, P., Lebel, L. and Furze, B. (2017). "The effectiveness of multi-stakeholder dialogues on water: reflections on experiences in the Rhine, Mekong, and Ganga-Brahmaputra-Meghna River basins." *International Journal of Water Governance*. 5, 39–60.

Hurlbert, M. and Gupta, J. (2015). "The split ladder of participation: a diagnostic, strategic, and evaluation tool to assess when participation is necessary." *Environmental Science & Policy*. 50, 100–113.

Kauffman, C. M. and Martin, P. L. (2012). "How courts are developing river rights Jurisprudence: comparing guardianship in New Zealand, Colombia, and India." *Journal of Environmental Law*, 547, 547–48.

Keskinen, M., Guillaume, J. H., Kattelus, M., Porkka, M., Räsänen, T. A. and Varis, O. (2016). "The water-energy-food nexus and the transboundary context: insights from large Asian rivers." *Water*, *8*(5), 193.

Köhne, M. (2014). "Multi-stakeholder initiative governance as assemblage: roundtable on sustainable palm oil as a political resource in land conflicts related to oil palm plantations." *Agriculture and Human Values*, *31*(3), 469–480.

Kooiman, J. (ed.) (1993). *Modern Governance: New Government-Society Interactions*. Sage.

Li, T. M. (2011) "Rendering society technical: government through community and the ethnographic turn at the World Bank in Indonesia." In: Mosse, D. (ed.). *Adventures in Aidland: The Anthropology of Professionals in International Development*. Oxford: Berghahn. pp. 57–80.

Margerum, R. D. (2011). *Beyond Consensus: Improving Collaborative Planning and Management*. Cambridge, MA: MIT Press.

Mees, H., Suykens, C. and Crabbe, A. (2017). "Evaluating conditions for integrated water Resource management at sub-basin scale. A comparison of the Flemish sub-basin boards and Walloon River contracts". *Environmental Policy and Governance*, 27, 59–73.

Meissner, R. and Funke, N. (2014). "The politics of establishing catchment management agencies in South Africa: the case of the Breede–Overberg Catchment Management Agency." In: Huitema, D. and Meijerink, S.V. (eds.). *The Politics of River Basin Organisations: Coalitions, Institutional Design Choices and Consequences*. Pp. 184–209.

Meissner, R., Funke, N. and Nortje, K. (2016). "The politics of establishing catchment management agencies in South Africa: the case of the Breede-Overberg catchment management agency". *Ecology and Society*, 21(3), 26. http://dx.doi.org/10.5751/ES-08417-210326

Meissner, R. and Warner, J. (2021). "Indigenous paradiplomacy and the Orokawe hydroelectric dam on the Kunene River." *Regions and Cohesion*, *11*(1), 21–48.

Munnik, V. (2020). *The Reluctant Roll-Out of Catchment Management Agencies: Assessing the Key Risks and Consequences of Delays in Finalising Institutional Arrangements for Decentralised Water Resource Management*. Pretoria: Water Research Commission. Report No. 2943/1/20.

Nadiruzzaman, M. and Wrathall, D. (2015). "Participatory exclusion–Cyclone Sidr and its aftermath." *Geoforum*, *64*, 196–204.

Ramirez, R. (2001). "Understanding the approaches for accommodating multiple stakeholders' interests." *International Journal of Agricultural Resources, Governance and Ecology*, *1*(3–4), 264–285.

Rap, E., Molle, F., Ezzat El-Agha, D. and Abou El Hassan, W. (2019). "The limits to participation: branch-canal water user associations in the Egyptian Delta." *Water International*, *44*(1), 31–50.

RSA (Republic of South Africa). 1998. *National Water Act No. 36 of 1998*. Pretoria: Government Printer.

Santbergen, L. (2016). "The International Zwin Commission: the beauty of a mayfly?" In: Warner, J. (ed.). *Multi-stakeholder Platforms for Integrated Water Management*. Aldershot: Ashgate, pp. 111–124.

Sarmiento Barletti, J. S., Larson, A. and Vigil, N. H. (2021). "Organizing for transformation? How and why organizers plan their multi-stakeholder forums." *International Forestry Review*, *23*(1), 1.

Sertas, M. (2018). *Do Multi-Stakeholder Platforms Work?* PhD Dissertation. Wageningen University. [Accessed on: 30 December 2022]. Available at https://edepot.wur.nl/459129

Steins, N. A. and Edwards, V. M. (1999). "Platforms for collective action in multiple-use common-pool resources." *Agriculture and Human Values*, *16*(3), 241–255.

Stel, N., Boer, D., Hilhorst, D., Haar, G., van der Molen, I., Douma, N. and Mostert, R. (2012). *Multi-Stakeholder Processes, Service Delivery and State Institutions*; Synthesis report. Peace, Security and Development Network.

Tremblay-Lévesque, L.-C., Warner, J., Gribanenkova, I. and Robbins, M. (2022). *The MSP Sourcebook: A Guide for Multi-Stakeholder Partnerships in Water Management. Global Water Partnership*. Stockholm, Sweden: Wageningen University and CapNet.

Turnhout, E., Van Bommel, S. and Aarts, N. (2010). "How participation creates citizens: participatory governance as performative practice". *Ecology and Society*, 15(4).

Uphoff, N., Meinzen-Dick, R. and Julien, N. S. (1985). "Improving Policies and Programs for Farmer Organization and Participation in Irrigation Water Management". *Water Management Synthesis II*. Ithaca, N.Y.: Cornell University Press.

Verhallen, A., Warner, J. and Santbergen, L. (2016). "Towards Evaluating MSPs for Integrated Catchment Management." In: Warner, J. (ed.). *Multi-Stakeholder Platforms for Integrated Water Management*. Aldershot: Ashgate, pp. 275–288.

Vink, M. (2015). *Navigating frames: a study of the interplay between meaning and power in policy deliberations over adaptation to climate change.* Doctoral dissertation, Wageningen University.

Warner, J. (2005). "Multi-stakeholder platforms: integrating society in water resource management?" *Ambiente & Sociedade*, *8*(2), 1–20.

Warner, J. (ed). (2016). *Multi-Stakeholder Platforms for Integrated Water Management.* Aldershot: Ashgate.

Warner, J. F. (2006). "More sustainable participation? Multi-stakeholder platforms for integrated catchment management." *Water resources development*, *22*(1), 15–35.

Warner, J. F. and Simpungwe, E. (2012). "La participation des usagers dans la gestion de l'eau: quelles perspectives pour les plateformes multiacteurs en Afrique Subsaharienne?" In: Julien, F. (ed.) *La gestion intégrée des ressources en eau en Afrique subsaharienne; Paradigme occidental, pratiques africaines.* Québec: Presses de l'Université de Québec. pp. 107–133.

Warner, J. F. and van Buuren, A. (2016). "Reframing long-term controversies in transboundary river management. The intermediate role of puzzling and powering in tackling wicked problems." *Futures*. 76, 18–29.

Wesselink, A., Paavola, J., Fritsch, O. and Renn, O. (2011). "Rationales for public participation in environmental policy and governance: practitioners' perspectives." *Environment and Planning A*, *43*(11), 2688–2704.

Zeitoun, M. and Warner, J. (2006). "Hydro-hegemony – a framework for analysis of trans-boundary water conflicts." *Water Policy*, *8*(5), 435–460.

18

CONFLICT AND COOPERATION OVER TRANSBOUNDARY WATERS

Jeroen Warner

Introduction

Water, so essential to life, has both political symbolism and strategic importance. The great ancient civilizations (such as Egypt, Mesopotamia, and India) developed around surface water resources (Nile, Euphrates-Tigris, and Ganges) that were shared across cultural, political, and administrative, borders. As they started to harness and develop their resources for irrigation, states inevitably impacted other actors' development opportunities and triggered potential conflicts of interest. As "the physical effects of our decisions spill across national frontiers" (WCED, 1987) the complex user interdependence of river, lake or aquifer riparians would be expected to incite a degree of coordination.

We cannot seem to agree on whether humans are fundamentally driven by self-interest or solidarity, and this is reflected in the water management literature. Water is considered to be "in crisis" (Gleick, 1993) and respected leaders have claimed that water is a catalyst for war (UN Secretary-Generals Serageldin and Ban-Ki Moon) or peace (e.g., South Africa's Kader Asmal). Given past droughts and floods and future common threats such as extreme floods and drought events, others would expect cooperation between upstream and downstream communities and leaders. But does water development actually promote cooperation and drive the creation of an international society? To tackle this conundrum, the present chapter provides an overview of different bodies of literature on transboundary interaction, engaging with both war and peace arguments while exploring alternative, more critical approaches, and ending in a reflection on whether water conflict and cooperation is really about water.

Water Peace?

Coined by Swedish *grande dame* of water Malin Falkenmark in 1998, the concept of hydro-solidarity institutionalizes interdependencies in river basins and calls for principles of ethics and equity. But this rejoinder is politically tricky as solidarity requires relinquishing freedom and territorial sovereignty (Falkenmark & Folke, 2002).

Hydro-solidarity echoes institutionalist and functionalist theories of regional integration such as the "English School" or liberal realists, led by Hedley Bull (Bull, 1979), predicting a spillover

DOI: 10.4324/9781003095545-21

from state collaboration from one domain into other domains, and from incidental to structural (regime) alignment into a "security community." The English School assumes that the anarchic international system would develop into regimes, creating mechanisms for coordination subscribed to voluntarily as a means of self-regulation in the international arena. The development of an international society of sovereign states interacting with and influencing each other will decrease the costs of cooperation while increasing them for those going it alone and/or starting entering into a conflict.

A related social constructivist perspective (Checkel, 1999; Wendt, 1999) applied to hydropolitics (Julien, 2012) highlights that state interests aren't cast in stone, as states can learn to redefine their interests and even their identity, say from a rogue state to a peacemaker. Focusing on norms and practices in the global arena, social constructivists see scope for change in how actors define their identities, their relations and problem definitions. This includes the meaning of "development" and "society." This may turn enemies into friends or non-committal passengers over time.

After the fall of the Berlin Wall, regime theory found water-sector application (Haftendorn, 2000; Jägerskog, 2003; Lindemann, 2008). Water regimes refer to patterned behaviour beyond treaties through the development of shared norms and common institutions lengthening the "shadow of the future." The Euphrates and Tigris rivers, for example, do not even have a comprehensive treaty, but show a high level of predictability. The International Law Commission developed principles of international water law, while the European Union has made great efforts to institutionalize the International Court of Justice (IJC) in The Hague to settle water disputes. During the 1990s the ICJ mediated between Hungary and Slovakia when the former defaulted on the joint Gabcikovo-Nagymaros dam project in 1989 but nothing much has happened since in the water domain.

Administrative and ecosystem boundaries do not coincide, costs and benefits from cooperation between states tend to be asymmetrical. Post-colonial states will strenuously assert their sovereignty over resources flowing within their territory.

A drive to promote global homogenization of water norms and governance (Gupta & Pahl-Wöstl, 2013) so far is not much in evidence, as those norms more often find themselves contested or ignored by neighbours engaged in dispute over one of the world's 276 transboundary rivers and lakes. It appears, however, that an immediate collective opportunity, "opportunitization" (Warner, 2004), can sometimes make states overcome their entrenched sovereignty concerns. Transboundary river basin organizations (Huitema & Meijerink, 2014) were established to support planned and coordinated regional resource development, such as the river La Plata in South America which serves the agro-export drive of its members: Argentina, Bolivia, Brazil, Paraguay, and Uruguay. But also previously, on the Asian river Mekong, the Mekong Basin Commission (MBC) was established as a Cold War ideological tool for joint hydropower development for Thailand, Lao PDR, Vietnam, and Cambodia (Sneddon, 2012). The prospect of joint benefits from and beyond the river (Sadoff & Grey, 2005), it is hoped, may offset power asymmetries and be a catalyst for regional cooperation beyond water (Wolf, 2004). Some countries have even invested in *joint risk-taking* projects for joint (regional) water development benefits such as the abortive Turco-Syrian Friendship Dam and India-Afghan Friendship Dam, emphasizing the historic commonalities between the neighbours. Small states such as Paraguay on the river La Plata, Lao PDR on the Mekong, and Lesotho on the Orange-Senqu have received investment and royalties from bigger neighbours to produce hydropower. However, the asymmetries between these neighbourly initiatives have led to rows over royalties and even military incursions (Meissner & Warner, 2017; Swatuk, 2012).

The fear of future water scarcity has triggered grandiose technical schemes, such as transboundary intra- and inter-basin transfers through long-distance "peace pipelines." Such pipeline projects may alleviate water scarcity, but do not only distribute water but also power to benefit some more than others (Turton, 2000). In the 1980s and 1990s, Turkish and Israeli engineers developed grandiose pipeline transfer schemes to provide the whole region with "excess" Turkish water. Recipients were not eager to be dependent on the vicissitudes of Turkish geopolitics. Likewise, Rhône water transfer to Barcelona got into political trouble, making Rinaudo and Barraqué (2015) wonder if the age of basin transfers might be over.

It is especially the common experience of a water crisis that appears to make states more prone to overcome their hesitation to coordinate their actions. Cooperation in risk mitigation has seen notable successes. Mostert (2005) has noted the importance of shock events such as disasters in international water regime formation, providing a window of opportunity for cooperation. The Rhine is an iconic example. The Rhine Action Plan was only implemented after the Sandoz chemical spill in Switzerland in November 1986 highlighted the interconnectivity of the river inciting joint emergency action. In Africa, prolonged droughts have precipitated interstate cooperation on the river Senegal (Kipping, 2009)[1] and Lake Victoria (Kagwanja, 2007). Taking this idea further, Ohlsson and Turton (2000) hypothesized that basin closure (the "turning of a screw") could lead to a reflexive turn towards demand management and cooperative institution building, provided states have sufficient "second-order" (human development) resources at their disposal (Balali, Keulartz & Korthals, 2009).

Trans-nationalists moreover, highlight the increasing role of non-state actors in network governance, predicting that complex interdependency, would increase interconnectedness. Where governments are not moving, multitrack diplomacy has sought to foster dialogues across borders. The Global Environmental Fund brokered a deal on the enormous, shared Guarani aquifer, and IUCN and WWF have facilitated conservation agreements on transboundary wetlands such as the Okavango.

Yet, domestic political change or traumatic national experiences such as economic crises can also make a state go move away from integrated water management. Ravaged by the 2008 financial crisis, the state of Portugal, well on its way towards integrated basin management, reversed course and fragmented and privatized river management, including transboundary rivers (Ioris, 2012). The accession of a populist party to the governmental coalition made the Netherlands renege on its water agreements with its neighbours on the rivers Rhine and Scheldt from 2010 to 2012. This points us to the importance of analysing transboundary relations as a two-level game (Weinthal, 2003).

Initiatives to make peace between states in conflict however wax and wane, often without noticeable results. Could it be that not everyone is all that interested in peacemaking and cooperation? The revival of "big (wo)man politics" running illiberal democracies (the United States, Brazil, Turkey, India, Bangladesh, Hungary, and Poland) seems to herald a partial retreat from internationalism and globalism. Meanwhile China, a one-party state, is successfully creating a globe-spanning Belt and Road initiative – although with some sensitive setbacks, such as Kachin ethnic nationalism putting a stop to the Chinese-Burmese Myitsone deal on the Irrawaddy (Kiik, 2016).

In the 21st century, an eco-centric New Water hydro-development paradigm has been pronounced by many, claiming that facing climate threats jointly would bring cooperation and multilateral institutions. This literature seems to overestimate the willingness of post-colonial states to share power with their riparian neighbours. For states pursuing a "hydraulic mission," a policy of extending safe and reliable water services to citizens and communities throughout its territory is part and parcel of their state building project. "Making the desert bloom" has been taken quite literally by Israel, Egypt, Saudi Arabia, and elsewhere (Molle & Floch, 2008). Water infrastructure is symbolic of prowess and pride, and it is therefore not just sovereignty but resource nationalism that holds countries back from the shared management of resources. Water development projects are material to state-building and national pride (Allouche, 2020; Menga, 2016) and then criticism of such projects is tantamount to criticism of the state. In such a context, the sharing of water or even water information can easily be non-negotiable.

While successful arrangements are negotiated and successful mediation is realized, there is little evidence for a thrust towards deepening water cooperation. If international treaties are signed, and there are over 3,600 of them, often they do not see much follow-up. Rather, they remain "shallow" and equipped with poor enforcement mechanisms (Daoudy & Kistin, 2008). Cooperation in treaties often usually stops with the exchange of technical data. Earle and Neal (2017: 5) point at the ubiquity of paper tigers, with "ecosystem degradation not being reversed, joint investments in water infrastructure not materialising and joint management organisations failing to attract significant long-term support from the respective basin states." One reason may be that such initiatives do not take off is asymmetric benefits. In transboundary flood management, flood risk asymmetry makes actors perceive flood risks differently (Becker, Aerts & Huitema, 2007). Muller, Chikozho and Hollingworth (2015) likewise warn that "hydro-centric" approaches promoting sustainable socio-economic development may be undermined when complementarity is limited: while there has been interstate coordination on water supply, irrigation and flood management in the Southern Africa region, water hasn't promoted integration, except for political symbolism.

Water Wars

So, if an unequivocal drive towards intensifying cooperation is not quite what we see in practice, does this then vindicate the dire predictions of scarcity leading to violent conflict, following Thomas Malthus' prediction of resource wars, and the intervention of a strong hegemon to contain them, after Thomas Hobbes (Warner, 2012)?

Environmental determinists keep predicting scarcity-induced violence, and in the water domain, wars have been predicted ever since Starr and Stoll (1987) sounded the alarm over violent conflict between states over scarce water. Blockbuster film scenarios (*Quantum of Solace*, *Tambien la lluvia*) reinforce the popular image of water wars about to break out, but also attracts military and policy attention, most recently from European Commissioner Timmermans predicting water and food wars (The Guardian, 2021). Recently the Hague-based Water, Peace and Security initiative developed an "early warning" tool to predict and detect water-related violent conflict and presented it to the UN Security Council (Iceland, 2019; The Guardian, 2020).

Already in the mid-1990s, however, Beaumont (1994), Allan et al. (1994), and Wolf (1995) started to push back against the water wars hypothesis, if based on very different reasonings.

Four research schools (in the cities of Toronto, Oregon, Oslo, and Geneva) conducting research over 30 years have so far failed to establish a direct relation between scarcity and violent conflict. Wolf, Yoffe and Giordano's (2003) Basin at Risk project at Oregon State University has developed an extensive database and found that conflict over water alone is extremely rare. Rather, cooperative events far outnumber conflictive events. Toronto's Peter Gleick and his team for example, publish a useful rolling conflict chronology of over 1,000 water-related disputes (www.worldwater.org/conflict/map) over water, but on closer scrutiny, these tend to concern water as a weapon rather than as a goal. They concern strategic acts of war building or destroying water infrastructure to hurt a rival.

Indeed, many transboundary problems that have not escalated into violence are also about new infrastructure. Put in to facilitate agricultural and industrial development or facilitate navigation, actual or feared downstream externalities incite diplomatic and, sometimes, military mobilization, such as in 1975 between Syria and Iraq over dam projects. A recent literature highlights how water and water infrastructure can be instrumentalized as a target or a lever for peaceful or bellicose ends (Daoudy, 2020). In modern times, we may note local fights and rebellions over wells, pipelines and/or institutional change (privatization) with global ramifications, such as the Cochabamba "Water war" in Bolivia, and water infrastructure targeted by displaces, or as a symbol of domination by insurgent groups such as Kurds targeting Ataturk Dam in 1984.

While the tripartite World Commission on Dams cast a critical light on high dams in light of their social and ecological impact and poor economic performance (Conca, 2006), megadams have been resurgent in Asia, Africa and Latin America, while in other regions they never really left. To legitimize multilateral funding or domestic displacement they are now often (re)labelled as "climate buffers" eligible to the climate fund and clean production, despite their obvious environmental setbacks (e.g., Fearnside, 2014).

States finding renewed confidence on the international scene, Ethiopia, Afghanistan and Tajikistan among them, maintain they need dams to catch up on the "hydraulic (development) mission" they had missed out on. The timing of the filling of dam reservoirs has brought strong tensions, these days notably between Ethiopia and Egypt over Ethiopia's Renaissance dam. Food or energy security is often "securitized" as a driver for such hydraulic development powering large-scale irrigation and electrification projects.

While upstreamers tend to claim they will use the water from megadams for peaceful and mutually beneficial ends, these projects inevitably put these states at loggerheads with downstreamers fearing the dams will be used as taps to be switched on (causing flooding) or off (causing scarcity) at will. The *potential* for water to be weaponized would seem to explain why downstream states tend to respond with such emotion over large upstream water projects. Egypt, as a downstreamer to ten coriparian states, is so dependent on the river Nile for irrigation that it has viewed any upstream hydraulic development as an act of war. A sudden "twist" of the tap can cause a flood or penury, badly impacting downstream development. It may be argued using dams as taps would be economically and strategically counterproductive to do in reality. However, there are historic feats of sacrificial flooding to preserve territory, including the Dutch "water line," a 85 km long, 3–5 km wide defence area that was manually inundated to resist invasions in the 17th and 18th century, and China's intentional dam break in 1938, which in order to resist the Japanese invasion, knowingly killed hundreds of thousands of Chinese civilians as well.

No War, No Peace

We have seen that the vogue for water peace-making does not appear to have reaped notable successes, but neither is there any evidence for the water war thesis. Many shared water basins remain in a status-quo, no-war, no-peace situation, oscillating between incendiary talk and cooperative initiatives. TWINS (Transboundary Water Interaction NexuS) analysis (Mirumachi & Allan, 2007), does not assume an inexorable linear thrust from conflict to cooperation between states, but accepts that the two alternate and coincide. Not only does this help sketch the bumpy path actors have negotiated over time, but also explicitly accepts that discourses of conflict or cooperation may not translate into durable friendship or enmity.

Is it that hegemonic power freezes or suppresses conflict? One way of understanding such "frozen conflicts" is to view them as the workings of basin hydro-hegemony keeping states in check (Zeitoun & Warner, 2006). Upstream hegemons can, to a degree, afford to ignore downstream concern (Lowi, 1993) so that deep-rooted differences between states never come to a confrontation. Adaptive hegemons denying conflict and mobilizing external legitimacy can show surprising longevity, especially when they provide collective (if in places unrequited) public goods and benefits (Petersen-Perlman & Fischhendler, 2018). In so doing, hegemons use a mix of hard (coercion) and soft power (generating consent) to consolidate their leading position in the basin. While upstream hegemons emphasize the joint benefits of their infrastructure, the fact that it is them who call the shots, and can withdraw their benevolence at will, can make downstreamers uncomfortable. Non-hegemons may, however, calculate it is more beneficial to dance to the hegemon's tune than to resist it, and the hegemon may be prepared to assume the extra burden of providing the collective good of basin stability. The resulting status quo, then, should not be taken for happy or even voluntary cooperation for sustainable regional development.

Such hegemonic relations moreover aren't cast in stone. The Nile is a spectacular example of a counter-hegemonic upset (Cascão, 2008). For decades upstream Ethiopia hesitated to challenge the hegemonic arrangement, a colonial legacy, long enforced by Egypt's military might and upstream disarray, contenting itself with building thousands of small dams with untraceable impact as Egypt created "facts on the ground" with a new irrigation project in the desert (Waterbury, 1979; Waterbury & Whittington, 1998). But Ethiopia gathered economic and political strength to "counter-hegemonically" defy downstream Egypt's claim to basin leadership. In other basins the predominance of one state may endure, but where the dominant actor is constantly resisted, as it has been on the river Euphrates and Tigris as well as the Jordan, there is hardly a question of consent and as a consequence, relations may more properly be termed dominance than hegemony. Downstream Syria and Iraq may be cooperative because they have no immediate interest in confronting Turkey which is upstream, but some may also have little to gain, for domestic reasons, in signing a peace treaty. As a result, we end up with a "frozen" status quo, of an imperfect peace (Kibarolu & Sayan, 2021; Vij et al., 2020). In this context it may be observed that states do not only seek to secure contemporary access but also scope for future water development and bide their time while creating conditions on the ground to strengthen their hand in basin negotiation.

Hydro-hegemony theory has been correctly critiqued for its "realist" state-centricity (Selby, 2007) and while non-state actors and transnational actors have since made themselves increasingly felt in hydrohegemonic analysis (Warner et al., 2017), it is fair to say it has remained eclectic. In other theoretical strands such as regime theory and hydrodiplomacy, it is recognized that while the state occupies critical nodes in the international system, the range of actors in (water) diplomacy is much wider (Hensengerth, 2015). This makes transboundary relations a form of network governance. Susskind and Islam (2012) see the resulting complexity as problematic, given heightened transaction costs. At the same time, the revolution in international and communication technology can bring openness to water data and transparency to what is traditionally a profession best suited to backrooms and corridors. While making water diplomacy potentially more inclusive, however, it also hands a megaphone to stir nationalist passions (Rigi & Warner, 2021).

Hydro-hegemony

The water diplomacy literature is wedded to "mutual gains" and problem-solving (Warner & De Man, 2020). It embraces interest-based rather than identity-based negotiation. This avoids a reiteration of well-rehearsed gripes, however it also risks reinforcing the status quo. To be truly transformative, conflict analysis will have to reach into the root causes of conflict which may not be water-related (Zeitoun, Mirimachi & Warner, 2020).

Should we instead attribute hydropolitical basin dynamics to the overlay of "hydro-hegemony" at the global level? Do geopolitical "great games" (Nagheeby & Warner, 2018), neoliberalism or state capitalism, exercised by large traders and supermarket chains (Sojamo et al., 2012) make a difference? If this should be the case, then it is a dynamic hegemony. The global scene is rapidly changing with the rise of Asian parastatals. A wider greening and localization of agriculture and trade, is countering the seemingly inevitable logic of globalization, while a counterhegemonic anti-globalist movement (e.g., Via Campesina) has claimed with some success the inviolability of indigenous, antiglobalist or Gaian worldviews next to Western worldviews, resisting the apparently Western tendency of cramming many different worlds into one.

Even successful examples of interstate peace-making may not necessarily benefit sub-state actors and ecological sustainability. Political ecologists, such as Sneddon and Fox (2006) have noted that developmentalist interstate cooperation paving the way for water (especially hydropower) development may not create a situation experienced as exploitation by intended beneficiaries. A radical critique of the "structural violence" of resource capture by the few pitting the rich against the poor (Swatuk, 2012) highlights the limits to hydro-solidarity. Ecofeminists such as Shiva (2016) have denounced (hydraulic) development itself as a masculine practice breaking the bond between humanity and nature, painting water development projects in terms of "water wars." State security clearly does not equal human security, state development is not tantamount to human development, and structural violence is far more widespread than inter-state water conflict.

Conclusion

The premise of the present book is that "water makes society." The present chapter explored whether water makes international/transboundary society. Rather than eternal struggle and an "allergy to the other" (Barnett, 2004), we see acts of pro-social transboundary "bridging" behaviour as well, especially in times of crisis. Wolf's (1995) rejection of the water wars thesis still stands – we do not see evidence of violent interstate conflict over water resources, and Katz (2011) is right to warn against attention-seeking hydropolitical hyperbole. Sudden shocks rather than absolute scarcity may create tensions and, in their view, can be contained through adaptive institutional frameworks.

Joint initiatives however have rarely proved transformative. The upward march of transboundary integration is far from deterministic. Rather than a ladder to be climbed one step at a time, cooperation is found to be non-linear and open-ended. (Bio)regional integrated water management tends to clash with the concerns of territorial sovereignty (Öjendal, Hansson & Hellberg, 2012).

Taking this a step further, so-called water conflicts (whether violent or nonviolent) rarely appear to be about water alone. Rather, they tend to highlight other conflictive issues such as longstanding territorial/border disputes, and regional hegemonic rivalry. Moreover, while open water wars are extremely rare to non-existent, critical analysts have highlighted how water projects and policies continue to create intentional or unintentional divisions and marginalize groups who depend on water.

As the water world appears to enter a new golden age of large dam building, tensions between affected neighbours and domestic substate groups continue to make themselves felt. Regional development integration was expected to overcome national sovereignties and pave the way for transboundary governance networks. Nevertheless, these are often halted by sovereign unilateral decisions, which tend to make a comeback when stakes are (seen as) high. Also, it has to be obvious to participants which problem is solved with an integrated river approach. In the Mekong Basin for example, this is unclear (Hirsch, 2012). While coordination would be superior, each country taking its own measures may still be superior to waiting for cooperation (Vij et al., 2020).

Note

1 The celebrated 1972 Senegal River Convention in West Africa, shared by Mali, Mauritania, and Senegal, within the Organisation pour la Mise en Valeur du Fleuve Sénégal, OMVS.

References

Allan, J.A., Mallat, C., Wade, S., Wild, J. and Elmusa, S.S. (1994). "Water in the Middle East: Legal, Political and Commercial Implications." *The Palestine Yearbook of International Law Online*, 8(1), 541–561.

Allouche, J. (2020). "Nationalism, Legitimacy and Hegemony in Transboundary Water Interactions." *Water Alternatives*, 13(2), 286–301.

Balali, M.R., Keulartz, J. and Korthals, M. (2009). "Reflexive Water Management in Arid Regions: The Case of Iran." *Environmental Values*, 8(1), 91–112.

Barnett, C. (2004). "Deconstructing Radical Democracy: Articulation, Representation, and Being-With-Others." *Political Geography*, 23(5), 503–528.

Beaumont, P. (1994). "The Myth of Water Wars and the Future of Irrigated Agriculture in the Middle East." *International Journal of Water Resources Development*, 10(1), 9–21.

Becker, G., Aerts, J.C.J.H. and Huitema, D. (2007). "Transboundary Flood Management in the Rhine Basin: Challenges for Improved Cooperation." *Water Science and Technology*, 56(4), 125–135.

Bull, H. (1979). "The State's Positive Role in World Affairs." Daedalus, 108(4), 111–123.

Cascão, A.E. (2008). "Ethiopia – Challenges to Egyptian Hegemony in the Nile Basin." *Water Policy*, 10(S2), 13–28.

Checkel, J.T. (1999). "Social Construction and Integration." *Journal of European Public Policy*, 6(4), 545–560.

Conca, K. (2006). *Governing water: contentious transnational politics and global institution building*. Boston: MIT Press.

Daoudy, M. (2020). "Water Weaponization in the Syrian Conflict: Strategies of Domination and Cooperation." *International Affairs*, 96(5), 1347–1366.

Daoudy, M. and Kistin, E.J. (2008). "Beyond water conflict: Evaluating the effects of international water cooperation." Paper presented at the ISA Annual Meeting, San Francisco.

Earle, A. and Neal, M.J. (2017). "Inclusive transboundary water governance." In Karar, E. (ed.) *Freshwater governance for the 21st century*. Cham: Springer. pp. 145–158.

Falkenmark, M. and Folke, C. (2002). "The Ethics of Socio-Ecohydrological Catchment Management: Towards Hydrosolidarity." *Hydrology and Earth System Sciences*, 6, 1–9.

Fearnside, P.M. (2014). "Impacts of Brazil's Madeira River Dams: Unlearned Lessons for Hydroelectric Development in Amazonia." *Environmental Science & Policy*, 3, 164–172.

Gleick, P. (1993). Water in crisis. *A guide to the world's freshwater resources. Pacific institute for studies in development, environment and security & Stockholm environment institute*. New York: Oxford University Press.

Gupta, J. and Pahl-Wöstl, C. (2013). "Global Water Governance in the Context of Global and Multilevel Governance: Its Need, Form, and Challenges." *Ecology and Society*, 18(4), 53.

Haftendorn, H. (2000). "Water and International Conflict." *Third World Quarterly*, 21(1), 51–68.

Hensengerth, O. (2015). "Where Is the Power? Transnational Networks, Authority and the Dispute over the Xayaburi Dam on the Lower Mekong Mainstream." *Water International*, 40(5–6), 911–928.

Hirsch, P. (2012). "8 IWRM as a participatory governance framework for the Mekong River basin?" In Öjendal, J., Hansson, S. and Hellberg, S. (eds.) *Politics and development in a transboundary watershed. The case of the lower Mekong basin*. Berlin: Springer. pp. 155–170.

Huitema, D. and Meijerink, S. (2014). "The politics of river basin organisations: institutional design choices, coalitions and consequences." In D. Huitema and S. Meijerink (eds.), *The politics of river basin organisations*. Northampton, MA: Edward Elgar.

Iceland, C. (2019) (5 December). *We predicted were violent conflicts will occur in 2020. Water is often a factor*. World Resources Institute. [Accessed on 28 January 2022] Available at https://www.wri.org/insights/we-predicted-where-violent-conflicts-will-occur-2020-water-often-factor#

Ioris, A.A.R. (2012). "The political geography of environmental regulation: implementing the Water Framework Directive in the Douro River Basin, Portugal." *Scottish Geographical Journal*, *128*(1), 1–23.

Jägerskog, A. (2003). *Why states cooperate over shared water: The water negotiations in the Jordan River Basin*. Doctoral dissertation, Linköping University Electronic Press.

Julien, F. (2012). "Hydropolitics Is What Societies Make of It (or Why We Need a Constructivist Approach to the Geopolitics of Water)." *International Journal of Sustainable Society*, 4(1–2), 45–71.

Kagwanja, P. (2007). "Calming the Waters: the East African Community and Conflict Over the Nile Resources." *Journal of Eastern African Studies*, 1(3), 321–337.

Katz, D. (2011). "Hydro-Political Hyperbole: Examining Incentives for Overemphasizing the Risks of Water Wars." *Global Environmental Politics*, 11(1), 12–35.

Kibaroglu, A. and Sayan, R.C. (2021). "Water and 'imperfect peace' in the Euphrates–Tigris river basin." *International Affairs*, 97(1), 139–155.

Kiik, L. (2016). "Nationalism and Anti-Ethno-Politics: Why 'Chinese Development' Failed at Myanmar's Myitsone Dam." *Eurasian Geography and Economics*, 57(3), 374–402.

Kipping, M. (2009). "Conflits et coopération liés à l'eau du fleuve Sénégal." *Géocarrefour*, 80(4), 335–347.

Lindemann, S. (2008). "Understanding Water Regime Formation—A Research Framework with Lessons from Europe." *Global Environmental Politics*, 8(4), 117–140.

Lowi, M. (1993). *Water and power: the politics of a scarce Resource in the Jordan River basin*. Cambridge, UK: Cambridge University Press.

Meissner, R. and Warner, J. (2017). "Hydro-hegemony or water security community? Collective action, cooperation and conflict in the SADC transboundary security complex." In Suhardiman, D., Nicol, A. and Mapedza, E. (eds.) *Water governance and collective action: multi-scale challenges*. London: Routledge. p. 46ff.

Menga, F. (2016). "Domestic and International Dimensions of Transboundary Water Politics." *Water Alternatives*, 9(3), 704–723.

Mirumachi, N. and Allan, J.A. (2007). Revisiting transboundary water governance: Power, conflict cooperation and the political economy. In *Proceedings from CAIWA international conference on adaptive and integrated water management: Coping with scarcity. Basel, Switzerland*, 1215.

Molle, F. and Floch, P. (2008). "Megaprojects and social and environmental changes: The case of the Thai "Water Grid"." *AMBIO: A Journal of the Human Environment*, *37*(3), 199–204.

Mostert, E. (2005). "How can International Donors Promote Transboundary Water Management?" *DIE discussion papers 2005/8, Bonn: Deutsches Institut für Entwicklungspolitik*

Muller, M., Chikozho, C. and Hollingworth, B. (2015). Water and Regional Integration: The Role of Water as a Driver of Regional Economic Integration in Southern Africa. *Water Research Commission Research Report* No. 2252-1-14, 150 pages.

Nagheeby, M. and Warner, J. (2018). "The Geopolitical Overlay of the Hydropolitics of the Harirud River Basin." *International Environmental Agreements: Politics, Law and Economics*, 18(6), 839–860.

Ohlsson, L. and Turton, A.R. (2000). "The turning of a screw: Social resource scarcity as a bottle-neck in adaptation to water scarcity." Occasional Paper #19, London Water Study Group, University of London.

Öjendal, J., Hansson, S. and Hellberg, S. (eds.) (2012). *Politics and development in a transboundary watershed. The case of the lower Mekong basin*. Berlin: Springer.

Petersen-Perlman, J.D. and Fischhendler, I. (2018). "The Weakness of the Strong: Re-Examining Power in Transboundary Water Dynamics." *International Environmental Agreements: Politics, Law and Economics*, 18(2), 275–294.

Richards, A. and Singh, N. (1997). "Two level negotiations in bargaining over water." In Parthasarathy., T., Dutta., B., Potters., J.A.M., Raghavan., T.E.S., Ray., D. and Sen, A. (eds.) *Game theoretical applications to economics and operations research*. Boston, MA: Springer. pp. 257–273.

Rigi, H. and Warner, J.F. (2021). "Pakistan's representation of transboundary water as a security issue." *International Journal of Water Resources Development*, 1–24. DOI: 10.1080/07900627.2020.1870937

Rinaudo, J.-D. and Barraqué, B. (2015). "Inter-basin transfers as a supply option: the end of an era?" In Grafton, Q., Daniell, K.A., Nauges, C., Rinaudo, J.-D. and Chan, N.W.W. (eds.) *Understanding and managing urban water in* transition. Berlin: Springer. pp. 175–200.

Sadoff, C.W. and Grey, D. (2005). "Cooperation on International Rivers: A continuum for Securing and Sharing Benefits." *Water International*, 30(4), 420–427.

Selby, J. (2007). "Beyond hydro-hegemony: Gramsci, the national, and the trans-national." In: *SOAS Occasional Paper*, *94*. London: University College London.

Shiva, V. (2016). *Water wars: privatization, pollution, and profit*. Toronto: North Atlantic Books.

Sneddon, C. (2012). "The 'sinew of development': Cold War Geopolitics, Technical Expertise, and Water Resource Development in Southeast Asia, 1954–1975." *Social Studies of Science*, 42(4), 564–590.

Sneddon, C. and Fox, C. (2006). "Rethinking Transboundary Waters: A Critical Hydropolitics of the Mekong Basin." *Political Geography*, 25(2), 181–202.

Sojamo, S., Keulertz, M., Warner, J. and Allan, J.A. (2012). "Virtual Water Hegemony: The Role of Agribusiness in Global Water Governance." *Water International*, 37(2), 169–182.

Starr, J. and Stoll, D. (1987). *Foreign policy on water resources in the middle East*. Washington, DC: Center for Strategic and International Studies.

Susskind, L. and Islam, S. (2012). "Water Diplomacy: Creating Value and Building Trust in Transboundary Water Negotiations." *Science & Diplomacy*, 1(3), 1–7.

Swatuk, L. (2012). "Water and security in Africa. State-centric narratives, human insecurities." In Schnurr, M.A. and Swatuk, L. (eds.) *Natural resources and social conflict*. London: Palgrave Macmillan. pp. 83–107.

The Guardian. (2020) (8 January). "Water wars: early warning tool uses climate data to predict conflict hotpots."

The Guardian. (2021) (30 April). "Climate crisis: our children face wars over food and water, EU deputy warns."

Turton, A.R. (2000). "Precipitation, people, pipelines and power in southern Africa: towards a 'virtual water'-based political ecology discourse. MEWREW occasional paper, water issues study group, school of oriental and African studies (SOAS). London: University of London.

Vij, S., Warner, J.F., Biesbroek, R. and Groot, A. (2020). "Non-Decisions Are Also Decisions: Power Interplay between Bangladesh and India Over the Brahmaputra River." *Water International*, 45(4), 254–274.

Warner, J. (2004). "Plugging the GAP-Working with Buzan: The Ilisu dam as a security issue." In: *SOAS Occasional Paper, 67*. London: University College London.

Warner, J. (2012). "Three Lenses on Water War, Peace and Hegemonic Struggle on the Nile." *International Journal of Sustainable Society*, 4(1–2), 173–193.

Warner, J. and de Man, R. (2020). "Powering hydrodiplomacy: How a broader power palette can deepen our understanding of water conflict dynamics." *Environmental Science & Policy, 114*, 283–294.

Warner, J., Mirumachi, N., Farnum, R.L., Grandi, M., Menga, F. and Zeitoun, M. (2017). "Transboundary 'hydro-hegemony': 10 Years Later." *Wiley Interdisciplinary Reviews: Water*, 4(6), e1242.

Waterbury, J. (1979). *Hydropolitics* of the *Nile* valley. Syracuse, NY: Syracuse University Press.

Waterbury, J. and Whittington, D. (1998). "Playing Chicken on the Nile? The Implications of Microdam Development in the Ethiopian Highlands and Egypt's New Valley Project." *Natural Resources Forum*, 22(3), 155–163.

WCED (1987). *Report of the world commission on environment and development: our common future*. New York: United Nations.

Weinthal, E. (2003). *State making and environmental cooperation: linking domestic and international politics in Central Asia*. Cambridge: MIT Press.

Wendt, A. (1999). *Social theory of international politics* (Vol. 67). Cambridge, UK: Cambridge University Press.

Wolf, A.T. (1995). *Hydropolitics along the Jordan River: scarce water and its impact on the Arab-Israeli conflict*. Tokyo: United Nations University Press.

Wolf, A.T. (2004). *Regional water cooperation as confidence building: water management as a strategy for peace*. Berlin: Adelphi Research.

Wolf, A.T., Yoffe, S.B. and Giordano, M. (2003). "International Waters: Identifying Basins at Risk." *Water Policy*, 5(1), 29–60.

Zeitoun, M., Mirimachi, N. and Warner, J. (2020). *Water conflicts: analysis for transformation*. Oxford, UK: Oxford University Press.

Zeitoun, M. and Warner, J. (2006). "Hydro-Hegemony – A Framework for Analysis of Trans-Boundary Water Conflicts." *Water Policy*, 8(5), 435–460.

19
STRATEGIES TOWARDS SDG 6 IMPLEMENTATION

Anik Bhaduri, Alexandre Teixeira, and Aditya Kaushik

Introduction

Access to water is inextricably related to sustainable development and is critical for socio-economic development, healthy ecosystems, economic growth, health, and welfare. Water, as an indispensable component of the Earth system flows through every dimension of human and natural development. In 2015, the United Nations (UN) adopted an ambitious agenda of 17 Sustainable Development Goals (SDGs), for global economic growth, social justice and environmental protection, leaving no one behind. Water is integral to this agenda as it is linked to all 17 SDGs. Goal 6-Ensure access to water and sanitation for all, focuses specifically on clean water, ensuring availability and sustainable water and sanitation management for all. SDG 6 is connected to most of the other SDGs as water is a critical input to meet many of the development and environmental targets and provides a critical foundation for the overall achievement of sustainability (Bhaduri et al., 2016). For instance, it plays a critical role for Zero hunger (SDG 2), Good health and wellbeing (SDG 3), Sustainable Cities and Communities (SDG 11), Responsible consumption and production (SDG 12), Climate Action (SDG 13), and Life on land (SDG 15). Hence, a delay in SDG 6 implementation can influence the progress of achieving other SDG goals.

The UN-Water's SDG 6 Global Acceleration Framework formulates that science has a clear role in supporting the implementation of SDG 6 and water governance. However, meaningful, stepwise change is needed at all levels to exchange and generate quality data, and optimize monitoring and assessment, to deepen disaggregation and analysis (UN-WATER, 2020c). The acceleration framework also recognizes that sharing data transparently within and between sectors and borders is crucial now, more than ever, in supporting good governance. This will help inform decision-making processes effectively, including drawing on coherent data and information systems, encouraging innovative solutions, and facilitating multi-stakeholder engagement, policy advice, and technical assistance. Timely assessment and innovative data technology are indispensable for water resource management and governance, particularly where hydrological variability may have enormous consequences on future developmental projects. It is essential to synthesize scientific results and data into knowledge and a fact-based decision-making process to make a difference in policymaking and produce expeditious knowledge transfer custom-made for countries. It will support in strategizing, planning, and accelerated implementation of a sustainable water governance

DOI: 10.4324/9781003095545-22

framework that takes synergies and trade-offs into consideration from all aspects of sustainable development – economically, socially, politically, and environmentally. The chapter summarizes the current state of SDG 6 implementation and discusses how scientific tools and assessments can help prioritize actions in SDG 6 implementation.

The Progress of SDG 6

It is unlikely that SDG 6 will be met by the year 2030, given the current rate of progress. As billions of people worldwide lack access to clean and safe water and adequate sanitation facilities, the COVID-19 crisis has also sharpened the linkage between water security and public health (Sadoff et al., 2020).

The delay in implementing SDG 6 could be costly and could undermine other dimensions of human development, including food security, achieving the Paris Agreement, and the post-2020 Global Biodiversity Framework. Further delay in SDG 6 implementation could have rebounding effects as access to the timely provision of water is also a key element in mitigating the coronavirus crisis and preventing future pandemics. As billions of people worldwide lack access to clean and safe water and adequate sanitation facilities, the COVID-19 crisis has also sharpened the linkage between water security and public health. Contrary to the water-related Millenium Development Goals (MDG), SDG 6 covers the entire water cycle, is globally applicable, and integrates social development, economic prosperity and environmental sustainability (UN-WATER, 2020b). The eight targets of SDG 6 include; (6.1) the provision of drinking water; (6.2) sanitation and hygiene services; (6.3) treatment and reuse of wastewater and ambient water quality; (6.4) water-use efficiency and scarcity; (6.5) integrated water resources management (IWRM), including through transboundary cooperation; (6.6) protecting and restoring water-related ecosystems; (6.7) international cooperation and capacity building; (6.A) and participation of local communities in water and sanitation management (6.B).

The United Nations (UN-WATER) provides information about the SDG-6 indicators in a global scale, in an effort with all countries of the world and custodian agencies. Table 19.1 illustrates the current state of the proportion of the population using safely managed services (SDG targets 6.1 and 6.2), extracted from UN-Water SDG-6 Data Portal (UN-Water, 2022). Rates of progress show that only on average, 1% annual progress has been achieved on expanding access to basic water supply and sanitation, instead of a minimum of 3% annual increases needed to ensure everyone has basic services by 2030.

Close to 2 billion people today lack access to improved drinking water services (UN-WATER, 2021). There is a considerable gap between the accessibility of services of urban and rural populations with regards to the proportion of the population using safely managed drinking water services. Even though the proportion of the global population using at least basic sanitation has increased substantially, 2.3 billion people still lack basic services, particularly in rural areas. Target 6.2 also highlights the importance of hygiene and calls for special attention to the needs of women and girls. Handwashing with soap and water is widely recognized as a top priority for reducing disease transmission, and hence target 6.2 is one of the most relevant indicators, particularly in the Covid pandemic period. Close to a billion people located particularly in the rural areas of Central Asia, Southern Asia, and Sub African areas still practise open defecation.

Target 6.2 is significantly linked with target 6.3, which focuses on the safe treatment of wastewater and attaining an acceptable level of ambient water quality. Data suggests that the risk levels of ambient water quality is increasing, which puts over 3 billion people at risk of disease (UN-WATER, 2021).

Table 19.1 Status of SDG Targets 6.1, 6.2, and 6.3 – year 2020.

Target/indicator	*World*	*Latin America*	*Europe and Northern America*	*Northern Africa and Western Asia*	*Sub-Saharan Africa*	*Eastern and South-Eastern Asia*	*Australia and New Zealand*	*Central and Southern Asia*
6.1.1 Proportion of population using safely managed drinking water services, by urban/rural (%)	74	75	96	79	30	89-	100	68
6.2.1a Proportion of population that uses a safely managed sanitation with basic handwashing facilities on premises (%)	54	34	78	42	21	60	82	47
6.2.1b Proportion of population has a handwashing facility with soap and water available at home (%)	71	NA	NA	91	26	NA	69	69
6.3.1 Proportion of domestic wastewater is safely treated (%)	56	41	80	63	28	65	79	26

Along with deteriorating water quality, many countries suffer from water stress. SDG target 6.4 aims to reduce the level of water stress through enhancing water use efficiency. Since 2015, water efficiency has increased by 10% (UN, 2018). A global measure of water stress masks the real effects of water scarcity. Even though the global average of water stress is only 11%, more than 25% of the population live in countries experiencing high stress. The highest stress levels occur in Northern and Sub-Saharan Africa and Western, Central and Southern Asia. Levels of water stress are likely to increase as populations and demand for water grow along with the intensification of climate change impacts. There exists a wide range of water use efficiency, where the values can vary from 2 US$/m^3 to 1,000 US$/m^3. This acts as a hindrance to defining detailed policy positions and making subsequent operational decisions to accelerate the implementation of the SDG target.

The estimation of global water scarcity conditions in the river basins is illustrated in Figure 19.1, derived from the standardized[1] Global Arid Index, calculated in Zomer et al. (2008), provided in the HydroAtlas (Linke et al., 2019). Vulnerable areas of water stress, due to aridity, are distributed throughout all continents. In Africa, such areas are mainly concentrated in the north, in regions including and surrounding the Sahara. In Australia, a significant part of the country's interior has acute water scarcity. Other clusters of aridity are observed in South America, particularly Chile (Atacama) and Argentina (Patagonia). There are also arid zones in Asia (China and Saudi Arabia, for example), Europe and North America (the United States and Mexico). This viewpoint however

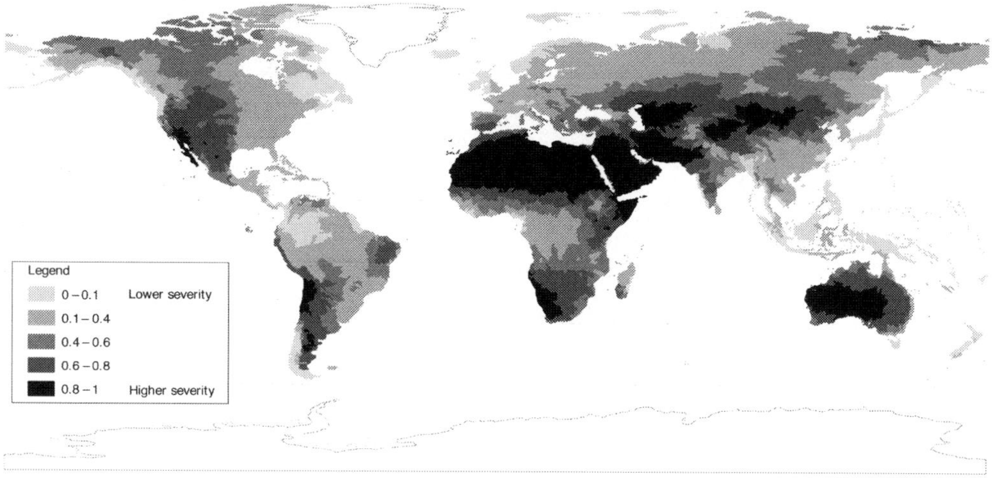

Figure 19.1 Global water scarcity.

is related to natural conditions and does not explain the capacity to supply the existing water demand within the water basins vis a vis the population density.

The spatial distribution of standardized global water availability per capita index (Figure 19.2), is related to the populational concentration in some regions, and signifies the water demand in comparison with the available superficial water. Therefore, areas with higher population density, such as micro basins in Japan, India, and China, are more vulnerable to water supply scarcity, than those (even arid regions) with a low population density. As a result of this, water efficiency actions in these areas could be prioritized to meet target 6.4.

SDG target 6.5 focuses on implementing IWRM, which requires consideration of linkages with socio-economic factors as well as possibilities of decisions designed for one sector affecting the water users in other sectors. IWRM involves multi-stakeholders in planning and implementing

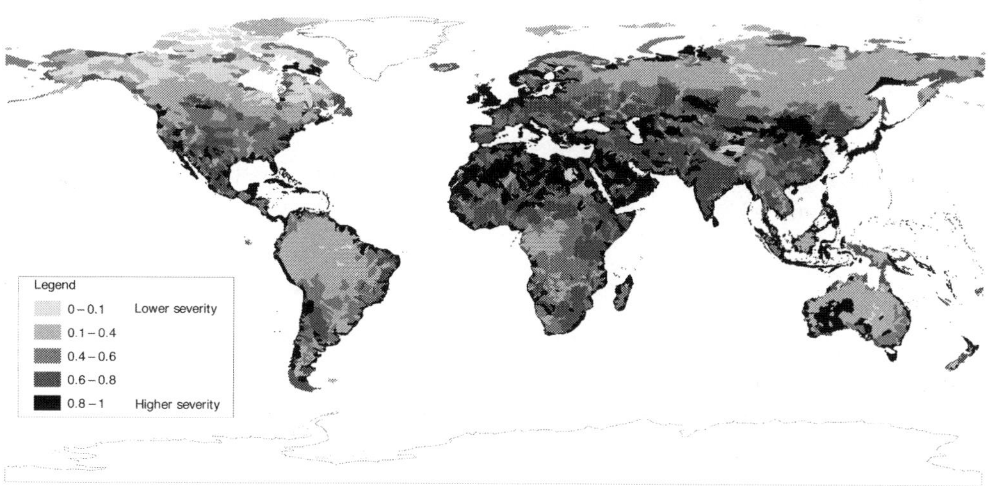

Figure 19.2 Water availability per capita.

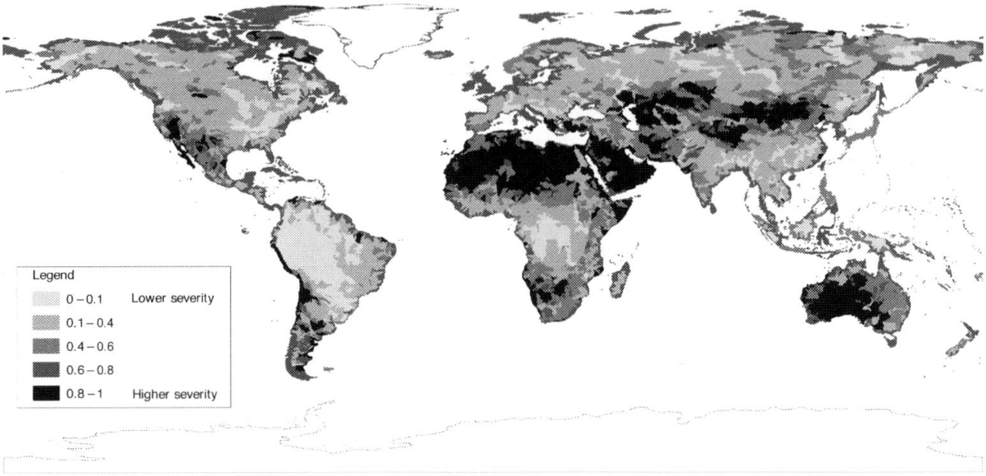

Figure 19.3 Minimum water flow.

water resources equitably and sustainably. The global average degree of the implementation of IWRM is only 48% (medium-low). However, there are significant variations among countries.

Cooperation for instance, is needed for riparian countries as transboundary river basins account for more than 60% of the global freshwater flow with close to 40% of the world's population being dependent on it. This indicates the interdependencies of natural, socio-economic, political, and environmental factors. The need for cooperation in water sharing is widely recognized, and it is reflected in having a dedicated target for cooperation in SDG 6.

SDG 6.6 focuses on protecting and restoring water-related ecosystems. The freshwater ecosystem provides the foundation for the implementation of SDG 6 and several other goals, particularly those relating to food and energy production, biodiversity, and land and sea ecosystems. Globally, 70% of natural wetlands has been lost over the last century, including significant loss of freshwater species. According to the United Nations (UN, 2018), freshwater aquatic ecosystems are the world's most biologically diverse environments and provide many services on which human wellbeing depends. Figure 19.3 shows the global standardized minimum flow, which is important to guarantee human and ecosystem water security in the river basins. Figure 19.3 indicates higher water availability for the Amazon River basin in South America with higher values of average flow in the world. On the other hand, regions in Northern Africa, at the centre of Australia, parts of Central and Eastern Asia, and North America, for example, present very lower values of minimum flow (null values in many basins), demonstrating a higher level of human and ecosystem water security threats.

Missing Links in SDGs Monitoring and Implementation

The complexity of infrastructure around water is constantly increasing, and there is a need to respond to growing demands and contribute to economic growth while addressing the SDGs. Tracking specific, measurable indicators can help assess sustainable water use (Markolf et al., 2018). However, current SDG indicators for water are ambiguous (e.g., what exactly constitutes "improved sanitation" or "appropriate transboundary cooperation") and thus may fail to provide a clear understanding of progress towards the goal (Bertule et al., 2018). For example, a drinking water service could be "safely managed" (SDG 6.1), but that does not reflect water's affordability and equitable distribution (Sadoff et al., 2020). The urban poor pay considerably more for a litre of water than their more affluent neighbours in some countries (Cooper, 2020). So, an indicator measuring the affordability of the drinking water supply might be a better measure of progress.

Furthermore, the definition and methodology of indicator 6.4.1 is very sensitive to changes in the proportion of water withdrawal accounted for by agriculture compared to industry and services. The indicator combines the economic values across all sectors, overlooking social and environmental values. The definition also completely ignores the effects of diminished return flows to the environment due to increased re-use of water (Hellegers & van Halsema, 2021).

Accelerated investments are urgently required to focus on target 6.2 to enhance the capacity of sanitation systems, including building an end-to-end robust information management system. Such information systems can help guide the assessment of sanitation data which is essential to understanding and illustrating the value of such investments in terms of health impacts. It is required to guide prioritize for the investments. There is also a need to identify the vulnerable and disadvantaged groups for a better understanding of the ways to increase access to improved drinking water services and reduce the water collection time. These data and knowledge gaps prevent better monitoring of SDG targets 6.1 and 6.2.

One of the reasons for higher risk in the implementation of target 6.3 is mainly because information on the water quality of their rivers, lakes, and groundwater is unknown. The extent of industrial pollution or the extent of domestic and industrial wastewater treated on-site, which are essential to guide investment priorities, is not known clearly. Such lack of data mainly stems from the lower in-country capacity to collect reliable water quality monitoring and analyse the data needed for a full assessment.

Further, non-point sources of pollution, such as polluted runoff from agricultural areas, are one of the major sources affecting the ambient water quality. The interlinkages and feedbacks related to the source, impacts, and solutions are unknown at a comprehensive level; for instance, the optimal placement of solutions, if not targeted well according to the source of the pollutants, will lower the value of investments. If conventional treatment is in place with a high incidence of industrial pollution, such a treatment process will not remove nutrients and hazardous substances like micro-pollutants. Untreated substances will severely affect human health and the environment (UN-WATER, 2020a).

As in other SDG targets mentioned above, little information is known for monitoring and assessing SDG 6.4, particularly on the interlinkages of water scarcity with socio-economic development, and the identification of key sectors with higher water use and lower water use efficiency. One of the challenges related to monitoring the SDG 6 implementation is the information availability at a disaggregated scale. A country-level indicator is important to compare regions, but sometimes does not reflect the main issues to SDG 6 achievements in a specific country. According to ANA (2019b), 60.9 million people (34% of the Brazilian urban population in 2017) live in cities with poor water supply. Despite this, Brazil can be classified as a good condition in terms of the target 6.4.2 (Level of Water Stress). It is due to the higher water availability in the Amazon Region, which impacts at an aggregated country level. Complementary indicators developed by Brazil (ANA, 2019a) at a disaggregated scale can be useful to support SDG 6 and the water resources policy implementation in Brazil.

SDG target 6.5 is very relevant in meeting other targets as the water crisis is largely a governance crisis (Pahl-Wostl et al., 2010; 2012) with limited cooperation. Water management and governance are crucial for the equitable distribution of resources to all stakeholders, including the environment. Water can lead to potential conflicts over its distribution; however, governance based on cooperation can transform a limited water quantity into a flexible resource (Mirumachi &

Allan 2007; Pahl-Wostl et al., 2012). There is no universal approach to implementing IWRM, and each country must develop its pathway based on political, social, environmental, and economic circumstances (UN-WATER, 2020c).

Meeting target 6.6 and other related targets requires overcoming strategic challenges related to protected areas, ecosystem-based solutions research, water observatories, and expanded technical readiness as one-fifth of the world's river basins are experiencing rapid changes in the area covered by surface waters (UN-WATER, 2021). The SDG 6 targets, which are key in reaching the desired human water security level, cannot be achieved on a truly sustainable basis without a well-functioning environment capable of supporting adequate resource quantity and quality. With ecological integrity in place, water-related ecosystem services provide a vast array of services, including the provision of clean water and climate change resilience.

There are several challenges in monitoring the implementation of SDG target 6.6. Global data on the freshwater ecosystem is available. However, there are challenges to gathering local-level data of higher granularity, which can help landscape planning and thus present an accurate understanding of SDG Interaction and its benefits. Remote sensing and Earth observations can complement local ground data, support national data acquisition, and alleviate the reporting burden.

One of the significant challenges in SDG implementation is the gap in increased financial resources. Over 80% of the countries reported insufficient funding even to meet the Water, Sanitation and Hygiene (WASH) targets. The official development assistance (ODA) in the water sector has remained constant at around 5% as a proportion of the total ODA disbursements.

Summarizing the critical challenges in SDG 6 implementation, we find that lack of disaggregated, time-series data and capabilities in monitoring and assessment at a desired spatial and temporal scale hinder the tracking of SDG progress. Most of the indicators of progress are monitored at the national level, which helps compare the country-wise status of the implementation of SDG. Monitoring is mainly based on in-situ data, and many countries lack data; for instance, in the Global Freshwater Quality Database (GEMSta), 71 out of the 110 river basins where we have data, have a density of 0.5 stations per 10,000 km^2 or less compared to minimum densities of around 1.5–4 stations per 10,000 km^2 of the river basin area in the United States and Europe (UNEP, 2016). Such a lack of data leads to biasedness in the assessment. The unavailability of monitoring data makes it difficult to determine rates of progress and limits the possibility of determining time trends.

Need for a Rigorous Water Security Assessment

There is a lack of understanding of the interlinkages between the different SDG targets. This leads to limitations in pinpointing the investment opportunities and identifying the challenges in mitigating water security threats. Such capabilities require an innovative combination of advanced scientific models, digital technologies, agile governance frameworks, and future-proof policy planning to help countries cope with these challenges. SDG implementation cannot be accelerated unless the progress is measured, monitored, and assessed on an acceptable temporal scale. Water security assessment is key to understanding and evaluating the impact of our socio-economic development (as drivers) on the progress of SDG 6 over time and space (Allen et al., 2019) including the disruption of the ecosystem services. Such assessment will help policymakers and planners devise investment strategies and reframing policies on where to address the gap and guide the impact of mitigation measures. It will also help to understand the system boundary, and help identify effective solutions to address trade-offs (Pahl-Wostl et al., 2013; Alcamo, 2019). A regular, consistent assessment function that provides a rigorous, systematic, timely, comprehensive assessment of SDG 6 on a global basis could be a principal contribution of the science community to anticipatory planning at all levels of government, a basis for investment and policy decisions related to water (Bhaduri et al., 2016). In the distant past, the assessment of water problems was generally restricted to national interests addressing national problems. However, with the advent of observing systems that view the whole earth and computers that can process large quantities of data and produce global analysis and predictions (Vörösmarty et al., 2013). The science community is ready to provide near real-time assessments. During this same period, water science also advanced. The study of water science considers the global impact of water management, the impact of water infrastructure, and the dependent elements such as biodiversity, ecosystems, and resources, such as energy and food. Also, water science frameworks explore, understands, measure, and assess the water cycle processes and its sensitivity to climate variability. Water science and its associated models and tools have allowed the community to address worldwide assessment challenges.

Water issues are too critical for society to be addressed with outdated technologies, poorly informed science, and incomplete information. Governments need reliable capabilities for doing leading-edge assessments. These assessments may be done in developed countries at the national level, but the infrastructure is not available for many countries to undertake such work.

For the last two decades, the water sciences community has been making significant advances towards understanding the complexities within the global water system and its dependencies with other systems such as food, health, and biodiversity, to name a few. Significant steps forward include the improvement in the resolutions and fidelity of atmospheric models in producing ever-reliable global-scale data sets, for example, high-resolution stream network system, spatial scale database on reservoirs and dams, satellite systems to monitor the depletion of groundwater, and global-scale geospatial assessments of the threat to the world's river systems from human and biodiversity perspectives. A well-developed knowledge base will enable policymakers and other stakeholders to accelerate the implementation of SDG targets by pinpointing opportunities and challenges. Digitalization is a key accelerator and a strategic enabler for the sustainable management of water and sanitation. Innovative ICT solutions can provide the necessary tools to assess and monitor water resources and meet the growing water demands worldwide. There are several areas where digitalization can make an immediate impact. The emergence of digital technologies has provided the necessary tools in water governance to strengthen capabilities through effective monitoring, decision-making, and forecasting. It can also contribute effectively to the implementation of Integrated Water Management (IWM), a key element for the success of SDG 6.

Further, insights from science seem not to have always been translated into path-breaking action to reverse these trends. It can also be attributed to the fact that past research has emphasized identifying problems rather than identifying solutions. Towards successful implementation of SDG, there is a need for an approach that deals with the increased demand for scarce water resources under environmental and socio-economic constraints, identify unrealized co-benefits of economic development and environmental protection; and embraces rapid emerging technology development in the field of Artificial Intelligence.

Local and global sources of knowledge and information are imperative to understand interconnected water security risks to humans, societies, and the environment and support water governance. Increasing the knowledge and information available may support decision-making processes and help avoid decisions that may lead to sub-optimal or even adverse outcomes. This is particularly relevant in Africa for the implementation of the African 2063 agenda and the global 2030 agenda for Sustainable Development, which are intrinsically multi-scale and characterized by complex interlinkages (Mugagga & Nabaasa, 2016). It is even more important for international donors willing to orient investments to effectively target continental priorities towards sustainable development to support the region in achieving its goals.

Improved quantitative, geospatial, and simulation of products, depicting the biophysical and human dimensions of freshwater across the sub-national scale, can expand the knowledge base and support decision-making by devising strategies that accelerate the development process and potentially reverse any negative environmental trends. Such assessments, framed within state-of-the-art scenario analyses of the possible futures, are needed to identify pathways for better decision-making in the water realm.

There is a critical need for baseline studies and good monitoring as an important foundation for planning and supervision and the systematic development of water service delivery. In Africa, for instance, collection and management of hydro-meteorological data have been marginal in many countries and have been in decline for the last two decades in terms of real-time data and long-term datasets for scenario-building (Houghton-Carr et al., 2006).

Water quality monitoring programmes and available databases are based on the laboratory analysis of water quality samples, collected at specific times and locations. These traditional techniques can be complemented by other data including remote sensing data available at spatial resolutions and covering river networks. The next generation of hyperspectral sensors will provide information that can be linked to a wide set of water quality parameters and that will require ground-truthing of the spectral signals. While satellite data could be used to fill some of these data gaps, it is important to facilitate the dialogue between knowledge providers and users and ensure the utilization of such data in policymaking and the implementation of action plans.

Integrated water resource management is imperative to ensure water security and consider quantity and water quality aspects as well as associated socio-economic dimensions in water management practices and infrastructure investment. This is also reflected in the concept of Water security, which is defined by United Nations as:

> The capacity of the population to safeguard sustainable access to adequate quantities of and acceptable quality water for sustaining livelihoods, human well-being, and socio-economic development, for ensuring protection against water-borne pollution and water-related disasters, and for preserving ecosystems in a climate of peace and political stability.
>
> *(UN-WATER 2013: 1)*

For instance, the regulation of non-point source pollution, including the contaminants in run-off from land, has been much trickier and poses a major challenge in implementing SDG effectively and managing water resources. Non-point sources of pollution stemming from the degradation of watersheds, either through land-use change or intensive agriculture, can affect sediment and nutrient runoff and require downstream users to invest more in water treatment. Around 85% of humankind currently relies on freshwater source areas that are moderately to severely threatened upstream and incur a rising cost to remedy the threats (Green et al., 2015). Protection of natural assets, including watersheds and aquatic ecosystems, and conversion to sustainable land practices watersheds and aquatic ecosystems are recognized as cost-effective means to improve water security by combining with engineered systems to provide services like safe drinking water, waste processing and dilution, erosion control, and flood risk reduction.

Such a combination of traditional engineering (such as building dams, and reservoirs) often referred to as grey infrastructure) with nature-based solution (such as reforestation, planting trees, and wetlands protection), referred to as green infrastructure, could be a cost-effective pathway to human water security. It represents a critical opportunity space within the overall sustainable development agenda (Vörösmarty et al., 2018).

Vörösmarty et al. (2021) explore the feasibility of such an approach in meeting water security threats and contributing to water-related sustainable development. The study finds that engineering expenditures in developing economies are projected to increase to $1.6 trillion US dollars by 2030 to meet crucial human water needs. On the other hand, preserving natural assets can play a crucial role in mitigating such threats and can avoid a potential cost of $2.3 trillion US dollars.

A blended approach may help to meet many SDG 6 targets like the provision of clean water for drinking (SDG 6.1), lowering the cost of building water treatment plants to attain an acceptable level of water quality (SDG 6.3), or reducing the hard infrastructure cost in building dams and reservoirs in the meeting (SDG 6.5). Such a blended approach supports high levels of water quality, improving various aspects of quality of life, such as reduced exposure to water-borne diseases in addition to cultural and spiritual wellbeing (Vörösmarty et al., 2015). Water management and efficient use play a fundamental role worldwide in climate mitigation and adaptation. However, with a delay in protecting natural assets, the loss of capital will increase the traditional cost of remedying the water security threats.

Further, there is a lack of knowledge in identifying areas of high environmental value where grey infrastructure could be avoided or areas where grey infrastructure, in combination with green infrastructure, can help support the maintenance of ecosystem services and benefit the population (e.g., agricultural productivity). It is important to communicate to policymakers the various gradation of risks (at different scales) in infrastructure planning and how to minimize such risks without compromising capacity to address human water security. There are gaps in spatial and temporal sequencing of green and grey investment that can ensure assets (physical and natural capital stock) deliver anticipated benefits over their operational lifetime, avoid premature obsolescence or costly retrofitting in the future, and reduce threats to human water security.

The next level of challenge is to operationalize the concept of water security with adaptive governance and management strategies and reduce the risk of increasing costs associated with controlling water security threats. It needs a complete rethinking of economic approaches to evaluate the economic feasibility of water-related investments; for instance, water investments need to be evaluated over longer periods instead of a snapshot cost-benefit analysis. Further, hydrological stationarity is dead today, and hence and future discounting should also involve uncertainty about water variability, climate risks. Such an approach should also consider multiple environmental and socio-cultural values while accounting for the natural assets. Otherwise, a narrow economic approach will favour traditional engineering approaches as the only remedy and may lead to suboptimal outcomes of higher threats and higher costs.

Implementing a circular economy approach can also reduce overconsumption and restore and regenerate ecosystems and natural capital. It requires designing new wastewater recycling and treatment systems at an appropriate level (centralized or decentralized) with the scope of supplying differentiating "waters" from different sources, with varied costs, qualities, and reliabilities.

Differentiated water can maximize the overall utility giving resource scarcity and capturing the opportunities offered by nature-based solutions to deliver water security.

Concluding Comments

Given that there is only a decade left to achieve the SDG targets, we urgently need to find ways to accelerate the implementation of SDG 6. We need accelerated changes at all levels to achieve the internationally agreed and time-bound targets. Timely assessment, innovative data technology, and governance approaches are indispensable for water resource management, particularly where hydrological variability may have enormous consequences on future development.

It is important to synthesize and translate scientific results and data into knowledge and eventually into fact-based decisions to guide policymaking priorities and produce such expeditious knowledge transfer custom-made for countries. Such an endeavour that considers synergies and trade-offs from all aspects of sustainable development – economically, socially, politically, and environmentally – is imperative to support and implement the agenda for sustainable water governance. The Earth system science, water engineering, digital transformation experts, and management communities can deliver these knowledge demands through high-resolution spatial datasets from remote sensing, simulation models, and allied ground-based information resources from networked observatories and socio-economic survey data. Such expertise, coupled with the outstanding experience of the UN water family in facilitating dialogues and producing knowledge products that bridge the gap between science and policymaking in the water sector, can upscale the impact, implementation, and use of the tools developed.

Note

1 Index values standardized in a 0–1 scale.

References

Alcamo, J. (2019). "Water quality and its interlinkages with the Sustainable Development Goals." *Current Opinion in Environmental Sustainability*, 36, 126–140.

Allen, C., Metternicht, G., and Wiedmann, T. (2019). "Prioritizing SDG targets: Assessing baselines, gaps and interlinkages." *Sustainability Science*, 14(2), 421–438.

ANA – NATIONAL WATER AND SANITATION AGENCY OF BRAZIL. (2019a). SDG 6 – ANA's Vision of the Indicators.

ANA – NATIONAL WATER AND SANITATION AGENCY OF BRAZIL. (2019b). Water Security National Plan. Brazil: Brasília. Available at: https://pnsh.ana.gov.br/pdf/ingles.pdf

Bertule, M., Glennie, P., Bjørnsen, P. K., Lloyd, G. J., Kjellen, M., Dalton, J., and Rieu-Clarke, A., et al. (2018). "Monitoring water resources governance progress globally: Experiences from monitoring SDG indicator 6.5. 1 on integrated water resources management implementation." *Water*, 10(12), 1744.

Bhaduri, A., Bogardi, J., Siddiqi, A., Voigt, H., Vörösmarty, C., Pahl-Wostl, C., Bunn, S. E., Shrivastava, P., Lawford, R., Foster, S., Kremer, H., Renaud, F. G., Bruns, A., and Osuna, V. R. (2016). "Achieving Sustainable Development Goals from a water perspective." *Frontiers in Environmental Science*, 4, 64. DOI: 10.3389/fenvs.2016.00064.

Cooper, R. (2020). Water for the Urban Poor and Covid-19. K4D Helpdesk Report 826. Brighton: Institute of Development Studies.

Green, P. A., Vörösmarty, C. J., Harrison, I., Farrell, T., Sáenz, L., and Fekete, B. M. (2015). "Freshwater ecosystem services supporting humans: Pivoting from water crisis to water solutions." *Global Environmental Change*, 34, 108–118.

Hellegers, P., and van Halsema, G. (2021). "SDG indicator 6.4. 1 'change in water use efficiency over time': Methodological flaws and suggestions for improvement." *Science of the Total Environment*, 801, 149431.

Houghton-Carr, H., Fry, M., and Wallingford, U. (2006). "The decline of hydrological data collection for development of integrated water resource management tools in Southern Africa." *International Association of Hydrological Sciences (IAHS) Publication*, 308, 51.

Linke, S., Lehner, B., Ouellet Dallaire, C., Ariwi, J., Grill, G., Anand, M., et al., and Thieme, M. (2019). "Global hydro-environmental sub-basin and river reach characteristics at high spatial resolution." *Scientific Data*, 6(1), 1–15.

Markolf, S. A., Chester, M. V., Eisenberg, D. A., Iwaniec, D. M., Davidson, C. I., Zimmerman, R., et al. and Chang, H. (2018). "Interdependent infrastructure as linked social, ecological, and technological systems (SETSs) to address lock-in and enhance resilience." *Earth's Future*, 6(12), 1638–1659.

Mirumachi, N., and Allan, J. A. (2007). "Revisiting transboundary water governance: Power, conflict cooperation and the political economy." In *Proceedings from CAIWA international conference on adaptive and integrated water management: Coping with scarcity*. Basel, Switzerland (Vol. 1215).

Mugagga, F., and Nabaasa, B. B. (2016). "The centrality of water resources to the realization of Sustainable Development Goals (SDG). A review of potentials and constraints on the African continent." *International Soil and Water Conservation Research*, 4(3), 215–223.

Pahl-Wostl, C., Holtz, G., Kastens, B., and Knieper, C. (2010). "Analyzing complex water governance regimes: The management and transition framework." *Environmental Science & Policy*, 13(7), 571–581.

Pahl-Wostl, C., Lebel, L., Knieper, C., and Nikitina, E. (2012). "From applying panaceas to mastering complexity: Toward adaptive water governance in river basins." *Environmental Science & Policy*, 23, 24–34.

Pahl-Wostl, C., Vörösmarty, C., Bhaduri, A., Bogardi, J., Rockström, J., and Alcamo, J. (2013). "Towards a sustainable water future: Shaping the next decade of global water research." *Current Opinion in Environmental Sustainability*, 5(6), 708–714.

Sadoff, C. W., Borgomeo, E., and Uhlenbrook, S. (2020). "Rethinking water for SDG 6." *Nature Sustainability*, 3(5), 346–347.

UN. (2018). Sustainable Development Goal 6 Synthesis Report on Water and Sanitation. New York, NY: United Nations.

UNEP. (2016). A Snapshot of the World's Water Quality: Towards a global assessment. United Nations Environment Programme, Nairobi, Kenya.

UN-WATER. (2013). Water Security & the Global Water Agenda: A UN-Water Analytical Brief. Hamilton: United Nations University.

UN-WATER. (2020a). Monitoring-water-and-sanitation-in-the-2030-Agenda.

UN-WATER. (2020b). The SDG 6 Global Acceleration Framework. Geneva: UN-Water.

UN-WATER. (2020c). The Sustainable Development Goal 6 Global Acceleration Framework. Geneva: UN-Water.

UN-WATER. (2021). Summary Progress Update 2021: SDG 6 — Water and Sanitation for All. Geneva: UN-Water.

UN-WATER. (2022). UN Water SDG-6 Data Portal. Available at: https://www.sdg6data.org/en/accessed on December 2022.

Vörösmarty, C. J., Hoekstra, A. Y., Bunn, S. E., Conway, D., and Gupta, J. (2015). "Fresh water goes global." *Science*, 349(6247), 478–479.

Vörösmarty, C. J., Osuna, V. R., Cak, A. D., Bhaduri, A., Bunn, S. E., Corsi, F., and Gastelumendi, J., et al. (2018). "Ecosystem-based water security and the Sustainable Development Goals (SDGs)." *Ecohydrology & Hydrobiology*, 18(4), 317–333.

Vörösmarty, C. J., Pahl-Wostl, C., Bunn, S. E., and Lawford, R. (2013). "Global water, the anthropocene and the transformation of a science." *Current Opinion in Environmental Sustainability*, 5(6), 539–550.

Vörösmarty, C. J., Stewart-Koster, B., Green, P. A., Boone, E. L., Flörke, M., Fischer, G., et al., and Stifel, D. (2021). "A green-gray path to global water security and sustainable infrastructure." *Global Environmental Change*, 70, 102344.

Zomer, R. J., Trabucco, A., Bossio, D. A., and Verchot, L. V. (2008). "Climate change mitigation: A spatial analysis of global land suitability for clean development mechanism afforestation and reforestation." *Agriculture, Ecosystems & Environment*, 126(1–2), 67–80.

PART IV

Themes and Issues

20

WATER, FOOD, AND IRRIGATION

Jaime Hoogesteger, Diana Suhardiman, Gert Jan Veldwisch, Juan Pablo Hidalgo-Bastidas, and Rutgerd Boelens

Introduction: Water and Food

This chapter explores the key role that irrigation plays in food production worldwide and points out the important challenges that this sector faces from a socio-environmental justice perspective. Water and food are intrinsically related. Water stands at the basis of all plant growth and as such forms the basis of all food production on earth. Agriculture also provides other needs of present-day society, such as raw material for clothes, building materials, fuel, flowers, medicine and many more. Broadly, there are two agricultural production modalities based on where the water used for crop growth comes from. These are rainfed and irrigated agriculture. Rainfed agriculture depends, as its name implies, solely on direct rainfall, while irrigated agriculture depends on water that is administered to the crops by humans through the use of hydraulic infrastructure such as buckets, weirs, canals, dams, tubes, and pumps amongst others. Irrigation water often complements rainfall, but in some cases, agricultural production solely depends on irrigation water that comes from surface or groundwater, or both. This process where people, nature and technology intertwine and co-shape each other forms the basis for food production and related livelihoods.

The continued expansion and intensification of irrigated as well as rainfed agriculture is often advocated as the only way to move forward in producing enough food, fibre, and other agricultural products for the growing world population and its demands. So far production has kept a-pace: "while the population has risen from 3 billion in 1960 to 7.5 billion in 2017 (and is expected to reach 9.2 billion by 2050), the proportion living near starvation conditions fell from between 25% and 30% in 1960 to just over 10% today" (Robinson, 2018: 140). Most undernourishment and near starvation comes from situations of violent conflict, state failure and the extremely uneven distribution of wealth worldwide, not because there is a lack of food at the global level.

Currently the highest rise in the demand for agricultural products does not come from population growth itself, but from more luxurious diets of the world's middle and upper classes (Borras & Franco, 2012; van der Ploeg, 2014), who have a higher per capita consumption of fish, meat, eggs and dairy as well as other often imported luxury food products such as fresh fruits and vegetables. All of these products require more water and land per produced calory than staple and traditional foods. Currently, livestock production alone uses 70% of the world's agricultural land (Robinson et al., 2011). Ironically, food abundance in many societies has become one of the most

DOI: 10.4324/9781003095545-24

important public health risks through widespread obesity, and millions of tons of produced food are thrown away yearly. At the same time, problems related to poverty, malnutrition, social exclusion, and related disparities in resource distribution in rural societies, particularly in the Global South prevail. The environmental costs of increased food production are also high and include soil degradation, water pollution, aquifer depletion, destruction of the world's major biomes, reduction of biodiversity, and increased greenhouse gas emissions from agriculture (Martinez-Alier, 2009).

Within this context, and placing irrigated agriculture within the broader context of food regime analysis (Holt-Gimenez & Shattuck, 2011), in this chapter we first present the development of the irrigation sector in the last century. Secondly, we explore the main challenges this sector is currently facing from a socio-environmental justice perspective. Finally, we present how peasant/smallholder production systems fit in this context and show some promising irrigation-based grassroots initiatives of smallholder producers that contribute to more sustainable food networks. The chapter closes with the conclusions.

Placing Irrigation in Global Food Production

Rainfed agriculture currently occupies between 75% and 80% of global arable land and produces around 60% of total agricultural production (Molden, 2007; see also FAO, 2020). Rainfed agriculture is prone to seasonal climatological variations that negatively affects crop production. This effect is expected to increase in the future due to global climate change (Ray et al., 2019). This makes production from rainfed agriculture in many areas of the world increasingly uncertain due to climatological extremes induced crop yield reductions and crop failure (FAO, 2020).

To overcome the uncertainties of rainfed agriculture, lengthen the growing season and enable multiple cropping seasons per year, for millennia people have diverted water from springs, streams, rivers, lakes and wells to irrigate their crops. Irrigation leads to higher and more secure agricultural production, while also enabling production in areas and in seasons in which these would normally not grow under rainfed conditions. As such irrigation has for centuries played a key role in food production and security of many societies around the world. As the basis for food production, irrigation has also played a key role in the cultural, political and economic structures of many societies.

Today many rural societies depend on irrigation for agricultural production for self-consumption and the market. Around 20–25% of the global agricultural area is irrigated, accounting for over 40% of total crop production; a value that is much higher when taking into account the economic value of irrigated production (Molden, 2007; FAO, 2020). According to Thenkabail et al. (2009) at the end of the last millennium worldwide there were around 300 million hectares (Mha) irrigated. This was comprised of 252 Mha from seasonal crops and 41 Mha from continuous year-round crops. According to these same authors, Asia accounts for 79% of all irrigated areas, followed by Europe (7%) and North America (7%). According to Faurès, Hoogeveen, and Bruinsma et al. (2021) irrigation expansion will continue as developing countries are expected to expand their irrigated areas from 202 Mha at the beginning of the 2000s to 242 Mha by 2030.

Irrigation Expansion in the Last Century: Controlling Waters to Feed the World

Since the 1950s, the irrigated area worldwide has grown exponentially. This was the result of enormous international and national investments to construct irrigation infrastructure and institutions to operate and manage them. The development of irrigation systems became an important means to develop rural areas, increase production, fight hunger, provide countries with food sovereignty and bring about capital accumulation and related socio-economic development (Molle, Mollinga, & Wester, 2009; Molle & Wester, 2009). Irrigation expansion took place on a three-pronged path which consisted of the development of large scale public irrigation schemes, investments in the expansion and modernization of smaller farmer-managed irrigation schemes, and the development of intensive groundwater use for agriculture as is explored below.

Large-Scale Public Irrigation Schemes

Surface water development has implied the construction of large scale infrastructure (dams, weirs, and canals) in most of the world's rivers to control and divert its waters to fertile lands aimed at the production of agricultural products. To this end a centralized bureaucracy specialized in the construction and management of these irrigation schemes was established in most countries (Molle et al., 2009). These powerful bureaucracies operated in a centralized top-down manner and in many areas forcefully created a new peasantry that was composed by a mix of local and re-settled families that were supported by state agencies to increase the production of specific crops (often staple foods), sometimes under strict state guidance and control (Veldwisch et al., 2012). The heavy financial burden that these bureaucracies held on national budgets coupled to the often poor performance of irrigation systems led to the decentralization of irrigation system management to organized water users associations around the world from the 1980s onwards (Veldwisch, 2013; Suhardiman et al., 2014; Hoogesteger et al., 2017; Cambaza, Hoogesteger, & Veldwisch, 2020). The results this decentralization had on the performance of irrigation management and production are variable (Senanayake, Mukherji, & Giordano, 2015) as they depended on context specific dynamics and outcomes (Suhardiman et al., 2014). One noticeable outcome of this process has been the increased control of large commercial farmers on irrigation management especially as many peasants/smallholders went bankrupt, sold their irrigated land and became labourers because they could not compete and remain productive in contexts of agricultural liberalization and globalization.

Farmer-Managed Irrigation Schemes

Private and collectively (farmer) managed irrigation systems play a key role in agricultural production worldwide. These decentred systems, that are often not recognized as irrigated area by state agencies and official data, existed in many regions of the world well before the larger-public irrigation systems. They are operated and maintained by farmers/users either privately or collectively through locally specific management forms (Boelens & Vos, 2014). External investments by state or non-governmental organizations, especially since the 1950s have often been used to expand and/or modernize these systems. However, these external interventions have in most cases also been accompanied by new technical, operational and management challenges (García-Mollá et al., 2020). Despite these challenges, these systems have shown great resilience to change as smallholders continue to sustain these systems (Hoogesteger et al., 2023a) as part of their peasant livelihoods as is further explored below.

Groundwater Irrigation

Groundwater irrigation took flight especially since the 1970s in many countries, due to advances in drilling technology and sharp drops in the cost of – often subsidized – pumps and the fuel or electricity to run them (Hoogesteger & Wester, 2015). Under these conditions, and paraphrasing

Shah et al. (2007: 409), groundwater became *available on site,* providing farmers with on-demand, just-in-time water for a very large variety of crops. This development was atomistic meaning that thousands and in some cases millions of farmers individually dug and controlled their own wells covering significant area extensions and bringing with it increased crop production, food security, poverty reduction and economic development to many rural families and societies (Hoogesteger, 2022). This has led to intensive (unsustainable) groundwater use and related aquifer overdraft in many parts of the world including India, the Western USA, the North China Plain, Spain, Iran, the Middle East and Northern Africa, the Peruvian and Chilean coast, and Mexico amongst others (de Graaf et al., 2019) with sever socio-environmental impacts as is further elaborated below.

Irrigation-Related Challenges

Though irrigated agriculture plays a central role in sustaining the food production levels worldwide, it is a sector that is also fraught with important challenges. Here we concentrate on two of the multiple domains of challenges and frictions: challenges within the irrigation systems, and environmental effects.

Challenges within Irrigation Systems

Many irrigation systems were built with public, private or donor investments and require high operational costs. In many irrigation systems, especially in the developing world, state agencies, and/or collectively managed water users associations are financially constrained and have managerial challenges resulting in prolonged break-downs, water delivery problems, low performance, and reduced cropped areas. In many systems the solution to managerial and water related challenges is sought through the modernization of irrigation infrastructure (new gates, lining canals, pressurizing irrigation, installation of sprinkler and drip, etc.). This is often seen as the silver bullet to increase water use efficiency and through it release water for new irrigated areas, other sectors or nature (Venot, Kuper, & Zwarteveen, 2017). However, this drive and blind belief in technology often oversees more simple, cost-efficient and effective measures to overcome the challenges of irrigation systems. In addition, "modernized" farmers tend to use "water savings" for the expansion of irrigation (Vos & Boelens, 2014). Through such modernization projects larger and often more capitalized and politically powerful producers tend to acquire more secure access to large shares of irrigation water than small producers (Franco, Mehta, & Veldwisch, 2013; Dell'Angelo, Rulli, & D'Odorico, 2018). This affects in particular peasant, indigenous and women farmer groups in irrigation systems.

In areas of intensive groundwater use, groundwater management and control by state agents, water users associations or through co-management arrangements has proven extremely challenging (Molle & Closas, 2020a; 2020b). The result is a race to the bottom of the aquifers in which economically powerful pumpers (commercial farmers, cities, industry and mining) outcompete smallholders that are not able to invest in the deepening and repositioning of wells or pay the ever-increasing pumping costs. The result is a silent process of water grabbing and accumulation in which smallholders and peasant producers gradually lose their access to groundwater and related livelihoods (Hoogesteger & Wester, 2015; Hoogesteger, 2018).

These "internal" challenges are exacerbated by external challenges related to water availability which is threatened by increased climatic variability related to global climate change and intersectoral competition. In many areas of the world great pressure is exerted on irrigation to liberate water (in particular from peasant land) for other water uses such as urban/domestic and industrial

(Molle et al., 2010). As Molle and Berkoff (2009) show, the financial and political clout of these sectors often leads to hidden and outright transfers of water from (peasant) irrigated agriculture to these more powerful sectors.

Broader Environmental Effects

The large infrastructure that is constructed in rivers to control and divert water for use in irrigation systems has drastically changed river environments, lakes and related water flows (Boelens et al., 2022). This affects – and in many cases severely destroyed – river, lake, and estuary fisheries and related ecosystems as well as the livelihoods and riverine cultures that depend upon them (e.g., Shah et al., 2019; Flaminio, 2021).

In many irrigated areas of the world salinization caused by a lack of appropriate drainage and increasing pollutants from pesticide and fertilizer residues in irrigation water have made millions of hectares of land unsuitable for agriculture (Ritzema et al., 2008). Salinization is often closely related to water quality deterioration. As rivers and thus irrigation water become increasingly polluted worldwide due to industrial, urban and agricultural pollutants (Strokal et al., 2019), new challenges arise for the production of healthy food crops and the sustenance of fertile soils.

In areas of intensive groundwater use, aquifer overdraft has become a serious socio-environmental threat. This affects the geo-hydrology and the contribution of aquifers to river flows (de Graaf et al., 2019) and related environments such as wetlands. Groundwater declines can also lead to salinity intrusion in coastal aquifers and land subsidence is a common occurrence in areas of intensive groundwater use. While the former has rendered some coastal aquifers too saline for use, land subsidence can lead to breaks in underground pipelines (sewerage, oil, gas, etc.), damages to houses and infrastructure and increased risk of flooding. Over-pumping can also have very serious impacts on groundwater quality. As aquifer levels decline, concentrations of arsenic, fluor and other toxic elements (nitrates, sulphates, heavy metals, etc.) increase in several aquifers leading to toxic concentrations of these elements in extracted groundwater (Knappett et al., 2020). All of these environmental changes tend to hit rural communities and smallholder farmers the hardest.

Within this broad context and generalized trends, there are very large differences in terms of how irrigated agricultural production is organized and who reaps its benefits or carries it's socio-environmental burdens. This is further explored in the next section.

Irrigation, Globalization, and the Expansion of Productivist Agriculture

Food trade is maybe as old as civilization but it has gained increased importance especially since the process of rapid globalization took flight in the 1980s. This entailed the liberalization of markets in most countries around the globe, the development of faster and more efficient food transport systems and the rise of transnational food chains and enterprises. The increased trade of agricultural goods has gone hand in hand with an increase in virtual water exports (that is, the water that is needed to produce the goods that are traded) mostly from the global south to the global north (Dalin et al., 2012).

Globalization and increased virtual water trade is closely associated with the rise of so-called productivist agriculture, characterized by an increased and increasing number of larger, well-capitalized (corporate) farms. These farming operations tend to access and accumulate the most fertile and productive lands and water for irrigation. Though a very broad spectrum of farming styles prevails the gap between productivist agriculture and smaller family/peasant-run farms is steadily increasing. This product of globalization is usually characterized by intensification, concentration, and specialization for which irrigation is often needed (Robinson, 2014, pp. 62–64). In the process capital often replaces labour as the production process transforms towards an increased reliance on new technologies, external specialized commodified inputs, mechanization, automation of production processes, and the application of advances in biotechnology.

Intensive use of technologies for production of high value crops such as vegetables, fruits, and flowers is done above all in irrigated agriculture and has gone hand in hand with a tendency of individual farms to specialize in the large-scale production of standardized outputs that can be easily processed and shipped to markets, both locally and worldwide. Specialization does not only take place at farm level but also at regional level. As a result, global hotspots of production and export of specific irrigated crops have emerged. These source a large portion of the world demand for one specific product from one or a few relatively small regions in the world. Examples are the production of roses and cut flowers in northern Ecuador, Colombia, Kenya and Tanzania; asparagus in the Ica Valley of Peru and central Mexico; grapes and other fruits in coastal Peru, Chile and South Africa; blackberries and avocados in western Mexico; fresh-cut vegetables for northern Europe in Murcia, Spain; pistachios in the province of Kerman, Iran and almonds in California, just to name a few examples in which production is irrigation based.

This specialization often goes hand in hand with the concentration of the means of production in fewer hands leading to less but larger specialized farming units that increasingly set of their production (sales) through food processing industries, many of which operate in global markets. These sales are often arranged through contract farming in which agricultural production is carried out via an agreement between the buyer (wholesaler, processor, and retailer) and the producer. It has also promoted the transformation of farm produce into inputs for the wider global food packaging, transport and manufacturing system. As a result producers and end-consumers of food have become increasingly removed from each other.

Many critical scholars have pointed out that the development of such specialized, large scale corporate agriculture underlies ongoing processes of land and water grabbing and accumulation by local elites and transnational agricultural companies in Africa, Asia and South America (Borras & Franco, 2012; Mehta, Veldwisch, & Franco, 2012). These processes of land and water accumulation often go hand in hand with the dispossession of access to these resources by peasant families and the transformation of rural livelihoods through increased proletarianization to poorly paid agricultural labour in productivist agriculture (Hoogesteger & Massink, 2021) and migration (Mena-Vásconez, Boelens, & Vos, 2016). These processes are compounded by outright and blatant dispossession through the use of force, but also through a slower process of de-peasantization which includes roping in small farmers into corporate modes of production through for instance contract farming schemes (Veldwisch, 2013; Hartman et al., 2022). These developments threaten the peasantry and the important role that it plays in local food production as is further explored below.

Peasant Production in Regional Food Systems

It is widely acknowledged that peasants see their household as a production unit that is strongly related to its own consumption and other needs, rather than by the capital profit maximization logic that dominates in commercial agriculture. This means that resources, labour and production do not only have an exchange value but also have a very important use value (van der Ploeg, 2014). The often-made distinction between "domestic/household" and "productive activities" generally blurs, as they combine and overlap. As a result, households are organized and sustained through a wide diversity of activities both within as well as outside of the production unit. In this sense, peasant households and communities are highly dynamic and integrated into the larger regional, national and transnational economies through their insertion in diverse labour markets, globalized production, trade and cyclical, or permanent migration (and related remittances) of household members (Veldwisch & Spoor, 2008; Hoogesteger & Rivara, 2021).

Peasant production, most of which combines irrigated and rainfed agriculture with extensive animal grazing as part of diversified production strategies, tends to focus on the cultivation of food crops for self-consumption and sale on local or regional food markets, though important exceptions exist (Hoogesteger & Rivara, 2021). It plays a key role in catering the regional and national food needs in many developing countries where corporate food companies, supermarkets and processed food are less dominant in the food sector. In many regions irrigated peasant production plays a fundamental role in the production of fresh vegetables, fruits and dairy products for the rural and growing urban populations.

Though in many areas of the world a process of gradual social differentiation and related de-peasantization (where less and less peasants remain in production) has been identified (Bernstein et al., 2018), renewed efforts to reproduce and re-peasantize food production through sustained or increased smallholder-based production exist (van der Ploeg, 2014). Some interesting processes in which smallholder/peasant production has increased or been re-vitalized based on irrigation are explored below.

Farmer-Led Irrigation Development

Farmers have been developing irrigation for centuries, largely without external support. Recently research has shown that smallholder farmers have continued expanding the irrigated area across a variety of agro-ecological contexts and tapping into different types of water resources, especially in Sub-Saharan Africa (Woodhouse et al., 2017). This process of relatively recent expansion of irrigation is called farmer-led irrigation development and has been defined as "a process whereby farmers drive the establishment, improvement, and/or expansion of irrigated agriculture, often in interaction with other actors (government agencies, NGOs, etc)" (Veldwisch et al., 2019: 2). Despite its substantial contribution to agricultural production and the expansion of irrigation over large areas (Bossenbroek, Van der Ploeg, & Zwarteveen, 2015; Hamamouche et al., 2018; Hartman et al., 2021), the process has remained little recognized by state and development organizations. These irrigating farmers largely grow for the nearby market, and have intensified their production processes, exemplified by the use of improved seeds, pesticides and fertilizers, by the degree of mechanization and by the hiring of labour (de Bont & Veldwisch, 2020).

Irrigated (Peri)Urban Agriculture

Irrigated agricultural production has been stable or growing in and around many cities in the global South (Drechsel & Keraita, 2014). This is mostly practiced by poor city dwellers, who frequently use urban water flows containing waste water, either treated, diluted or raw (Crush, Hovorka, & Tevera, 2011). Regardless of health risk for both farmers and consumers substantial areas are irrigated in this way.

Most countries have approaches to urban irrigation that aim to enforce strict standards and rules regarding the use of urban space and urban water sources. These policies are often combined with promulgating high-tech sanitation and water treatment solutions. This approach that combines an emphasis on rule-enforcement with faith in technological solutions has largely failed, with untreated and partially treated wastewater mixing with natural flows and continue to be used as a stable, year-round, flow of water that by default re-cycles nutrients. Ghana forms an exception and has taken up a more expedient approach that is very promising (Ayambire et al., 2019).

Agro-ecology and Local Organic Solidarity and Production Networks

Around the world, over the last 20 years there has been a resurgence of local (mostly organic) farmers markets and producers cooperatives. In these, smallholder producers sell their products directly to the cooperative or the end consumers without intermediaries. This enables producers to make higher revenues from production while bringing consumers local fresh produce (Hoogesteger et al., 2023b). Most of these local markets depend on, and are the result of, extended solidarity networks that are often historically initiated by peasant communities and federations, sometimes supported and facilitated by non-governmental organizations and local governments. Most members of these networks produce organically or follow principles of agro-ecology and permaculture. This type of agriculture is based on the enforcement of agrobiodiversity, zero agrochemical usage (pesticides, industrialized fertilizers, etc.), empowerment of solidarity networks (Vos et al., 2020), and enhancement of socio-ecological resilience capacity of production systems (Wezel et al., 2009).

Conclusions: Are We Feeding the World in a Socio-environmentally Sustainable Manner?

Over the past century worldwide food production has kept a-pace with population growth, ensuring widespread food security and enabling a transition to richer and more luxurious diets for many around the globe. Irrigation has played a key role in this process, but from a socio-environmental justice perspective important questions can be raised about its socio-environmental costs and current processes. The construction of irrigation infrastructure in most rivers of the world has destroyed river dependent ecosystems and related livelihoods. Intensive groundwater use has led to the dissection of river base flows and wetlands as aquifers are over-exploited and groundwater quality deteriorates in many areas of the world.

Socially the development of surface and groundwater irrigation initially brought food security and economic development to many rural areas, forming the basis for the development of new-peasantries and smallholder based rural development. Especially since the late 1970s liberalization, globalization and commodification of production through (inter)national food chains has increasingly threatened peasant irrigation and related livelihoods. The rise and expansion of productivist agriculture at the expanse of peasant agriculture is leading to fewer, specialized larger production units that focus on economically lucrative agro-export crops such as flowers, fruits, vegetables, or dairy. These productive units accumulate access to land and water resources through legal and illegal mechanisms of resource grabbing often at the cost of peasants' access to these resources. The result is increased resource and capital accumulation in the hands of a small elite, while rural livelihoods increasingly come to depend on migration and proletarianization. This process of de-peasantization, which is facilitated in many countries by liberal agricultural, land and water policies, is troubling from a socio-environmental justice perspective in terms of resource access equity and when recognizing the important role that peasant agriculture plays in local and regional food supply and food security systems.

The promising developments of farmer-led irrigation development in Africa, peri-urban agriculture based food production and local agro-ecological and organic production networks in many parts of the world offer promising, more sustainable and socially equitable, alternatives to the productivist imperative in the irrigation sector. Although these initiatives are generally not recognized nor supported by governmental policies, they show that more sustainable, equal and just food production systems that build on peasant and smallholders agriculture are viable in different contexts. Therefore, global discussions on water and food, irrigated agriculture and socio-environmental sustainability need to incorporate peasant agriculture as a key building block for the (re)shaping of global food regimes. For policy makers there is much scope to recognize, support, and upscale such grassroots initiatives. Finally, we as consumers can decide what kind of agriculture and producers we want to support through our choices about where and what foods we buy and consume on a daily basis.

References

Ayambire, R. A., Amponsah, O., Peprah, C. and Takyi, S. A. (2019). "A review of practices for sustaining urban and peri-urban agriculture: Implications for land use planning in rapidly urbanising Ghanaian cities." *Land Use Policy*, 84, 260–277.

Bernstein, H., Friedmann, H., van der Ploeg, J. D., Shanin, T. and White, B. (2018). "Fifty years of debate on peasantries, 1966–2016." *The Journal of Peasant Studies*, 45(4), 689–714.

Boelens, R., Escobar, A., Bakker, K., Hommes, L., Swyngedouw, E., Hogenboom, B., ... and Wantzen, K. M. (2022). "Riverhood: Political ecologies of socionature commoning and translocal struggles for water justice." *The Journal of Peasant Studies*, 1–32.

Boelens, R. and Vos, J. (2014). "Legal pluralism, hydraulic property creation and sustainability: The materialized nature of water rights in user-managed systems." *Current Opinion in Environmental Sustainability*, 11, 55–62.

Borras, S. M. Jr and Franco, J. C. (2012). "Global land grabbing and trajectories of Agrarian change: A preliminary analysis." *Journal of Agrarian Change*, 12(1), 34–59.

Bossenbroek, L., Van der Ploeg, J. D. and Zwarteveen, M. (2015). "Broken dreams? Youth experiences of agrarian change in Morocco's Saïss region." *Cahiers Agricultures*, 24(6), 342–348.

Cambaza, C., Hoogesteger, J. and Veldwisch, G. J. (2020). "Irrigation management transfer in Sub-Saharan Africa: An analysis of policy implementation across scales." *Water International*, 45(1), 3–19.

Crush, J., Hovorka, A. and Tevera, D. (2011). "Food security in Southern African cities: The place of urban agriculture." *Progress in Development Studies*, 11(4), 285–305.

de Bont, C. and Veldwisch, G. J. (2020). "State engagement with farmer-led irrigation development: Symbolic irrigation modernisation and disturbed development trajectories in Tanzania." *The Journal of Development Studies*, 56(12), 2154–2168.

de Graaf, I. E., Gleeson, T., van Beek, L. R., Sutanudjaja, E. H. and Bierkens, M. F. (2019). "Environmental flow limits to global groundwater pumping." *Nature*, 574(7776), 90–94.

Dell'Angelo, J., Rulli, M. C. and D'Odorico, P. (2018). "The global water grabbing syndrome." *Ecological Economics*, 143, 276–285.

Drechsel, P. and Keraita, B. (eds.). (2014). *Irrigated urban vegetable production in Ghana: Characteristics, benefits and risk mitigation*. IWMI, Colombo: Sri Lanka.

FAO. (2020). *The state of food and agriculture 2020. Overcoming water challenges in agriculture*. Rome, Italy: FAO.

Faurès, J. M., Hoogeveen, J. and Bruinsma, J. (2002). *The FAO irrigated area forecast for 2030*. Rome, Italy: FAO, 1–14.

Flaminio, S. (2021). "Modern and nonmodern waters: Sociotechnical controversies, successful anti-dam movements and water ontologies." *Water Alternatives*, 14(1), 204–227.

Franco, J., Mehta, L. and Veldwisch, G. J. (2013). "The global politics of water grabbing." *Third World Quarterly*, 34(9), 1651–1675.

García-Mollá, M., Ortega-Reig, M., Boelens, R. and Sanchis-Ibor, C. (2020). "Hybridizing the commons. Privatizing and outsourcing collective irrigation management after technological change in Spain." *World Development*, 132, 104983.

Hamamouche, M. F., Kuper, M., Amichi, H., Lejars, C. and Ghodbani, T. (2018). "New reading of Saharan agricultural transformation: Continuities of ancient oases and their extensions (Algeria)." *World Development*, 107, 210–223.

Hartman, S., Chiarelli, D. D., Rulli, M. C. and D'Odorico, P. (2021). "A growing produce bubble: United States produce tied to Mexico's unsustainable agricultural water use." *Environmental Research Letters*, *16*(10), 105008.

Hartman, S., Farfán, M., Hoogesteger, J. and D'Odorico, P. (2022). "Mapping the expansion of berry greenhouses onto Michoacán's ejido lands, México." *Environmental Research Letters*, *17*(11), 115004.

Holt-Gimenez, E. and Shattuck, A. (2011). "Food crises, food regimes and food movements: Rumbling of reforms or tides of transformation?" *The Journal of Peasant Studies*, 38, 109–144.

Hoogesteger, J. (2018). "The ostrich politics of groundwater development and neoliberal regulation in Mexico." *Water Alternatives*, 11(3), 552–571.

Hoogesteger, J. (2022). "Regulating agricultural groundwater use in arid and semi-arid regions of the Global South: Challenges and socio-environmental impacts." *Current Opinion in Environmental Science & Health*, *27*, 100341.

Hoogesteger, J., Bolding, A., Sanchis-Ibor, C., Veldwisch, G. J., Venot, J. P., Vos, J. and Boelens, R. (2023a). "Communality in farmer managed irrigation systems: Insights from Spain, Ecuador, Cambodia and Mozambique." *Agricultural Systems*, *204*, 103552.

Hoogesteger, J., Konijnenberg, V., Brackel, L., Kemink, S., Kusters, M., Meester, B., ... and Sanchis-Ibor, C. (2023b). "Imaginaries and the commons: insights from irrigation modernization in Valencia, Spain." *International Journal of the Commons*, *17*(1), 109–124.

Hoogesteger, J. and Massink, G. (2021). "Corporate labour standards and work quality: Insights from the agro-export sector of Guanajuato, Central Mexico." *Third World Quarterly*, 42(6), 1196–1212.

Hoogesteger, J. and Rivara, F. (2021). "The end of the rural/urban divide? Migration, proletarianization, differentiation and peasant production in an ejido, Central Mexico." *Journal of Agrarian Change*, 21(2), 332–355.

Hoogesteger, J., Tiaguaro-Rea, Y., Rap, E. and Hidalgo, J. P. (2017). "Scalar politics in sectoral reforms: Negotiating the implementation of water policies in Ecuador (1990–2008)." *World Development*, 98, 300–309.

Hoogesteger, J. and Wester, P. (2015). "Intensive groundwater use and (in)equity: Processes and governance challenges." *Environmental Science and Policy*, 51, 117–124.

Martinez-Alier, J. (2009). "Social metabolism, ecological distribution conflicts, and languages of valuation." *Capitalism Nature Socialism*, 20(1), 58–87.

Strokal, M., Emiel Spanier, J., Kroeze, C., Koelmans, A. A., Flörke, M., Franssen, W., Hofstra, N., Langan, S., Tang, T., van Vliet, M. T. H., Wada, Y., Wang, M., van Wijnen, J. and Williams, R. (2019). "Global multi-pollutant modelling of water quality: Scientific challenges and future directions." *Current Opinion in Environmental Sustainability*, 36, 116–125.

Mehta, L., Veldwisch, G. J. and Franco, J. (2012). "Introduction to the special issue: Water grabbing? Focus on the (re)appropriation of finite water resources." *Water Alternatives*, 5(2), 193–207.

Mena-Vásconez, P., Boelens, R. and Vos, J. (2016). "Food or flowers? Contested transformations of community food security and water use priorities under new legal and market regimes in Ecuador's highlands." *Journal of Rural Studies*, 44, 227–238.

Molden, D. (ed.) (2007). *water for food, water for life: A comprehensive assessment of water management in agriculture*. London, UK: Earthscan.

Molle, F. and Berkoff, J. (2009). "Cities vs. agriculture: A review of intersectoral water re-allocation." *Natural Resources Forum*, 33(1), 6–18.

Molle, F. and Closas, A. (2020a). "Comanagement of groundwater: A review." *Wiley Interdisciplinary Reviews: Water*, 7(1), e1394.

Molle, F. and Closas, A. (2020b). "Why is state-centered groundwater governance largely ineffective? A review." *Wiley Interdisciplinary Reviews: Water*, 7(1), e1395.

Molle, F., Mollinga, P. P. and Wester, P. (2009). "Hydraulic bureaucracies and the hydraulic mission: Flows of water, flows of power." *Water Alternatives*, 2(3), 328–349.

Molle, F., Venot, J. P., Lannerstad, M. and Hoogesteger, J. (2010). "Villains or heroes? Farmers' adjustments to water scarcity." *Irrigation and Drainage*, 59(4), 419–431.

Molle, F. and Wester, P. (2009). *River Basin trajectories: Societies, environments and development* (Vol. 8). Reading, UK: C.A.B. International.

Knappett, P. S. K., Li, Y., Loza, I., Hernandez, H., Avilés, M., Haaf, D., Majumder, S., Huang, Y., Lynch, B., Piña, V., Wang, J., Winkel, L., Mahlknecht, J., Datta, S., Thurston, W., Terrell, D. and Nordstrom,

K. (2020). "Rising arsenic concentrations from dewatering a geothermally influenced aquifer in central Mexico." *Water Research*, 185, 116257.

Ray, D. K., West, P. C., Clark, M., Gerber, J. S., Prishchepov, A. V. and Chatterjee, S. (2019). "Climate change has likely already affected global food production." *PLoS ONE*, 14(5), e0217148.

Ritzema, H., Satyanarayana, T., Raman, S. and Boonstra, J. (2008). "Subsurface drainage to combat waterlogging and salinity in irrigated lands in India: Lessons learned in farmers' fields." *Agricultural Water Management*, 95(3), 179–189.

Robinson, G. M. (2018). "Globalization of agriculture." *Annual Review of Resource Economics*, 10, 133–160.

Robinson, T. P., Thornton, P. K., Franceschini, G., Kruska, R. L., Chiozza, F., Notenbaert, A., Cecchi, G., Herrero, M., Epprecht, M., Fritz, S., You, L., Conchedda, G. and See, L. (2011). *Global livestock production systems*. Rome, Italy: FAO and ILRI.

Senanayake, N., Mukherji, A. and Giordano, M. (2015). "Re-visiting what we know about Irrigation Management Transfer: A review of the evidence." *Agricultural Water Management*, 149, 175–186.

Shah, E., Vos, J., Veldwisch, G. J., Boelens, R. and Duarte-Abadía, B. (2019). "Environmental justice movements in globalising networks: A critical discussion on social resistance against large dams." *The Journal of Peasant Studies*, 48(5), 1–25.

Suhardiman, D., Giordano, M., Rap, E. and Wegerich, K. (2014). "Bureaucratic reform in irrigation: A review of four case studies." *Water Alternatives*, 7(3), 442–463.

Thenkabail, P. S., Biradar, C. M., Noojipady, P., Dheeravath, V., Li, Y., Velpuri, M., Gumma, M., Gangalakunta, O. P. R., Turral, H., Cai, X., Vithanage, J., Schull, M. A. and Dutta, R. (2009). "Global irrigated area map (GIAM), derived from remote sensing, for the end of the last millennium." *International Journal of Remote Sensing*, 30(14), 3679–3733.

van der Ploeg, J. D. (2014). "Peasant-driven agricultural growth and food sovereignty." *The Journal of Peasant Studies*, 41(6), 999–1030.

Veldwisch, G. J. (2013). "Local governance issues after irrigation management transfer: A case study from Limpopo province, South Africa." In Perret, S., Farolfi, S. and Hassan, R. (eds.) *Water governance for sustainable development*. London, UK: Earthscan. pp. 75–91.

Veldwisch, G. J., Mollinga, P., Hirsch, D. and Yalcin, R. (2012). "Politics of agricultural water management in Khorezm, Uzbekistan." In Martius, C., Rudenko, I., Lamers, J.P.A. and Vlek, P.L.G. (eds.) *Cotton, water, salts and soums: Economic and ecological restructuring in Khorezm, Uzbekistan*. Springer, Vol. 9789400719637. pp. 127–140.

Veldwisch, G. J., Venot, J. P., Woodhouse, P., Komakech, H. C. and Brockington, D. (2019). "Re-introducing politics in African farmer-led irrigation development: Introduction to a special issue." *Water Alternatives*, 12(1), 1–12.

Venot, J. P., Kuper, M. and Zwarteveen, M. (eds.). (2017). *Drip irrigation for agriculture: Untold stories of efficiency, innovation and development*. London, UK and New York, NY: Earthscan/Routledge.

Vos, J. and Boelens, R. (2014). "Sustainability standards and the water question." *Development and Change*, 45(2), 205–230.

Vos, J., Boelens, R., Venot, J. P. and Kuper, M. (2020). "Rooted water collectives: Towards an analytical framework." *Ecological Economics*, 173, 106651.

Wezel, A., Bellon, S., Doré, T., Francis, C., Vallod, D. and David, C. (2009). "Agroecology as a science, a movement and a practice. A review." *Agronomy for Sustainable Development*, 29(4), 503–515.

Woodhouse, P., Veldwisch, G. J., Venot, J. P., Brockington, D., Komakech, H. and Manjichi, Â (2017). "African farmer-led irrigation development: Re-framing agricultural policy and investment?" *The Journal of Peasant Studies*, 44(1), 213–233.

21
GROUNDWATER

Susann Baez Ullberg and Henrik Josefsson

Groundwater Basics

In the 21st century, the global demand for freshwater is increasing in order to supply agriculture, industry, and domestic needs. Representing 97% of the world's available freshwater resources, groundwater is often preferred over surface water because of expectations in terms of both quality and quantity. The definition of groundwater varies but often refers to the water found underground in what are called aquifers, that is, subterranean cracks and spaces in soil, gravel, sand, sandstone and fractured rock. In large parts of the world, aquifers are extensive collections of water underneath the surface, where it can be stored from tens to tens of thousands of years. Globally, a smaller proportion of groundwater resources are found in smaller entities, often referred to as groundwater bodies or small aquifers. Water in aquifers is brought to the surface naturally through a spring or can be discharged into lakes and streams. Groundwater can also be extracted through a well drilled into the aquifer. In contrast to surface water, groundwater's underground existence and movement take place largely out of human sight and control. This subterranean condition constitutes what we call a phenomenological challenge as it turns aquifers into unknown terrain and groundwater into a hidden and unpredictable resource. The perceptual challenge and a lack of public awareness create limited political and legal responses, which risk turning groundwater use into a "tragedy of the commons" (Dellapenna, 2012). Notwithstanding, more than 50% of the world's population use groundwater for drinking purposes and more than 40% of all groundwater is used for irrigation (Siebert et al., 2010). With global annual withdrawal rates between 800 and 1,500 km^3 per year, groundwater is allegedly the most extracted raw material in the world (Godfrey et al., 2019: 1). It is fundamental to ecosystems, as it maintains the base flow of rivers and prevents land subsidence and seawater intrusion. Groundwater is vital to human societies in sustaining livelihoods and is a critical solution for people without access to other sources of freshwater (Re et al., 2022: xvii). Expectations regarding groundwater's quality and quantity (Gleeson et al., 2020; Lapworth et al., 2020) make it a preferred resource in many places. The increased global consumption of groundwater since the 1970s has however resulted in the depletion of many of the world's aquifers today and careless land and surface water use has contributed to the contamination of small and large groundwater bodies (Bierkens & Wada, 2019; Dalin et al., 2019; Guppy et al., 2018).

DOI: 10.4324/9781003095545-25

Groundwater, just like surface water, connects different parts of society and acts as a bridge between scientific disciplines and political commitments such as Agenda 2030 of the United Nations and its 17 sustainable development goals (SDGs) and the Paris Agreement, as agriculture, industry, and households depend on this vital and shared resource for survival and development in the Anthropocene (Gleeson et al., 2020). Despite their importance, aquifers and groundwater have been largely invisible as public objects of concern (United Nations, 2022), indexing its hidden nature underground. While groundwater research is an established field of research within the realm of the natural and engineering sciences, it is a relatively understudied area in social sciences and humanities, especially if we compare it to surface water, as this handbook testifies. This chapter makes no claim to being exhaustive in describing this field, rather it draws on an eclectic collection of key works and lesser known research to provide an overview of the literature with a global scope.

Groundwater Uses

Groundwater epistemologies in the past and the present

The knowledge and use of groundwater is probably as old as humanity itself as people have settled close to springs, formed when groundwater flows onto the surface, to secure their water supply (Re et al., 2022: xvii). For example, the ancient Minoan towns of Knossos, Tylissos, and Syme on Crete developed and thrived around 2000 B.C. through their access to springs on the island (Mays, 2010: 3). In the Colca Valley of Peru's southern highlands, the principal livelihood is agriculture, which depends on irrigation. This is largely supplied by *puquios*, springs that villagers believe are controlled by superhuman forces who demand the offering of gifts to release its subterranean water for human use (Paerregaard, 2022). This research makes clear that groundwater is more than a natural substance; it is a hydrosocial phenomenon and a resource that different epistemic communities understand in varying ways. To render a subterranean and hidden world that can never be seen, remote sensory technologies are often used by experts nowadays, not only to monitor and chart aquifers, but to conceptualize and make them knowable (Ballestero, 2019a; 2019b). Numeric models on geologic strata, time, and groundwater extraction are used to narrate a form of balance between nature and culture (Brooks, 2017), and to tell the story and make meaning of imbalance when aquifers are depleted (Bessire, 2021).

Technologies of groundwater detection and extraction

In many places, especially where surface water is scarce, wells provide a reliable and ample supply of groundwater for domestic uses, farming and industries (cf. Re et al., 2022). The relative invisibility of groundwater prompts water users to detect it by way of different technologies. The terrain and the vegetation of a landscape provide clues to where groundwater can be found. Dowsing or water witching is an ancient practice of divination to locate subterranean water by using a forked stick, rod, or pendulum, found in pre-modern Europe and in other parts of the world through colonialization (Kalb, 2020; Krautwurst, 1998). In modern times, geophysical prospecting and drilling technology have enabled the detection of groundwater. Since the 1970s, space technologies such as remote sensing, geographic information systems (GIS), and the global positioning system (GPS) have been playing an increasing role on a global scale in the assessment, monitoring, and use of groundwater. Artificial intelligence (AI) is currently expected to complement such technologies when it comes to interpreting data such as water quality maps and models for sustainable groundwater management (Bhunia et al., 2021).

A majority of households in countries like India and Bangladesh, and in many sub-Saharan countries and Southeast Asian and Pacific nations self-supply drinking water by extracting groundwater from scoop holes, dug wells, or boreholes, well fields or springs belonging to individual proprietors, local communities or NGOs (Grönwall & Danert, 2020: 2). The building and drilling of wells is an ancient and worldwide practice as far as archaeology can tell. The oldest known well in the world is a wooden well dated to 5600 B.C. found in the village of Hemudu in South East China (Jiao, 2007). Another ancient groundwater extraction technology is the *qanat*.[1] This system consists of a subterranean tunnel that uses gravity to convey water from the water table to the surface of lower lands. This technology was developed in Persia in the years 3000 B.C. and later spread to Asia, North Africa, and the Mediterranean[2] (Mays, 2010: 3–4). The oldest *qanats* have been found in the northern part of Iran and tens of thousands are at still use there in present day (Eslamian et al., 2016: 124). A more recent practice of urban and commodified water supply in for example in India, Uganda, and Bangladesh is to connect the groundwater borehole to "water ATMs," automated vending machines that dispenses clean drinking water in public spaces with contaminated or scarce water (Grönwall & Danert, 2020).

Sharing groundwater: Conflict, collaboration, and transboundary governance

Groundwater moves in largely unknown ways beneath the social and political-administrative boundaries that exist above ground. This poses a particular challenge to the communities that share aquifers when groundwater is to be extracted. Ostrom (1990) demonstrated through empirical research that communities using the same groundwater resource could develop local self-sustaining alliances, so called groundwater commons, even if self-regulation would be more likely to work for smaller scale groundwater systems than larger ones (cf. Curtis et al., 2016). Recent research shows that a lack of access to groundwater is associated with a higher risk of communal violence, even if the risk is lower in areas with a strong presence of the state (Döring, 2020). This research also suggests that seizing groundwater access from other communities is more likely to occur when water property rights disputes (due to scarcity for example) are not effectively mitigated by political institutions (Döring, 2020). State intervention in uses of local groundwater can also act in detriment to communities' right to self-supply from groundwater sources, however (Grönwall & Danert, 2020), turning water users into activists who have to become experts on technical and legal matters to make their claim of water as a human right over the paradigm of water as a commodity (Ballestero, 2019b).

Transboundary groundwater governance at local and regional levels

Legal frameworks can play a crucial role for effective groundwater governance. They flank and support water policy and provide users and the administration with rights and obligations to use, manage, and protect groundwater resources to achieve sustainable use. It is often said that legal frameworks for groundwater must be adapted to each country's condition, such as institutional capabilities, customs, cultures, political systems, and prevailing practices (Mechlem, 2016). This becomes a challenge in places where groundwater is transboundary, as it often is, not only crossing national borders but also limits between political administrative territories on lower levels. Consider for example, the case of shared groundwater resources on the border between the states of Chihuahua, Coahuila, Nuevo Leon, and Tamaulipas in northeastern Mexico and the US state of Texas. This a politically sensitive region due to migratory politics. It is suggested that in order to develop more effective cross-border relations and enhance knowledge, collaborative management, and sustainability of the region's shared aquifers, stakeholders should pursue local and regional arrangements that focus primarily on water quality and environmental issues (Sanchez & Eckstein, 2020). For the results of such efforts to be permanent and sustainable on a local level, authors suggest that they must consider both formal and informal long-term cooperative models to achieve stronger systemic impacts and funding commitments to collaborative groundwater governance (Sanchez & Eckstein, 2017). Corresponding aquifer management and governance schemes must encompass local needs with regional, national, and global perspectives, which means that there are similarities in the demands and needs between translocal, transregional, and transnational aquifers.

International and transnational groundwater agreements

While groundwater and surface water form part of the same hydrological system and are largely managed according to the same policy regime in most countries, this practice has been questioned due to the unique characteristics of groundwater. For example, groundwater is typically more vulnerable to pollution and other forms of contamination than surface water because it generally flows at much slower rates than surface water. While one of the first transnational legal acts on groundwater focused on limiting groundwater pollution was established within the EU in 1980,[3] it was not until the 21st century that the international community begun to pay attention to the topic of transboundary aquifers (Eckstein, 2017). This resulted in the Draft Articles on the Law of Transboundary Aquifers, which were annexed to the UN General Assembly Resolution 63/124. Over the years, the Articles have been discussed four times before the UN General Assembly 6th Committee, yet not much more has happened. Given this lack of progress of adopting the Draft Articles, the opinion is that there is no will to establish international laws that may be stronger than multilateral or bilateral agreements. One alleged reason for this is a concern about having to cede riparian rights (Alfonso & Candela, 2018). Another reason is the idea that while the management of transboundary groundwater is critical for the success of the UN 2030 Agenda, international law is not a suitable framework to regulate transboundary aquifers. Instead, it is argued that this type of legislation must be tailored to each state (Alfonso & Candela, 2018). The difficulties in fully respecting national ideas of, for example, groundwater rights and developing international law on groundwater indicate that the Draft Articles may stay as drafts for a long time.

With the exception of the EU Water Framework Directive (WFD) and Groundwater Directive (which we will discuss in more detail next), no international groundwater law is yet in place. Instead, other legal options are transnational agreements. In some cases, these are used to regulate, for example, equitable and sustainable use, and pollution prevention. There are at least 592 transboundary aquifers and groundwater bodies in the world, yet only seven of them are subject to a transboundary agreement binding the parties to its content, and a few others are covered by less formalized transboundary arrangements that are not binding. One such agreement that has been in place for 40 years and is often claimed to be a success is the Genevese groundwater agreement, which regulates the management of the Genevese aquifer between communities belonging to the EU (e.g., France) and a Swiss canton (de los Cobos, 2018). Another well-known agreement is the one covering the Guaraní aquifer, the second-largest known aquifer system in the world, located under Argentina, Brazil, Paraguay, and Uruguay. The Guaraní Aquifer Agreement, signed by the governments of these countries in 2010 and ratified in 2020, is one of the few treaties signed in a precautionary context (Villar et al., 2011). Despite many conflicts regarding surface waters in the La Plata Basin and in spite of hard pressure on the groundwater due to extraordinary droughts in the region in 2015 and in 2019–2022, the transboundary agreement has not yet truly been put to the test. However, it is likely that climate change will soon put more pressure on this aquifer, creating unprecedented challenges for its governance.

European Union Law on Groundwater

One very notable exemption to the legal void in transnational groundwater legislation that seldom is discussed in international law circuits is the EU WFD and Groundwater Directive, which includes both transboundary aquifers and national groundwater bodies. In contrast to most other countries that lack legal management structures for transboundary aquifers or do not engage in establishing agreements, the EU has implemented a joint assessment method for groundwater status and require member countries to secure a good groundwater status before 2027. While the problems of quantity and quality of groundwater may seem to be greater in other parts of the world, EU Countries also suffer from the over-exploitation of groundwater. Since this legislation is often, or at least occasionally, referenced in discussions on the establishment of an international groundwater law, it is relevant to describe it here in more detail and discuss its possibilities and constraints.

The EU law on water is one of the more complex strands of EU law and has developed over a number of years. To bring coherence and consistency to this area of law, the EU WFD was established in the year 2000 and a few years later, in 2006, a "daughter directive" on groundwater was established.[4] The main focus of the 2006 Groundwater Directive is on the prevention and control of groundwater pollution, with a view to ensuring the protection of drinking water sources and dependent ecosystems.[5] The Groundwater Directive builds on the WFD and the two cannot be separated. What makes the WFD and Groundwater Directive interesting and rather unique is that they focus on the groundwater environment at large and not only on the supply of drinking water. The latter is instead specifically covered by the newly revised Drinking Water Directive.[6] Among the key requirements of the WFD and the Groundwater Directives is that member states are required to establish groundwater monitoring networks and river basin management plans to provide a comprehensive overview of groundwater status in terms of both quality and quantity.[7] The aim of these requirements concerning groundwater is to prevent or limit the input of pollutants and the deterioration of groundwater quality; to protect, enhance, and restore all bodies of groundwater and ensure a balance between extraction and recharge of aquifers; and to reverse any significant and sustained upward trend in the concentration of any pollutant resulting from the impact of human activity in order to progressively reduce the pollution of groundwater.[8]

The environmental objectives are binding for the member states, and they must achieve them.[9] This means that member states that apply the normal legal remedies to groundwater pollution (that is, control of the discharge and the process from which the discharge originates, or programmes/best available practices for diffuse sources) and still do not achieve the environmental objectives are breaching EU law. In legal terms, this is called a combined approach.[10] The EU Commission (2019) considers that the Directives align with the EU's international obligations related to groundwater governance, such as the 2030 Sustainable Development Goal Agenda. For example, SDG 6 "Clean water and sanitation" includes six targets that address matters of water quality and quantity that has bearing on the status of groundwater. Groundwater levels of nitrates and pesticides must not exceed the EU-wide standards[11] in order to be classified as of good quality. In terms of quantity, the level of groundwater must not exceed the long-term annual average rate of extraction, for example. The above mentioned objectives of the Directives are slowly under achievement in Member State countries EU Commission (2019) EU Commission (2019). EU member states sharing a body of groundwater must coordinate their activities in monitoring, the setting of threshold values and the identification of relevant hazardous substances. They must ensure that the establishment of threshold values for transboundary groundwater bodies is coordinated among the member states concerned. Where a body of groundwater extends beyond the territory of the EU, the member states concerned shall "endeavour" to establish threshold values in coordination with the relevant non-EU countries.[12] There is also an option for member states to use structures originating from international agreements in their governance system,[13] meaning that they may entrust an international commission established under international law with the enforcement of the Directives' requirements regarding a transboundary aquifer.

Two decades after implementing the WFD and the Groundwater Directive in the EU, have they served the member states to achieve their environmental objectives? With respect to groundwater, achievements regarding good groundwater chemical status is 74% and good groundwater quantitative status is on 89% of the area of groundwater bodies in 2019 (EU Commission 2019). The figures need to be considered with some caution, however. Due to the stark differences in size of aquifers and groundwater bodies, any strict comparison based on number of water bodies is not viable. Furthermore, the identification of pollutants that characterize bodies of groundwater as being at risk differs substantially between member states, ranging from 10 to 90 identified substances between different countries. Recent research indicates that this is a growing problem (Hartman et al., 2021).

Groundwater Challenges and Research Gaps

Groundwater has played a crucial role for the development of societies and ecosystems in the past, and will increasingly do so in a future with profound societal and ecological changes. This entails many challenges. Scholars have recently argued that to meet these challenges by drawing on groundwater in sustainable ways, it is crucial to strengthen the relationship between water users and managers, and to promote the sharing of groundwater information (Re et al., 2022: xviii). Policy makers at local, national, regional, and international levels have a major responsibility here. Equal and sustainable policy decisions about aquifer resource management need to be based on sound scientific knowledge. It is often argued that water legislation follows the logic of the natural sciences, yet both the EU WFD and the Groundwater Directive indicate that this is an incorrect assumption. Instead, these agreements – and the political discourses they stem from – affect the development of sciences. This indicates the need for laws on groundwater use and protection to take the social sciences as its starting point to make natural scientific knowledge usable in socio-legal contexts. In addition, given that many aquifers are transboundary, the data must come from similar assessment structures in each country to ensure that the information is comparable. As the EU law and the Genevese groundwater agreement show, this will only happen if there is a politico-legal agreement on this.

Another challenge is the issue of groundwater ownership. Revisions to water rights may be an initial stage in the implementation of groundwater policies. This is a fundamental legal issue that requires attention on a regular basis because the introduction of protection measures by the State usually involves the enactment of control mechanisms that apply to groundwater or to activities on the ground above the aquifer. Such control mechanisms may infringe on existing practices and rights. It may be necessary to modify or withdraw the existing rights of individual groundwater users in order to enact the necessary changes. More research is needed on groundwater governance, especially when it comes to transboundary aquifers, but it is likely that for groundwater use and protection policies to be effective, democratic, and sustainable within and across national borders, they must take account of not only the interrelationship of quantity and quality of groundwater, but of the institutional and social environment in the countries. The financial viability and social acceptance of any proposed measure for protection should be part of the equation, since what is considered reasonable in some countries may not be acceptable in others. Therefore, when policy proposals build on models and experiences from other regions, it is essential to analyse them in the light of local conditions and to include all public and private stakeholders in the discussions at the earliest possible stage.

Groundwater is a vital yet hidden element that exists beneath our feet and much out of our sensory scope, and behaves in ways that we still do not know much about. Much of the hydrological, ecological and public health effects on groundwater from human activities in terms of extraction and pollution are matters we still do not fully understand, not only because of their complexity, but also because underground research is a quest. This phenomenological challenge calls for more research into the ontological and epistemological assumptions that inform historical and contemporary relationships with groundwater in different geo hydrosocial settings. Notwithstanding the ecological, legal, and political dimensions of groundwater that this review has shed light on, we argue that a phenomenological approach can provide an ecocritical perspective to balance an anthropocentric framing of groundwater as a management problem or a mere technological and economic resource that is instrumental to global development.

Notes

1 *Qanat* is a Semitic word meaning "to dig."
2 This technology is known by other names such as *karez* (Afghanistan and Pakistan), *kanerjing* (China), *falaj* (United Arab Emirates), and *foggara* and *fughara* (North Africa).
3 Directive 80/68/EEC on the protection of groundwater against pollution caused by certain dangerous substances.
4 Required in Water Framework Directive (WFD) Article 17.
5 Art. 4(1)(b) and 7 WFD.
6 Directive (EU) 2020/2184 of the European Parliament and of the Council of 16 December 2020 on the quality of water intended for human consumption (recast).
7 The river basin management plans must include, among other things, a thorough analysis of pressures and impacts of human activity on groundwater status, a map of monitoring results, a summary of the economic analysis of water use, a summary of protection programs, control, or remediation measures.
8 Art. 4(1)(b) WFD and art. 5 Groundwater Directive (GWD).
9 This has been concluded by the Court of Justice of the European Union, see C-461/13, Bund für Umwelt und Naturschutz Deutschland (2015) and C-535/18 – Land Nordrhein-Westfalen (2020).
10 In more detail, this means that the Directives first require the implementation of emission controls based on best available techniques, or the relevant emission limit values, or in the case of diffuse impacts the controls including, as appropriate, best environmental practices. And secondly, it requires that if there is a quality objective or quality standard, whether established pursuant to WFD or any other Union legislation, requires stricter conditions than those which would result from the application of previous mentioned measures, more stringent emission controls shall be set accordingly. There are two main rules of exemption, but these also require that the deterioration is stopped and that best possible groundwater status is achieved by an implementation of all reasonable measures.
11 Relevant conditions are specified in Annex V.
12 GWD Article 3, paras 3 and 4.
13 WFD, Article 3, paras 5 and 6.

References

Alfonso, R. and Candela, L. (2018). "Fifteen-year experiences of the internationally shared aquifer resources management initiative (ISARM) of UNESCO at the global scale." *Journal of Hydrology: Regional Studies*, 20: 5–14.

Ballestero, A. (2019a). "Touching with light, or, how texture recasts the sensing of underground water." *Science, Technology and Human Values*, 44(5): 762–785.

Ballestero, A. (2019b). *Future History of Water*. Durham: Duke University Press.
Bessire, L. (2021). *Running Out: In Search of Water on the High Plains*. Princeton and Oxford: Princeton University Press.
Bhunia, G. S., Shit, P. K., Gupta, H. D. and Adhikary, P. P. (2021). "Groundwater and Space Technology: Issues and Challenges." In: Shit, P. K. Bhunia, G. S., Adhikary, P. P. and Dash, Ch. J. (eds). *Groundwater and Society*. Cham: Springer. pp. 83–102.
Bierkens, M. F. and Wada, Y. (2019). "Non-renewable groundwater use and groundwater depletion: A review." *Environmental Research Letters*, 14(6): 063002.
Brooks, E. (2017). "Number narratives: Abundance, scarcity, and sustainability in a California water world." *Science as Culture*, 26(1): 32–55.
Curtis A., Mitchell M., & E. Mendham. (2016). "Social Science Contributions to Groundwater Governance." In Jakeman A. J., Barreteau, O., Hunt, R. J., Rinaudo, J. D. and Ross, A. (eds). *Integrated Groundwater Management*. Cham: Springer. pp. 477–492.
Dalin, C., Taniguchi, M. and Green, T. R. (2019). "Unsustainable groundwater use for global food production and related international trade." *Global Sustainability*, 2(12): 1–11.
Dellapenna, J. W. (2012). "A primer on groundwater law." *Idaho Law Review* 49(2): 265.
de los Cobos, G. (2018). "The Genevese transboundary aquifer (Switzerland-France): The secret of 40 years of successful management." *Journal of Hydrology: Regional Studies*, 20: 116–127.
Döring, S. (2020). "Come rain, or come wells: How access to groundwater affects communal violence." *Political Geography*, 76(1): 102073.
Eckstein, G. (2017). *The International Law of Transboundary Groundwater Resources*. London: Earthscan, Routledge.
Eslamian, S., Davari, A. and Reyhani, M. N. (2016). "Iranian Qanāts: An Ancient and Sustainable Water Resources Utilization." In: Angelakis, A. N. Chiotis, E., Eslamian, S. and Weingartner, H. (eds.). *Underground Aqueducts Handbook*. London: Routledge. pp. 123–150.
EU Commission. (2019). *Report from The Commission to The European Parliament and The Council on the implementation of the Water Framework Directive (2000/60/EC) and the Floods Directive (2007/60/EC) Second River Basin Management Plans First Flood Risk Management Plans*. 26 February. COM (2019) 95 final. [Accessed February 02, 2023]. Available from https://ec.europa.eu/environment/water/water-framework/impl_reports.htm
Gleeson, T., Cuthbert, M., Ferguson, G. and Perrone, D. (2020). "Global groundwater sustainability, resources, and systems in the Anthropocene." *Annual Review of Earth and Planetary Sciences*, 48: 431–463.
Godfrey, S., Hailemichael, G. and Serele, C. (2019). "Deep groundwater as an alternative source of water in the Ogaden Jesoma sandstone aquifers of Somali Region, Ethiopia." *Water*, 11(8): 1735.
Grönwall, J. and Danert, K. (2020). "Regarding groundwater and drinking water access through a human rights lens: Self-supply as a norm." *Water* 12(2): 419.
Guppy, L., Uyttendaele, P., Villholth, K. G. and Smakhtin, V. (2018). "Groundwater and Sustainable Development Goals: Analysis of Interlinkages." *UNU-INWEH Report Series*, 4. Hamilton: United Nations University Institute for Water, Environment and Health.
Hartman, A., Jasechko, S., Gleeson, T., Wada, Y., Andreo, B., Barberá, J. A., Brielmann, H., Bouchaou, L., Charlier, J.-B., Darling, W. G., Filippini, M., Garvelmann, J., Goldscheider, N., Kralik, M., Kunstmann, H., Ladouche, B., Lange, J., Lucianetti, G., Martín, J. F., Mudarra, M., Sánchez, D., Stumpp, C., Zagana, E. and Wagener, T. (2021). "Risk of groundwater contamination widely underestimated because of fast flow into aquifers." *Proceedings of the National Academy of Sciences*, 118(20): e2024492118.
Jiao, J. J. (2007). "A 5,600-year-old wooden well in Zhejiang Province, China." *Hydrogeology Journal*, 15: 1021–1029.
Kalb, M. (2020). "The dowsing debate: Water, science and colonialism in German Southwest Africa." *German History*, 38(4): 568–593.
Krautwurst, U. (1998). "Water-witching modernist epistemologies and dowsing rationality: Exporting models of non-rationality through colonial and development discourses." *Political and Legal Anthropology Review*, 21(2): 71–82.
Lapworth, D. J., MacDonald, A. M., Kebede, S., Owor, M., Chavula, G., Fallas, H., Wilson, P., Ward, J. S. T., Lark, M., Okullo, J., Mwathunga, E., Banda, S., Gwengweya, G., Nedaw, D., Jumbo, S., Banks, E., Cook, P. and Casey, V. (2020). "Drinking water quality from rural handpump-boreholes in Africa." *Environmental Research Letters*, 15(6): 064020.
Mays, L. (ed.) (2010). *Ancient Water Technologies*. Heidelberg, London, and New York: Springer.

Mechlem, K. (2016). "Groundwater governance: The role of legal frameworks at the local and national level-established practice and emerging trends." *Water*, 8(8): 347.

Ostrom, E. (1990). *Governing the Commons: The Evolution of Institutions for Collective Action*. Cambridge: Cambridge University Press.

Paerregaard, K. (2022). "Lubricating water metabolism: How mountain offerings contribute to water sustainability in the Peruvian Andes." *kritisk etnografi – Swedish Journal of Anthropology*, 4(2), 83–98.

Re, V., Manzione, R. L., Abiye, T. A., Mukherji, A. and MacDonald, A. (2022). "Introduction: Groundwater, Sustainable Livelihoods and Equitable Growth." In: Re, V., Manzione, R. L., Abiye, T. A., Mukherji, A. and MacDonald, A. (eds). *Groundwater for Sustainable Livelihoods and Equitable Growth*. Boca Raton, London, and New York: CRC Press. pp. xvii–xxiv.

Sanchez, R. and Eckstein, G. (2017). "Aquifers shared between Mexico and the United States: Management perspectives and their transboundary nature." *Groundwater*, 55(4): 495–505.

Sanchez, R. and Eckstein, G. (2020). "The path towards groundwater management in the borderlands of Mexico and Texas." *Wiley Interdisciplinary Reviews: Water*, 7(1): e1399.

Siebert, S. J., Burke, J., Faures, J. M., Frenken, K., Hoogeveen, J., Doll, P. and Portmann, F. T. (2010). "Groundwater use for irrigation – A global inventory." *Hydrology and Earth System Science*, 14(10): 1863–1880.

United Nations. (2022). *The United Nations World Water Development Report 2022: Groundwater: Making the Invisible Visible*. Paris: UNESCO.

Villar, Pilar C. and Wagner C. Ribeiro. (2011). *The Agreement on the Guarani Aquifer: a new paradigm for transboundary groundwater management?* Water International, 36(5): 646-660.

22
WATER STRESS AND SCARCITY

Zafar Adeel

Introduction

The international dialogue and discussion around the availability of water,[1] or lack thereof, has been receiving considerable attention in traditional and social media (Graffy, 2006; Felter & Robinson, 2021). A variety of definitions of water stress are used to compare water withdrawals against annually renewable resources; a situation that leads to a divergent description of the size of the global population living under water stress, ranging from a few hundred million to 2.3 billion to over 4 billion (Vörösmarty et al., 2000; Mekonnen & Hoekstra, 2016; UN-Water, 2021). Ostensibly, there is a tendency to use numbers that depict the largest possible impact of water scarcity; implicit in this approach is the argument that larger the size of population living under water stress, the greater the policy response and magnitude of investments sought. Therefore, the matter of how water stress and scarcity are defined gains considerable importance in determining how the international community, national and sub-national governments, and communities should respond.

As is discussed in the subsequent sections, the topics of water stress and scarcity – however we choose to define them – are frequently discussed in research publications, media outlets, and social media platforms (Felter & Robinson, 2021; Encyclopaedia Britannica, 2022); the corresponding public attention does not always correlate to a greater emphasis in addressing water scarcity in a political or policy context. While one would expect a proportionately increased investment of financial and human resources to address water scarcity, the political response is often not so uniform or coherent (UN-Water, 2021).

There is also often a disconnect between the global dialogue on water scarcity and how it might be perceived at a local level. This disconnect operates at both geographical scales and different societal levels. Most of the worldwide analyses of water scarcity provide national-scale estimates of water stress or scarcity; this approach tends to mask the variability within the countries. It is important for policy-formulation purposes for water availability and/or scarcity data be available at the lowest scale of jurisdiction, which can be an intensive and expensive exercise. The societal disconnects also become obvious in terms of what is considered "adequate" water consumption. Data tend to show that greater the level of economic prosperity, greater the level of water consumption on a per capita basis (Mekonnen & Hoekstra, 2016; Emile et al., 2022).

DOI: 10.4324/9781003095545-26

Definitions of water stress and scarcity

Overall, water scarcity is defined as having insufficient water resources available for meeting societal and environmental needs within a geographic location (Savenije, 2000). But there remain some challenges and ambiguities in how this definition is created and applied. Let us consider the most significant of these definitional challenges here.

First, there are many ways in which "sufficiency" of water is defined because the societal demand on water resources is a stretchable quantity that is highly dependent on the local social and economic context (Savenije, 2000; Emile et al., 2022). In many countries around the world, the water demand has been significantly reduced through better management and minimizing water losses. Historically, the availability of over 1,000 m^3/capita/year of water was considered to be the threshold of water scarcity and 500 m^3/capita/year as the "absolute water scarcity" (Falkenmark, 1997; UN, 2020). However, these thresholds are somewhat arbitrary and outdated, and often lack contemporary scientific reasoning or rigorous data analysis.

A second ambiguity in these approaches towards defining water scarcity is that of describing "environmental" water needs (Savenije, 2000; Vörösmarty et al., 2005). Water availability in any river- or lake-basin is typically fluctuating with seasons and the ecosystems dependent on this water have adapted themselves to this variability. Water planners, however, had to come up with a scheme of abstracting water that can, or appears to, protect the ecosystem. The response to that dilemma is to use the average lowest water availability in a basin as the absolute minimum required to sustain the ecosystem (Vörösmarty et al., 2005); this argument is obviously flawed from a scientific standpoint because it could be utilized to deplete the available water resources to their lowest possible level. Ostensibly, this approach supports the rationale that anthropogenic demands eventually trump any ecosystem demands.

There is a third type of ambiguity: how does one define the "area" for which water scarcity or stress is measured? While all hydrological systems – such as river and lake basins – are agnostic to politically defined geographical boundaries, practically all computations pertaining to scarcity are performed in the context of political boundaries. This approach becomes problematic for transboundary water resources – river basins and lakes that cut across international borders. It is quite rare to find reliable data about annual water availability per capita on a basis that evenly covers an entire transboundary watershed.

Finally, the approaches to describe and quantify water scarcity rarely incorporate the quality of water. Water that is "available" for consumption in a watershed may actually not be suitable in terms of its quality for that specific consumptive use – whether it happens to be direct human consumption or its use in agriculture or other industries. For the purposes of our discussion in this chapter, one may conclude that the delineation and description of water security is probably overestimating the water availability because it is not accounting for water that is either polluted or contains excess minerals like sodium, magnesium, or metals like arsenic and iron or various organic contaminants.

There are a number of definitions of water stress, as shown in Table 22.1. This list of definitions is not meant to be exhaustive, but to demonstrate two common hallmarks of all approaches related to water stress and scarcity: first, they tend to get more complicated over time in efforts to encompass the many dimensions of water usage, including balancing out the needs of various stakeholders against those of the natural ecosystems. Second, they provide significantly different pictures at various geographic scales of what water scarcity looks like. Consequently, very different population figures are often cited about how many people around the world are living with water scarcity or stress and how those numbers are forecasted to change in future as both

Table 22.1 Commonly used water stress and scarcity indicators.

Commonly used title	*Category of indicator*	*Quantitative measurement*	*Geographical scale*	*Reference*
Falkenmark indicator	Water shortage	Per capita water availability	Sub-national (presented at global scale)	Falkenmark et al. (1989, 2009)
Criticality ratio	Water stress	Ratio of water use to availability	Sub-national (presented at global scale)	Falkenmark (1997); Alcamo et al. (2000); Vörösmarty et al. (2000)
IWMI indicator	Physical and economic water scarcity	Proportion of water supply that is water availability, accounting for water infrastructure	Sub-national (presented at global scale)	Seckler et al. (1998)
Water Poverty Index		Weighted average of five components: water availability, access, capacity, use, and environment	National	Sullivan et al. (2003)
Water footprint	Water consumption based assessment	The ratio of water footprint to water availability	–	Hoekstra et al. (2011)
Water stress	Physical water scarcity	The ratio of water withdrawal for all economic activities to the total renewable water resources available	National	UN-Water (2021)

Source: Adapted from Liu et al. (2017).

water demand and supply change. We should also note that the water scarcity indicators shown in Table 22.1 simply do not take water quality into account, which has consequences for determining the actual scarcity of water resource.

Hydrologic water scarcity is driven by nature, for the most part (that is, if we don't consider the impacts of climate change on the hydrological cycle), and is derived from the supply side of water resources. The annual precipitation in many parts of the world, typically those labelled as hyper-arid, arid, or semi-arid, is relatively low and correspondingly the evapotranspiration rates are high. These "drylands" are scientifically characterized through the measurement of Aridity Index.[2] An overlay of Aridity Index with population densities in drylands countries shows that significant water-stressed population densities can be found in arid and semi-arid regions (Uriel & Adeel, 2005). The ecosystems in these zones have adapted to the Aridity Index over eons, but often are unable to remain viable when additional, and often intense, anthropogenic demands are placed on them. These excessive demands on water resources can lead to what is often described as economic water scarcity.

Economic water scarcity is a demand-side phenomenon in which the societal demand for water resources exceeds their capacity. Such exceedance in demand can occur due to three broad reasons: either the population density is too high to sustain on scarce water resources, or sufficient water storage and delivery infrastructure does not exist to meet the demands of a society, or the existing water resources and infrastructure are not managed adequately. Poor management *inter alia* stems from the inability to modulate water demands by different users (Liu et al., 2022).

Another reason for economic water scarcity could be that water is economically undervalued as a finite natural resource, often for political reasons (Savenije, 2000). The economic value of water is dependent on its use within the societal context. That is, drinking water, water for industrial production, water for agricultural production, and water to sustain aquatic and terrestrial ecosystems can all be perceived and valued very differently within and across societies. The most common examples of water mismanagement can be found in countries that rely heavily on agricultural production to maintain their local or national economies – about 70% of global water withdrawals are for agricultural purposes (DNI, 2012; United Nations, 2020). In many developing countries, such excessive water consumption for crop or cattle farming is typically subsidized or provided at a flat rate (Hoekstra, 2014) – a situation that is driven by efforts to achieve food security but de-emphasizes water conservation.

Hydrologic versus economic water scarcity

Human-induced climate change is bringing about changes to the global and regional hydrologic cycles in a way that many parts of the world will face a long-term increase in the Aridity Index (i.e., lower average annual precipitation and higher average temperatures) (Hoekstra, 2014; Liu et al., 2022). It may, therefore, be argued that watersheds in different regions are already locked into hydrologic water scarcity for the coming decades. The trends in increased aridity over the past few decades are already observable; areas that are particularly impacted include the Sahel region, Middle East and North Africa, Central and South Asia, and parts of China (Adeel et al., 2005). However, hydrologic water scarcity does not automatically lock a society into economic water scarcity.

Economic water scarcity, on the other hand, can be addressed in multiple ways; the most prominent of these approaches is integrated water resources management (Liu et al., 2022). Successful water management approached include investments in water storage and delivery infrastructure, demand management in key economic sectors, minimizing water losses during delivery, and re-use of (waste)water for multiple purposes. An important consideration for these management approaches is the presence of robust institutions (Schuster-Wallace et al., 2015).

Global distribution of water stress

The global figures about water scarcity and water stress vary considerably (Encyclopaedia Britannica, 2022). The most commonly cited figure is that about 4 billion people live under conditions of severe water scarcity for at least one month each year (Mekonnen & Hoekstra, 2016); it should be noted that this estimate is based on "blue water" resources which include both groundwater from aquifers and surface water from lakes and rivers. This estimate is depicted graphically on the global map with subnational data in the 2020 World Water Development Report (UN, 2020), as shown in Figure 22.1. The same report also states that about 1.6 billion people face economic water shortage.

Another depiction of water stress in the latest version of the UN World Water Development Report (UN, 2022) shows the "blue water scarcity indicator" at the global scale (Figure 22.2). It shows the ratio between net surface water and groundwater withdrawal that is not returned to the source, and the available surface water and groundwater (calculated as the sum of run-off generated in that area and received from upstream, minus the water footprint in all upstream contributing areas). Areas in Figure 22.2 shown in grey or black depict locations in which water reservoirs, including both surface and groundwater resources, are being depleted significantly. A

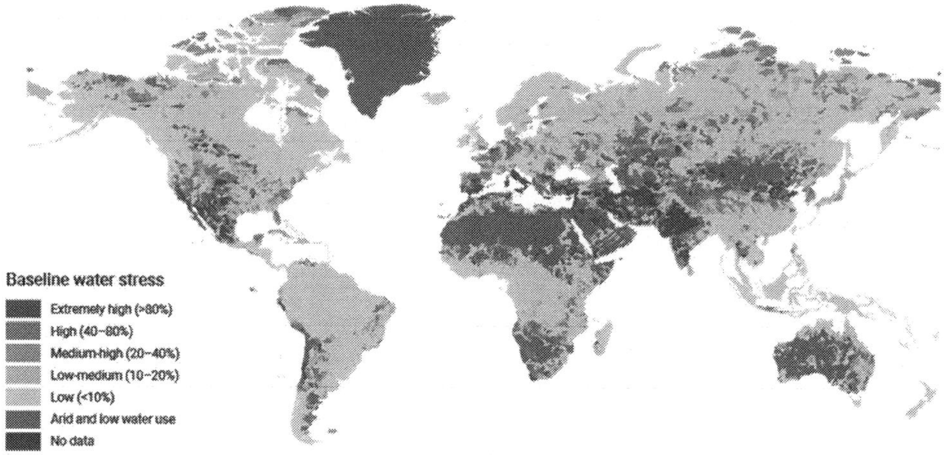

Note: Baseline water stress measures the ratio of total water withdrawals to available renewable water supplies. Water withdrawals include domestic, industrial, irrigation and livestock consumptive and non-consumptive uses. Available renewable water supplies include surface and groundwater supplies and considers the impact of upstream consumptive water users and large dams on downstream water availability. Higher values indicate more competition among users.

Source: WRI (2019). Attribution 4.0 International (CC BY 4.0).

Figure 22.1 Depiction of "Annual baseline water stress" in the World Water Development Report (2020).

quick comparison of Figures 22.1 and 22.2 reveals that while presenting similar concepts based on the work of Mekonnen and Hoekstra (2016), the surface footprint of water-stressed area appears quite different, which might be confusing for a lay person.

The UN-Water progress update in implementation of the Sustainable Development Goal (SDG) #6 to the United Nations includes a map of water-stressed countries (UN-Water, 2021). It claims that when water withdrawal by all economic activities in an area exceeds 25% of the total renewable water resource available, it is identified as "water-stressed." These data are assimilated at the national level, which typically masks areas with order-of-magnitude diversity in water

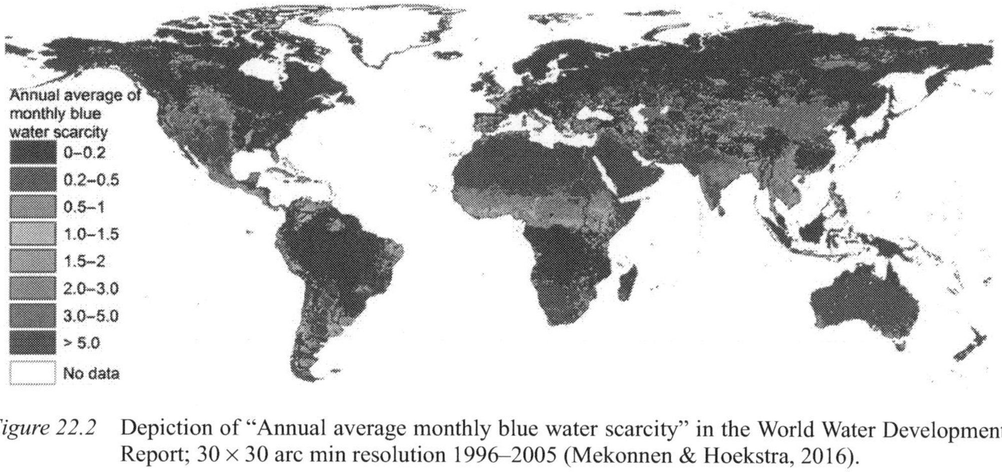

Figure 22.2 Depiction of "Annual average monthly blue water scarcity" in the World Water Development Report; 30 × 30 arc min resolution 1996–2005 (Mekonnen & Hoekstra, 2016).

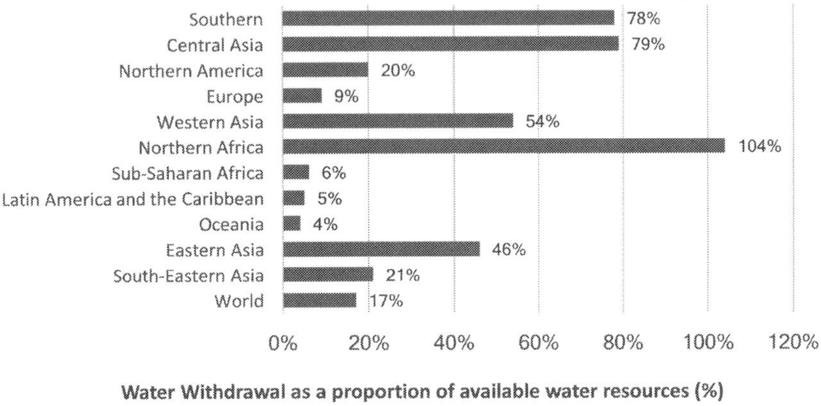

Figure 22.3 Level of water stress, by region, presented as a proportion of water withdrawal as a proportion of available water resources; adapted from UN-Water (2021).

stress levels within countries. Nonetheless, as shown in the corresponding Figure 22.3, 5 out of 11 regions show water stress levels above 25%. Importantly, the same report notes that data accumulation and reporting from some 84 countries is not performed in a consistent way. While this discrepancy in data reporting should be highlighted in the form of confidence in estimation of water scarcity at national level, no such qualification of the data is presented in most of the international reports, or in the subsequent media coverage.

The Water Poverty Index (WPI) is another composite water stress indicator that incorporates a measure of water availability together with access (Lawrence et al., 2002). Specifically, WPI incorporates the average scores pertaining to resources, access, capacity, use, and environment in a given country. Figure 22.4 presents a depiction of WPI at the global scale, demarcated at the national level. It can be argued that WPI is measuring more of economic water scarcity than hydrologic water scarcity. When juxtaposed against each other Figures 22.1 and 22.4 depict

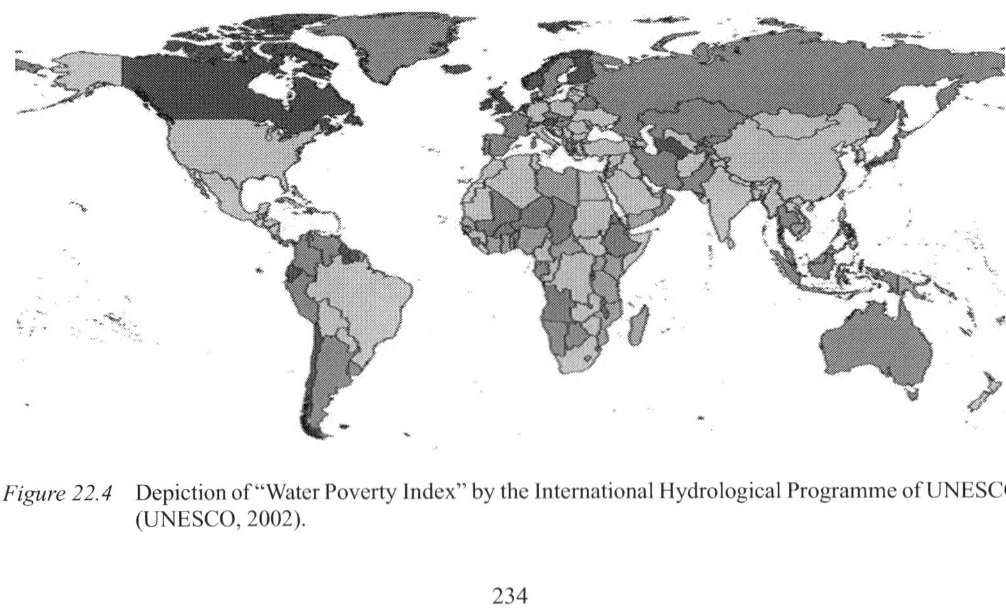

Figure 22.4 Depiction of "Water Poverty Index" by the International Hydrological Programme of UNESCO (UNESCO, 2002).

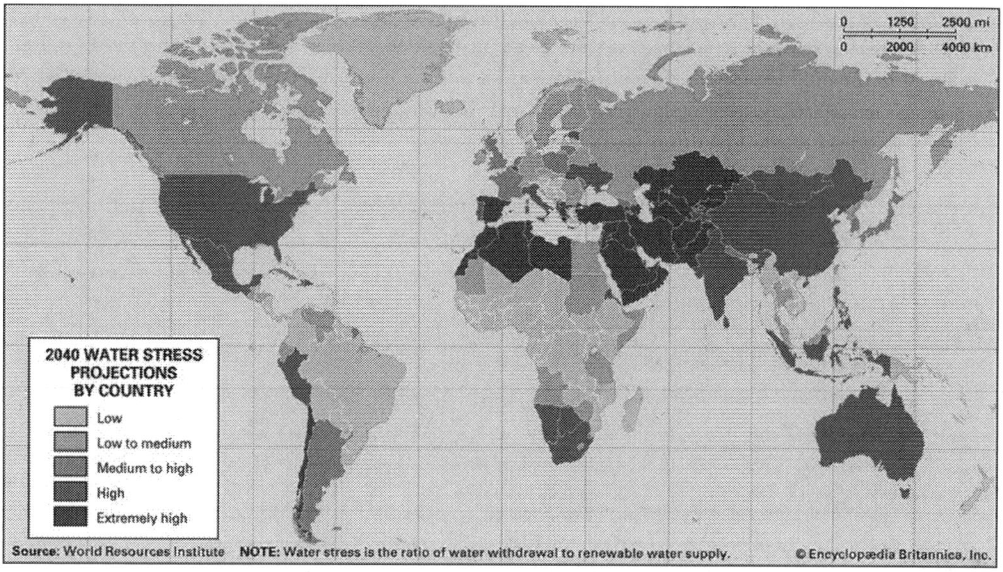

Figure 22.5 Depiction of forecasted water stress in the year 2040 (Encyclopaedia Britannica, 2022; data source: World Resources Institute).

significant differences in which areas are shown as water stressed; this difference is quite pronounced for Sub-Saharan Africa, for example.

Another challenge in depicting water scarcity is to provide meaningful and usable forecasts for how it might change over time. Figure 22.5 shows a country-by-country projection of water stress to be encountered in 2040. While this geographical depiction may be marginally useful in describing how water stressed a country will be in the year 2040, it does not offer information that is precise or usable for policy formulation purposes.

Analysis of Approaches to Address Water Scarcity

The notion of achieving water security for all users worldwide is closely linked to overcoming water scarcity (Adeel, 2017). Approaches to addressing water scarcity are vastly different and depend on the following key factors:

- *Available renewable water resources:* the most fundamental factor in addressing water scarcity is the amount of renewable water resources available; these resources are frequently constrained by geography and hydrology.
- *Ability to utilize transboundary water resources:* when water demands in an area or country exceed the available renewable water resources, one option is to look elsewhere – typically either diverting from upstream riparian areas or occasionally from adjacent watersheds – to bring in the additional water resources. Such water transfers or sharing arrangements, however, always bring significant political, societal, financial, and environmental costs.
- *Level of water expertise:* it is essential to have the requisite expertise and know-how to enhance water management approaches and effectively address water scarcity. While the shortage of such expertise is universal, the gaps in human resource capacity are particularly pronounced in developing countries.

- *Strength of water governance mechanisms and institutions:* the robustness of governance mechanisms is critical to addressing water scarcity in the context of changing social and economic conditions. Government institutions play a pivotal role in overall water governance, and degradation of these institutions due to political turmoil, armed conflict, or corruption typically brings serious impediments to addressing water scarcity.
- *Sufficient political will and policy framing:* while it borders on cliché, political attention on water scarcity can play a significant role in overcoming roadblocks and mobilizing public opinion.
- *Level of economic and financial resources:* national and local affluence, or lack thereof, can influence the options available to address the factors listed above. While water delivery systems can generate their own revenue, political considerations often trump the ability of governments to price water for full cost recovery. Such political sensitivities around water pricing can further limits application of sustainable long-term solutions.

Approaches and solutions to water management differ considerably, based on the level of economic and social development in a country. Therefore, it is instructive to separately review the approaches used in developing and developed countries.

Addressing water scarcity in developing countries

A quick glance at any of the global maps shown earlier, particularly that in Figure 22.1, makes it clear that a large fraction of the countries in the global south are under some level of water stress. This is particularly true for countries in the Sahel region and southern Africa, West Asia and North Africa, Central and South Asia, and Southeast Asia. Specifically, a number of river basins in these regions have been identified as facing multiple dimensions of stress: Indus, Jordan, Mekong, Nile, Tigris-Euphrates, Amu Darya, and Brahmaputra (DNI, 2012). One key common factor in these basins is that the riparian countries are significantly dependent on agricultural production for their economies, which in translates into a high degree of vulnerability to short- and long-term water stresses (DNI, 2012). Studies focused on these regions have identified considerable water inefficiencies and losses related to poor condition of the related infrastructure and inadequate management approaches (Seckler et al., 1998; Vorosmarty et al., 2005; Felter & Robinson, 2021; UN-Water, 2021).

Not surprisingly, water policies in developing countries are often driven by food security, energy security, and the need for economic growth. With a few exceptions, water resources management does not rank high in the political agenda at the national level. This becomes obvious when reviewing the progress to implement SDG #6 (UN-Water, 2021).

For example, the water scarcity situation in Cape Town, South Africa has been well-known for decades but was largely ignored at the political level until the media picked up on the notion of "Day Zero" in early 2017 – referring to the potential of no water being supplied to Cape Town residents as a result of prolonged, multi-year drought. This situation triggered an emergency response by the various levels of government, which included severe restriction on water use by the Cape Town residents. Some have argued that even these extreme measures might have proven ultimately insufficient had rains not returned to that part of the world (Pascale et al., 2020).

Addressing water scarcity in developed countries

Most developed countries place a significant emphasis on environmental conservation and water policies, and often attempt to incorporate climate change impacts. Economic affluence often means that transfer of economic and social resources from less water-stressed areas of these countries is a viable option. Such transfer of resources often means that there is no differential impact of water stress on social and economic development; the western, water-scarce states in the United States are an example of this outcome, in which no significant differences in well-being indicators could be found when compared with more humid, eastern states (Uriel & Adeel, 2005).

Impacts of Water Scarcity on Sustainable Development

Scarcity of water resources plays a significant role in challenges related to societal and economic development, perhaps much more so than the credit it usually gets. Water scarcity sits at the apex of a downwards spiral observed in many countries (Adeel et al., 2005; Uriel & Adeel, 2005): water scarcity is linked to expansion of cropped areas in efforts to achieve food security, which in turn leads to reduction in natural vegetation cover and expansion of irrigation systems that trigger soil erosion and salinization, respectively. When combined with climate driven factors such as extended and more frequent droughts, overall land productivity declines. All of these outcomes of water scarcity have been shown to contribute to a number of societal outcomes: persistent poverty of those dependent on land resources, migration away in search of better livelihoods, reduced well-being of population, and overall political and economic instability (Adeel et al., 2005). It has been argued that the combination of these factors leads to the greatest threats to those who are most vulnerable (King-Okumu, 2021).

Having water scarcity, whether it is hydrological or economic, does not automatically lead to all the adverse outcomes listed above. Water conservation approaches at different scales, combined with healthy water systems that espouse good water quality and sustainability of ecosystems, can lead to increased resilience against water scarcity and droughts, and eventually improve human well-being (King-Okumu, 2021). There are two different lenses through which these interactions can be discussed and analysed: using the 2030 Sustainable Development Agenda as a policy and action framework, and engaging key sectors of economy in re-envisioning water efficiency and conservations, thus integrating water security into economic development trajectories.

Overcoming water scarcity to achieve SDGs

Extensive work has been undertaken to identify that overcoming water scarcity and fostering sustainable water resources are the cornerstones of achieving most of the SDGs (Adeel, 2017). SDG 6 is most directly concerned with sustainable management of water resources, including those cutting across international borders. SDG 14 on aquatic life and SDG 15 on land life also intersect with water conservation and sustainability of water resources in a fairly fundamental way. Importantly, however, the most impactful linkages can be made to the following SDGs: poverty reduction (SDG 1, through enhanced economic activity), food security (SDG 2, through increase productivity and water use efficiency in the food sector), good health (SDG 3, through reduction of gastro-intestinal diseases), quality education (SDG 4, through a focus on youth and access to safe water and sanitation), and gender equality (SDG 5, through reduction of water stewardship burdens on women). Achievement of other SDG targets also partly dependent on access to sustainable water resources for affordable clean energy (SDG 7, through hydropower), economic growth (SDG 8), infrastructure (SDG 9), sustainable cities (SDG 11), responsible consumption (SDG 12), and climate action (SDG 13).

Based on the brief description of the intersections between water resources (or their scarcity) and achievement of most SDG targets, it might be argued that the 2030 Agenda for Sustainable Development provides a broad policy framing for the former. Conceptually, the theory of change describes that as governments invest in achievement of SDGs, some of those resources would end up in addressing water scarcity. In practice, however, there is little evidence that such investment of resources is taking place. This conclusion can be inferred from the UN-Water report that states that progress along most of the targets for SDG 6 remains stagnant at best, resulting in 2.3 billion people living in water stressed areas (UN-Water, 2021). A key stumbling block is the mobilization of sufficient financial resources, which are estimated to be in the range of about US$2 trillion a year (Schuster-Wallace et al., 2015). The out-of-box thinking to mobilize resources to implement the approaches discussed in *Analysis of Approaches to Address Water Scarcity* section of this chapter is largely absent; for example, the entire water, sanitation and hygiene sector received US$9.25 billion in 2019 in overseas development assistance – a figurative drop in the bucket when compared to the scale of the challenge (UN-Water, 2021).

Sectoral responses to water scarcity

In terms of challenges presented by water stress and scarcity, it can be argued that the international business community – representing a broad spectrum of economic sectors – has been better at understanding nature of the challenge and implementing responsive approaches. An excellent insight into this thinking can be obtained by observing how the global leaders gathered at the World Economic Forum in Davos, Switzerland, have repeatedly ranked water challenges as issues of highest concern. For a number of years, the Global Risks Report has highlighted in its top ten most severe global risks as including natural resources crises, extreme weather events (including floods, hurricanes, and droughts), and human-caused environmental damage (WEF, 2022). This awareness is to a large extent driven by the impact of these drivers on the financial bottom-line; for example, companies reported over US$38 billion in losses as a result of water-related challenges (Stinson & Nelson, 2020).

There is also a growing realization that addressing water, energy, and food are strongly interdependent and treating them as isolated sectors of the economy often leads to ineffective and mutually counterproductive policy decisions. As an example, agricultural sector utilizes 69% of the global water resources and 30% of the global energy consumption is used for food production and supply (Liu et al., 2022). The notion of working on the water-energy-food "nexus" has gained considerable traction over the past two decades (Carr et al., 2020). Achieving integration of these three policy streams has, however, proven challenging so far, mostly stymied by bureaucratic inertia of existing institutions and silo-ed approaches to policy formulation.

Overcoming Water Scarcity in the Coming Decades

There is considerable uncertainty in forecasting the most significant and impactful global processes: population growth, economic development and shocks, climate change impacts, mass movement of peoples across international borders and continents, and improvements in determinants of human well-being. Further, there are unexpected events such as the COVID-19 pandemic (2019-ongoing) and the Russia-Ukraine War (2022-ongoing) that can further skew any forecasts. It is, therefore, in the realm of possibilities that the water stress and scarcity we observe today might get significantly worse in parts of the world, even more so than the relatively reliable forecasts emerging from the international research community. That possibility is a good reason to ponder the consequences for mismanaging water scarcity and take timely and proportionate action.

It is understandable that the scale and scope of water scarcity would vary in different parts of the world, and accordingly, the response should also be scaled proportionately. The uncertainty in the variable usage of water stress indicators notwithstanding, national governments can play a significant role in fully understanding the nature of the challenge facing them and creating responsive and proactive actions. The international community must play a central role in supporting these national efforts by providing financial, institutional, and human-resource support.

The private sector's role is two-fold. First, it should continue the trajectory of improving its own water efficiency and reducing water consumption. Such approaches have been shown to directly benefit the fiscal bottom-line. Second, it needs to play a greater role in engaging with the public sector by mobilizing financial resources and bringing cutting-edge management approaches to the table.

The international community must move in a decisive manner to address water scarcity with location-specific, culturally safe, locally acceptable, and scalable solutions. The 2030 Agenda and SDG framing provide a good mechanism through which investments can be channelled and outcomes can be monitored. National governments need support and capacity building to fully understand their challenges and create effective responses; they must be supported by the United Nations, bilateral aid agencies, development banks, and the private sector.

Notes

1 The term "water" in the context of this chapter is meant to represent freshwater, whether abstracted from surface water resources such as river or lake or from groundwater resources such as an aquifer.

2 Aridity Index (AI) is the ratio of precipitation to the potential evapotranspiration in an area. Areas with AI values less than 1 are considered to be "drylands."

References

Adeel, Z. (2017). "A renewed focus on water security within the 2030 agenda for sustainable development." *Sustainability Science*. 12, 891–894.

Adeel, Z., Safriel, U., Niemeijer, D., White, R., de Kalbermatten, G., Glantz, M., Salem, B., Scholes, R., Niamir-Fuller, M., Ehui, S. and Yapi-Gnaore, V. (2005). *Millennium Ecosystem Assessment – Ecosystems and Human Well-Being: Desertification Synthesis*. World Resources Institute, Washington DC.

Alcamo, J., Henrichs, T. and Rösch, T. (2000). *World Water in 2025-Global Modeling and Scenario Analysis for the World Commission on Water for the 21st Century, Kassel World Water Series*. Univ. of Kassel, Kassel.

Carr, G., Barendrecht, M. H., Debevec, L., Kuil, L. and Crossmark, G. B. (2020). "People and water: understanding integrated systems needs integrated approaches." *Journal of Water Supply: Research and Technology-Aqua*. 69(8), 819–832. DOI: https://doi.org/10.2166/aqua.2020.055

DNI. (2012). Global Water Security – Intelligence Community Assessment, Office of the Director of National Intelligence (DNI), USA. [Accessed 18 September 2022]. Available from https://www.dni.gov/files/documents/Special%20Report_ICA%20Global%20Water%20Security.pdf

Emile, R., Clammer, J. R., Jayaswal, P. and Sharma, P. (2022). "Addressing water scarcity in developing country contexts: a socio-cultural approach." *Humanities and Social Sciences Communications*, 9, 144. DOI: https://doi.org/10.1057/s41599-022-01140-5

Encyclopaedia Britannica. (2022). Water Scarcity, Encyclopaedia Britannica, Inc. [Accessed on 6 May 2022]. Available from https://www.britannica.com/topic/water-scarcity

Falkenmark, M. (1997). "Meeting water requirements of an expanding world population." *Philosophical Transactions of the Royal Society*, 352, 929–936. DOI: https://doi.org/10.1098/rstb.1997.0072

Falkenmark, M., Lundqvist, J. and Widstrand, C. (1989). "Macro-scale water scarcity requires micro-scale approaches." *Natural Resources Forum*. 13, 258–267. DOI: https://doi.org/10.1111/j.1477-8947.1989.tb00348.x

Falkenmark, M., Rockström, J. and Karlberg, L. (2009)." Present and future water requirements for feeding humanity." *Food Security*. 1, 59–69. DOI: https://doi.org/10.1007/s12571-008-0003-x

Felter, C. and Robinson, K. (2021). "Water stress: a global problem that's getting worse." *Council on Foreign Relations*.

Graffy, E. A. (2006). "Expert forecasts and the emergence of water scarcity on public agendas." *Society and Natural Resources*, 19(5), 465–472. DOI: 10.1080/08941920600561173

Hoekstra, A. Y. (2014). "Water scarcity challenges to business." *Nature Climate Change*. 4, 318–320.

Hoekstra, A. Y., Chapagain, A. K., Aldaya, M. M. and Mekonnen, M. M. (2011). *The Water Footprint Assessment Manual: Setting the Global Standard*. Earthscan, Washington, DC.

King-Okumu, C. (2021). A *R*apid *R*eview of *D*rought *R*isk *M*itigation *M*easures – Integrated *D*rought *M*anagement. FAO, Rome. https://doi.org/10.4060/cb7085en

Lawrence, P., Meigh, J. and Sullivan, C. (2002). "The Water Poverty Index: An International Comparison, Keele Economics Research Papers (2002/19)." [Accessed on 3 November 2022]. Available from https://www.researchgate.net/publication/235737531_The_Water_Povery_Index_an_International_Comparison

Liu, X., Liu, W., Tang, Q., Liu, B., Wada, Y. and Yang, H. (2022). "Global agricultural water scarcity assessment incorporating blue and green water availability under future climate change." *Earth's Future*. 10, e2021EF002567. DOI: https://doi.org/10.1029/2021EF002567

Liu, J., Yang, H., Gosling, S. N., Kummu, M., Flörke, M., Pfister, S., Hanasaki, N., Wada, Y., Zhang, X., Zheng, C., Alcamo, J. and Oki, T. (2017). "Water scarcity assessments in the past, present, and future." *Earth's Future*. 5, 545–559. DOI: https://doi.org/10.1002/2016EF000518

Mekonnen, M. M. and Hoekstra, A. Y. (2016). "Four billion people facing severe water scarcity." *Science Advances*. 2(2), e1500323. DOI: https://doi.org/10.1126/sciadv.1500323

Pascale, S., Kapnick, S. B., Delworth, T. L. and Cooke, W. F. (2020). "Increasing risk of another cape town 'Day Zero' drought in the 21st century." *Earth, Atmospheric, and Planetary Sciences*. 117(47), 29495–29503. DOI: https://doi.org/10.1073/pnas.2009144117

Savenije, H. H. G. (2000). "Water scarcity indicators; The deception of The numbers." *Physics and Chemistry of the Earth Part B Hydrology Oceans and Atmosphere*. 25(3), 199–204. DOI: https://doi.org/10.1016/S1464-1909(00)00004-6

Schuster-Wallace, C. J., Sandford, R., Dickin, S. K., Vijay, M., Laycock, K. and Adeel, Z. (2015). *Water in the World We Want*. United Nations University Institute for Water, Environment and Health, Canada and United Nations Office for Sustainable Development, South Korea.

Seckler, D., Amarasinghe, U., Molden, D. J., de Silva, R. and Barker, R. (1998). *World Water Demand and Supply, 1990 to 2025: Scenarios and Issues*. International Water Management Institute (IWMI), Colombo, Sri Lanka.

Stern, D. I. (2018). "The environmental Kuznets curve." *Earth Systems and Environmental Sciences*. DOI: https://doi.org/10.1016/B978-0-12-409548-9.09278-2

Stinson, C. and Nelson, J. (2020). "3 Actions Business Leaders can Take to Help Solve Our Water Crisis." World Economic Forum, Davos, Switzerland. [Accessed on 7 November 2022]. Available from https://www.weforum.org/agenda/2020/01/3-actions-business-leaders-can-take-to-tackle-the-worlds-water-crisis/

Sullivan, C. A., Meigh, J. R. and Giacomello, A. M. (2003). "The Water Poverty Index: development and application at the community scale." *Natural Resources Forum*. 27, 189–199. DOI: https://doi.org/10.1111/1477-8947.00054

United Nations. (2020). *The United Nations World Water Development Report 2020. Water and Climate Change*. UNESCO, Paris.

United Nations. (2022). *The United Nations World Water Development Report 2022: Groundwater: Making the Invisible Visible. UNESCO*, Paris.

UN-Water. (2021). "*Summary Progress Update 2021 – SDG 6 – Water and Sanitation for all*." Version: 1 March 2021, Geneva, Switzerland.

Uriel, S. and Adeel, Z. (2005). "Dryland Systems." In Hassan, R., Scholes, R. and Ash, N. (eds.). *Millennium Ecosystem Assessment: Current Status and Trends*. Island Press, Washington DC.

Vörösmarty, C. J., Green, P., Salisbury, J. and Lammers, R. B. (2000). "Global water resources: vulnerability from climate change and population growth." *Science*, 28, 284–288. DOI: https://doi.org/10.1126/science.289.5477.284

Vorosmarty, C. J., Leveque, C. R. and Revenga, C. (2005). "Fresh Water." In: Hassan, R., Scholes, R. and Ash, N. (eds.). *Millennium Ecosystem Assessment: Current Status and Trends*. Island Press, Washington, DC.

WEF. (2022). *The Global Risks Report 2022, 17th Edition*. World Economic Forum, Davos.

23

WATER, MIGRATION, AND DEVELOPMENT

Ander Jagerskog and Ashok Swain

Introduction

During the last decade the world has witnessed an unprecedented level of human migration – both within countries as well as across international borders. Conflicts and unrest are key reasons for the increase in migration but environmental concerns such as water scarcity, droughts, and floods are key in this respect. With increasing water scarcity globally, coupled with global climate change which increases the prospects for disruption of the weather patterns that often are underpinning people's livelihoods, the challenges for people, especially the poor and vulnerable, are bound to increase. In a context where the increasing focus is being put on emphasizing linkages between different issue areas and avoiding 'silo' approaches it is relevant to try to see how migration can be seen to be situated and better understood in for example the water, climate, energy, and food discussions.

This chapter aims to provide an overview of the current state of research on the links between water migration and development. Hitherto, there has been an assumed link between the topic of water and migration, judging by headlines in the international press where often blame is put on drought and climate change while complexity and deeper analysis of issues such as poor governance is shied away from (Jägerskog & Swain, 2016; Swain & Jägerskog, 2016; Selby et al., 2017; Zaveri et al., 2021). There are often many reasons (including underlying push and pull factors) as to why people flee or migrate. We hold the view that deciding to move is a complex and multifaceted decision informed by a range of considerations, including those relating to climate change, environment, water, and people in search of jobs and improved livelihoods. Thus, it is rarely the result of considering mainly one aspect but is rather an outcome of a complex process. We aim to provide a critical perspective where causality is cautioned against and where water's role is more closely analysed. Fortunately, research using large numbers of migration data has started to emerge that provide a closer look as well as improved understanding, also at an aggregate level, of the links between water, development and migration (Zaveri et al., 2021). Furthermore, from a science-policy perspective understanding the reasons behind and possible links between water and migration is imperative for international organizations. In order to be able to put in place proper policies to address the root causes of the challenges related to the water, migration and development nexus an improved understanding of the issue is key.

DOI: 10.4324/9781003095545-27

Human migration has always existed and represents, in many ways, one of the key features underpinning development. A new influence, knowledge, and ideas leading to progress in development is linked to people moving throughout history. However, new challenges have been brought through the increasing water scarcity, coupled with development and population growth. This has led researchers as well as international organizations to try to improve the understanding of the links between water, migration, and development. Today, more than 40% of the global population is suffering from water scarcity. Climate change is expected to seriously aggravate the problem, particularly in Asia, Africa, and the Middle East. As water supply and demand patterns change, existing water-sharing arrangements between countries and regions are likely to be increasingly challenged and prone to tension and conflict. The World Bank (Zaveri et al., 2021) have analysed the biggest data set on internal migration available containing over half a billion people from 189 population censuses in 64 countries from 1960 to 2015, which found that water scarcity of water is contributing to 10% of the increase in global migration. It notes that the decision to migrate is a complex one but that water does play a role and furthermore that it is the lack of water, rather than excessive amounts of water, that shows a stronger link to migration. Overall, rainfall variability – driven by climate change – is also experienced predominantly in the developing world. The report notes that around 85% of the people that experience the challenges brought by rainfall variability live in low- or middle-income countries. In middle and low-income countries it is, furthermore, the poorest that are often unable to migrate in search of better prospects in life. The World Bank finds that people living in poor countries are four times less likely to be able to migrate in comparison with those in rich countries (Zaveri et al., 2021).

What Do We Know Today about Water, Migration, and Development?

When discussing migration it is important to distinguish between different types of migration. There are several different types of migration: internal migration within a country that is not forced; internal migration that is forced; and cross-border migration that is not forced and cross-border migration of forcibly displaced (refugees). Migration is a complex phenomenon which ranges from chosen migration to coerced or forced displacement. There is rarely one reason to migrate but voluntary human migration underscores the economic motives arguing that people move to take advantage of better economic prospects for income and employment. Forced migration or displacement is largely the results of civil war, political unrest and ethnic oppression (Swain & Jägerskog, 2016). In this chapter we are looking at: (i) the overall relationship between water, migration and development; (ii) causes of migration (with a focus on water) and how that effects development; and (iii) the effects of migration on sending as well as receiving communities and countries.

According to the International Organization for Migration (IOM) global estimate there were around 281 million international migrants in the world in 2020, excluding refugees (IOM, 2020). Most of these migrants are moving as a result of economic prospects being (or being presumed to being) better elsewhere and in many ways represents labour migration. Internal migration (people who are moving inside the border of a country but are not forcibly displaced) is estimated to be around 760 million people globally (IOM, 2020). Accounting for all migration globally the number is well over one billion people – or roughly every 7th person in the world.

The number of internally displaced and refugees, according to UNHCR has reached 89.3 million people at the end of 2021. Never before in the history has the world witnessed such a high number of forcibly displaced. Out of those 89.3 million, around 21.3 million were defined as refugees and around 53.2 million internally displaced. In addition there are close to 5.8 million

Palestinians that are displaced and falls under the mandate of United Nations Relief and Works Agency for Palestine Refugees in the Near East (UNRWA) and 4.6 million asylum seekers and 4.4 million Venezuelans displaced abroad (UNHCR[1]). Around two thirds of all refugees (people that are forcibly displaced and cross international borders) come from just five countries: Syria (6.8 million), Venezuela (4.6 m), Afghanistan (2.7 m), South Sudan (2.4 m), and Myanmar (1.12 m). The number of internally displaced persons have risen to 53.2 million by 2021 and are people that have been forced to flee their homes by armed conflict and generalized violence, and were living in displacement within the borders of their own country. This is a significant increase from around 21 million in 2000. Around 75% of the internally displaced people live in just ten countries: Syria (6.5 million), Colombia (5.6 m), Democratic Republic of Congo (5.5 m), Yemen (3.6 m), Afghanistan (3 m each), Somalia and Nigeria (2.2 m each), Sudan (2.1 m) and Iraq (1.6 m), and Ethiopia (1.4 m) (International Organisation of Migration, 2020[2]).

While we know that part of the migration contributes to development in both the receiving country, who benefit from labour, as well as the sending country who are often receiving remittances that the migrant workers are transmitting back to their home country. However, international migration can also create tension and conflict. Large-scale transborder migration has several dimensions for inducing tension between the receiver and sender states regardless of the underlying reasons behind them (Swain & Jägerskog, 2016). In some cases, giving permission to the migrants to enter into its own territory may strain the relationship between the receiving state and the sender country. The tension may arise from the exposure of the sender state's inability to provide basic livelihood to its own citizens. The other possibility is that the migrants after being settled in the host country may indulge themselves in anti-government activities against their native government, whom they will perceive as the perpetrator of their relative deprivation. In some cases, the migrants may be encouraged or be manipulated by the host state in their effort to take revenge because of existing political differences between the host and the sender states (Swain, 2015).

Traditionally, environmental factors and in particular water have not been receiving too much attention in the academic debate focusing on migration although in recent years a lot of more attention has been paid to climate change as well as water and other environmental factors such as loss of fertile land (Gleick, 2014; Kelley et al., 2015; Swain & Jägerskog, 2016). In many ways the climate change materializes through water. This is either through increasing or decreasing water availability, and increased floods or increased droughts. From a development perspective it is noted that this is increasing vulnerability, especially for poor people. What the data shows is that it is more often people that are better off that are able to move while those that are poor often become 'stuck' in their place of origin even if water availability and other environmental factors take a turn for the worse (Zaveri et al., 2021).

Researchers and international organizations have long warned that climate change may be a major reason behind human migration. Thirty years ago, in its 1990 assessment, the UN's Intergovernmental Panel on Climate Change (IPCC) warned that large-scale human migration could be the single most significant impact of climate change. However, climate change has raised the possibility of mass migration on a scale the world has never witnessed before. In 2021, the World Bank, in its updated *Groundswell* report (Clement et al., 2021) highlights that climate change could lead to 216 million people needing to migrate within countries by 2050, with the following division: Sub-Saharan Africa 86 million internal climate migrants; East Asia and the Pacific, 49 million; South Asia, 40 million; North Africa, 19 million; Latin America, 17 million; and Eastern Europe and Central Asia, 5 million. Significant hot spots for internal migration are identified where developments are expected to take place significantly earlier than 2050. Around 2030 major migration hotspots may experience significantly increased internal migration, including the

Middle East and North Africa where water availability (or lack thereof) will be the main driver for people to move. As bleak as this may sounds it is at the same noted that it is not too late to slow some of these developments. The reports notes that if:

> Early and concerted action to cut global greenhouse gas emissions and to ensure inclusive and resilient development can reduce the scale of internal climate migration by as much as 80%.
> *(Clement et al., 2021)*

In the North Africa region in particular, water availability is highlighted as at risk of being the main driver of migration. Certain areas such as the northeastern coast of Tunisia and Northwestern coast of Algeria, coupled with western and southern Morocco and the western part of the Nile Delta are identified as out-migration hotspots and climate change continue to affect water availability. On the other side of the spectrum a number of in-migration hotspots are identified such as important urban centres including Cairo, Algiers, Tunis, and Tripoli where water is, in relative terms, more abundant. Another region that is expected to suffer from climate changed induced water availability challenges is Central Asia. For example, for the Ferghana Valley in-migration is expected as it is attractive as an economically productive areas for those that are meeting increased climate variability and decreased water availability. As such, a movement from urban areas to rural areas is expected.

The United Nations University has proposed a three dimensional framework for understanding the water-migration nexus where water quality, water quantity, and water extremes are highlighted as 'forces' that can led to loss of income and livelihood, disease burden and food insecurity and as such can lead people to move, either voluntarily or involuntarily (Nagabhatla et al., 2020). While these factors are important, it is often the ways in which these are managed that decide whether a person will move or not. If proper measures to address these challenges are present and incorporated in a functioning governance system and which countries have a certain degree of resilience built even if pressures are increased it may be possible to mitigate their impacts. However, in situations where the level of development is low and/or conflict is present, the likelihood of people needing to move will increase.

In addition, aspects relating to gender and inclusion becomes imperative in relation to water, migration and development. Women are often the ones who are taking the main responsibility for provision of water as well as for the household. At the same time, migration data shows that men are more likely to migrate than women, both for economic reasons as well as in situations of conflict (Miletto et al., 2017; Jobbins et al., 2018). This often means women are left to deal with an increasingly challenging situation in places of origin in which they will need to take on some of the men's traditional responsibilities (e.g. farming) to support the livelihood of the family in addition to the traditional female roles such as parenting and cooking (Nagabhatla et al., 2020).

The Case of the Middle East and North Africa – A Precursor for Improved Understanding of the Nexus of Water, Migration, and Development

The case of the Middle East and North Africa is important in and of itself but it can also be seen as an offering broader lessons for other regions in primarily two aspects. Firstly, it offers lessons on how water and migration intersects from a forced displacement perspective (international and across international borders). Secondly, it offers a glimpse of a situation which other regions may face at a later stage. This is because the region has the highest levels of water scarcity per capita and is, on top of that, facing significant changes in the climate variability, manifested through hotter climate (driving water scarcity further) and also increased fluctuations in rainfall between both seasons and years. As such it offers important lessons for water and development globally. For research on the topic of water, migration and development in other regions this offers a cautionary tale and a potential glimpse of what may come and also underscores the need for deep contextual analysis to be undertaken before drawing overly simplistic conclusions.

As identified by many, the Middle East and North Africa region is already today and is on the trajectory to face severe challenges relating to water and migration in the future, not least in view of climate change which will affect the region significantly, not least in terms of water availability and variability (Falkenmark & Jägerskog, 2010; Gleick, 2014; Weinthal et al., 2015; Swain & Jägerskog, 2016; Nagabhatla et al., 2020; Borgomeo et al., 2021; Clement et al., 2021; Zaveri et al., 2021). While the economic migration can be seen to assist development in the region, it is largely other types of migration, namely forced displacement, that is, the hallmark of the region. As such the migrants often face both vulnerabilities and development risks. This is both in the sending region as well as in the receiving region (Borgomeo et al., 2021). What this often means from an international relations perspective is that the topic becomes 'securitized.' In essence a 'securitization' of new issue areas such as water means it is not approached primarily as an environmental or development but as a security issue which enables the issue to move up in political importance (Buzan, 1991). During the last few decades it is apparent that a securitization of the water issue has taken place in the region and water has moved into the national security scene. As such, the topic of water and migration in the region is likely to be viewed through a security lens. The securitization of water in the region is likely there to stay given an increasing water scarcity driven by population growth and migration (in receiving countries such as Jordan and Lebanon) a high level of transboundary water reliance and a general conflictual situation in the region. This means that water, which often is best treated as a development issue, is increasingly treated as security issue. This could for example, mean that instead of taking regional approaches to water to address floods and droughts, countries resort to unilateral measures. Thus, undermining collective action that would have been beneficial to all (Swain & Jägerskog, 2016).

Historically migration and water in the region related to, by and large, nomads and pastoralists looking for water and feed for their animals. Nomads at the African continent, in the Middle East and Asia often lived a life where they moved in accordance with where they could sustain a livelihood for themselves. The new debate tries to see linkages between water scarcity (which includes natural variability), migration (and refugee flows), and conflict and as such its effects on overall development in both receiving as well as sending regions. We argue that it is important not to draw hasty conclusions as regards causal linkages in this respect. The availability of fresh water has also fallen short of adequately meeting its increased demand in most parts of Asia, Africa, and the Middle East. In addition, the region has a very high proportion of transboundary waters (around 60%) with very few agreements on these shared water resources, thus adding to the conflict aspect. The loss of livelihood due to increasing water scarcity and variability, whether 'natural' or man-made could force people to migrate. Furthermore, the water scarcity is becoming much more problematic due to global climate change, which often leads to increased variability of flow thereby fuelling vulnerability. Climate change can also potentially alter the usual rainfall pattern, which may lead to increased flooding, drought, and soil erosion in tropical and arid regions of the world. In that sense climate change is exacerbating already existing phenomena such as natural climatic variability. In the MENA region, as well as in regions in conflict, gender aspects are also relevant to understand better. Both in places of origin, as well as places of destination the vulnerability for women and girls (either in country of origin or as refugees) increase significantly both to gender-based violence related to unsafe water and sanitation facilities in host communities and refugee camps as well as by the overall risk that being forcibly displaced means (Borgomeo et al., 2021).

The fact that the region is facing ever larger climatic variation as a result of climate change will (according to most projections) lead to higher temperature (meaning more evaporation) and less rainfall and thereby on the whole an even greater water scarcity. This is also likely be the pattern in some other regions over the next decades. While securitization of water or climate is understandable, especially in a region like the Middle East but also in other regions, it runs the risk of encouraging inappropriate government responses. White (2011) also notes that a securitization of environmental challenges enables the security sector to be the main actor in formulating the debate and responding policies. This leads to a focus on defending borders, enabling 'deeper' security cooperation and coordination with states that are transit places while policies to address the environmental challenges, which are exaggeratedly deemed to be causing the migration, gets little attention. White holds that:

> 'Getting tough' – responding in a militarized fashion – is an easy, cynical step in a warming world [...] Building a fence is easier than changing lifestyles. Yet the injection of security imperatives into climate-induced migration is unethical and unworkable.
>
> *(2011: 7)*

Not only does the securitization play into the consideration and adoption of inadequate policy measures, it may also serve to 'explain' failure of governments to address the underlying challenges relating for example to water governance in Syria. De Châtel (2014), and Weinthal et al. (2015) have all noted that it was the significant neglect of the Syrian regime of putting in place proper water governance and irrigation systems in the southern parts of Syria that led to an unpreparedness when the drought hit harder than usual, in the years before the Syrian civil war. While the drought led to major internal migration the reasons were not the drought in itself but were rather 'part of a broader pattern of rural neglect' (Weinthal et al., 2015, p. 1).

As security and development are increasingly treated as closely interlinked, which means that in water scarce situations like the Middle East and North Africa water becomes securitized. Therefore understanding these links becomes increasing important for understanding regional *discourses* on water, migration and development. The Middle East and North Africa region, again, offers insights into this question. For a farmer in the region, faced with the brute reality of climate change, meaning for example increasing variability in terms of rainfall and increased temperature may have no other resort than to move. However, what has happened, in terms of media reporting as well as part of the research community (Kelley et al., 2015), is that this has been intermingled and sometimes confused as a reason for conflict although that link is far from well established (Swain & Jägerskog, 2016; Selby et al., 2017; Borgomeo et al., 2021).

What Measures Can Be Taken?

As noted at the outset of the chapter migration is an integral part of development, moving both labour as well as ideas within countries as well as across international borders. As such migration is positive and it contributes both to the local economy as well as is a large source of important remittances supporting communities and livelihoods in areas of the origin of the migrant. Remittances account for large parts of GDP in many countries – during the year of 2020, for example, 32% in Lebanon, 27% in Tajikistan, and 10% in Ukraine.However, most of these migrants are neither forcibly displaced nor have had to move as an outcome of climatic changes and the related water challenges. People needing to move as a result of water related challenges are around 10%. This is according to data from the World Bank (Zaveri et al., 2021) which covers the last few decades and while these migrants are invariably contributing to positive developments, it may also lead to strains and challenges for societies – both within countries as well as in other countries. However, there are policies and measures that can be put in place to address the challenges that rainfall shocks bring.

The World Bank (Zaveri et al., 2021) have highlighted a number of policy measures. Putting in place *people-centred investments* which includes safety nets that can protect livelihoods for communities that may face significant rainfall deficits. Alongside flexible and active labour market policies and investments in training and capacity building for workers basic investments in water supply and sanitation will also contribute to protect people. In addition, putting in place investments in livelihoods in places where migrants are originating is key and will include additional water storage and supplemental irrigation infrastructure in areas projected to be vulnerable. A range of investments can be envisaged – ranging from green infrastructure to traditional grey infrastructure. These can create a stronger resilience for communities at risk. In addition, as many migrants are moving from rural areas to cities it will be important to both maintain and improve urban water services provision and infrastructure. Finding adequate and adaptable governance mechanisms that are able to deal with trade-offs between groups and managing a scarce supply will imperative as well (Zaveri et al., 2021).

The MENA case discussed in this chapter highlights a number of additional measures with relevance not only for that region but for areas where forced displacement is prevalent but also of relevance as being the region in which water scarcity is most pronounced. Among key interventions from a migration and water perspective would be to increase the focus on long-term water resilience to future shocks. In a situation where conflict, water scarcity and forced displacement is prevalent it is not enough to focus on water from a development perspective. On the contrary, humanitarian efforts by agencies such as UNICEF and ICRC need to be complemented by regular development efforts and actions coordinated. However, in cases where conflict is persistent coordination with security actors may also be necessary and improve the situation for the people forcibly displaced as well as in the places of origin (Borgomeo et al., 2021). The above should be the characteristics of the approach in a crisis situation when a conflict is ongoing but as a situation moves into a more stable situation and recovery and development starts in a more tangible manner efforts focusing on people and area-based interventions should be more central (Zaveri et al., 2021).

Conclusion

The aim of this chapter has been to provide an overview of the state of the knowledge of the area of water, migration, and development. This nexus between is complex and the decision to move is a multifaceted decision for a migrant where a number of considerations come into play. The research to entangle this nexus has evolved significantly over the last decades to one in which not only qualitative analysis dominate but also substantial quantitative analysis has advanced. Going forward, there will be a need to further explore both qualitative and quantitative aspects related to water and migration, not least as climatic changes intensify and makes peoples futures less predictable.

This chapter attempts to identify the main threads and development of knowledge over the last few years noting that there is scope to advance further. That said, it is indeed good that migration and notably water, and the linkages between the issue areas are being discussed, researched, and increasingly well understood. It is of utmost importance to critically analyse the linkages. Overly simplistic attempts to assign 'casual' linkages between development, water scarcity, climate change, and variability and migration will not be useful and may encourage policy responses that are not adequate. While it was previously observed that there existed a tendency to jump to conclusion about how there almost exists a causality between water scarcity, climate change, and migration the debate has evolved noting that simple linkages are rarely found. Indeed, the decision to migrate is a complex one that is often affected by a range of considerations. Therefore, water should be seen as one of many areas that can contribute to migration together with a range of other issues. In the end it represents one of many challenges that can be addressed in a systematic manner (or not).

As noted in the World Bank's *Ebb and Flow: Water, Migration, and Development* report (Zaveri et al., 2021) around 10% of internal migration is explained by rainfall deficits, drawing on Machine Learning analysing vast amounts of data with input from over 150 countries over a 30-year period. What is clear also is that it is not the poorest that migrate but the, relatively speaking, more well of people that have means and capacity to move. Indeed, the poorest are 80% less likely to migrate as a result of the rainfall shocks (Zaveri et al., 2021).

Going forward the interlinkages between water challenges and climate change will be critical to understand and as noted in the Groundswell report (Clement et al., 2021) the scenarios for climate change migration (some of it water related) are significant over the next decades although it will, at least partly, depend on what policies are put in place both from a mitigation and adaptation perspective. The climate changes, as the relate to migration are manifested in the form of for example increased variability and uncertainty are often not the main causes of large-scale population migration but is considered a 'push factor' multiplier. Indeed, as noted around a 10% increase in migration can be water related. Social, economic, and political factors will also affect the vulnerability or resilience of communities. In many regions, the ability to cope with climate change and water scarcity decreases, and the likelihood of migration increases, as a result of factors that include poverty, low levels of education, lack of skills, weak institutions, limited infrastructure, lack of technology and information, limited access to health care, poor access to resources, overexploitation of resources, etc. Certain groups are more negatively affected and vulnerable from a water, migration, and development perspective. Women and girls, either in the place of origin if they stay (as more vulnerable groups often are doing) or if they are moving, they are likely to be more vulnerable than men and boys, either as they take on extra responsibilities for water, the household and livelihoods in places of origin or as they face increasing gender-based violence risks relating to water and sanitation facilities (as well as other risks of violence).

Notes

1 See: https://www.unhcr.org/figures-at-a-glance.html
2 See: https://worldmigrationreport.iom.int/wmr-2020-interactive/

References

Borgomeo, E. and Jägerskog, A. Zaveri, E., Russ, J., Khan, A., Damania, R., (2021). *Ebb and Flow, Volume 2: Water in the Shadow of Conflict in the Middle East and North Africa*. Washington, DC: World Bank.

Buzan, B. (1991). *People, States & Fear – An Agenda for International Security Studies in the Post-Cold War Era*. London: Harvester Wheatsheaf.

Clement, V., Rigaud Kumari, K., de Sherbinin, A., Jones, B., Adamo, S., Schewe, J., Sadiq, N. and Shabahat, E. (2021). *Groundswell Part 2: Acting on Internal Climate Migration*. Washington, DC: World Bank.

De Châtel, F. (2014). "The Role of Drought and Climate Change in the Syrian Uprising: Untangling the Triggers of Revolution." *Middle Eastern Studies*, 40(4), 521–535.

Falkenmark, M. and Jägerskog, A. (2010) "Sustainability of Transnational Water Agreements in the Face of Socio-Economic and Environmental Change." In Earle, A., Jägerskog, A. and Öjendal, *Transboundary Water Management: Principles and Practice*. London: Earthscan. pp. 157–170.

Gleick, P. (2014). "Water, Drought, Climate Change and Conflict in Syria." *Weather, Climate and Society*, 6(3), 331–340.

International Organization of Migration (IOM). (2020). *World Migration Report.* United Nations.

Jägerskog, A. and Swain, A. (2016) "Water, Migration and How They Are Interlinked." *SIWI Working Paper*, 27. Stockholm: Stockholm International Water Institute.

Jobbins, G., Langdown, I. and Bernard, G. (2018). *Water and Sanitation, Migration and the 2030 Agenda for Sustainable Development.* [Accessed 21 December 2022]. Available from https://www.odi.org/sites/odi.org.uk/files/resource-documents/12299.pdf

Kelley, C. P., Mohtadi, S., Cane, M. A., Seager, R. and Kushnir, Y. (2015). "Climate Change in the Fertile Crescent and Implications of the Recent Syrian Drought." *Proceedings of the National Academy of Sciences*, 112(11), 3241–3246.

Miletto, M., Caretta, M. A., Burchi, F. and Zanlucchi, G. (2017). *Migration and Its Interdependencies With Water Scarcity, Gender and Youth Employment.* Paris: UNESCO.

Nagabhatla, N., Pouramin, P., Brahmbhatt, R., Fioret, C., Glickman, T., Newbold, K. B. and Smakhtin, V. (2020). "Water and Migration: A Global Overview." *UNU-INWEH Report Series*, Issue 10. Hamilton: United Nations University Institute for Water, Environment and Health.

Selby, J., Dahi, O. S., Fröhlich, C. and Hulme, M. (2017). "Climate Change and the Syrian Civil War Revisited." *Political Geography*, 60, 232–44.

Swain, A. (2015). "Climate Change: Threat to National Security," In Bearfield, D. A. and Dubnick, M. J. (eds.) *Encyclopedia of Public Administration and Public Policy*, Third Edition, London: CRC Press.

Swain, A. and Jägerskog, A. (2016). Emerging Security Threats in the Middle East: The Impact of Climate Change and Globalization. Lanham: Rowman and Littlefield Publishers.

Weinthal, E., Zawahari, N. and Sowers, J. (2015). "Securitizing Water, Climate and Migration in Israel, Jordan and Syri", *International Environmental Agreements*. [Accessed 21 December 2022]. Available from DOI: 10.1007/s10784-015-9279-4.

White, G. (2011). *Climate Change and Migration: Security and Borders in a Warming World.* Oxford: Oxford University Press.

Zaveri, E., Russ, J., Khan, A., Damania, R., Borgomeo, E. and Jägerskog, A. (2021). *Ebb and Flow, Volume 1: Water, Migration, and Development.* Washington, DC: World Bank.

24
WATER AND CLIMATE CHANGE

Deliang Chen and Hui-Wen Lai

Water and Development under Climate Change

Water is necessary for all life forms and for maintaining a functional human society and ecosystem. It plays a critical role in sustainable development, as it is closely linked to the social, economic, and environmental dimensions of sustainability. An adequate freshwater supply (i.e., a net gain between precipitation minus evapotranspiration) is vital for sustaining agriculture, industry, and a domestic water supply that can meet society's and ecosystems' needs. Hydropower is also an important source of renewable energy.

Global warming has been intensifying the global hydrological cycle, and regional water availability will be reshaped in both space and time, seriously impacting food security, economic prosperity, and societal well-being.

In 2015, the United Nations established 17 Sustainable Development Goals (SDGs) (United Nations, 2015). These SDGs included 169 targets that were designed to improve the quality of human lives and natural ecosystems. The sixth SDG was designed to ensure the availability and sustainable management of water and sanitation worldwide. Irannezhad et al. (2022) argued that SDG 6 is an essential prerequisite for achieving 62 out of the 163 targets related to all other SDGs.

Global water availability and security are expected to be strongly affected by climate change (e.g., Liu et al., 2018). Furthermore, the human-induced warming of the climate system in the future is an indisputable issue that is attracting increasing worldwide concern. The Sixth Assessment Report of IPCC (AR6) concluded that human influence has caused the globally averaged surface temperature between 2010 and 2019 to increase by 1.1°C when compared to that between 1850 and 1900. This warming is projected to continue in the coming decades (IPCC, 2021).

Predicting how natural and human systems will change in the future is difficult and carries significant uncertainties. This is especially true for climate change, which is controlled by a range of complex processes. In essence, the future climate will be shaped by past and future human activities, with the dominant control being the rates at which greenhouse gases are emitted into the atmosphere (IPCC, 2021). Since accurately predicting future human activities is very difficult – if not entirely impossible – and there are many uncertainties, the necessary climate variables for the future are dealt with by creating and exploring several different outcomes for climate states which

DOI: 10.4324/9781003095545-28

consider a series of assumed emissions. This creates a set of possible future climate projections that are called future climate scenarios.

These scenarios allow the exploration of possible futures, the assumptions they depend upon and the courses of action that could change them. Therefore, climate change scenarios are necessary and powerful tools that can be used to understand climate change, chart response strategies, and make new climate policies.

Key Concepts and Methods Used for Climate Projections

Before showing how a future climate change scenario is produced and what uncertainties it involves, we must first describe several key concepts and methods.

Weather and climate

Weather describes the day-to-day state of the atmosphere and its short-term (i.e., hours, days, or weeks) variations, including temperature, precipitation, cloud cover, and wind speed. By contrast, the climate is a statistical summary of weather information that synthesizes weather variations in a specific area during a specific period of time. Typically, the climate is based on the weather that a geographic locality experiences over a period of 30 years. Climate, therefore, records the mean conditions throughout the period considered and can also provide statistical information about extreme weather events. Extreme weather conditions can be called extreme climate events in the context of climate change. These extremes include unexpected, unusual, severe, or unseasonal weather events, which are often associated with serious impacts in the form of natural hazards, physical damage to property or infrastructure, and/or the loss of human life.

Future climate change is expected to alter the intensity, frequency, and extent of extreme weather or extreme climate events. Indeed, extreme events often have larger impacts than changes to mean conditions, and an increasing amount of attention has recently been given to studying how and why these extreme events occurred. Thus, long-term weather data from the past and projection for the future, including extreme events, is needed to study climate change and the impacts that various scenarios may have.

Most data concerning historical climate change patterns are derived from observations collected from meteorological stations, whereas creditable future climate projections can only be provided by numerical models forced by scenarios of climate drivers, such as changing greenhouse gas emissions.

Climate variability and climate change

Climate variability refers to all variations in the climate that have a longer duration than individual weather events; however, the term climate change is reserved for variations that persist for a much longer period of time (e.g., a decade or longer).

Climate variation and change are characterized as deviations from a reference climate state called the baseline. For example, a commonly used baseline represents the climate during the period 1961–1990, although other periods usually longer than 10 years can also be used. When the baseline changes, climate variability and change also adjust. Therefore, the baseline should always be clearly stated when a certain climate change or climate variation is presented.

Climate variability refers to how a climatic variable differs from its long-term mean, and natural variability usually refers to how much the climate varies in the absence of anthropogenic influences; for example, climate variations between years or between seasons are dominated by natural variability. Prior to the industrial revolution, climate variations and changes were dominated by natural factors. Variation in the climate is an inherent characteristic of the climate, given that some years a region may be characterized by a wet summer with a large amount of rain, but other years it may have a dry summer. Natural variability can change depending on the time scales of interest; for example, different circulations or teleconnection patterns in the oceans can strongly influence climate variability for several decades, such as the North Atlantic oscillation (NAO) and El Niño. Natural variability can sometimes be large compared to anthropogenic climate change signals, which typically represent long-term changes. This is especially true at high latitudes where natural variability can be significant. In particular, it can be difficult to identify climate change caused by human activities in regions with strong natural variability, meaning that near-future climate projections have large uncertainties due to the limited time available for the climate to change. In that case, climate change cannot be easily distinguished from natural variation.

Climate change refers to any changes in climate over time, whether due to natural factors or anthropogenic forces. The Earth's climate has changed on a variety of time scales throughout its history due to a wide range of natural factors, such as changes in solar radiation and the effects of continental drift. The current period of global warming is of particular significance, given that it has been caused by human activities since the pre-industrial era. Recent warming of the climate first became significant during the mid-20th century and has proceeded at an unprecedented rate. When future climate changes are discussed, it typically refers to climate changes from the present day to 2100, with respect to the pre-industrial era, which is taken by AR6 to be from 1850 to 1900 (Chen et al., 2021).

Emission scenarios with a focus on SSP

To quantify the anthropogenic influence on climate, the extent of social and economic activities related to energy and land use must first be constrained, which themselves determine the amount of greenhouse gases and aerosols added to the atmosphere by human activities.

Figure 24.1 shows that social and economic activities are the starting point to assess climate change and its impact. The activities include many factors, such as population size, economic activity, urbanization, education, social equality, consumption patterns, lifestyles, and institutions. These factors determine how human society evolves, its habits, and how energy and land are used.

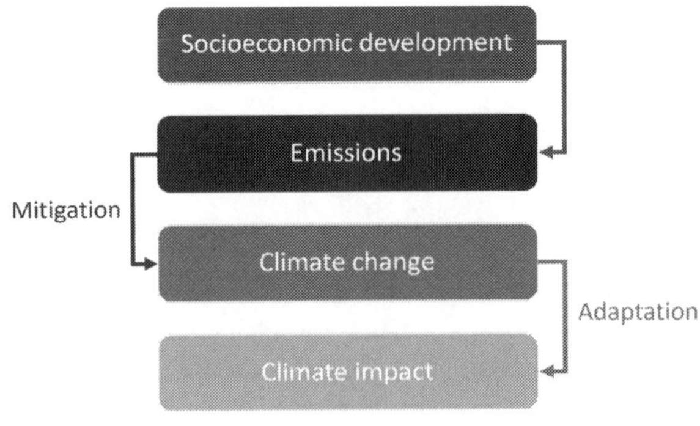

Figure 24.1 Human activities as drivers of climate change and its impacts. Knowledge of climate change is needed for both mitigation and adaptation measures.

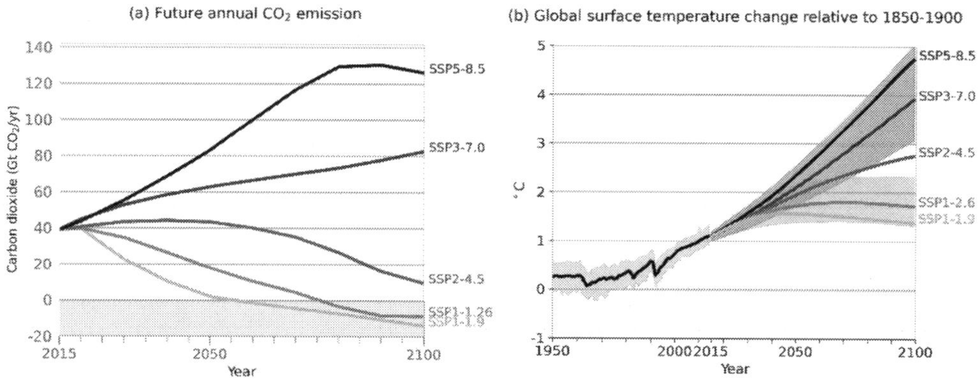

Figure 24.2 The five illustrative emission scenarios used by IPCC's sixth report (a) and global historical and projected temperature changes relative to 1850–1900 driven by each emission scenario (b).

Scenarios have been used frequently by the IPCC as a method for systematically examining possible futures. Over the years the IPCC has released several sets of emission scenarios used to drive global climate models (GCMs) towards developing climate change scenarios. In 2021, a new set of illustrative scenarios that cover a range of possible future developments for anthropogenic drivers of climate change reported in the literature, derived from shared socioeconomic pathways (SSPs), was used to synthesize knowledge across the physical sciences and impact, adaptation, and mitigation research (Chen et al., 2021). These scenarios were based on the SSPs that provided a consistent set of assumptions about societal, technical, cultural, and economic developments during the 21st century (O'Neill et al., 2017). The five SSP scenarios used in AR6 comprised SSP1-1.9, SSP1-2.6, SSP2-4.5, SSP3-7.0, and SSP5-8.5, which cover a broad range of emissions pathways. They started in 2015 and have scenarios ranging from high and very high greenhouse gas emissions (SSP3-7.0 and SSP5-8.5) to scenarios with very low and low greenhouse gas emissions (SSP1-1.9 and SSP1-2.6). Emissions vary between scenarios depending on socioeconomic assumptions, levels of climate change mitigation, aerosols and non-methane ozone precursors, and air pollution controls.

Based on socioeconomic scenarios, integrated assessment models simulating complex energy–land–economy–climate interactions alongside socioeconomic assumptions to produce energy, land use, and emissions scenarios (Riahi et al., 2017). Figure 24.2a shows the emission scenarios for carbon dioxide (CO_2) under the five AR6 illustrative SSPs.

Global climate models and future climate projections

A GCM, sometimes also known as an Earth System model, is a mathematical representation of physical, chemical, and biological processes occurring within the Earth's climate system, which includes the atmosphere, ocean, land, biosphere, and cryosphere.

All climate models are based on the same basic scientific understanding of the climate system; however, their results can still differ, given that the processes that operate in the climate system can be described in different ways and so may have model-specific details. Models can include a large or small number of processes, use different mathematical methods to solve the equations, and cover either global or regional domains. In addition, models are intrinsically imperfect replications of real systems, such that it is useful to compare the results from different models to see how and why they differ. Such a comparison allows measurement of the importance of different model descriptions for simulated climate and quantification of the associated model uncertainty.

If driven by natural forcings, a GCM describes the natural variability. Climate models may not match the real climate at short time scales. Nonetheless, a well-functioning climate model can reliably reproduce climate conditions, such as mean values and variability over a long period (e.g., the correct number of wet and dry summers over a 30-year period, although these modelled summers may occur in a different order than in the observed climate).

Changes in land use and the composition of the atmosphere drive climate change. Thus, GCMs can be used to reconstruct past climate conditions or project future climate changes based on historical emissions or emission scenarios used as future forcings (O'Neill et al., 2016). These models calculate many different properties of the climate, including atmospheric temperature, precipitation, pressure, wind, and humidity. For example, Figure 24.2b shows projections of global mean surface temperature changes under the five illustrative emission scenarios using GCMs from the Coupled Model Intercomparison Project (CMIP) phase 6 (CMIP6; see *Climate change scenarios based on GCMs and RCMs* section), starting in 2015 and ending at 2100. Shaded regions for the initial time period leading up to 2015 show how the CMIP6 climate models describe historical climate changes (i.e., 1951–2015), while the black line represents observations made during that period. The early stages of future projections demonstrate the difficulty in distinguishing signals of climate change from internal variability; However, differences in climate changes soon become clear. Here, the choice of model has great significance for the result and the choice of emission scenario has the greatest importance for the simulated climate towards the end of the century.

Climate models provide information on climate changes in both space and time. Figure 24.3 shows the spatial distribution of relative changes in the number of dry days and amount of rain under worst- and best-case emission scenarios.

Downscaling for future climate change scenarios

A meaningful impact assessment requires past and future climate information at appropriate temporal and spatial scales; all of these data, especially those for the future, come from GCM simulations. Figure 24.4 illustrates the steps needed to obtain the right information to perform such an assessment.

To date, GCMs are the best available tools to project future climate changes. Figure 24.3 shows that modelled climate characteristics change according to location. If the resolution of the GCM used – which is typically 100–200 km – is good enough for the application in hand, the simulated past and future climates can be used directly in an impact model, shown graphically as a jump from step 4 to step 6 in Figure 24.4; otherwise, downscaling (i.e., step 5 in Figure 24.4) must be performed. Downscaling is the process of translating coarse-resolution GCM outputs to a finer spatial scale, such as for a ground station or a small drainage basin (Xu, 1999).

Various downscaling techniques have been developed to provide information at finer scales (Benestad et al., 2008), with these methods generally being divided into dynamical (Nikulin et al., 2011) and statistical or empirical techniques (Hanssen-Bauer et al., 2005). Dynamical downscaling generates higher-resolution climate change scenarios by using a high-resolution regional climate model (RCM) forced by GCM outputs. Typically, RCMs have a resolution of about 25 km for long-term climate projection (Nikulin et al., 2011), although short-term simulations use a resolution as fine as a few kilometres. Dynamical downscaling using a RCM with a resolution of a few

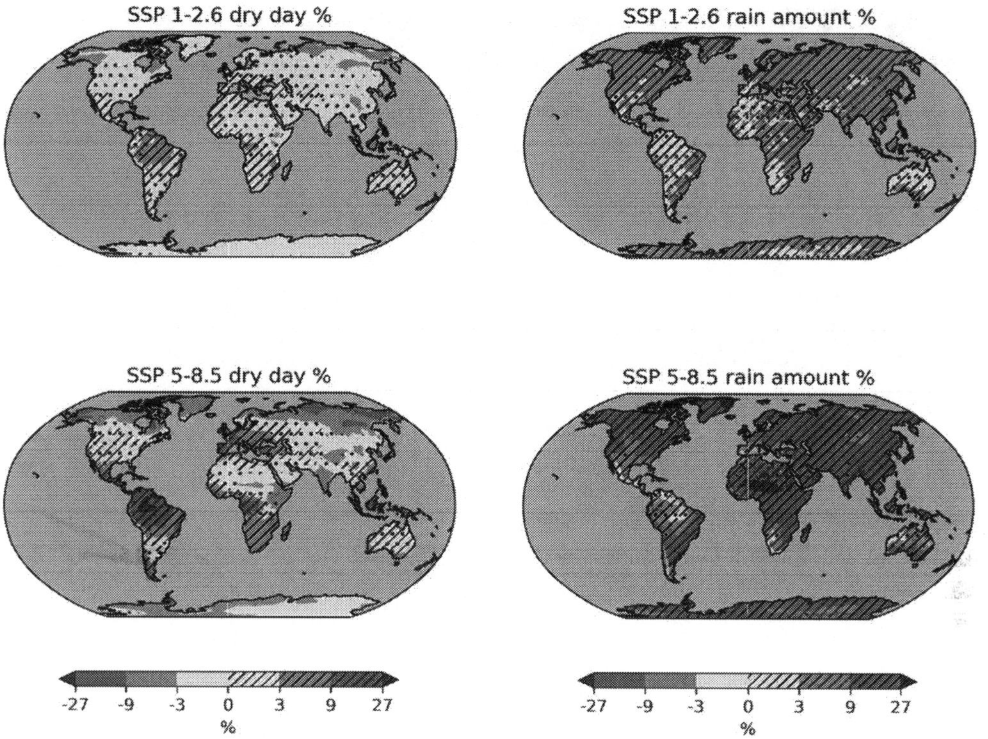

Figure 24.3 Global maps of projected annual mean relative changes (%) in the number of dry days (i.e., days with less than 1 mm of rain) and daily precipitation intensity (mm/day), estimated as the mean daily precipitation amount during wet days (i.e., days with rain intensity greater than 1 mm/day) averaged across CMIP6 models in the SSP1-2.6 and SSP5-8.5 scenarios respectively. Uncertainty is represented using a simple approach: whereby no overlay indicates regions with a strong model agreement, defined by ≥80% of models agreeing on the sign of change, and dots indicate regions with a poor model agreement, defined by <80% of models agreeing on the sign of change. Diagonal lines indicate positive values.

kilometres is rare, and creating a GCM with a similar resolution for long–period climate simulation will remain impractical for the foreseeable future; however, statistical downscaling offers an attractive and useful alternative for some applications that shows comparable skills to dynamic downscaling (e.g., Hellström et al., 2001). In statistical downscaling, observed long-term series are used to establish an empirical relationship between local and large-scale variables (Benestad et al., 2008).

Akin to GCM simulations, an attractive feature of dynamical downscaling is that it does not require historical observations for the locations in question, which makes the approach possible for data-scarce regions or areas without observations. *Historical climate observations* section describes a global effort to coordinate regional downscaling based on this method.

Despite these benefits, both RCMs and GCMs have the limitation that their simulations of historical climate are often biased compared to observed data, as they have inappropriate resolutions and intrinsic model deficiencies. As such, future simulations are expected to be biased too. This logic explains why the difference between simulations of the past and future is considered more reliable than simulations of the past and the future. Because of these biases, RCM or GCM

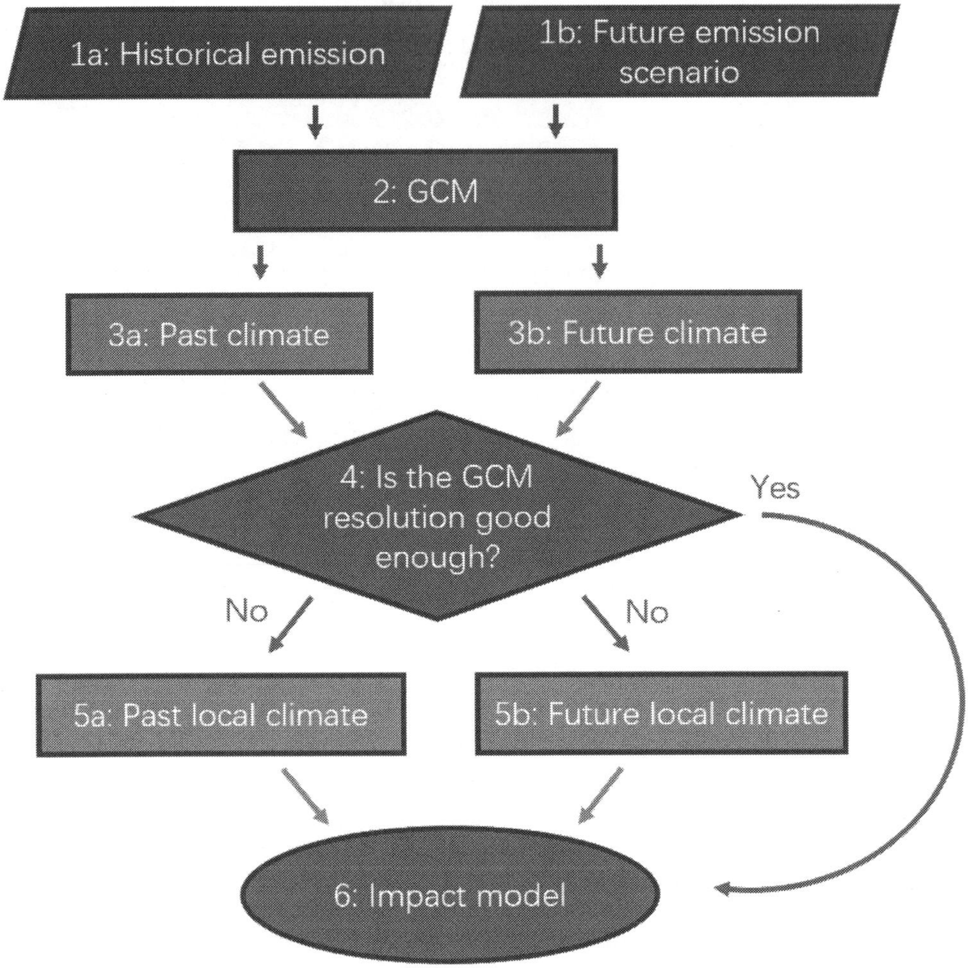

Figure 24.4 Steps involved in the provision of climate change information for impact models.

outputs are not always directly used in climate change impact assessments (e.g., Kotlarski et al., 2005). This has motivated the development of algorithms that combine outputs from GCMs and/or RCMs with time series information obtained from weather stations, which together create more realistic climate change scenarios. Indeed, the application of statistical downscaling algorithms to dynamically downscaled outputs is becoming increasingly common in the community (e.g., Teutschbein & Seibert, 2010).

The delta method illustrated in Figure 24.5 is a simple approach that has been widely used in the community since it effectively avoids bias-related problems and simultaneously respects current (and future, to some extent) fine-scale climate variation (Gleick, 1986; Hay et al., 2000). The delta method builds on the principle that changes in climate simulated by a GCM are only poorly influenced by model biases related to simulations of current and future climates since the same model would produce similar biases for different climates. Further, fine-scale observations of the current climate are used to represent the current climate.

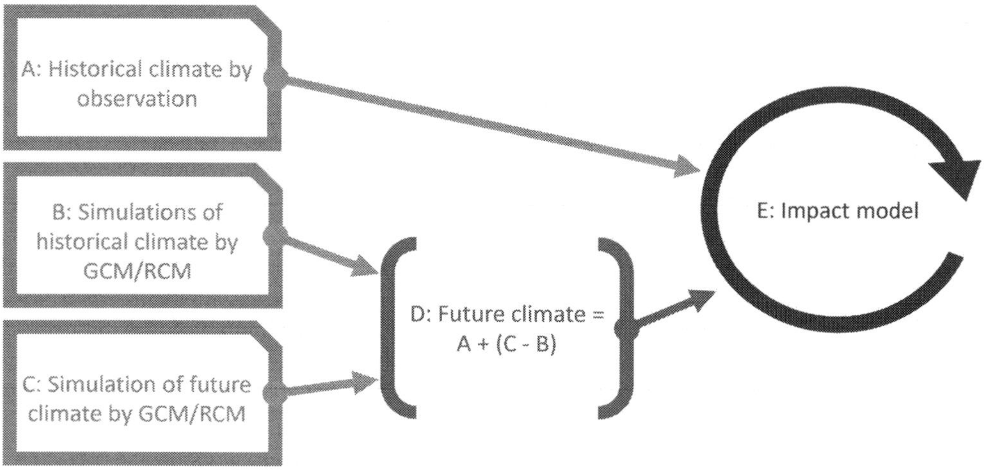

Figure 24.5 The principle of the delta method.

As this approach does not explicitly downscale the GCM results but does preserve fine-scale observations of the present climate, the modelled future climate also has a finer-scale structure that is considered fairly close to reality.

While the delta method represents the simplest way to remove overall model bias for monthly or annual data, more advanced methods are required to bias-correct variables with shorter time intervals, such as daily data (Guo et al., 2020). Successful application of these bias-correction methods relies on the availability and quality of observational data from historical periods, as well as reliable climate change signals simulated by the GCM and/or RCM that is employed. In addition, some skills in developing, adopting, and using statistical downscaling methods are required. *Resources for Climate Data and Scenarios* section provides resources for implementing downscaling needed for particular applications.

Uncertainty in climate change scenarios

One important feature of future climate change scenarios is their inherent uncertainty. Typically, these scenarios are presented for a range of plausible pathways or targets that capture the relationships between human choices, greenhouse gas emissions, atmospheric greenhouse gas concentrations, and temperature changes. Nonetheless, given the broad uncertainties in predicted emission rates, climate models, and internal variability of the climate system, many different estimations of future climate can be produced.

The range of emissions and climate changes reflects the uncertainty inherent in quantifying human activities (including technological development) and their impact on climate. In addition to these forcings, climate change scenarios also depend on climate models and eventually the downscaling methodologies used. Uncertainties in the forcings, climate simulations, and downscaling methods are discussed below to help users of climate change scenarios better appreciate the value and limitations of these scenarios.

One effective way to deal with uncertainty about the future is to make several climate scenarios based on different emission scenarios to show many possible future climates. An ensemble is a collection of climate scenarios that are individually different from each other due to different climate models used or different initial conditions of the same climate model. The differences between members thus provide a useful overview of the uncertainties associated with models or internal variability of the climate system. Therefore, the spread of an ensemble can be seen as a measure of the reliability of the results. If many different climate scenarios give similar results, then the results can be considered more reliable than if all scenarios predict very different outcomes.

An important result of the published climate scenarios is that the choice of emission scenario has very little impact on warming within the next few decades. This is because internal variability is larger compared to anthropogenic climate change in the near future. It is only after the middle of the century that the choice of emission scenario becomes an important factor for climate development.

Key variables for water resource studies

Surface air temperature (T) and global warming levels. The surface air temperature, often called surface temperature, is the temperature of air measured at 2 m above the ground or ocean surface. Surface temperature is one of the most important climatic variables, with the global mean surface temperature used as an indicator of global warming. Changes in global surface temperature with respect to preindustrial (1850–1900) levels are called global warming levels. An important finding of AR6 is that the global warming level is often closely related to a range of hazards and regional climate impacts, making it an ideal indicator for policy communication, as a reference point within and across climate change impact and adaptation studies, and for mitigation-related research. The Paris Agreement sets out a global framework to avoid dangerous climate change by limiting global warming levels to less than 2°C and pursuing efforts to limit it to 1.5°C.

Precipitation (P) describes the condensation of atmospheric water vapor to form clouds, which then fall to the Earth's surface under the influence of gravity either in solid (snow, hail, etc.) or liquid form (rain). While surface temperature represents thermal conditions at the Earth's surface, precipitation is the most important variable for the water balance of the surface, as it describes the wetness or dryness of the climate.

Evapotranspiration (ET) is the process that transfers water from the land to the atmosphere. This occurs by evaporation of water from soils and other surfaces, and by transpiration from plants. If the water supply in a region is not limited to evaporation and/or transpiration, evapotranspiration becomes maximized and is then termed potential evapotranspiration (PET).

Water availability (P–ET). Since precipitation and evapotranspiration are the main processes to increase or decrease the land surface water budget respectively, the accessibility of water resources for human consumption and ecosystems depends to a large extent on the spatio-temporal distribution of both precipitation and evapotranspiration (e.g., Liu et al., 2018).

The aridity index (AI = P/PET) is a common indicator for the dryness of the climate in a given location, although other forms of aridity indices exist.

Resources for Climate Data and Scenarios

There are numerous sources of historical climate/weather data, future climate projections using simulations from GCMs and RCMs, and statistical downscaling methods available in the community. Below we briefly introduce the most commonly used resources; however, we note that this is not a complete list of all resources and those mentioned remain under development, such that future updates are expected.

Climate change scenarios based on GCMs and RCMs

The World Climate Research Programme (WCRP) Working Group on Coupled Modelling (WGCM) CMIP was formed to compare sets of GCM simulations. The CMIP focuses on systematically modelling and comparing results produced by several climate modelling centres worldwide. The CMIP aims to better understand past and future climate changes driven by natural and/or anthropogenic forces. The IPCC Fifth Assessment Report (IPCC, 2013) employed the ensemble model results from phase 5 of the CMIP (CMIP5). The latest results (i.e., CMIP6) have a higher horizontal and vertical spatial resolution than CMIP5, meaning that it can better represent the global hydrological cycle and other processes. The main focuses of CMIP6 are to explore how the Earth system responds to forcing and how future climate projections vary between different scenarios, and to assess models' performances (Eyring et al., 2016).

The WCRP Coordinated Regional Climate Downscaling Experiment (CORDEX) was launched to evaluate model performances and produce climate projections that explore climate information (Giorgi et al., 2009). The CORDEX coordinates regional climate downscaling simulations from reanalyses, CMIP5 and CMIP6. Modelling domains in CORDEX cover all of Africa, Asia, Australia, South America, North America, Europe, and adjacent oceans. Both empirical statistical and downscaling are employed in CORDEX simulations, and the downscaling results provide high-resolution information for regional climate changes and future extremes. CORDEX downscaling simulations have been used in the AR6 digital atlas for assessing regional climate information. The CORDEX information and datasets are available at the WCRP CORDEX website (https://cordex.org/).

The IPCC digital atlas (https://interactive-atlas.ipcc.ch) makes climate data and scenarios for any region in the world much easier to access, display, and analyse. It can be used to obtain past climate observations and future climate change scenarios for any regions, time periods, and variables of interest.

Historical climate observations

Historical observations for past climate have been made with help of land surface stations, ocean surface measurements from ships and buoys, underwater instrumentation, satellite, and surface-based remote sensing, and in situ atmospheric measurements from airplanes and balloons. For impact studies, land surface stations and satellite retrievals play a special role, as they are useful to validate climate models, as well as establish and calibrate statistical downscaling models or bias-corrected GCM/RCM simulations of past climate.

Surface observations are available as station data or gridded data. Many research centres in the world provide a large amount of climate and one of them is the National Center for Atmospheric Research (NCAR) which has a rich collection of climate data.[1]

In addition to gridded or stated observation, there is another type of product called reanalysis. Reanalysis is comprehensive data that combines short-term weather forecasts from numerical models with observations through data assimilation. Since the observations are usually limited in space, reanalysis fills the gaps and provides global or regional gridded outputs in space and time. Several research institutes develop and support high-resolution global reanalyses, such as ERA5 from European Centre for Medium-Range Weather Forecasts (ECMWF) (Hersbach et al., 2020), MERRA version 2 from NASA Global Modeling and Assimilation Office (GMAO) (Gelaro et al., 2017), and CFSR from National Centers for Environmental Prediction (NCEP) (Saha et al., 2010).

Bias-corrected historical climate data and statistically downscaled scenarios

Several research projects and centres offer downscaled and bias-corrected historical and future climate data. One such example is WorldClim (https://www.worldclim.org/data/index.html#), which offers high spatial resolution global weather and climate data. These gridded climate and weather data cover historical (1970–2000) and future climate conditions by using future climate projections based on CMIP6. A simple downscaling method was applied to these data, which were also bias-corrected using WorldClim v2.1 (Fick & Hijmans, 2017) as a reference.

Conclusions

Water is not only the most essential component of life on Earth but also a critical enabler for food and energy security, human health and well-being, and ending hunger and poverty. Water plays a central role in sustainable development which includes social, economic, and environmental dimensions. Climate change can have impacts through worsening floods, rising sea levels, wildfires, and droughts. On the other hand, water can be useful in fighting climate change, as sustainable water management is central to building the resilience of societies and ecosystems and to reducing carbon emissions. Assessing water's positive and negative effects under future climate change requires climate projections and baseline at appropriate temporal and spatial scales, which requires GCM simulations and probably also downscaling products. This chapter provides practical guidance to the principles, procedures, and methods to deliver this actionable knowledge.

Note

1 See https://climatedataguide.ucar.edu/climate-data

References

Benestad, R. E., Hanssen-Bauer, I. and Chen, D. (2008). *Empirical-Statistical Downscaling*. World Scientific Publishing Co. Pte. Ltd., Singapore. p. 300.

Chen, D., M. Rojas, B. H., Samset, K., Cobb, A., Diongue Niang, P., Edwards, S., Emori, S. H., Faria, E., Hawkins, P., Hope, P., Huybrechts, M., Meinshausen, S. K., Mustafa, G-K., Plattner, and A.-M. Tréguier. (2021). "Framing, Context, and Methods." In Masson-Delmotte, V., P. Zhai, A. Pirani, S. L. Connors, C. Péan, S. Berger, N. Caud, Y. Chen, L. Goldfarb, M. I. Gomis, M. Huang, K. Leitzell, E. Lonnoy, J. B. R. Matthews, T. K. Maycock, T. Waterfield, O. Yelekçi, R. Yu and B. Zhou (eds.). *Climate Change 2021: The Physical Science Basis*. Contribution of Working Group I to the Sixth Assessment Report of the Intergovernmental Panel on Climate Change. Cambridge University Press, Cambridge and New York, NY, pp. 147–286.

Eyring, V., Bony, S., Meehl, G. A., Senior, C. A., Stevens, B., Stouffer, R. J. and Taylor, K. E. (2016) "Overview of the coupled model intercomparison project phase 6 (CMIP6) experimental design and organization." *Geoscientific Model* Development, 9, 1937–1958.

Fick, S. E. and Hijmans, R. J. (2017). "WorldClim 2: new 1km spatial resolution climate surfaces for global land areas." *International Journal of Climatology*, 37(12), 4302–4315.

Gelaro, R., McCarty, W., Suárez, M. J., Todling, R., Molod, A., Takacs, L., Randles, C. A., Darmenov, A., Bosilovich, M. G., Reichle, R., Wargan, K., Coy, L., Cullather, R., Draper, C., Akella, S., Buchard, V., Austin Conaty, A., da Silva, A. M., Gu, W., Kim, G.-K., Koster, R., Lucchesi, R., Dagmar Merkova, D., Nielsen, J. E., Partyka, G., Pawson, S., Putman, W., Rienecker, M., Schubert, S. D., Sienkiewicz, M. and Zhao, B. (2017). "The modern-era retrospective analysis for research and applications, version 2 (MERRA-2)." *Journal of Climate*, 30(14), 5419–5454.

Giorgi, F., Jones, C. and Asrar, G. R. (2009). "Addressing climate information needs at the regional level: the CORDEX framework." *World Meteorological Organization (WMO) Bulletin*, *58*(3), 175.

Gleick, P.H. (1986). "Methods for evaluating the regional hydrologic impacts of global climatic changes." *Journal of Hydrology*, 88(1–2), 97–116.

Guo, L., Jiang, Z., Chen, D., Treut, H. L. and Li, L. (2020) "Projected precipitation changes over China for global warming levels at 1.5°C and 2°C in an ensemble of regional climate simulations: impact of bias correction methods." *Climatic Change*, 162, 623–643.

Hanssen-Bauer, I., Achberger, C., Benestad, R., Chen, D. and Førland, E. (2005). "Statistical downscaling of climate scenarios over Scandinavia." *Climate Research*, 29(3), 255–268.

Hay, L. E., Wilby, R. L. and Leavesley, G. H. (2000). "A comparison of delta change and downscaled GCM scenarios for three mountainous basins in the United States." *Journal of the American Water Resources Association*, 36(2), 387–397.

Hellström, C., Chen, D., Achberger, C. and Räissänen, H. (2001). "A comparison of climate change scenarios for Sweden based on statistical and dynamical downscaling of monthly precipitation." *Climate Research*, 19(1), 45–55.

Hersbach, H., Bell, B., Berrisford, P., Hirahara, S., Horányi, A., Muñoz-Sabater, J., Nicolas, J., Peubey, C., Radu, R., Schepers, D., Simmons, A., Soci, C., Abdalla, S., Abellan, X., Balsamo, G., Bechtold, P., Biavati, G., Bidlot, J., Bonavita, M., De Chiara, G., Dahlgren, P., Dee, D., Diamantakis, M., Dragani, R., Flemming, J., Forbes, R., Fuentes, M., Geer, A., Haimberger, L., Healy, S., Hogan, R. J., Hólm, E., Janisková, M., Keeley, S., Laloyaux, P., Lopez, P., Lupu, C., Radnoti, G., de Rosnay, P., Rozum, I., Vamborg, F., Villaume, S. and Thépaut, J. N. (2020). "The ERA5 global reanalysis." *Quarterly Journal of the Royal Meteorological Society*, 146(730), 1999–2049.

IPCC. (2013). "Summary for Policymakers. In: Climate Change 2013: The Physical Science Basis." In Stocker, T. F., D. Qin, G.-K. Plattner, M. Tignor, S. K. Allen, J. Boschung, A. Nauels, Y. Xia, V. Bex and P. M. Midgley (eds.). *Fifth Assessment Report of the Intergovernmental Panel on Climate Change*. Contribution of Working Group I to the Fifth Assessment Report of the Intergovernmental Panel on Climate Change. Cambridge University Press, Cambridge, UK and New York, NY.

IPCC. (2021). "Summary for Policymakers. In: Climate Change 2021: The Physical Science Basis." In Masson-Delmotte, V., P. Zhai, A. Pirani, S.L. Connors, C. Péan, S. Berger, N. Caud, Y. Chen, L. Goldfarb, M.I. Gomis, M. Huang, K. Leitzell, E. Lonnoy, J.B.R. Matthews, T.K. Maycock, T. Waterfield, O. Yelekçi, R. Yu and B. Zhou (eds.). *Sixth Assessment Report of the Intergovernmental Panel on Climate Change*. Contribution of Working Group I to the Sixth Assessment Report of the Intergovernmental Panel on Climate Change. Cambridge University Press, Cambridge and New York, NY, pp. 3–32.

Irannezhad, M., Ahmadi, B., Liu, J., Chen, D. and Matthews, J. H. (2022). "Global water security: a shining star in the dark sky of achieving the sustainable development goals. Sustainable Horizons." 1, 100005.

Kotlarski, S., Block, A., Böhm, U., Jacob, D., Keuler, K., Knoche, R., Rechid, D. and Walter, A. (2005). "Regional climate model simulations as input for hydrological applications: evaluation of uncertainties." *Advances in Geosciences*, 5, 119–125.

Liu, W., Lim, W. H., Sun, F., Mitchell, D., Wang, H., Chen, D., Bethke, I., Shiogama, H. and Fischer, E. (2018). "Global freshwater availability below normal conditions and population impact under 1.5 and 2°C stabilization scenarios." *Geophysical Research Letters*, 45(18). DOI:10.1029/2018GL078789.

Nikulin, G., Kjellström, E., Hansson, U., Jones, C., Strandberg, G. and Ullerstig, A. (2011). "Evaluation and future projections of temperature, precipitation and wind extremes over Europe in an ensemble of regional climate simulations." *Tellus A: Dynamic Meteorology and Oceanography*, *63*(1), 41–55.

O'Neill, B. C., Kriegler, E., Ebi, K. L., Ebi, K. L., Kemp-Benedict, E., Riahi, K., Rothman, D. S., van Ruijven, B. J., van Vuuren, D. P., Joern Birkmann, Kok, K., Levy, M. and Solecki, W. (2017). "The roads ahead: narratives for shared socioeconomic pathways describing world futures in the 21st century." *Global Environmental Change*, 42, 169–180.

O'Neill, B. C., Tebaldi, C., Van Vuuren, D. P., Eyring, V., Friedlingstein, P., Hurtt, G., Knutti, R., Kriegler, E., Lamarque, J.-F., Lowe, J., Meehl, G. A., Moss, R., Riahi, K. and Sanderson, B. M. (2016). "The scenario model intercomparison project (ScenarioMIP) for CMIP6." *Geoscientific Model Development*, 9(9), 3461–3482.

Riahi, K., Van Vuuren, D. P., Kriegler, E., Edmonds, J., O'Neill, B. C., Fujimori, S., Bauer, N., Calvin, K., Dellink, R., Fricko, O., Lutz, W., Popp, A., Cuaresma, J. C., Leimbach, K. C., Jiang, M., Kram, L., Rao, T., Emmerling, S., Ebi, J., Hasegawa, K., Havlik, T., Humpenöder, P., Da Silva, F., Smith, L. A., Stehfest, S., Bosetti, E., Eom, V., Gernaat, J., Masui, D., Rogelj, T., Strefler, J., Drouet, J., Krey, L., Luderer, V., Harmsen, G., Takahashi, M., Baumstark, K., Doelman, L., Kainuma, J. C., Klimont, M., Marangoni, Z., Lotze-Campen, G., Obersteiner, H., Tabeau, A. and Tavoni, M. (2017). "The shared socioeconomic pathways and their energy, land use, and greenhouse gas emissions implications: an overview." *Global Environmental Change*, 42, 153–168.

Saha, S., Moorthi, S., Pan, H., Wu, X., Wang, J., Nadiga, S., Tripp, P., Kistler, R., Woollen, J., Behringer, D., Liu, H., Stokes, D., Grumbine, R., Gayno, G., Wang, J., Hou, Y., Chuang, H., Juang, H. H., Sela, J., Iredell, M., Treadon, R., Kleist, D., Van Delst, P., Keyser, D., Derber, J., Ek, M., Meng, J., Wei, H., Yang, R., Lord, S., van den Dool, H., Kumar, A., Wang, W., Long, C., Chelliah, M., Xue, Y., Huang, B., Schemm, J., Ebisuzaki, W., Lin, R., Xie, P., Chen, M., Zhou, S., Higgins, W., Zou, C., Liu, Q., Chen, Y., Han, Y., Cucurull, L., Reynolds, R. W., Rutledge, G. and Goldberg, M. (2010). "The NCEP climate forecast system reanalysis." *Bulletin of the American Meteorological Society*, 91(8), 1015–1058.

Teutschbein, C. and Seibert, J. (2010). "Regional climate models for hydrological impact studies at the catchment scale: a review of recent modeling strategies." *Geography Compass*, 4(7), 834–860.

United Nations. (2015). *Transforming Our World: The 2030 Agenda for Sustainable Development*. UN Publishing, New York, NY.

Xu, C.-Y. (1999). "From GCMs to river flow: a review of downscaling methods and hydrologic modelling approaches." *Progress in Physical Geography*, 23(2), 229–249.

25
DROUGHT

Elisa Savelli

Introduction: Drought and Development

The word drought often evokes images of the prolonged absence of precipitation, expanses of parched soil, dried-up agricultural fields or depleted riverbeds. For a general audience, a drought might occur when the water reservoirs of a city are unable to supply water to the urban population. Indeed, drought is all of these things together, intertwined, interdependent, and difficult to disentangle. As of today, natural scientists have not agreed on a universal definition of drought (Cook et al., 2018; Lloyd-Hughes, 2014; Mishra & Singh, 2010; Wilhite, 2000). The large diversity of available definitions is due to the complexity of such a phenomenon across several disciplines, in addition to the fact that drought can mean many different things to different audiences depending on their specific interests (Cook et al., 2018; Mishra & Singh, 2010). Many scientists go as far as to describe drought as the most complex and least understood natural hazard (Cook et al., 2018; Lloyd-Hughes, 2014; Wilhite, 1993).

Besides surpassing other hazards in terms of complexity, droughts are also classified amongst the most extreme hydro-climatological phenomena, as they exceed all other events in terms of number of people affected (Meza et al., 2020). Since the earliest civilizations, drought has been a recurring hazard, producing the largest population movements, and the most dire famines, and the greatest societal collapses ever recorded in human history (Lucero et al., 2017; McLeman et al., 2010; Sear et al., 2020). Over time, the relationship between drought and society has considerably changed due to the expansion and intensification of anthropogenic pressure manifested amongst others in phenomena such as urbanization, deforestation, unsustainable water consumption, and human-induced climate change. Specifically, human activities have increasingly altered natural hydrological conditions to the point of leaving an imprint on almost all the water sources existing on Earth (Linton & Budds, 2014; Rockström et al., 2014; Savenije et al., 2014). Such changes have intensified the interactions between water and social systems and, as a result, increased the frequency, duration, and intensity of droughts (Guerreiro et al., 2018; Spinoni et al., 2019; Leng et al., 2015; Van Lanen et al., 2013). At the same time, this intense interplay between water and society has increased the vulnerability of systems affected by droughts, and in particular, exacerbated the condition of the most disadvantaged groups or individuals. In turn, drought projections accounting for climate change, economic

DOI: 10.4324/9781003095545-29

development, and population growth dynamics paint alarming pictures in many regions of the world.

Today, the world seems still unprepared to face future droughts, much less address their implications. Generally, it is still difficult to foresee when and for how long droughts are likely to occur, not to mention what their impacts will be (Nature, 2019). In turn, droughts constitute some of the most compelling and serious challenges that society must address (IPCC, 2022; UNDRR, 2021). This chapter argues that the reasons for this impasse lies in the fact that neither natural nor social scientists alone have been able to fully apprehend the socioenvironmental complexity of drought. On the one hand, physical and engineering sciences' tend to homogenize society into a unified whole, and consider social processes as apolitical. On the other hand, critical social sciences have not seriously appreciated the heterogeneity and multitude of the biophysical processes coproducing drought. In response to these research gaps, this chapter engages with both natural and social sciences as a way to account for the environmental as well as societal complexities of drought. Ultimately, such an interdisciplinary approach claims to offer a more effective way to address future drought.

Conceptualizing Drought Risk, Hazard, and Vulnerability

Physical and engineering sciences conceptualizations of drought are crucial to understand the components of drought risk and their long-term coevolution. In general, these disciplines define drought risk as the combination of the probability of occurrence of a drought, and the potential negative consequences that drought entails. The first component of risk represents the hazard, whilst the second component reflects the vulnerabilities of the systems exposed to the hazard.

The term 'drought hazard' defines the intensity and duration of a drought throughout its propagation from the atmosphere to the ground. Many scholars agree that drought hazard does not represent a static condition but rather constitutes a physical process which material property changes over time and across space (AghaKouchak et al., 2021). A drought hazard is usually set off by prolonged atmospheric conditions induced by climate variability and further enhanced by meteorological factors, and anthropogenic climate change (Mishra & Singh, 2010). Thus, the first manifestation of a drought consists of a *meteorological drought*, which encompasses a prolonged condition of below normal precipitation. This can (though not always) evolve into a *soil moisture (or agricultural) drought*, which expresses the impact of precipitation shortfall on the reduction of soil moisture and vegetation growth. During an agricultural drought the soil becomes dryer than normal for a prolonged amount of time and thus unable to absorb or maintain the water required by the surrounding ecosystem. Subsequently, the latter conditions can (though not always) unfold into a *hydrological drought*, manifesting into reduced surface water and groundwater levels as a result of the previous meteorological and agricultural dryness. Finally, with *socio-economic drought* natural scientists mostly describe the deficit of water demand for socio-economic goods and services relative to the amount of water supplied (Wilhite, 2000).

Earlier definitions of drought hazard considered it as a component of risk that society could not control, but only measure or predict (Salvador et al., 2021). Thus, the propagation of drought hazard constituted a unidirectional process mostly influenced by meteorological features and other physical properties such as hydrogeology and vegetation cover. Human processes remained confined at the downstream end of such propagation and were responsible of reshaping only vulnerabilities and impacts of the system affected by drought hazard. More recently, literature has showed that society can directly influences the propagation of the hazard from the atmosphere to the ground (de Freitas, 2020; Jaeger et al., 2019; Lange et al., 2017; Omer et al., 2020; Van Loon et al., 2016). For instance, physical and engineering scientists found that human activities can intensify the aridity of the soil, reduce the amount of water available in hydrological systems, and induce shortages of water supply for different socio-economic activities. Socio-hydrological studies have calculated that anthropogenic activities can under certain conditions intensify the severity of drought hazard by 500% (Wada et al., 2013). Other analyses found that land cover changes can lead to drought hazards that are twice as severe and three-times more frequent relative to hazards that would have occured under pristine conditions (Taufik et al., 2020). Ultimately, physical and engineering scientists seem to find it extremely challenging to distinguish between the diverse hydro-climatological, ecological, and social forces that reshape the production of drought hazard (Van Loon et al., 2016).

Whilst developing this classification of drought hazard, physical and engineering scientists explain that the level of risk that a meteorological, agricultural, hydrological, or socio-economic drought produces, does not merely result from the intensity and severity of the hazard itself (Wilhite, 2000). Instead, drought hazard becomes a more or less serious risk depending on the vulnerability of the socioenvironmental systems exposed to such hazard. This means that a more vulnerable social group or ecosystem would face higher drought risk and more destructive impacts relative to a less vulnerable one even if the two were exposed to the same drought hazard. Accordingly, vulnerability represents the degree to which a socioenvironmental system is susceptible to and/or unable to cope with the adverse effects of drought events (Bhavani et al., 2017; Dumitraşcu et al., 2018; McNeeley, 2014; Wang et al., 2020). According to physical and engineering sciences, vulnerability results from the combination of diverse socio-economic factors which range between wealth, education, access to resources and infrastructure, or relate more specifically to types of crops and livestock in agricultural settings (Savelli et al., 2022a).

Overall, when conceptualizing drought risk, physical and engineering studies have achieved crucial scientific advances by first recognizing and then quantifying the role that humans play in the production drought hazard and vulnerabilities. Explicitly or not, their studies helped conceiving drought risk as socioenvironmental rather than just hydroclimatic in that its manifestation becomes as much a result of anthropogenic activities, as it is of climatic or geophysical processes (Taylor et al., 2009; Van Dijk et al., 2013). However, in their quest to measure and forecast physical transformations, these disciplines have overlooked the complexity of the social processes which intersect with drought risk. What characterises most of their investigations on drought hazard and vulnerability is that physical and engineering studies mostly consider society as homogenous and social processes apolitical.

By privileging quantitative methods, physical and engineering works on drought have produced reductionist interpretations of society which lack the potential to capture the power relations underlying the production of drought hazard and vulnerabilities. One of the reasons behind this resistance in engaging with social power stems from the formative experience of most physical and engineering scientists, which is based on positivist theories and beliefs. Positivism considers as valid knowledge that which can be observed, measured or experimented (Lane, 1996). Yet, not all social processes can be quantified or experimented. Nor can society be considered homogenous inasmuch as every individual or social group has different abilities to interact with and reshape drought-related processes (Sanderson, 2018). In turn, physical and engineering investigations on drought can end up ignoring the sociotechnical and political-economic dynamics that drive human activities whilst also obscuring the disproportionate agency of privileged subjects relative to the most vulnerable (Acha, 2020; Swyngedouw, 2019).

To avoid similar simplifications or oversights of social processes, physical and engineering scientists should engage more actively with critical social sciences (Savelli et al., 2022a). This interdisciplinary effort would expose the power dynamics that intersect with the production of drought hazard whilst at the same time, link broader patterns of inequality with the uneven distribution of drought vulnerability and impacts. Ultimately, more politically aware explorations of droughts would offer alternative ways to deal with their causes, rather than with their symptoms.

The Social Complexity of Drought

Amongst critical social scientists, political ecologists, and human geographers have been particularly active in criticizing engineering and physical scientists for neglecting the social power that intersect with every socioenvironmental transformation (Loftus, 2007). Building on this idea, critical scholars have been able to apprehend the power imbalances and dynamics which govern social complexities (Mollinga, 2008). Specifically, most political ecologists and human geographers have focused their attention on the relationships that intertwine ecology, power, and capital, thereby challenging existing power structures and problematizing those social conditions of domination and oppression that risk being taken for granted (Heynen et al., 2006; Swyngedouw, 2009, 1997). This focus has brought about theorizations of the socio-economic and political processes that reshape society and its heterogeneous relations with ecology. With this awareness, critical scholars are crucial to understand the way humans intersect with drought risk, both whilst coproducing the hazard as well as when reshaping vulnerability.

First, political ecologists and human geographers internalize the relationships between water and society as opposed to physical and engineering visions that consider water and society as separate entities. By doing so, critical scholars are able to emphasise the influence of socio-political processes in the physical characteristics of water. Political ecology in particular, fuses together water and social power in inseparable ways and makes of what once was the hydrological cycle a *hydrosocial* one (Linton & Budds, 2014; Swyngedouw, 2009). This means that flows of water become *socionatural*[1] processes in which water and society shape and reshape each other over time and across space. For instance, the water flowing through a hydropower dam constitutes both the material transformation of water from precipitation to monthly inflow into the dam, as well as the material agency of powerful actors who advocated for the dam's construction to suit their political and economic benefit. This hybrid conceptualization proposed by the *hydrosocial cycle* demonstrates that the material circulation and biophysical characteristics of water result from particular configurations of social power, and concurrently, that produced water reconfigures social relations. Although rarely focused on drought, this hybrid conceptualization can be useful to further understanding of drought hazard. Indeed, as much as flows of water are socionatural processes, shortages of water, dried-up river beds or parched soils do not merely fall within the realm of nature, but can be conceived as hybrid products of *socionatural* transformations. For instance, Bakker's (2000) examination of the 1995 Yorkshire drought reveals that the water scarcity experienced in that period, was largely the outcome of meteorological modelling, demand forecasting, and neoliberal restructuring of the water companies, rather than a mere consequence of rainfall reduction.

Second, critical scholars explain that drought risk is as much a natural as it is historical and context specific. This means that drought risk is deeply rooted in distinctive socio-political processes that, throughout history, have materially and discursively produced different drought experiences and impacts. Amongst other socio-political processes, critical scholars have retraced in particular, colonial processes of accumulation by dispossession, capitalist exploitations of labour, neoliberal policies and governmentalities, along with other race, gender and class-based exclusion/inclusion processes (Ahlers, 2010; Bakker, 2003; 2000; Kaika, 2006; 2003; Loftus, 2007; Mehta, 2010; 2003; Millington, 2018; Musemwa, 2009; Otero et al., 2011; Savelli et al., 2021; 2022b; Walker, 2014). According to Musemwa (2009), for instance, the unequal experiences of the 1982–1992 drought in Bulawayo, Zimbabwe follow a class dynamic which is deeply rooted in colonial and post-colonial policies. In practice, Bulawayo's urban poor bore the brunt of the drought by experiencing the highest levels of water insecurity relative to the rest of the city in the same way they did during and after colonial times (Musemwa, 2009). Like Musemwa (2009), critical scholars mostly conceive drought as water scarcity socially produced by dominant political-economic regimes and prevailing power structures related to class, ethnicity, gender and other intersectional axes.

The contingency of drought phenomena to historical socio-political processes reveals a third important contribution that critical social sciences can bring to the study of drought: the temporality of drought. In other words, one could conceive droughts as conjunctures that emerge from certain dynamics which have endured historically and will eventually persist in the future (Scheba & Millington, 2018). Thus, on the one hand, the temporality of drought can help explain the genesis of drought risk whilst at the same time, reveal how the manifestation of a drought reshapes the future. For instance, the manner in which a city experiences drought might create ruptures from the past or materially and discursively reproduce pre-existing *socionatural* relations. Kaika (2003) provides a matching example of drought temporality in her examination of the drought that hit Athens, Greece between 1989 and 1991. She argues that the 170 days of water crisis that resulted from the drought were crucial for the government to mask a political crisis with a natural one. The drought and its manifestation enabled the government to acquire public consensus and in turn, facilitated the uncontested implementation of an emergency act which strengthened the privatization of the water sectors with serious social, economic, political and environmental consequences for the city (Kaika, 2003). Alternatively, drought could also engender civil reactions and demands for socio-political changes. In Cape Town for instance, the struggle of the recent Day Zero water crisis triggered trade unions and activists to reclaim their rights and advance their political agenda against the ongoing water privatization and enduring neoliberal policies (Alzate González & Peñaloza Lanza, 2019).

Lastly, by focusing on social power, critical scholars are able to expose the heterogeneity of society and with it, reveal the uneven distribution of disproportionate burdens and interests existing across different social groups or individuals. This heterogeneous understanding is relevant to advance understandings of drought risk in two ways. First, such critical lens links broader patterns of inequality with the uneven distributions of drought vulnerabilities and impacts. Second, by unravelling the inequalities and injustice that shape peoples' access to, use, and control over water, land and other resources, critical scholars are able to expose the responsibility that distinctive social groups or individuals have in reshaping drought hazard (Savelli et al., 2021). Critical scholars have studied social heterogeneity at different spatial scales: from global injustices and urban inequalities, to the uneven water (in)security experienced by different bodies (Doshi, 2017). Every spatial scale reveals distinctive power dynamics and socio-political processes at the origins of such inequalities. For instance, investigating inequalities in a city can help retrace how certain development policies have unequally redistributed land and water resources across different social groups (Tiwale et al., 2018). A focus on the body, disentangles the intersectional axes of gender, race, religion, or class that reshape the manner in which drought is experienced and perceived (Sultana, 2011; Truelove, 2019). Focusing on the urban scale, Millington (2018) and Savelli et al. (2021) describe the water crises that affected respectively São Paulo and Cape Town, as a result of severe droughts. They both argue that the unequal experiences of drought amongst the cities' inhabitants, mostly resulted from inequalities existing long before the droughts. Indeed, the unequal experience of drought was contingent upon the type of water infrastructure as well as the different abilities to pay for or access to different water sources (Enqvist & Ziervogel, 2019; Hellberg, 2020). In turn, during the drought, whilst low-income residents experienced recurring water shortages, wealthier areas in the city centre had access to continuous supply of water (Millington, 2018; Savelli et al., 2021). In addition, the work of Truelove (2011) is helpful for studying experiences of drought at a body scale. Focusing on women's bodies, Truelove's (2011) examination of water insecurity in Delhi, explains that even when water supply is seemingly improved across an urban area, women continue to struggle in their daily water access and uses, therefore remaining disadvantaged relative to the rest of the household (Truelove, 2011). Together, by directly linking social inequalities with the production of drought risk, these examples reveal how inequalities matter in drought research.

Despite not necessarily engaging with concepts of risk, hazard, and vulnerability, critical social sciences have proved crucial in enriching physical and engineering sciences' conceptualizations of drought both in terms of hazard and vulnerabilities. The problem with these critical conceptualizations is they overemphasise the socially constructed nature of drought thereby concealing the multitude of ecological processes that coproduce drought hazard, i.e., the soil moisture variability, the vegetation, hydrogeological changes, and micro-climatic alterations (AghaKouchak et al., 2021). In general, critical social scientists seem to have merged the heterogeneity of these biophysical processes into the unique category of water scarcity (Bakker & Bridge, 2006). However, As Kallis (2008: 104) aptly illustrates, "one cannot presume that a drought will not happen if politics are different without adequately engaging with the hydrometeorological and infrastructural specificities of the case". Therefore, *dismissing the ecology* (Walker, 2005) of drought, these scholars can fail to understand the genesis of drought risk and its distribution.

The Environmental Complexity of Drought

To more seriously consider the biophysical processes that intersect with socioenvironmental transformation, critical scholars have started decentring humans and reconceptualizing them in their entanglements with non-human processes (Lima Costa et al., 2017). Post-human (or more-than-human) geographers have developed a relational approach which investigates the agency of animals, gardens, mushroom and other biophysical entities (Bingham, 2006; Panelli, 2010; Power, 2005; Tsing, 2012). Other scholars have focused on *materiality* to emphasise the material conditions and the way they can enable or constrain social transformations (Bakker & Bridge, 2006). Throughout the disclosure of these relations, what was before the social-production of nature, now becomes the co-production of nature, a process in which humans and non-humans coevolve and mutually reshape each other.

Thus, decentring human agency and focusing on the material relationalities of drought can potentially serve as a way to overcome the impasse of the social construction of water scarcity. However, a focus on materiality alone, would not suffice to adequately apprehend the agency of biophysical elements coproducing drought (Savelli et al., 2022b). Instead, capturing the co-productive capacity of non-human elements, requires also engaging with engineering and physical sciences. As shown throughout this chapter, these disciplines serve to account for the way and extent to which distinctive processes of vegetation, soil moisture variability, hydrogeological changes, and micro-climatic alterations reshape the propagation of drought hazard.

For instance, Savelli et al. (2022b) combines political ecology with hydroclimatological understandings of drought hazard, to explain the manner in which a drought can push a socioenvironmental system to the verge of collapse. Focusing on Ladismith, an agricultural area in the Western Cape Province of South Africa, Savelli et al. (2022b) account for both the biophysical and political transformations that co-produced the drought in the area. To do so, the authors expand the temporal and spatial scales of their analysis beyond the event itself. The study goes back until the

Colonial era and dissect the manner in which Ladismith's soil, vegetation, hydrology and microclimate have set the basis for distinctive socioeconomic transformations such as the establishment of commercial agriculture and the accumulation of capital by the most successful white farmers. Ultimately, the study finds that these socio-economic transformations ended up co-producing the drought by degrading the soil and vegetation, overexploiting the water resources, altering the microclimatic conditions, along with exacerbating the social vulnerability of less privileged inhabitants. Thus, Ladismith's case study demonstrates that the social production of water scarcity alone cannot fully explain the politics of drought and its injustice. Instead, capturing the power dynamics and the political changes that intersect with drought also requires retracing the agency of its non-human processes. Ultimately, without considering the agency of biophysical processes it would be difficult if not impossible to retrace the origins of a drought.

Conclusion: Unravelling the Socioenvironmental Genesis of Drought

Most natural scientists conceive humanity as a major force able to influence distinctive environmental processes and, in turn, exacerbate the occurrence of future droughts in many regions of the world. Concurrently, critical social sciences explain that any manifestation of drought is both a consequence and a trigger of unequal power dynamics. This chapter showed that although crucial, these conceptualizations alone are insufficient to apprehend the manner in which drought risk is produced and redistributed. These disciplines either merge questions of capitalism, power, class, race and gender, or overlook the distinctive biophysical processes that co-produce drought. By doing so, both natural and social sciences risk perpetuating the unethical divide between nature and society and in turn, hamper scientific understanding of future droughts and their unfolding. Conversely, this chapter engaged with both natural and social sciences understandings to consider what Moore (2016) would call *the web of life*, or otherwise the endless and complex entanglements of nature and society. In turn, this interdisciplinary perspective conceives droughts as dynamically produced by interwoven socioenvironmental processes. Such a perspective expands scientific investigations of drought by enabling finer understandings about *how* and *why* future droughts will unfold.

Note

1 This section employs the word *socionatural* instead of the more common socioenvironmental because the former is term that commonly used by critical social scientists conceptualize human interactions with ecology.

References

Acha, M. R. (2020). "Climate justice must be anti-patriarchal or it will not be systemic." In edited by Rocío Silva Santisteban *Indigenous Women & Climate Change*, 105–112, Chapter 6. https://iwgia.org/images/publications/new-publications/Indigenous_Women_and_Climate_Change_IWGIA.pdf#page=10

AghaKouchak, A., Mirchi, A., Madani, K., Di Baldassarre, G., Nazemi, A., Alborzi, A., Anjileli, H., Azarderakhsh, M., Chiang, F., Hassanzadeh, E. and Huning, L. S. (2021). "Anthropogenic drought: definition, challenges and opportunities." *Review of Geophysics*. 59, e2019RG000683. https://doi.org/10.1029/2019RG000683

Ahlers, R. (2010). "Fixing and nixing: the politics of water privatization." *Review of Radical Political Economics*, 42(2), 213–230.

Alzate González, L. D. and Peñaloza Lanza, R. A. (2019). "Day Zero: the role of social movements in the face Cape Town's water crisis" (Linnaeus University Master Thesis, 2019).

Bakker, K. J. (2000). "Privatizing water, producing scarcity: the Yorkshire drought of 1995." *Economic Geography*, 76(1), 4–27.

Bakker, K. J. (2003). "A political ecology of water privatization." *Studies in Political Economy*, 70(1) 35–58.
Bakker, K. J. and Bridge, G. (2006). "Material worlds? Resource geographies and the matter of nature." *Progress in Human Geography*, 30(1), 5–27.
Bhavani, P., Roy, P. S., Chakravarthi, V. and Kanawade, V. P. (2017). "Satellite remote sensing for monitoring agriculture growth and agricultural drought vulnerability using long-term (1982–2015) climate variability and socio-economic data set." *Proceedings of the National Academy of Sciences, India Section A: Physical Sciences*, 87(4), 733–750
Bingham, N. (2006). "Bees, butterflies, and bacteria: biotechnology and the politics of nonhuman friendship." *Environment and Planning A*, 38(3), 483–498.
Cook, B. I., Mankin, J. S. and Anchukaitis, K. J. (2018). "Climate change and drought: from past to future." *Current Climate Change Reports*, 4(2), 164–179.
de Freitas, G. N. (2020). "São Paulo drought: trends in stream flow and their relationship to climate and human-induced change in Cantareira watershed, Southeast Brazil." *Hydrology Research*, 51(4), 750–767.
Doshi, S. (2017). "Embodied urban political ecology: five propositions." *Area*, 49(1), 125–128.
Dumitraşcu, M., Mocanu, I., Mitrică, B., Dragotă, C., Grigorescu, I. and Dumitrică, C. (2018). "The assessment of socio-economic vulnerability to drought in Southern Romania (Oltenia Plain)." *International Journal of Disaster Risk Reduction*, 27, 142–154
Enqvist, J. P. and Ziervogel, G. (2019). "Water governance and justice in Cape Town: an overview." *Wiley Interdisciplinary Reviews: Water*, 6(4), e1354
Guerreiro, S. B., Dawson, R. J., Kilsby, C., Lewis, E. and Ford, A. (2018). "Future heat-waves, droughts and floods in 571 European cities." *Environmental Research Letters*, 13(3), 034009.
Hellberg, S. (2020). "Scarcity as a means of governing: challenging neoliberal hydromentality in the context of the South African drought." *Environment and Planning E: Nature and Space*, 3(1), 186–206.
Heynen, N., Kaika, M. and Swyngedouw, E. (2006). *In the nature of cities: urban political ecology and the politics of urban metabolism*, vol. 3. London and New York: Taylor & Francis.
IPCC. (2022). *Climate Change 2022: Impacts, Adaptation and Vulnerability. IPCC Sixth Assessment Report.* Intergovernmental Panel on Climate Change. [Accessed 05/05/2022]. Available at https://www.ipcc.ch/report/ar6/wg2/
Jaeger, W. K., Amos, A., Conklin, D. R., Langpap, C., Moore, K. and Plantinga, A. J. (2019). "Scope and limitations of drought management within complex human–natural systems." *Nature Sustainability*, 2(8), 710–717.
Kaika, M. (2003). "Constructing scarcity and sensationalising water politics: 170 days that shook Athens." *Antipode*, 35(5), 919–954.
Kaika, M. (2006). "The political ecology of water scarcity: the 1989–1991 Athenian drought." In Heynen, N., Kaika, M., and Swyngedouw, E. (eds.). *In the nature of cities the political ecology of water scarcity: the 1989–1991 Athenian drought*, vol. 3. London and New York: Taylor & Francis. pp. 172–187.
Kallis, G. (2008). "Droughts." *Annual Review of Environment and Resources*, 33, 85–118.
Lane, R. (1996). "Positivism, scientific realism and political science; Recent development in the philosophy of science." *Journal of Theoretical Politics*, 8(3), 361–382.
Lange, B., Holman, I. and Bloomfield, J. P. (2017). "A framework for a joint hydrometeorological-social analysis of drought." *Science of the Total Environment*, 578, 297–306.
Leng, G., Tang, Q. and Rayburg, S. (2015). "Climate change impacts on meteorological, agricultural and hydrological droughts in China." *Global Planetary Change*, 126, 23–34.
Lima Costa, C. D., Cavalcanti, I. and Hara, J. (2017). "On the posthuman." *Ilha do Desterro*, 70(2), 9–14.
Linton, J. and Budds, J. (2014). "The hydrosocial cycle: defining and mobilizing a relational-dialectical approach to water." *Geoforum*, 57, 170–180.
Lloyd-Hughes, B. (2014). "The impracticality of a universal drought definition." *Theoretical and Applied Climatology*, 117(3), 607–611.
Loftus, A. (2007). "Working the socio-natural relations of the urban waterscape in South Africa." *International Journal of Urban and Regional Research*, 31(1), 41–59.
Lucero, L. J., Harrison, J., Larmon, J., Nissen, Z. and Benson, E. (2017). "Prolonged droughts, short-term responses, and diaspora: the power of water and pilgrimage at the sacred cenotes of Cara Blanca, Belize." *Wiley Interdisciplinary Reviews: Water*, 4(4), e1148.
McLeman, R., Herold, S., Reljic, Z., Sawada, M. and McKenney, D. (2010). "GIS-based modeling of drought and historical population change on the Canadian Prairies." *Journal of Historical Geography*, 36(1), 43–56.

McNeeley, S. M. (2014). "A 'toad's eye' view of drought: regional socio-natural vulnerability and responses in 2002 in Northwest Colorado." *Regional Environmental Change*, 14(4), 1451–1461.
Mehta, L. (2003). "Contexts and constructions of water scarcity." *Economic and Political Weekly*, 5066–5072.
Mehta, L. (2010). "The social construction of scarcity: the case of water in western India." In Peet, R., Robbins, P., and Watts, M. (eds.) *Global political ecology*. London and New York: Routledge Taylor & Francis Group. pp. 385–400.
Meza, I., Siebert, S., Döll, P., Kusche, J., Herbert, C., Eyshi Rezaei, E., Nouri, H., Gerdener, H., Popat, E., Frischen, J. and Naumann, G. (2020). "Global-scale drought risk assessment for agricultural systems." *Natural Hazards & Earth System Sciences*, 20(2), 695–712.
Millington, N. (2018). "Producing water scarcity in São Paulo, Brazil: the 2014–2015 water crisis and the binding politics of infrastructure." *Political Geography*, 65, 26–34.
Mishra, A. K. and Singh, V. P. (2010). "A review of drought concepts." *Journal of Hydrology*, 391(1–2), 202–216.
Mollinga, P. P. (2008). "Water, politics and development: framing a political sociology of water resources management." *Water Alternatives*, 1(1), 7.
Moore, J. W. (2016). *Anthropocene or capitalocene?: Nature, history, and the crisis of capitalism*. Oakland, CA: PM Press.
Musemwa, M. (2009). "Coping with water scarcity: the social and environmental impact of the 1982–1992 droughts on Makokoba township, Bulawayo, Zimbabwe." In Locatelli, F., and Nugent, P (eds.) *African Cities: competing claims on urban spaces*. Brill, Leiden. pp. 157–185.
Nature. (2019). "End the drought in drought research." *Nature Editorial*, 573, 310. DOI: 10.1038/d41586-019-02782-3. [Accessed 16/11/2020]. Available at https://www.nature.com/articles/d41586-019-02782-3
Omer, A., Zhuguo, M., Zheng, Z. and Saleem, F. (2020). "Natural and anthropogenic influences on the recent droughts in Yellow River Basin, China." *Science of the Total Environment*, 704, 135428.
Otero, I., Kallis, G., Aguilar, R. and Ruiz, V. (2011). "Water scarcity, social power and the production of an elite suburb: the political ecology of water in Matadepera, Catalonia." *Ecological Economics*, 70(7), 1297–1308.
Panelli, R. (2010). "More-than-human social geographies: posthuman and other possibilities." *Progress in Human Geography*, 34(1), 79–87.
Power, E. (2005). "Human–nature relations in suburban gardens." *Australian Geographer*, 36(1), 39–53.
Rockström, J., Falkenmark, M., Allan, T., Folke, C., Gordon, L., Jägerskog, A., Kummu, M., Lannerstad, M., Meybeck, M. and Molden, D. (2014)."The unfolding water drama in the Anthropocene: towards a resilience-based perspective on water for global sustainability." *Ecohydrology*, 7, 1249–1261.
Salvador, C., Nieto, R., Linares, C., Díaz, J., Alves, C. A. and Gimeno, L. (2021). "Drought effects on specific-cause mortality in Lisbon from 1983 to 2016: risks assessment by gender and age groups." *Science of The Total Environment*, 751, 142332.
Sanderson, M. R. (2018). "Everything flows… unevenly: social stratification in coupled socio-ecological systems." *Current Opinion in Environmental Sustainability*, 33, 51–57
Savelli, E., Rusca, M., Cloke, H. and Di Baldassarre, G. (2021). "Don't blame the rain: social power and the 2015–2017 drought in Cape Town." *Journal of Hydrology*, 594, 125953.
Savelli, E., Rusca, M., Cloke, H. and Di Baldassarre, G. (2022a). "Drought and society: scientific progress, blind spots, and future prospects." *Wiley Interdisciplinary Reviews: Climate Change*, e761.
Savelli, E., Rusca, M., Cloke, H., Flügel, T. J., Karriem, A. and Di Baldassarre, G. (2022b). "All dried up: the materiality of drought in Ladismith, South Africa." *Environment and Planning E: Nature and Space*, 25148486221126617.
Savenije, H. H., Hoekstra, A. Y. and van der Zaag, P. (2014). "Evolving water science in the Anthropocene." *Hydrology and Earth System Sciences*, 18(1), 319–332.
Scheba, S. and Millington, N. (2018). "Crisis temporalities: intersections between infrastructure and inequality in the Cape Town Water Crisis." *International Journal of Urban & Regional Research*. [Accessed on 05/05/2022]. Available at https://www.ijurr.org/spotlight-on/parched-cities-parched-citizens/crisis-temporalities-intersections-between-infrastructure-and-inequality-in-the-cape-town-water-crisis/
Sear, D. A., Allen, M. S., Hassall, J. D., Maloney, A. E., Langdon, P. G., Morrison, A. E., Henderson, A. C., Mackay, H., Croudace, I. W., Clarke, C. and Sachs, J. P. (2020). "Human settlement of East Polynesia earlier, incremental, and coincident with prolonged South Pacific drought." *Proceedings of the National Academy of Sciences*, 117(16), 8813–8819.

Spinoni, J., Barbosa, P., De Jager, A., McCormick, N., Naumann, G., Vogt, J. V., Magni, D., Masante, D. and Mazzeschi, M. (2019). "A new global database of meteorological drought events from 1951 to 2016." *Journal of Hydrology: Regional Studies*, 22, 100593.
Sultana, F. (2011). "Suffering for water, suffering from water: emotional geographies of resource access, control and conflict." *Geoforum*, 42(2), 163–172.
Swyngedouw, E. (1997). "Power, nature, and the city. The conquest of water and the political ecology of urbanization in Guayaquil, Ecuador: 1880–1990." *Environment and Planning A*, 29(2), 311–332.
Swyngedouw, E. (2009). "The political economy and political ecology of the hydro-social cycle." *Journal of Contemporary Water Research & Education*, 142(1), 56–60.
Swyngedouw, E. (2019). "The perverse lure of autocratic postdemocracy." *South Atlantic Quarterly*, 118(2), 267–286.
Taufik, M., Minasny, B., McBratney, A. B., Van Dam, J. C., Jones, P. D. and Van Lanen, H. A. J. (2020). "Human-induced changes in Indonesian peatlands increase drought severity." *Environmental Research Letters*, 15(8), 084013.
Taylor, V., Chappells, H., Medd, W. and Trentmann, F. (2009). "Drought is normal: the socio-technical evolution of drought and water demand in England and Wales, 1893–2006." *Journal of Historical Geography*, 35(3), 568–591.
Tiwale, S., Rusca, M. and Zwarteveen, M. (2018). "The power of pipes: mapping urban water inequities through the material properties of networked water infrastructures – the case of Lilongwe, Malawi." *Water Alternatives*, 11(2), 314–335.
Truelove, Y. (2011). "(Re-) Conceptualizing water inequality in Delhi, India through a feminist political ecology framework." *Geoforum*, *42*(2), 143–152.
Truelove, Y. (2019). "Rethinking water insecurity, inequality and infrastructure through an embodied urban political ecology." *Wiley Interdisciplinary Reviews: Water*, 6 (3), e1342.
Tsing, A. (2012). "Unruly edges: mushrooms as companion species, for Donna Haraway." *Environmental Humanities*, 1(1), 141–154.
UNDRR. (2021). "Global Assessment Report on Disaster Risk Reduction: Special Report on Drought 2021." *United Nation Office for Disaster Risk Reduction Official report*, SBN/ISSN/DOI 9789212320274. [Accessed on 22/07/2021]. Available from https://www.undrr.org/publication/gar-special-report-drought-2021
Van Dijk, A. I., Beck, H. E., Crosbie, R. S., de Jeu, R. A., Liu, Y. Y., Podger, G. M., Timbal, B. and Viney, N. R. (2013). "The millennium drought in southeast Australia (2001–2009): natural and human causes and implications for water resources, ecosystems, economy, and society." *Water Resources Research*, 49(2), 1040–1057.
Van Lanen, H. N., Wanders, L., Tallaksen, A. and Van Loon. (2013). "Hydrological drought across the world: impact of climate and physical catchment structure." *Hydrology Earth System Sciences*, 17, 1715–1732.
Van Loon, A. F., Stahl, K., Di Baldassarre, G., Clark, J., Rangecroft, S., Wanders, N., Gleeson, T., Van Dijk, A. I., Tallaksen, L. M., Hannaford, J. and Uijlenhoet, R. (2016). "Drought in a human-modified world: reframing drought definitions, understanding, and analysis approaches." *Hydrology and Earth System Sciences*, 20(9), 3631–3650.
Wada, Y., Van Beek, L. P., Wanders, N. and Bierkens, M. F. (2013). "Human water consumption intensifies hydrological drought worldwide." *Environmental Research Letters*, 8(3), 034036.
Walker, G. (2014). "Water scarcity in England and Wales as a failure of (meta)governance." *Water Alternatives*, 7(2), 388–413.
Walker, P. A. (2005). "Political ecology: where is the ecology?" *Progress in Human Geography*, *29*(1): 73–82.
Wang, Y., Duan, L., Liu, T., Li, J. and Feng, P. (2020). "A Non-stationary Standardized Streamflow index for hydrological drought using climate and human-induced indices as covariates." *Science of The Total Environment*, 699, 134278.
Wilhite, D. A. (1993). "The enigma of drought." In Wilhite, D. A. (ed.) *Drought assessment, management, and planning: theory and case studies*. Springer, Boston, MA. pp. 3–15
Wilhite, D. A. (2000). "Chapter 1 drought as a natural hazard: concepts and definitions." In Wilhite, D. A. (ed.). *Drought: a global assessment*, vol. I. Routledge, London. pp. 3–18.

26
WATER-ENERGY NEXUS

Aiko Endo

Introduction

Academia, policy-makers, and industries have recognized the interrelationship, interdependency and tradeoff relationships of water, energy, and food resource systems and sectors. The water-energy-food (WEF) nexus has been acknowledged as one of the emerging discourses in the international academic community. International organizations have particularly begun to recognize a set of water and energy issues as the water-energy nexus concept, such as the United Nations (UN) International Decade for Action "Water for Life" (2005–2015). A recent review on the WEF nexus found that the focus on the water-energy nexus within WEF nexus studies has increased over the last three years between 2018 and 2020 (Endo et al., 2020). Triggering events and several driving factors have enhanced this trend, such as: (1) the occurrence of extreme climate, weather events, and disasters, including droughts, heatwaves, freezes, and floods; (2) social changes, including migration and urbanization; (3) technological breakthroughs, including hydraulic fracking for shale gas development; (4) enhancements to regional energy and economic cooperation through the integration of energy and water security; (5) large-scale infrastructure developments for water and energy, such as hydropower in developing countries initiated by the donor-driven top-down approach; (6) identification of tradeoffs among water, energy, and low-carbon policies at different levels and scales; and (7) targeting multiple global goals, such as Sustainable Development Goals (SDGs). These events and factors are intricately interrelated, have accumulated from the past to the present, and will continue into the future with uncertainty.

For example, the amount of energy used in the water sector, such as desalination, wastewater treatment, large-scale water transfer, water supply, reuse, and distribution, is projected to more than double from the present to 2040 (IEA, 2020a). Desalination and wastewater treatment are energy-intensive water supply sources; desalination will have the highest increase in electricity consumption in the water sector from 5% in 2014 to 15% in 2040, followed by large-scale water transfer and wastewater treatment (IEA, 2020a). The United States consumes more electricity in the water sector than any other region or country, and China is a close second (IEA, 2016).

On the other hand, global water withdrawal by the energy sector, such as that used by power plants, is predicted to decrease from 338 billion cubic meters (bcm) in 2016 to 272 bcm in 2030 due to a shift to less water-intensive renewable energy sources, such as solar photovoltaic (PV)

DOI: 10.4324/9781003095545-30

and wind, and away from coal power generation. While water consumption will increase by more than 50% from 47 bcm in 2016 to 72 bcm in 2030, owing to a shift to water-intensive, low-carbon technologies, such as concentrating solar power (CSP), carbon capture sequestration (CCS), or nuclear power enhanced by a low-carbon policy (IEA, 2020a).

The followings are significant concerns and challenges in the water-energy nexus, such as improving water and energy efficiency, developing technology for cooling power plants, treating desalination and wastewater using the embedded energy in waste and wastewater, together with achieving low-carbon emissions, promoting cost-effectiveness, reducing environmental impacts on ecosystems, and synergizing energy and water policies.

The purpose of this chapter is (1) to understand academic trends in the water-energy nexus from the perspective of these triggering events and driving factors and (2) to discuss the contributions and potential roles of the social sciences in the water-energy nexus.

Triggering Events and Driving Factors

Extreme climate and weather events

Extreme heatwaves occurred in France in 2003, 2006, and 2009, which led to increased power demand. As a result, many nuclear power plants had to reduce their output or shut down to avoid violating thermal pollution limits set by the European Fish Directive. That heatwaves stretched across Europe: in Germany in 2003 and 2006, in Spain in 2006, and in Switzerland in 2015, affecting electric power generation (Gjorgiev & Sansavini, 2018; Webber, 2017). In 2007, 2008, and 2012, the United States experienced severe drought, which resulted in record electricity demands. Although power plants in Texas were sufficiently flexible at the plant level to adapt by switching to less water-intensive technologies (Scanlon et al., 2013), limited water availability caused by severe drought revealed the vulnerabilities associated with power generation, inducing reductions in electricity production or shutdowns to protect the aquatic environment due to the Clean Water Act. In China, severe droughts in 2010 and 2011 led to vast water shortages and electricity generation losses from hydro-powered sources (Al-Saidi & Elagib, 2017). This has enhanced studies on the water-energy nexus, i.e., analysing water withdrawal and consumption for thermal electric power generation; and assessing different sources of water (i.e., surface or groundwater, conventional or non-conventional water), sources of energy (i.e., coal, natural gas, hydroelectric, renewables including wind, solar PV, CSP, nuclear, or geothermal), cooling technologies (i.e., once-through, closed-cycle, or dry-air cooling), electricity storage, pricing, and policy constraints (Gjorgiev & Sansavini, 2018; Scanlon et al., 2013; Webber, 2017). Efforts to increase the use of non-freshwater sources for cooling, especially in coal-fired power plants, have been observed in China and the United States (IEA, 2016).

Migration and urbanization

The world population is expected to increase from 7.8 billion in 2020 to 9.7 billion in 2050 and peak at nearly 11 billion ca. 2100 (UN, 2019). Fifty-five per cent of the world's population lived in urban areas in 2018. This proportion is expected to increase to 68% by 2050 (UN, 2018). Urban areas consume between 67% and 76% of global energy and generate approximately three-quarters of global carbon emissions (IPCC, 2014), where a large amount of water is consumed for energy supply. Sovacool and Sovacool (2009) discussed the impacts that droughts and water shortages have on thermoelectric capacity additions by the electric industry, highlighting population growth and demographic shifts, as the population in arid areas, such as the southwestern United States, is likely to continue to increase (US DOE, 2014). The urban water-energy nexus in the metabolism framework is a popular topic in the water-energy nexus academic community, as water and energy are the main obstacles to urban development (Chen & Chen, 2016; Wang & Chen, 2016). A recent study shows that population growth is not the primary driving force of water consumption in China; rather, changes in energy intensity translate into more significant influences on water consumption (Duan & Chen, 2020).

Technological breakthrough

Previous and emerging technological innovations and industrial revolutions in the energy sector have positively and negatively influenced the energy market, social development, and environmental sustainability. For example, recent technological advancements in combined horizontal well drilling and hydraulic fracturing allowed the shale gas industry in Texas, United States, to produce commercial volumes by the year 2000. The top seven shale deposit countries in 2013 were China, Argentina, Algeria, the United States, Canada, and Mexico. Shale gas production is projected to increase by 2040 in the United States (USEIA, 2013). The majority of water-energy nexus articles on shale gas development are derived from American production experience on the following topics: (1) analysing water resource sustainability using quantitative modelling and simulation; (2) assessing environmental impacts on the aquifer; (3) examining wastewater management, including the reuse, export, and import of wastewater and injection disposal; and (4) discussions on the lack of integrated water and energy policies and regulations. Wastewater management for shale gas development has experienced tradeoffs in environmental impacts and social equity and environmental justice between on-site wastewater treatment areas and waste-transported areas, and the studies based on American production experience can provide a reference for shale gas development in other potential areas (Rahm et al., 2013).

Regional energy cooperation addressing water-energy nexus security

The primary purpose of regional energy cooperation is to ensure regional energy security by creating an internal competitive energy market, developing technologies for energy efficiency and low-carbon emissions, and creating job opportunities. The European Union (EU) Energy Union (EU, 2019, 2020a), formulated in 2015, the Greater Mekong Subregion Economic Cooperation Program of the Asian Development Bank (ADB) in Southeast Asia, established in 1992 (ADB, 2017), the Krabi Initiative of the Association of Southeast Asian Nations (ASEAN), started in 2010 (ASEAN, 2020), and the EU4Energy Programme, concentrating on 11 countries in the Caucasus, Eastern Europe, and Central Asia, initiated in 2016 (EU, 2020b), have incorporated the water-energy nexus framework.

Mabhaudhi et al. (2016) discussed the water-energy nexus challenges faced by the Southern African Power Pool (SAPP), which is the regional cooperation body established in 1955, consisting of 12 Southern African Development Community (SADC) member countries in the energy sector. The SAPP member countries are interconnected through a grid to export and import power, providing relief to possible spatial and temporal shortfalls in the electrical capacity by aiming at providing affordable and environmentally friendly energy and increasing accessibility to rural communities (SAPP, 2021). In the background that the demand for power is expected to increase by 44% from 2014 to 2025, where coal, with high carbon emissions, generates 75% of the power in the SADC, and 85% of the SADC region's water resources are transboundary, the SAPP serves as a buffering mechanism for climate-induced river basin-scale electricity insecurity (Conway et al., 2015; Mabhaudhi et al., 2016). Although South Africa has already examined the water-energy nexus at the national level, the SAPP requires a transboundary water-energy nexus policy, studies on enhancing institutional arrangements for watercourses, and development of the modelling capacity, along with addressing data scarcity.

Donor-driven large-scale infrastructures development

Currently, 789 million people live without access to electricity, and 3 billion people, especially rural populations, rely heavily on wood, coal, charcoal, or animal waste for cooking and heating (UN, 2021). The UN SDG 7 aims to ensure access to affordable, reliable, sustainable, and modern energy, specifically focusing on improving access to clean and safe cooking fuels and technologies, though progress has been made in energy access, energy efficiency, and renewable energy (UN, 2021).

In the Mekong region, where millions still live without basic amenities, such as clean drinking water and electricity (MRC, 2021), the discourses in both the Integrated Water Resources Management and WEF nexus approaches have been applied in politics, research, funding, and institutions (Dore et al., 2012) to address those issues. As of 2012, activities in the MRC were supported by a group of 18 donors (World Bank, 2012). The following current bilateral and multilateral cooperation and development projects are ongoing as of May 2021: the Mekong-IWRM Project (World Bank), the ASEAN-Mekong Basin Development Cooperation, Australia-Mekong Water Facility (Australia), the Great Mekong Subregion (GMS) programme (ADB), the Network of Asian River Basin Organizations (ADB), the Asia-Europe Meeting (EU), the Lancang-Mekong Cooperation (China), and Japan-Mekong Cooperation.

Large-scale water infrastructure development in developing countries has produced tradeoffs in enhancing economic development based on national policies. Käkönen & Kaisti (2012) argued that the launch of World Bank projects had dual objectives, targeting a win-win scenario in Lao PDR: a large-scale hydropower development project was promoted as an on-grid solution, while a rural electrification project, known as the solar home system (SHS), was promoted as an off-grid solution for poverty alleviation and expansion of low-carbon renewable energy. The authors revealed that the project created winners and losers as well as unsolved challenges and uncertainties for the sustainability of the proposed alternative livelihoods, including fisheries in resettled villages. Most of the electricity produced has been exported to neighbouring countries such as Thailand. Additionally, the implementation of the SHS did not consider the local culture and practices. The Siemenpuu Foundation (2021) focuses on sustainable energy issues, including the impacts that hydropower projects have on ecosystems and local livelihoods in the Mekong region from climate and energy justice perspectives.

Tradeoffs and synergies among water, energy, and low-carbon policies

The energy sector is a key contributor to climate change, accounting for more than two-thirds of global greenhouse gas (GHG) emissions (IEA, 2020b). As electricity production increased by 7.6 times between 1990 and 2012, with coal contributing to approximately 80% of electricity generation, each country must regulate its climate targets, water sustainability, and environmental protection at different levels to meet UN SDG 13 (Climate action) globally.

In the United States, Webber (2017) pointed out that the amended 1972 Clean Water Act, which regulates thermal discharge from point sources to reduce environmental concerns, including protections for fish and aquatic life, has accelerated the transformation of cooling technology in thermoelectricity power plants from open-loop, with high water withdrawal, to closed-loop, with a cooling tower, which requires less water withdrawal. DeNooyer et al. (2016) noted that a shift from open- to closed-loop cooling could decrease withdrawal with the tradeoff of increased water consumption. Moreover, the authors revealed that a shift from coal to nuclear, thermal solar, and clean coal energy, following the U.S. Environmental Protection Agency's Clean Power Plan and the proposed Low-Carbon Portfolio Standard in Illinois, could lead to a net increase in water requirements. Furthermore, CCS technology for climate change mitigation will be able to eliminate GHG emissions from power plants in addition to requiring more water and energy (Feng et al., 2014).

In 2020, China accounted for one-third of global CO_2 emissions and announced to attempt achieving carbon neutrality by 2060 (IEA, 2021). Feng et al. (2014) analysed the total life-cycle CO_2 emissions of eight electricity generation sources, including coal, oil, natural gas, biomass, solar PV, wind, nuclear, and hydropower in China between 2000 and 2010, as well as the total life-cycle water consumption for these eight sources. The results showed that increasing the share of renewable energy in China's electricity fuel mix could not only curb CO_2 emissions but also reduce pressure on the local environment, especially in already water-scarce regions. Specifically, wind power electricity generation, which can create synergy, is estimated to save more than 79% of life-cycle CO_2 emissions and consume up to 83% less water than fossil-fuel-based electricity generation, such as coal.

Previous studies have also identified challenges to existing water-energy nexus policies at different geographical locations, levels, and sectors. The primary challenges are the lack of policies to (1) support the incorporation of the nexus approach into decision-making processes; (2) accommodate tradeoffs between water, energy, and low-carbon policies and regulations; (3) collect, access, and integrate data to analyse the water-energy nexus in quantitative manners for modelling and simulation, (4) consider regional differences, local cultures, and behavioural changes related to human water and energy use; and (5) incentivize private investment and the utilization of various technological options in developing countries (Craig, 2013; Larsen & Drews, 2019; Mabhaudhi et al., 2016; Nair et al., 2014; Rahm et al., 2013).

Contributions of Social Sciences to Water-Energy Nexus

Water-energy nexus studies are at present dominated and accelerated by engineering disciplines (e.g., life-cycle engineering, process systems engineering, and biological engineering), and emerging engineering technologies (e.g., membrane engineering) by developing low-carbon, low-cost, low-environmental impact, and high-energy and water-efficient technologies. Meanwhile, natural science developed quantitative modelling and future scenarios incorporating social-economic data such as GDP per capita and population from the view of hydrological discipline. The natural science community, especially the Earth science community addressing the planetary system, initially promoted the WEF nexus discourse to tackle global-scale environmental problems that have become more interrelated, complicated, and uncertain by proactively inviting social scientists (Leck et al., 2015) and by incorporating transdisciplinary concept.

Socio-economic data integration

The concepts, methods, and data for the social sciences, such as economics, finance, environmental law, and political science (including international relations and public policies), are solely or conjointly applied in water-energy nexus studies. As examples of techno-economic optimization models, previous studies have used different technologies to apply capital expenditure and operational expenditure for desalination (Muhammad Wakil et al., 2017). A mixed-integer linear fractional programming model was developed to maximize profit for the shale gas production supply chain network (Gao et al., 2015). A multi-period mixed-integer nonlinear programme was applied to attain the maximum annual profit of an entire system mixing solar energy, thermal storage, and fossil fuel for shale gas production (Al-Aboosi et al., 2018). An optimal demand-side management model was developed for co-optimizing water-energy distribution networks (Li et al., 2019). Please refer to Vakilifard et al. (2018) for more information on the water-energy nexus optimization model, which considers economic constraints.

As for other techno-economic approaches, previous studies have performed economic analyses of retrofitting in different cooling technologies in power plants (DeNooyer et al., 2016). Discourses on the circular economy have been used to assess different biotechnologies for energy recovery in wastewater treatment (Puyol et al., 2017). Furthermore, utility-scale solar energy has been discussed from socio-economic and environmental perspectives (Hernandez et al., 2014).

Besides techno-economic analysis, Zhang and Anadon (2014) examined China's domestic inter-provincial virtual water trade using input-output analysis. Duan and Chen (2020) applied a structural decomposition analysis based on input-output analysis to illustrate energy intensity by sector, including water production and supply, between 1990 and 2014 in China.

Behavioural and cultural studies

From the point of water-use perspective, Wada et al. (2016) discussed management options of the water-energy-climate nexus and revealed that very few assessments of global water use for model validations had been built on the shared socio-economic pathways (SSP); however, future scenarios of global water use will be influenced by social, economic and political factors. The authors summarized three state-of-the-art models (H08, PCR-GLOBWB, WaterGAP) in water use, including domestic, industrial including energy and manufacturing, and agriculture from the point of the SSP such as population, GDP, and water withdrawal. The authors also assumed future scenarios with three SSP storylines using hydro-economic classification at the country level. They discovered that all projections indicated significant increases in future domestic use and highlighted that future sustainability of water use accounts for the nature of human water use behaviour envisaged population growth and socio-economic variables despite data limitations. The authors also suggested that scenario development should be derived from expert and/or stakeholder consultation.

Duan and Chen (2020) analysed water and energy consumption in China under the water-energy-nexus framework consisting of three tiers, consuming fresh water and primary energy sources withdrawn from the environment, intermediate consumers in 123 industrial sectors, and final consumers such as households and government along with their purchasing and consuming behaviours. The authors discovered that the household is the greatest water consumer in China. Therefore, reducing the water consumption in households relies on changes in consumption habits and propaganda for an ecofriendly lifestyle.

Nair et al. (2014) reviewed water-energy-GHG nexus studies in the urban context from energy for water perspective based on the previous studies, which indicate that (1) water conservation at the household level is the most effective way to reduce carbon emissions originating from the water system, (2) water heating, laundry, and shower use are the most energy-intensive water uses at the household level, (3) energy-consumption in water-end-use is heavily dependent on the behavioural and cultural factors and therefore, (4) behavioural change plays a vital role in reducing water-related energy use and carbon emissions by at least 50%. Moreover, three previous studies found that studying for motivating behavioural and cultural changes at the household level is one of the keys to reducing water and energy use and GHG.

Multi-level and multi-scale difference between water and energy

Scott et al. (2011) point out that researchers frequently address national and global demand for resources and rarely pay adequate attention to local-scale consequences and impacts on environmental quality and human health. Local decision-makers may also have limited capacity or authority to address such impacts when energy development is pursued by powerful or global commercial interests. The global and national-level demands for energy resources may have local impacts on water resources and quality of life, and the process of energy-development permitting that is administered by federal or state agencies means that local resources may not be managed by local authorities. These multi-level and multi-scale challenges partially derive from the dissimilarities between the water and energy resources and sectors: (1) energy is a transformable and more transportable resource than water; therefore, (2) the interdependencies among source regions and demand sites for energy are very different compared to source and demand links for water (Scott et al., 2011); (3) the energy sector has regulated international markets for an extended period while water is a more recent initiative; (4) energy data is available for most countries worldwide in various formats, whereas there are considerably fewer data available on water; (5) there are a greater number of energy businesses involved with larger financial industries than that of water; therefore, these differences influence policies and decision-making (Walsh et al., 2015). To implement a multi-tiered intuitional arrangement, identifying, and filling scaler-related gaps between various kinds of water and energy resource systems and sectors should be studied from a socio-economic perspective.

Potential lost economy effect

Puls and Sanders (2017) discussed the energy boom and bust cycles created by technological development in hydraulic fracturing of unconventional gas production by focusing on sustainable development in communities. Macke and Gardner (2012) divided the community boom-bust cycles into three phases, preboom, boom phase, and boom contraction. The process of the cycles started from a rapid rise in population and job opportunities, increased cost of goods and land, and insufficient infrastructure in the community. The public and private agents make long-term investments in infrastructure such as water, sewer, roads, housing, schools, hospitals, etc. When market glut and over-exploitation of the limited local natural resources result in a bust, the public and private agents have to manage the aftermath of the boom. Furthermore, the communities will encounter challenges called the lost economy effect, such as carrying this over-investment cost in the form of higher taxes as well as retraining a labour force for new opportunities and potentially increased crime and health crisis.

In the case of shale gas development boom, when a drilling boom begins, (1) there is a rush to hold leases for mineral rights, (2) sweet spots where production is highest are identified, (3) production rises, but the production per well declines, (4) more wells are drilled, and lateral lengths may be extended to maintain overall production, (5) the area is shifted outside the sweet spots where the production per well reduces, and (6) overall production of the area begins to decline with increased drilling. In addition to reductions in the price of gas, drought or lack of water resources, and other environmental, economic, and social factors, there would be a bust following the boom. Although it is true that oil and gas resource extraction activities use less than 1% of all total fresh water consumption at a global, national, and regional level; however, it may not be true at the local level as most discussions do not evaluate the scaler gaps (Puls & Sanders 2017). The adverse impacts of hydraulic fracturing on local communities and people should be studied by evaluating the costs of water, obtaining a water permit, and wastewater treatment and disposal, along with the duration and expansion of energy boom-bust cycles in the process of planning for building socio-economic and environmentally sustainable communities.

Conclusion

Several triggering events since the beginning of the 2000s have enhanced the water-energy nexus movement, and engineering and natural science studies have currently led and accelerated water-energy nexus studies incorporating social and economic data. Social science, including data, methods, concepts, and theory, has addressed the water-energy nexus by applying disciplines such as economics, finance, environmental law and policy, political economy, rural development, international relations, public policies, and environmental or energy justice into the nexus approach. However, taking interdisciplinary and transdisciplinary approaches such as collecting or creating socio-economic data and indicators to integrate them into the process of quantitative models and simulation, and multiple future scenarios cooperating with diverse groups of stakeholders, and institutionalizing them into decision- and policy-making systems remain a challenge to address the complexity and uncertainty of the linkages among water and energy resources and sectors in the human-nature system in different scales and levels. Further highlighting social science perspectives of the water-energy nexus will contribute to promoting inter-and-transdisciplinary approaches in the context of the WEF nexus.

References

ADB. (2017). "Water–Energy Nexus in the People's Republic of China and Emerging Issues." Asian Development Bank Report. [Accessed 10 February 2023]. Available at https://www.adb.org/sites/default/files/publication/384291/water-energy-nexus-prc.pdf

Al-Aboosi, F. Y., Mahmoud, M. and El-Halwagi, M. M. (2018). "An integrated approach to water-energy nexus in shale-gas production." *Processes* 6(5), 52.

Al-Saidi, M. and Elagib, N. A. (2017). "Towards understanding the integrative approach of the water, energy and food nexus." *Science of the Total Environment* 574, 1131–1139.

ASEAN. (2020). "ASEAN STI from Krabi Initiative." [Accessed: 25 January 2021]. Available at https://www.unescap.org/sites/default/files/1Opportunities%20for%20ASEAN%20Cooperation%20in%20the%20Water-Energy-Food%20Nexus%20by%20STI.pdf

Chen, S. and Chen, B. (2016). "Urban energy–water nexus: A network perspective." *Applied Energy* 184, 905–914.

Conway, D., van Garderen, E. A., Deryng, D., Dorling, S., Krueger, T., Landman, W., Lankford, B., Lebek, K., Osborn, T., Ringler, C., Thurl, J. T. J., Zhu, T. J. and Dalin, C. (2015). "Climate and Southern Africa's water-energy-food nexus." *Nature Climate Change* 5, 837–846.

Craig, R. K. (2013). "Hydraulic fracturing (fracking), federalism, and the water-energy nexus." *Idaho Law Review* 49, 241–264.

DeNooyer, T. A., Peschel, J. M., Zhang, Z. and Stillwell, A. S. (2016). "Integrating water resources and power generation: The energy–water nexus in Illinois." *Applied Energy* 162, 363–371.

Dore, J., Lebel, L. and Molle, F. A. (2012). "Framework for analysing transboundary water governance complexes, illustrated in the Mekong region." *Journal of Hydrology* 466–467, 23–36.

Duan, C. and Chen, B. (2020). "Driving factors of water-energy nexus in China." *Applied Energy* 257, 113984.

Endo, A., Yamada, M., Miyashita, Y., Sugimoto, R., Ishii, A., Nishijima, J., Fujii, M., Kato, T., Hamamoto, H., Kimura, M., Kumazawa, T. and Qi, J. (2020). "Dynamics of water–energy–food nexus methodology, methods, and tools." *Current Opinion in Environmental Science and Health* 13, 46–60.

EU. (2019). *Water-Energy Nexus in Europe*. EU, Luxembourg.
EU. (2020a). "Report on the State of the Energy Union pursuant to Regulation (EU) 2018/1999 on Governance of the Energy Union and Climate Action." [Accessed 24 January 2021] Available at https://eurlex.europa.eu/legalcontent/EN/TXT/?qid=1602743359876&uri=COM:2020:950:FIN
EU. (2020b). "EU4Energy Eastern Europe Energy Efficiency and Environment Partnership." [Accessed 25 January 2021]. Available at https://www.euneighbours.eu/en/east/stay-informed/projects/eu4energy-eastern-europe-energy-efficiency-and-environment-partnership
Feng, K., Hubacek, K., Yim Ling Siu, Y. L. and Li, X. (2014). "The energy and water nexus in Chinese electricity production: A hybrid life cycle analysis." *Renewable and Sustainable Energy Reviews* 39, 342–355.
Gao, J., Fengqi and You, F. (2015). "Optimal design and operations of supply chain networks for water management in shale gas production: MILFP model and algorithms for the water-energy nexus." *Process Systems Engineering* 61(4), 1184–1208.
Gjorgiev, B. and Sansavini, G. (2018). "Electrical power generation under policy constrained water-energy nexus." *Applied Energy* 210, 568–579.
Hernandez, R. R., Easter, S. B., Murphy-Mariscal, M. L., Maestre, F. T., Tavassoli, M., Allen, E. B., Barrows, C. W., Belnap, J., Ochoa-Hueso, R., Ravi, S. and Allen, M. F. (2014). "Environmental impacts of utility-scale solar energy." *Renewable and Sustainable Energy Reviews* 29, 766–779.
IEA. (2016). "*Water Energy Nexus: Excerpt from the World Energy Outlook*". OECD/IEA, Paris.
IEA. (2020a). "Introduction to the water-energy nexus." [Accessed 23 January 2021] Available at https://www.iea.org/articles/introduction-to-the-water-energy-nexus
IEA. (2020b). "Climate Change." [Accessed 22 January 2021] Available at https://www.iea.org/topics/climate-change
IEA. (2021). "*An Energy Sector Roadmap to Carbon Neutrality in China*." IEA, Paris.
IPCC. (2014). "Climate Change 2014: Mitigation of Climate Change: Working Group III contribution to the Fifth Assessment Report of the Intergovernmental Panel on Climate Change." Chap 12, Geneva, Switzerland. pp. 923–1000.
Käkönen, M. and Kaisti, H. (2012). "The World Bank, Laos and renewable energy revolution in the making: Challenges in alleviating poverty and mitigating climate change." *Forum for Development Studies* 39(2), 159–184.
Larsen, M. A. D. and Drews, M. (2019). "Water use in electricity generation for water-energy nexus analyses: The European case." *Science of the Total Environment* 651, 2044–2058.
Leck, H., Conway, D., Bradshaw, M. and Rees, J. (2015). "Tracing the water–energy–food nexus: Description, theory and practice." *Geography Compass* 9(8), 445–460.
Li, Q., Yu, S., Al-Sumaiti, A. S. and Turitsyn, K. (2019). "Micro water–energy nexus: Optimal demand-side management and quasi-convex hull relaxation." *IEEE Transactions on Control of Network Systems* 6(4), 1313–1322.
Mabhaudhi, T., Mpandeli, S., Madhlopa, A., Modi, A. T., Backeberg, G. and Nhamo, L. (2016). "Southern Africa's water–energy nexus: Towards regional integration and development." *Water* 8(6), 235.
Macke, D. and Gardner, D. (2012). *Policy Education White Paper – Western North Dakota Energy Project*. Paper presented to Vision West Consortium. RUPRI Center for Rural Entrepreneurship, Lincoln, NE.
Mekong River Commission (MRC). (2021). "People." [Accessed 26 January 2021] Available at https://www.mrcmekong.org/about/mekong-basin/people
Muhammad Wakil, S., Muhammad, B., Li, A. and Kim Choon, N. (2017). "Energy-water-environment nexus underpinning future desalination sustainability." *Desalination* 413, 52–64.
Nair, S., George, B., Malano, H. M., Arora, M. and Bandara Nawarathna, B. (2014). "Water–energy–greenhouse gas nexus of urban water systems: Review of concepts, state-of-art and methods." *Resources, Conservation and Recycling* 89, 1–10.
Puls, R. and Sanders, L. D. (2017). "Water-energy nexus and environmental aspects of oil and gas production." In *Competition for Water Resources*. Pp. 160–177. https://www.sciencedirect.com/book/9780128032374/competition-for-water-resources#book-description
Puyol, D., Batstone, D. J., Hülsen, T., Astals, S., Peces, M. and Krömer, J. O. (2017). "Resource recovery from wastewater by biological technologies: Opportunities, challenges, and prospects." *Frontiers in Microbiology* 7, 2106.
Rahm, B. G., Bates, J. T., Bertoia, L. R., Galford, A. E., Yoxtheimer, D. A. and Riha, S. J. (2013). "Wastewater management and Marcellus shale gas development: Trends, drivers, and planning implications." *Journal of Environmental Management* 120, 105–113.

SAPP. (2021). "About SAPP." [Accessed 24 January 2021]. Available at http://www.sapp.co.zw/

Scanlon, B. R., Duncan, I. and Reedy, R. C. (2013). "Drought and the water–energy nexus in Texas." *Environmental Research Letters* 8(4), 045033.

Scott, C. A., Pierce, S. A., Pasqualetti, M. J., Jones, A. L., Montz, B. E. and Hoover, J. H. (2011). "Policy and institutional dimensions of the water–energy nexus." *Energy Policy* 39, 6622–6630.

Siemenpuu Foundation. (2021). "Climate and Energy Justice." [Accessed 26 January 2021]. Available at https://www.siemenpuu.org/en/funding/energy

Sovacool, B. K. and Sovacool, K. E. (2009). "Identifying future electricity water tradeoffs in the United States." *Energy Policy* 37, 2763–2773.

United Nations. (2018). "News." [Accessed 19 January 2021]. Available at https://www.un.org/development/desa/en/news/population/2018-revision-of-world-urbanization-prospects.html

United Nations. (2019). "Revision of World Population Prospects." [Accessed 19 January 2021]. Available at https://www.un.org/en/sections/issues-depth/population/

United Nations. (2021). "Sustainable Development Goal 7 Affordable and Clean Energy." [Accessed 23 January 2021]. Available at https://www.un.org/sustainabledevelopment/energy/

US Department of Energy (DOE). (2014). "The Water-Energy Nexus: Challenges and Opportunities Overview and Summary." https://www.energy.gov/sites/default/files/2014/07/f17/Water%20Energy%20Nexus%20Executive%20Summary%20July%202014.pdf

US Energy Information Administration (EIA). (2013). "Technically Recoverable Shale Oil and Shale Gas Resources: An Assessment of 137 Shale Formations in 41 Countries Outside the United States." Available at https://eia.gov

Vakilifard, N., Anda, M., Bahri, P. A. and Ho, G. (2018). "The role of water-energy nexus in optimising water supply systems – Review of techniques and approaches." *Renewable and Sustainable Energy Reviews* 82, 1424–1432.

Wada, Y., Flörke, M., Hanasaki, N., Eisner, S., Fischer, G., Tramberend, S., Satoh, Y., van Vliet, M. T. H., Yillia, P., Ringler, C., Burek, P. and Wiberg, D. (2016). "Modeling global water use for the 21st century: The Water Futures and Solutions (WFaS) initiative and its approaches." *Geoscientific Model Development* 9, 175–222.

Walsh, P., Murray, S. and O'Sullivan, D. T. J. (2015). "Water energy nexus, an ISO50001 water case study and the need for a water value system." *Water Resources and Industry* 10, 15–28.

Wang, S. and Chen, B. (2016). "Energy–water nexus of urban agglomeration based on multiregional input–output tables and ecological network analysis: A case study of the Beijing–Tianjin–Hebei region." *Applied Energy* 178, 773–783.

Webber, M. E. (2017). "Water for Electricity Generation in the United States." In Competition for Water Resources. Elsevier, pp. 212–232. https://www.sciencedirect.com/book/9780128032374/competition-for-waterresources#book-description

World Bank. (2012). "Implementation Completion Report Review. East Asia and Pacific-4M–Mekong Water Utilization." [Accessed 23 January 2021]. Available at http://documents.worldbank.org/curated/en/641311474492235787/pdf/000146597-20140619165326.pdf

Zhang, C. and Anadon, L. D. (2014). "A multi-regional input–output analysis of domestic virtual water trade and provincial water footprint in China." *Ecological Economics* 100, 159–172.

27

WATER INEQUALITIES

Maria Rusca

Introduction

Adequate access to clean water is essential for human well-being and for a dignified and healthy life. Because of the fundamental nature of drinking, practicing hygiene, cooking, and washing clothes, multiple negative conditions arise from inadequate access to water quantity and quality. There are clear indications that inadequate access to water can exacerbate existing well-being and gender inequalities, and undermine the dignity of underserved population that struggle with performing and complying with local hygiene norms (Brewis et al., 2019; Harris et al., 2017; Rusca et al., 2017; Buechler & Hanson, 2015). Epidemiological studies have shown the correlation between cholera and other diarrheal diseases and the consumption of contaminated water (Prüss-Ustün et al., 2019; Escamilla et al., 2013), whilst recent work has highlighted the social stigma associated with these diseases (Rusca et al., 2021). Concurrently, anthropological interventions have noted that insecure access to water is associated with physical stress and negatively affects mental health (Brewis, Choudhary, & Wutich, 2019; Wutich & Ragsdale, 2008).

It is, therefore, not surprising that improving access to drinking water has long been at the centre of the international development agenda. The launch of the Water and Sanitation Decade (1980–1990) marked the first systematic attempt at improving access to water for the large unserved or underserved population in the global South. This effort was followed by several global campaigns (UN International Year of Freshwater, 2003; the 2005–2015 UN International Decade for Action on Water for Life), the recognition of the Human Rights to Water and Sanitation by the UN General Assembly and the Human Rights Council (UNGA, 2010), and the Millennium Development Goals' (MDGs) pledge to halve population without access to water by 2015. Today, the International Decade for Action, "Water for Sustainable Development" 2018–2018 and the inclusion of water and sanitation as a stand-alone goal (SDG6) in the Sustainable Development Goals Agenda (2015–2030) mark the increased attention to water access. By emphasising universal access, SDG 6 also acknowledges emerging water struggles in the global North and sets a more ambitious target of ensuring access to water for all by 2030.

Whilst considered by many as promising steps towards universal access to water, these policy instruments are open to interpretation and abuse by power holders and tend to promote

DOI: 10.4324/9781003095545-31

depoliticised and technocratic interpretations of water inequalities (Sultana, 2021b; Hope et al., 2020). This prevents from bringing about the transformative change required to significantly reduce and progressively eliminate water inequalities. Eschewing these technocratic explanations, I draw on critical water studies to reflect on the political nature of water inequalities. The chapter proceeds as follows. Section "Global Inequalities in Water Access" maps global inequalities in access, highlighting distinct scales at which multiple patterns of water inequalities are (re)produced. Sections "The Production of Inequalities and Uneven Waterscapes: The Power-Money-Water Nexus" and "Articulating Water Inequalities through an Everyday Lens," specifically consider the multiple articulations of power – from structural processes to everyday water negotiations – that (re)produce multiple dimensions of water inequalities. The chapter concludes with a discussion on possibilities for more equitable and inclusive water access.

Global Inequalities in Water Access

Despite the increasing developmental attention devoted to ensuring universal access to water, stark inequalities persist in many regions. Over 2.2 billion people worldwide lack access to "safely managed" drinking-water services, and 785 million still rely on untreated and unsafe water streams. For monitoring purposes, the SDGs consider "improved" all sources – i.e., tubewells and boreholes, piped water, protected springs and dug wells, and rainwater – that potentially ensure access to safe drinking water. If on the premises, the service is defined as *Safely managed*, when access requires more than 30 minutes roundtrip it is classified as *Limited*, below this time it is considered *Basic* (WHO/UNICEF, 2017). If this ladder aims, as stated by the WHO/UNICEF (2017), at progressively reducing inequalities, it also works to institutionalise them. Indeed, the entire architecture of the SDG is grounded on the fundamental acceptance that universal access to water will be achieved by normalising stark inequalities across service levels.

As several scholars have noted, marked disparities in conditions of access to water across regions, rural–urban populations, intra-urban spaces, income groups, as well as across identities (e.g., gender, race) persist. Between 2000 and 2017, for instance, sub-Saharan Africa has made the greatest progress in ensuring access to at least basic water services, but the region still has the highest percentage of unserved population (WHO/UNICEF, 2019). Within countries, the urban-rural divide remains significant, with rural populations disproportionally affected by water infrastructure and services deficits. Although access to improved water sources has grown more rapidly in rural than urban areas, improved sources remain significantly lower than in urban areas (WHO/UNICEF, 2019). Whilst urban populations enjoy higher quality services, with relatively high levels of coverage of safely managed sources (e.g., piped water on the premises), rural areas mostly rely on community wells (Bain et al., 2014). Although both sources are considered "improved," urban residents are exposed to lower risks of faecally contaminated water than rural populations (Bain et al., 2014). Moreover, handpumps are subject to frequent breakage. To illustrate, it is estimated that each year 60,000 hand pumps are installed in sub-Saharan Africa, but one-third is not functional at each moment in time (Fisher et al., 2015). In many cases, therefore, coverage does not ensure access.

Although, comparatively, urban populations are likely to have better access levels than rural areas, stark inequalities in access across urban spaces are the norm. Several scholars have observed that in cities in the Global South, water utilities only serve 40%–70% of urban residents, whilst the rest of the urban population relies on a variety of small-scale water providers operating for profit or as social enterprises, or on unimproved sources (WHO/UNICEF, 2019; Ahlers et al., 2014; Marston, 2014). Moreover, whilst there is a tendency of relating domestic water inequalities to the global South, the (re)emergence of racialised water inequalities in post-industrial cities in the US and elsewhere, has led scholars to question the North/South, served/unserved geographic dichotomy (Meehan et al., 2020; Silver, 2019; Ranganathan & Balazs, 2015).

A vast literature has demonstrated that, across urban and rural spaces, multiple identities, including class, status, gender race, religion and ethnicity are tied to and mediate conditions of access to water. These inequalities are mutually reinforcing and generate multiple forms of discrimination, spatial segregation and exclusion from water access (Kabeer, 2010). To illustrate, whilst higher income residents enjoy *Safely managed* services, vulnerable low-income urban dwellers residing in informal settlements or slums are unserved or rely on *Basic* or *Limited services* (Meehan et al., 2020; Tiwale, Rusca, & Zwarteveen, 2018; Truelove, 2016). Socio-economic status intersects with racial and ethnic segregation processes that work to spatially distance and exclude minorities from adequate access to basic services (Meehan et al., 2020; Filčák, Szilvasi, & Škobla, 2018). This has cascading effects on gendered inequalities, as women are responsible for providing water for the household in areas with *Basic* or *Limited* access to water (Harris et al., 2017; Birkenholtz, 2013; Sultana, 2006).

The classed, gendered, racialised nature of water access reveals that water inequalities are a work of history and power. In the next section, I turn to critical water studies to examine the production and politics of water inequalities, often conceptualised as *uneven waterscapes*.

The Production of Inequalities and Uneven Waterscapes: The Power-Money-Water Nexus

As socio-spatial inequalities in water access persist, a greater understanding of the processes generating uneven access to water becomes increasingly urgent. Political ecology has been pathbreaking in examining the social relations of water inequalities and in excavating the role of power and politics in shaping infrastructural configurations and uneven water flows across urban and rural spaces (e.g., Loftus, March, & Purcell, 2019; Birkenholtz, 2013; Heynen, Kaika, & Swyngedouw, 2006; Swyngedouw, 1997, 2004; Gandy, 2004; Bakker, 2003). This critical perspective is crucial as overlooking the power structures that define society and shape water access and distribution can only lead to perpetuating water injustices (Loftus, 2015).

A central concern in this literature is the relation between capital, water flows and power, and the production of urban configurations and outcomes that are unevenly experienced by residents. Swyngedouw's examination of the processes of urban transformation and marginalisation that worked to exclude urban poor from public water supplies in Guayaquil, Ecuador, pioneered work on uneven waterscapes (Swyngedouw, 1997). Between 1880 and 1990, subsequent "ecological conquests" and capital accumulation from agricultural land and oil continuously (re)structured social power relations and water flows across the city. Through this work, he demonstrates that water circulation interiorises power relations at multiple nested geographical scales – from the body to the urban political ecology of Guayaquil and global development processes – that make the uneven waterscape.

Subsequent studies have examined the impact of variegated neoliberal reforms and market-based pricing of water on its circulation and access (Bakker, 2003, 2013; Bond, 2010). Rusca and Schwartz (2018) identify three phases in the process of neoliberalisation of water services: the privatisation decade (1993–2003), during which many countries (e.g., Mozambique, 1999; Namibia, 2000; Kenya, 2002; Tanzania, 2002) reformed the sector to allow for and incentivise private sector participation; commercialisation of water providers, entailing the adoption of private sector ethos (prioritising cost recovery and efficiency) by a public provider (see also

McDonald, 2016; Bakker, 2003) and the recent reappraisal of small-scale water provides, previously dismissed as informal/illegal providers and now praised for embodying several private sector characteristics (Ahlers et al., 2014). Importantly, promoters of these reforms do not explicitly disregard equity concerns. On the contrary, they argue, private sector participation and full cost recovery are a prerequisite to extend coverage to the urban poor as they provide capital and incentives (i.e. market opportunities) for the utility to expand services (Franceys & Gerlach, 2012; Berg & Mugisha, 2010).

These interpretations have been critiqued on several grounds. First, Castro (2007, p. 756) claims that neoliberal reforms never intended to address water inequalities they claimed to provide a solution to: "the pro-poor rhetoric was added to these policies since the 1990s, probably as a result of increasing citizen unrest." In this light, the emphasis on cost-recovering tariff reflects a more fundamental shift in state-society relations, which emphasise efficiency and individualism over equity and universalism, making inclusion to water services conditional to customers' ability to pay (Rusca & Schwartz, 2018). Moreover, several studies have demonstrated that in spaces regulated by market relations, the material link between power, water and money shapes circulation and ultimately produces socially stratified forms of water provision and access, and the marginalisation of lower-income groups (McDonald, 2016; Castro, 2007; Loftus, 2005; Swyngedouw, 2004; Bakker, 2003). For instance, providers may opt for cutting on maintenance and expanding services based on the consumers' ability to pay. Another perverse outcome generated from market-based principles is that in the attempt of reducing the costs of service provision, utilities may opt for outsourcing the provision of water services to intermediary providers, such as community-based water users associations (WUAs) or small private sector firms, tasked with managing publicly owned water supply infrastructures (Rusca & Schwartz, 2018). In many cases, however, this leads to a paradox. These intermediaries operate as a business and apply cost recovery principles more rigorously than the public provider, which is under greater pressure to balance economic efficiency and social mandates. As a result, low-income residents served by intermediaries end up paying a higher tariff for lower-quality services (Rusca & Schwartz, 2018). As aptly illustrated by Bakker, these reforms produce ultimately highly uneven waterscapes: "for the urban elite, water supply is often relatively abundant, and relatively cheap. For the urban poor, the scarcity of potable water is a daily hardship" (2003, p. 333).

The more recent expression of the power-money-water nexus is the financialisaton of water infrastructures, which have become tradable commodities in global markets (Loftus, March, & Purcell, 2019; Bayliss, 2014). If the privatisation decade has been the trojan horse for the emergence of a broader and more diverse range of investors in the water sector (March & Purcell, 2014), financialisaton reflects the persistent presence and influence of the private sector in the water policy arena (Bayliss, 2014). For Loftus and colleagues (2019, p. 1) processes of financialisaton have led to a radical reconfiguration of the institutional arrangements of water delivery and have turned "infrastructure into a wealth extraction mechanism." These arrangements, Furlong (2020) and Bayliss (2014) have argued, are not compatible with the social mandate of water providers. In Medellin, Colombia, for instance, debt financialisaton has led to an ever-increasing focus on full-cost recovery which, in turn, has exacerbated users' indebtedness, and increased disconnections and pre-paid metering, thereby intensifying water inequities across the city (Furlong, 2020).

These analyses aptly illustrate the mutual constitution of power, money, infrastructural configurations, and water inequalities. Concurrently, over the past years, geographic interventions have called for situated analyses that account for the everyday practices and hydraulic relations generated by and transformative of these hegemonic projects (e.g., Sultana, 2020; Truelove, 2011; Ekers

& Loftus, 2008). In the next section, I focus on the everyday to analyse processes of inclusion or exclusion, and the hybrid arrangements constituting uneven waterscapes.

Articulating Water Inequalities through an Everyday Lens

Attending calls for a situated political ecology (Lawhon, Ernstson, & Silver, 2014), critical water studies have taken an everyday lens to understand how water inequalities are experienced, contested and reproduced through everyday practices and life. Situated analyses have shed light on dispersed forms of power and hybrid governance arrangements generating uneven waterscapes (Rusca & Cleaver, 2022). Work by Truelove (2011, 2016), for instance, contributes to illuminate water inequalities by revealing how they are reproduced or transformed in everyday spaces of work, streets and households, as well as in the body itself. Highly unequal infrastructural configurations and waterflows are navigated through localised political informal or illegal arrangements that reveal how social relations, status and everyday (re)negotiations shape conditions of access and, in turn, water inequalities across spaces. Research on heterogenous service configurations within the centralised water supply network in Lilongwe, Malawi, sheds light on how local elites mediating access to water in informal and traditional settlements of the capital, simultaneously enhance their status and exacerbate water inequalities for lower income residents (Rusca et al., 2015). Here, status is the crucial mediating mechanisms in shaping hierarchies within WUAs to which the public, corporatised utility outsourced the responsibility of managing water kiosk, and in establishing "benefits" (honoraria, salaries, funeral allowances, etc.) elite members are entitled to. These costs are passed on to low-income consumers, who pay 4–5 times more for poorer conditions of access to water. WUAs must reinvest their profits into the community and whether they opt for developing their own headquarters, paying for a fence or a funeral, or building additional kiosks, these activities (paid for by the consumers' tariffs) further empower them. Thus, heterogeneous infrastructures do not only embody but also shape social relations of power (see also Hellberg, 2014, 2018). Yet, everyday practices can also work to resist and transform hegemonic projects, reduce water inequalities, and incrementally produce more just waterscapes. Non-market-based water gifting practices from water-rich to water-poor households (Wutich et al., 2018; Zug & Graefe, 2014), as well as occasional practices of field engineers redirecting water to water-stressed homes rather than water-rich ones (Velzeboer, Hordijk, & Schwartz, 2018) are examples of counter-hegemonic everyday practices.

Analyses attentive to the everyday have also shed light on the relationality of water access, and the multiple ways in which water inequalities and citizenship are tied to multiple identities (e.g., Sultana, 2020; Truelove, 2011). In Delhi and Lilongwe, for instance, women are responsible for fetching water, a labour-intensive and psychologically challenging practice, subject to everyday risks of violence and theft, shame, stigmatisation and, in some cases, criminalisation that simultaneously reinforces gendered inequalities and uneven access to water (Sultana, 2020; Rusca et al., 2017; Truelove, 2011). In Delhi these dynamics extend to domestic workers' experience of accessing water in middle-class households, where they are asked for additional hours of work in exchange of a few buckets of water. Similarly, Sultana's work (2006) has illuminated the gendered experience of water scarcity and embodied nature of everyday practices of accessing water in rural Bangladesh, where poisoned sources exacerbated women's burden of care and fetching water. Here, the spatiality of arsenic contamination, which varies from well to well, intersects with the technology and economic ability to develop safer sources (i.e., deeper tube wells) to produce uneven gendered outcomes and unequal exposure to water contamination.

Last, everyday analyses have revealed the dynamics of infrastructures "in use," showing how operation and maintenance intersect with infrastructural configurations and contribute to produce highly differentiated water distribution across urban spaces (Alda-Vidal, Kooy, & Rusca, 2018; Björkman, 2015; Anand, 2012). Through everyday informal practices of operating the network, field engineers often redirect water to "important" customers, thereby exacerbating water stress in low-income areas. These practices symbolically and materially link large-scale water infrastructures to end-users (Tiwale, Rusca, & Zwarteveen, 2018): dams are discursively promoted as a way of quenching the thirst of *all* citizens. Yet, network's material properties mediate unequal access by enabling larger water flows to high-income areas, and everyday practices reinforce inequalities by redirecting flows to already better served areas, ultimately disconnecting some users from the dams and producing spatially differentiated experiences of water scarcity.

Conclusions

Despite the positive global trends on water coverage and access (WHO/UNICEF, 2019) and UN's claim that "structural changes in the world economy" and "new models" are needed to meet the SDGs (2013, p. 5), water inequalities persist across regions, urban-rural populations, intra-urban spaces, income groups, and identities. As a result, approximately 485,000 people still die each year from preventable diseases caused by contaminated drinking water (WHO/UNICEF, 2019), and gendered, classed and racialised inequalities continue to be (re)produced and reinforced. Today, the COVID-19 pandemic and the recent urban water crises in Cape Town (2015–2017) and São Paulo (2014–2015) have shown that pre-existing inequalities in water access increase vulnerabilities to public health emergencies and climate change (Neely & Lopez, 2022; Sultana, 2021a, 2021b; Savelli et al., 2021; Ekumah et al., 2020; Millington & Scheba, 2020; Millington, 2018), thereby reinforcing the urgency of ensuring universal access to water.

This chapter has argued that crucial mechanism producing water inequalities persist or have taken new forms through financialisation. The chapter also brought attention to how everyday practices performed to compensate for structural inequalities (re)produce classed, status and gendered differentiation and variegated citizenship, and entrench water inequalities. It also suggests that everyday practices can work to incrementally erode and subvert established authoritative patterns producing water inequalities and to create spaces of possibility for more equitable and inclusive water access. As aptly illustrated by Loftus, "those involved in making urban environments through their day-to-day interactions with one another are well positioned […] for generating a transformative politics" (Loftus, 2012, p. xxiv). This potential is crucial, especially in times of entrenched capitalism, when structural transformations appear more difficult to achieve (Truelove & Cornea, 2021; Lawhon, Ernstson, & Silver, 2014; Pieterse, 2013).

References

Ahlers, R., Cleaver, F., Rusca, M. and Schwartz, K. (2014). "Informal Space in the Urban Waterscape: Disaggregation and Co-Production of Water Services." *Water Alternatives*, 7(1), 1–14.

Alda-Vidal, C., Kooy, M. and Rusca, M. (2018). "Mapping Operation and Maintenance: An Everyday Urbanism Analysis of Inequalities within Piped Water Supply in Lilongwe, Malawi." *Urban Geography*, 39(1), 104–121.

Anand, N. (2012). "Municipal Disconnect: On Abject Water and Its Urban Infrastructures." *Ethnography*, 13(4), 487–509.

Bain, R., Cronk, R., Wright, J., Yang, H., Slaymaker, T. and Bartram, J. (2014). "Fecal Contamination of Drinking-Water in Low-and Middle-Income Countries: A Systematic Review and Meta-Analysis." *PLoS Medicine*, 11(5), e1001644.

Bain, R. E. S., Wright, J. A., Christenson, E. and Bartram, J. K. (2014). "Rural: Urban Inequalities in Post 2015 Targets and Indicators for Drinking-Water." *Science of the Total Environment*, 490, 509–513.

Bakker, K. (2003). "Archipelagos and Networks: Urbanization and Water Privatization in the South." *Geographical Journal*, 169(4), 328–341.

Bakker, K. (2013). "Neoliberal Versus Postneoliberal Water: Geographies of Privatization and Resistance." *Annals of the Association of American Geographers*, 103(2), 253–260.

Bayliss, K. (2014). "The Financialization of Water." *Review of Radical Political Economics*, 46(3), 292–307.

Berg, S. V. and Mugisha, S. (2010). "Pro-Poor Water Service Strategies in Developing Countries: Promoting Justice in Uganda's Urban Project." *Water Policy*, 12(4), 589–601.

Birkenholtz, T. (2013). "'On the Network, off the Map': Developing Intervillage and Intragender Differentiation in Rural Water Supply." *Environment and Planning D: Society and Space*, 31(2), 354–371.

Björkman, L. (2015). *Pipe Politics, Contested Waters*. Duke University Press.

Bond, P. (2010). "Water, Health, and the Commodification Debate." *Review of Radical Political Economics*, 42(4), 445–464.

Brewis, A., Choudhary, N. and Wutich, A. (2019). "Low Water Access as a Gendered Physiological Stressor: Blood Pressure Evidence from Nepal." *American Journal of Human Biology*, 31(3), e23234.

Brewis, A., Wutich, A., du Bray, M. V., Maupin, J., Schuster, R. C. and Gervais, M. M. (2019). "Community Hygiene Norm Violators Are Consistently Stigmatized: Evidence from Four Global Sites and Implications for Sanitation Interventions." *Social Science & Medicine*, 220, 12–21.

Buechler, S. and Hanson, A.-M. S. (2015). *A Political Ecology of Women, Water and Global Environmental Change*. Routledge.

Castro, J. E. (2007). "Poverty and Citizenship: Sociological Perspectives on Water Services and Public–Private Participation." *Geoforum*, 38(5), 756–771.

Ekers, M. and Loftus, A. (2008). "The Power of Water: Developing Dialogues between Foucault and Gramsci." *Environment and Planning D: Society and Space*, 26(4), 698–718.

Ekumah, B., Armah, F. A., Yawson, D. O., Quansah, R., Nyieku, F. E., Owusu, S. A., Odoi, J. O. and Afitiri, A.-R. (2020). "Disparate On-Site Access to Water, Sanitation, and Food Storage Heighten the Risk of COVID-19 Spread in Sub-Saharan Africa." *Environmental Research*, 189, 109936.

Escamilla, V., Knappett, P. S., Yunus, M., Streatfield, P. K. and Emch, M. (2013). "Influence of Latrine Proximity and Type on Tubewell Water Quality and Diarrheal Disease in Bangladesh." *Annals of the Association of American Geographers*, 103(2), 299–308.

Filčák, R., Szilvasi, M. and Škobla, D. (2018). "No Water for the Poor: The Roma Ethnic Minority and Local Governance in Slovakia." *Ethnic and Racial Studies*, 41(7), 1390–1407.

Fisher, M. B., Shields, K. F., Chan, T. U., Christenson, E., Cronk, R. D., Leker, H., Samani, D., Apoya, P., Lutz, A. and Bartram, J. (2015). "Understanding Handpump Sustainability: Determinants of Rural Water Source Functionality in the Greater Afram Plains Region of Ghana." *Water Resources Research*, 51(10), 8431–8449.

Franceys, R. and Gerlach, E. (2012). *Regulating Water and Sanitation for the Poor: Economic Regulation for Public and Private Partnerships*. Routledge.

Furlong, K. (2020). "Trickle-down Debt: Infrastructure, Development, and Financialisation, Medellín 1960–2013." *Transactions of the Institute of British Geographers*, 45(2), 406–419.

Gandy, M. (2004). "Rethinking Urban Metabolism: Water, Space and the Modern City." *City*, 8(3), 363–379.

Harris, L., Kleiber, D., Goldin, J., Darkwah, A. and Morinville, C. (2017). "Intersections of Gender and Water: Comparative Approaches to Everyday Gendered Negotiations of Water Access in Underserved Areas of Accra, Ghana and Cape Town, South Africa." *Journal of Gender Studies*, 26(5), 561–582.

Hellberg, S. (2014). "Water, Life and Politics: Exploring the Contested Case of EThekwini Municipality through a Governmentality Lens." *Geoforum*, 56, 226–236.

Hellberg, S. (2018). *The Biopolitics of Water: Governance, Scarcity and Populations*. Routledge.

Heynen, N., Kaika, M. and Swyngedouw, E. (2006). *In the Nature of Cities: Urban Political Ecology and the Politics of Urban Metabolism*. Taylor & Francis.

Hope, R., Thomson, P., Koehler, J. and Foster, T. (2020). "Rethinking the Economics of Rural Water in Africa." *Oxford Review of Economic Policy*, 36(1), 171–190.

Kabeer, N. (2010). *Can the MDGs Provide a Pathway To Social Justice? The Challenge of Intersecting Inequalities – World*. IDS. [Accessed April 22, 2021]. Available from https://reliefweb.int/report/world/can-mdgs-provide-pathway-social-justice-challenge-intersecting-inequalities

Lawhon, M., Ernstson, H. and Silver, J. (2014). "Provincializing Urban Political Ecology: Towards a Situated UPE Through African Urbanism." *Antipode*, 46(2), 497–516.

Loftus, A. (2005). "Free Water as Commodity: The Paradoxes of Durban's Water Service Transformations." *The Age of Commodity: Water Privatization in Southern Africa*, 189–203.
Loftus, A. (2012). *Everyday Environmentalism: Creating an Urban Political Ecology*. University of Minnesota Press.
Loftus, A. (2015). "Water (in)Security: Securing the Right to Water." *The Geographical Journal*, 181(4), 350–356.
Loftus, A., March, H. and Purcell, T. F. (2019). "The Political Economy of Water Infrastructure: An Introduction to Financialization." *WIREs Water*, 6(1), e1326.
March, H. and Purcell, T. (2014). "The Muddy Waters of Financialisation and New Accumulation Strategies in the Global Water Industry: The Case of AGBAR." *Geoforum*, 53, 11–20.
Marston, A. (2014). "The Scale of Informality: Community-Run Water Systems in Peri-Urban Cochabamba, Bolivia." *Water Alternatives*, 7(1), 72–88.
McDonald, D. A. (2016). "To Corporatize or Not to Corporatize (and If so, How?)." *Utilities Policy*, 40, 107–114.
Meehan, K., Jurjevich, J. R., Chun, N. M. J. W. and Sherrill, J. (2020). "Geographies of Insecure Water Access and the Housing–Water Nexus in US Cities." *Proceedings of the National Academy of Sciences*, 117(46), 28700–28707.
Millington, N. (2018). "Producing Water Scarcity in São Paulo, Brazil: The 2014–2015 Water Crisis and the Binding Politics of Infrastructure." *Political Geography*, 65, 26–34.
Millington, N. and Scheba, S. (2020). "Day Zero and The Infrastructures of Climate Change: Water Governance, Inequality, and Infrastructural Politics in Cape Town's Water Crisis." *International Journal of Urban and Regional Research*, 45(1), 116–132.
Neely, A. H. and Lopez, P. J. (2022). "Toward Healthier Futures in Post-pandemic Times: Political Ecology, Racial Capitalism, and Black Feminist Approaches to Care." *Geography Compass*, 16(2), e12609.
Pieterse, D. E. (2013). *City Futures: Confronting the Crisis of Urban Development*. Zed Books Ltd.
Prüss-Ustün, A., Wolf, J., Bartram, J., Clasen, T., Cumming, O., Freeman, M. C., Gordon, B., Hunter, P. R., Medlicott, K. and Johnston, R. (2019). "Burden of Disease from Inadequate Water, Sanitation and Hygiene for Selected Adverse Health Outcomes: An Updated Analysis with a Focus on Low- and Middle-Income Countries." *International Journal of Hygiene and Environmental Health*, 222(5), 765–777.
Ranganathan, M. and Balazs, C. (2015). "Water Marginalization at the Urban Fringe: Environmental Justice and Urban Political Ecology across the North–South Divide." *Urban Geography*, 36(3), 403–423.
Rusca, M., Alda-Vidal, C., Hordijk, M. and Kral, N. (2017). "Bathing without Water, and Other Stories of Everyday Hygiene Practices and Risk Perception in Urban Low-Income Areas: The Case of Lilongwe, Malawi." *Environment and Urbanization*, 29(2), 533–550.
Rusca, M. and Cleaver, F. (2022) "Unpacking Everyday Urbanism: Practices and the Making of (Un) Even Urban Waterscapes." *Wiley Interdisciplinary Reviews: Water*, 9(2), e1581.
Rusca, M., Gulamussen, N. J., Weststrate, J., Nguluve, E. I., Salvador, E. M., Paron, P. and Ferrero, G. (2021). "The Urban Metabolism of Waterborne Diseases: Variegated Citizenship, (Waste) Water Flows, and Climatic Variability in Maputo, Mozambique." *Annals of the American Association of Geographers*, 112(4), 1159–1178.
Rusca, M. and Schwartz, K. (2018). "The Paradox of Cost Recovery in Heterogeneous Municipal Water Supply Systems: Ensuring Inclusiveness or Exacerbating Inequalities?" *Habitat International*, 73, 101–108.
Rusca, M., Schwartz, K., Hadzovic, L. and Ahlers, R. (2015). "Adapting Generic Models through Bricolage: Elite Capture of Water Users Associations in Peri-Urban Lilongwe." *The European Journal of Development Research*, 27(5), 777–792.
Savelli, E., Rusca, M., Cloke, H. and Di Baldassarre, G. (2021). "Don't Blame the Rain: Social Power and the 2015–2017 Drought in Cape Town." *Journal of Hydrology*, 125953.
Silver, J. (2019). "Decaying Infrastructures in the Post-Industrial City: An Urban Political Ecology of the US Pipeline Crisis." *Environment and Planning E: Nature and Space*, 4(3), 756–777.
Sultana, F. (2006). "Gendered Waters, Poisoned Wells: Political Ecology of the Arsenic Crisis in Bangladesh." *Fluid bonds: Views on Gender and Water*. [Accessed November 16, 2020]. Available from https://experts.syr.edu/en/publications/gendered-waters-poisoned-wells-political-ecology-of-the-arsenic-c
Sultana, F. (2020). "Embodied Intersectionalities of Urban Citizenship: Water, Infrastructure, and Gender in the Global South." *Annals of the American Association of Geographers*, 110(5), 1407–1424.
Sultana, F. (2021a). "Political Ecology II: Conjunctures, Crises, and Critical Publics." *Progress in Human Geography*, 45(6), 1721–1730.

Sultana, F. (2021b). "Climate Change, COVID-19, and the Co-Production of Injustices: A Feminist Reading of Overlapping Crises." *Social & Cultural Geography*, 22(4), 447–460.

Swyngedouw, E. (1997). "Power, Nature, and the City. The Conquest of Water and the Political Ecology of Urbanization in Guayaquil, Ecuador: 1880–1990." *Environment and Planning A*, 29(2), 311–332.

Swyngedouw, E. (2004). *Social Power and the Urbanization of Water: Flows of Power*. Oxford, New York: Oxford University Press.

Tiwale, S., Rusca, M. and Zwarteveen, M. (2018). "The Power of Pipes: Mapping Urban Water Inequities through the Material Properties of Networked Water Infrastructures-The Case of Lilongwe, Malawi." *Water Alternatives*, 11(2), 314–335.

Truelove, Y. (2011). "(Re-)Conceptualizing Water Inequality in Delhi, India through a Feminist Political Ecology Framework." *Geoforum*, 42(2), 143–152.

Truelove, Y. (2016). "Incongruent Waterworlds: Situating the Everyday Practices and Power of Water in Delhi." *South Asia Multidisciplinary Academic Journal*, (14).

Truelove, Y. and Cornea, N. (2021). "Rethinking Urban Environmental and Infrastructural Governance in the Everyday: Perspectives from and of the Global South." *Environment and Planning C: Politics and Space*, 39(2), 231–246.

UN. (2013). *A New Global Partnership: Eradicate Poverty and Transform Economies through Sustainable Development*. New York: United Nations. [Accessed December 4, 2021]. Available from https://sustainabledevelopment.un.org/content/documents/8932013-05%20-%20HLP%20Report%20-%20A%20New%20Global%20Partnership.pdf

UNGA. (2010). *RES/15/9 Human Rights and Access to Safe Drinking Water and Sanitation/RightDocs – Where Human Rights Resolutions Count* [Accessed April 14, 2021]. Available from https://www.right-docs.org/doc/a-hrc-res-15-9/

Velzeboer, L., Hordijk, M. and Schwartz, K. (2018). "Water Is Life in a Life without Water: Power and Everyday Water Practices in Lilongwe, Malawi." *Habitat International*, 73, 119–128.

World Health Organization & United Nations Children's Fund (UNICEF). (2017). Safely Managed Drinking Water: Thematic Report on Drinking Water 2017. World Health Organization. [Accessed April 22, 2021]. Available from https://apps.who.int/iris/handle/10665/325897.

WHO/UNICEF. (2019). *Progress on Household Drinking Water, Sanitation and Hygiene 2000–2017: Special Focus on Inequalities*. New York: United Nations Children's Fund (UNICEF) and World Health Organization. [Accessed April 16, 2021]. Available from http://www.who.int/water_sanitation_health/publications/jmp-report-2019/en/

Wutich, A., Budds, J., Jepson, W., Harris, L. M., Adams, E., Brewis, A., Cronk, L., DeMyers, C., Maes, K., Marley, T., Miller, J., Pearson, A., Rosinger, A. Y., Schuster, R. C., Stoler, J., Staddon, C., Wiessner, P., Workman, C. and Young, S. (2018). "Household Water Sharing: A Review of Water Gifts, Exchanges, and Transfers across Cultures." *WIREs Water*, 5(6), e1309.

Wutich, A. and Ragsdale, K. (2008). "Water Insecurity and Emotional Distress: Coping with Supply, Access, and Seasonal Variability of Water in a Bolivian Squatter Settlement." *Social Science & Medicine*, 67(12), 2116–2125.

Zug, S. and Graefe, O. (2014). "The Gift of Water. Social Redistribution of Water among Neighbours in Khartoum." *Water Alternatives*, 7(1), 140–159.

28

GENDERED INTERSECTIONS IN WATER AND DEVELOPMENT

Gaylean Davies, Evelyn Arriagada Oyarzún, and Leila M. Harris

Introduction

Statistics regarding global "water and sanitation" appear to be straightforward: 2.2 billion people lack safely managed drinking water; 4.2 billion lack safely managed sanitation; and 3 billion lack a soap and water handwashing facility at home (UN Water, 2021). Yet such summary statistics belie considerable complexity. In particular, there remain unanswered questions regarding gender equity as linked with water-related development, or how water-related development might differentially affect men, women, or gender relations more broadly. In this chapter, we first seek to outline how gender, water, and development are linked conceptually and in practice. Second, we offer counter-narratives to common ways gender is considered in water and development contexts. We do so through a focus on key themes of empowerment, participation, activism, and resistance. It is important to highlight at the outset that none of the concepts we engage are fixed. Whether it is gender (see Butler, 1990), development (see Escobar, 2012), or even water itself (see Linton, 2010) – all are contested, geographically specific, and charged in uneven terrains of power-knowledge. At times, we discuss gender as relational and deeply connected to norms, roles, and power dynamics. At other moments, we attend to the ways that water development projects seek to target and engage women. Apart from a critical conceptual overview and broad synthesis on these themes, a key analytical contribution centres on our discussion of participation and activism vis-a-vis empowerment debates. Towards the conclusion, we also highlight emergent themes due for further consideration, including: intersectional inequities; focus on the Global North; LGBTQI2+ communities; climate change; and embodiment and emotions.

Gender as a Contested Concept

It is imperative to begin with some of the broad theoretical currents regarding gender (see the feminist theory chapter, in this volume). Gender as a concept is commonly used to classify people according to the male/female binary, and relatedly, dominant understandings of femininity and masculinity (MacGregor, 2017). However, definitions of gender vary across time and space, are deeply contested, and are constantly in flux. There has been a considerable movement away from

DOI: 10.4324/9781003095545-32

the assumption that men and women are "self-evident and useful categories" (Özerol & Harris, 2020: 4). Indeed, many theorists understand gender as dynamic, varying along a spectrum, and as an outcome of specific operations of power-knowledge. The concept of "gender performativity" (Butler, 1990), offers that sex difference or gender meanings are not fixed or given, but constantly reproduced and reinforced through everyday practices. Such insights have been taken up in studies of water and development, particularly through the work of feminist political ecologists who understand "gender as a process" sedimented and made to appear as fixed through iterative citational practices (Harris, 2006).

Other contributions from decolonial, postcolonial, queer, feminist, and Indigenous researchers have historized and challenged epistemological and ontological assumptions related to gender (and other binaries). Such contributions have defied key dualities that frame our world: e.g., mind and body; environment and society; male and female (see Ahlers & Zwarteveen, 2009, for a discussion of Andean cultures and dualism as complementarity). In other contexts, there might be recognition of multiple genders, at times along a continuum, e.g., the Hijra in India, Fa'afafine in Samoa, *quariwarmi* (men-women) shaman in the Andes, and Two-Spirit members of many Indigenous communities. Vinyeta et al. (2016) highlight the considerable harms associated with the imposition of binary gender categories on Indigenous communities in North America. All told, gender cannot be considered universal, but, rather, needs to be situated in context. As Harris offers: "gender is not only historically and geographically variable, but linked to specific practices (politics, knowledges) [...] at different times and spaces" (2006: 188).

A final conceptual shift worth noting concerns "intersectionality" (see Crenshaw, 1990). Intersectionality posits that gender always intersects with other axes of social differentiation (i.e., class, race, or ability), which creates overlapping and co-constituted experiences of marginality. The intersectional approach helps to nuance the analysis of power relations and social hierarchies, moving beyond essentialized and uniform views about femininity and masculinity, to include analysis of how gender, class, race, and other aspects of social difference are mutually imbricated (cf. Mollett & Faria, 2013). In the water and development realm, the caste system in South Asia offers a salient example. For instance, women of lower castes are often not permitted to touch a water source (a spring or well) and might rely on higher caste women to fill their vessels. As such, a water project which installs taps into the homes of higher status households may inadvertently intensify marginalization, leaving women of lower castes to wait longer for their vessels to be filled (Leder et al., 2017). Another notable example, learning from the arsenic crisis in Bangladesh, shows the differentiated experiences of water contamination as well as the ways that gender, caste, and class (among other factors) condition how households are differentially able to navigate this poisoned waterscape, with some being left to rely on contaminated wells or engage in difficult social interactions to secure water (Sultana, 2011). In the following sections, we aim to elucidate the connection between gender, development, and water, drawing on illustrative examples to underscore these complex intersections.

Development Praxis, Politics, and Histories of a Contested Relationship with Gender

The integration of gender into development thinking, planning, and project implementation has a long and storied history. As described by Connelly and Barriteau (2000) and Harcourt (2016), early development policies and interventions often focused on "modernizing" what were considered traditional or backwards societies – following teleologies of "progress" (largely defined by Western/white scholars and practitioners). Early on (e.g., immediate post-WWII era), women were generally not highlighted as in need of particular attention, although notions of primitivism or backwardness were often classified in gendered terms, e.g., by focusing on the status of women, marital practices, natality, or sexuality (see Harris, 2008, for example of gendered notions of underdevelopment in Turkey). Gendered consequences of development interventions eventually were highlighted, whether the consequences of focusing on industry and men's productive work (Connelly et al., 2000), or the consequences of the green revolution on shifting farming practices and gendered labour (Kilby, 2019).

From 1970, various frameworks highlighted the need to focus on women and gender in development (Reddock, 2000; Coles et al., 2015). The women in development (WID) framework focused on women's economic participation, but was criticized for failing to attend to strategic gender goals – e.g., those relating to women's political emancipation and liberation (Reddock, 2000). This was followed by the women and development (WAD) model, which sought the creation of "women only" development projects and spaces and a greater attention to women's strategic goals and collective organization (Connelly et al., 2000). Finally, the gender and development (GAD) framework highlighted unequal relations of power, relational dynamics between men and women, and eventually, a focus on intersectional differences between different groups of women (see Harcourt, 2016). Tensions remain between the WAD and GAD models, both of which are in evidence today: there are those who see a specific focus on women as leading to gendered conflict, and others who consider that a broad focus on gender minimizes women's specific needs (Coles & Wallace, 2005; de Jong, 2016). Recent institutional focus on gender mainstreaming (e.g., through the World Bank and United Nations) requires that gender be considered in all aspects of development practice, with the goal of achieving gender equality. However, this too has been roundly criticized for depoliticizing gender, often relying on simplistic notions of inclusion, as we discuss below in the context of water development (Kabeer, 2005; Porter & Sweetman, 2005; de Jong, 2016).

Gender, Water, and Development

Turning now to the complex intersections between gender, water and development, it is often implied that there are core linkages between these categories. However, the evidentiary basis for these connections remains unclear. Does gender equality lead to development, or the other way around? Does water access lead to gender equality? Does improved water security allow women to spend more time in productive work? Or for girls to attend school? Some have suggested that gender equality is important for the longevity of water projects and programmes (Ivens, 2008); others that water infrastructure and development underpin gender equality goals (Fisher et al., 2017). Numerous studies have shown a decrease in women and girls' time poverty connected with access to water sources, but where this time might go is not obvious. There is some evidence that reducing the time necessary to collect water increases girls' school attendance (in Ghana, see Nauges & Strand, 2013); while other evidence from Ghana suggests this also reduces the probability of women being involved in market work (Costa et al., 2009). Still other research suggests that water and sanitation interventions increase men's leisure time, but not women's (in an assessment of Water Aid's projects in Ethiopia, Ghana, India, and Tanzania, see Wallace & Wilson, 2005), or exactly the opposite from work in Kenya (Bisung & Elliott, 2018). As well, even when latrine or sanitation projects directly target women as beneficiaries, gender empowerment outcomes remain mixed (see O'Reilly, 2010 on interventions in India). What is clear is that the associations between water development, gender equality, and women's well-being are frequently inferred. Results are, however, complicated and context-dependent, resulting in the difficulty of providing a summary assessment of such interventions.

That said, in general terms the water sector has largely tracked the WID-WAD-GAD paradigms (Sweetman & Medland, 2017), with women and gender frequently highlighted in policy interventions. The centrality of women in water policy was enshrined in the 1992 Dublin Principles (UNEP, 1992), which noted that "Women play a central part in the provision, management, and safeguarding of water." However, these principles simultaneously considered water a "finite and vulnerable resource" which should be managed as an "economic good." Scholars have pointed to the contradictions inherent in these principles as it is often low-income women who are least able to ensure they have the economic means to pay for water (Ahlers, 2005; Sweetman & Medland, 2017). The associated shifts, therefore, away from public provision of water, towards privatization, and user-pay models, have potentially significant consequences for many women, as well as for broad gender equity goals (Harris, 2009).

In terms of other interventions, many institutions (the United Nations, World Bank, International Monetary Fund, together with NGOs and non-profits) have focused their work in line with the Millennium Development Goals (MDGs, 2000–2015), and Sustainable Development Goals (SDGs, 2015–2030). Both frameworks include targets related to women and water, and both aim to mainstream gender. For the SDGs, civil society, feminist, and women's organizations worked to ensure that gender equality targets were embedded within each goal, with consideration of power relations, socio-political, and economic dynamics to address perceived weaknesses of the earlier MDG framework (Esquivel & Sweetman, 2016). The SDGs also include a specific gender equality goal (SDG 5), and a standalone goal related to *equitable* access to and distribution of drinking water and sanitation (SDG 6.1) with specific mention of the impact of water and sanitation on women and girls (SDG 6.2) (United Nations, 2019). The SDGs and the development sphere more broadly have their detractors, particularly those who critique an ever-present individual and economic focus, and the downplaying of the role of collective organizing to challenge societal inequities (Esquivel & Sweetman, 2016; Harcourt, 2016).

Apart from tracking how development interventions or water infrastructures differentially affect men, women, or other marginalized individuals and communities (e.g., Aladuwaka & Momsen, 2010), it has generally been suggested that gender mainstreaming lacks a critical edge, and as such often might reify gendered divisions (Regmi, 2005). Regarding participation specifically, women are increasingly included in water development projects but often remain excluded from *meaningful* participation, leadership roles, decision-making, and providing strategic input (Smith, 2008; Adams et al., 2018). Research in Egypt by Barnes (2014) showed that while women were included in water user associations, they were often only able to speak about domestic water uses in ways that retrenched gendered expectations. Work in Malawi (Adams et al., 2018), and Kenya (Abu et al., 2019) reached similar conclusions about women's involvement in formal water associations – finding they are often excluded from substantial participation in decision-making.

It is important to note that gender dynamics, power relations, and sociocultural norms are deeply embedded, and thus are resistant to change. In addition, theorists have increasingly refined their understanding of gender as produced through social and power relations, involving complex and intersectional dynamics (discussed above). Such a perspective is relatively more difficult to address, resulting in clear tensions with the persistent focus on women, quotas, or technical and numerical measurement in many policy arenas (Arora-Jonsson, 2014; Esquivel & Sweetman, 2016). Reliance on apolitical and technical understandings of gender, or tokenistic participation of women, in turn may increase women's burden of work rather than materially improving their circumstances (Regmi, 2005). Some of these tensions are clear in the next section where we delve further into key distinctions between "participation," "empowerment," "activism," and "resistance."

Empowerment and Participation versus Activism and Resistance

Considering these debates, it is clear that there is a bifurcation between notions of empowerment (commonly expressed in development circles) and activism as evidenced in social movements literatures – leaving a gap where the development literature and policy community generally pay scant attention to women's collective activism. Here, we offer some notes on the concepts of empowerment and participation, in relation to evidence of collective activism and resistance in the water realm.

In formal development policy and praxis, discussions of women's "participation" and "empowerment" abound. However, what this means in practical terms requires critical analysis. Who is able to participate in water projects? How do they participate? Is their participation meaningful or does it increase an already heavy work burden? When a woman, Indigenous community member, or other historically marginalized person speaks, are they heard and taken seriously? Such questions are key to any intersectional or gendered analysis of water-related development. In this vein, only a small number of empirical studies have engaged a critical feminist approach to evaluate women's own experiences of water projects and, in particular, whether or not they experience empowerment through participation (Aladuwaka & Momsen, 2010; Dickin et al., 2021). For example, it is clear that women of different economic or caste status may have different abilities to harness the benefits of water projects, or to engage in decision-making (see work in Nepal by Leder et al., 2017).

While many examples speak to whether or not women or marginalized community members are able to participate in development and water-related institutions, there is a tendency to ignore and downplay other types of engagement that do not fit as easily within gender mainstreaming paradigms. As such, many forms of collective organizing and action are often ignored, despite their importance to addressing societal inequities (Esquivel & Sweetman, 2016; Harcourt, 2016), and despite persistent language of "gender empowerment" (Sweetman & Medland, 2017). For issues of water access, use, control, and management, resistance and activism are clearly important to understanding gendered responses. For example, Arora-Jonsson (2014) emphasizes how women's water groups may catalyse individual and community empowerment through collective organizing. As well, work in Mexico and other Latin American countries has suggested that women's participation in water governance opened up opportunities to be more involved in their communities and politics, more generally (see Ennis-McMillan, 2005 and other chapters in Bennett et al., 2005).

The literature on water activism also highlights other movements that have challenged hegemonic narratives and practices. Varied examples include female activists and organizations protesting against dams (Drew, 2014; Yaka, 2017); mining and extractivist activities (Caretta et al., 2020); privatization of water resources (Bustamante et al.,2005); as well as mobilization to protect diverse water bodies (Hayman et al., 2015). Often, leaders of water organizations and movements around the world are Indigenous, peasant, poor and/or racialized women – providing a stark point of contrast with the literature on "participation" of women in formal water governance and development arenas in which the focus often centres on representation and quotas. As Arora-Jonsson (2014) notes, development projects may unintentionally produce "dissident political subjects" who push for transformative change, particularly in postcolonial and settler colonial contexts.

We can hypothesize that water development literature has tended to overlook these processes of collective and political empowerment because they focus instead on what is considered to be "legitimate" forms of participation (Morinville & Harris, 2014) – often technically and numerically defined, tokenistic, and apolitical – putting less emphasis on forms of organizing that challenge core precepts of many development efforts. In some cases, women's groups explicitly challenge mainstream definitions and practices of development, or notions of women's empowerment, extending their activism in innovative ways and reclaiming the authority to speak, theorize, and dream for themselves (see for instance the work of Sangtin Writers Collective and Nagar, 2006). Whether the water development field is able to recognize this tension and learn from such examples as key illustrations of robust and empowered engagement remains to be seen.

Future Directions/Novel Contributions/Alternative Perspectives

While we have aimed to highlight some key themes, and tensions related to gender, water and development, there is significant terrain that we have not addressed. Here we signal several themes as important for follow up work, including: gendered water issues in the Global North; plural water ontologies; climate change; embodied and affective experiences; masculinities; and LGBTQ2+ issues.

Global North: While many examples in water and development literature relate to the Global South, these issues are also important in high-income countries, particularly if we take a broad intersectional perspective that attends to gender, Indigeneity, housing precarity, income, race, and so forth (Meehan et al., 2020). All such issues are critical in terms of who experiences water insecurity, and relatedly, who is able to participate meaningfully in water governance. An example of this is the well-known case of Flint, Michigan, which looms large in conversations of racialized and spatialized water insecurity at present. The Flint water crisis exposes environmental injustice and democratic disenfranchisement, revealing intersectional vulnerabilities that emerge as a function of race, income, and democratic failures (Pauli, 2019). Some attention has also been given to women's experiences of the crisis, mirroring many issues already raised: excess time spent waiting for water; or the financial, emotional, and physical burdens associated with water insecurity (Radonic & Jacob, 2021).

Indigenous and Plural Ontologies: Other relevant literature considers plural water ontologies, with particular reference to Indigenous ontologies – highlighting diverse understandings of what water *is* and how humans relate to it (Yates et al., 2017). For many Indigenous peoples, water is considered a living being in a complex relationship with humans (ibid., Hayman et al., 2015; Vinyeta et al., 2016). The diverse social, cultural, and spiritual values associated with multiple ontologies of water are poorly reflected in water governance mechanisms and development projects – most often governed by colonial or neoliberal institutions which view water as a resource to be technically managed (Yates et al., 2017). Indigenous women are also seen to have a unique relationship to water: "Water is a life giver, and because women also give life they are the keepers of the water" (Nibi Walk organizers, quoted in Barker, 2019). These gendered relationships and ways of seeing and relating to water have been poorly integrated into the literature on gender, water, and development (Barker, 2019; Georgeson & Hallenbeck, 2018; Vinyeta et al., 2016, see also theory chapter, this volume).

Climate Change: Understanding gendered and intersectional aspects of water insecurity and development will be particularly important as regions face the intensified effects of climate change – likely exacerbating vulnerabilities of women and other marginalized groups. For example, women in Bangladesh (Sultana, 2018), India (Singh & Singh, 2015), Mexico (Buechler, 2009), and Indigenous women in the United States (Whyte, 2014; Vinyeta et al., 2016) are frequently highlighted in terms of particular vulnerabilities related to climate change and associated changes in water availability and quality. Again, it is important to emphasize the intersectional dimensions in terms of specific vulnerabilities as a function of race, class, ethnicity, or location (Arora-Jonsson, 2011; Sultana, 2018). We expect that debates regarding gender, water and development will continue to take new forms in the context of these ongoing shifts and challenges. It is also likely that women (or Indigenous women) might be particularly knowledgeable about the strategies needed to adapt to and mitigate the worst effects of climate change (Whyte, 2014; Singh & Singh, 2015; Sultana, 2018), suggesting that community-based initiatives focused on women, water, and climate change are especially needed moving forward (Perkins & Figueiredo Walker, 2015).

Embodied and Affective Experiences: The literature increasingly calls attention to the embodied and affective experiences of water, as a counterpoint to the common focus on macro scales and material dimensions. As such, it is imperative not only to centre the implications of water-related changes for economies and livelihoods, but also other effects for gendered and sentient bodies. Feminist political ecologists, in particular, have highlighted the importance of attention to the body, everyday politics, and emotional-affective lives (Truelove, 2019) to address differentiated responses to water-related changes (Sultana, 2009, 2011; Wutich et al., 2020). Scholars have also noted that women's water activism is rooted in varied bodily practices, knowledges, and affections (Caretta et al., 2020; Yaka, 2017). Additionally, we can draw inspiration from feminist and queer theory to challenge hegemonic notions of water, reconnecting to more-than-human others while also re-imagining alternatives to enable transformative changes (Neimanis, 2013).

Masculinities: Focus on masculinities offer another path forward to more adequately connect gender, water and development. Some work has focused on practices of "hegemonic masculinity" that reproduce or intensify gender inequalities at different scales. As a key example, Zwarteveen (2011) makes visible the dominance of male engineers in water and irrigation agencies in South Asia, connecting to gendered and colonial hierarchies. Meanwhile, Braun's (2011) work on South Africa shows how large-scale water projects are dependent on global hierarchies of male labour (entangled with race and class) while also maintaining local gender inequalities (see also recent attention to water insecurity and gender-based violence, Cole et al., 2020).

LGBTQI2+: A final area for exploration relates to LGBTQI2+ experiences of water and sanitation, as well as involvement in water governance, a topic relatively understudied given common reliance on binary gender frameworks. Recent studies have highlighted the value of understanding gender as a continuum or bringing in non-binary and transgender peoples and perspectives. Specifically, Boyce et al. focused on transgender communities in their assessment of sanitation programming in India and Nepal, finding that the stigma and discrimination of this community hinges in part on their inability to access "safe and dignified sanitation" (2018: 112). A further trio of scholars is also advocating for the gender continuum to be considered in water development, particularly to highlight the ways that water and sanitation insecurities affect non-binary, trans, and intersex peoples (cf. recent articles and reports by Mukherjee et al., 2020, 2021). The important report by Vinyeta et al. (2016) also connects issues of non-binary gender understandings to resilience of Indigenous communities in settler colonial contexts, again illustrating the relevance of queer and non-binary approaches moving forward (see also Brewis et al, forthcoming for overview of water insecurity and retheorizing gender building on such perspectives).

Concluding Thoughts

All told, there is a considerable body of evidence, but also much still to be learned and understood regarding complex water, gender, and development intersections. These interlinkages are complex, context-specific, and important. Moreover, researchers and policy communities must do more than simply highlight differences between men and women. Instead, we have sought to offer multiple lenses and starting points to analyse broader gendered and power dimensions of water and development. Doing so must also occur across scales and build on insights from diverse geographical locations. As well, it is important to understand patterns and processes across the Global North and South, to address why some silences and gaps remain (such as those we highlighted regarding empowerment and activism or LGBTQI2+ experiences). Finally, not only are all of these issues even more acute in the context of climate change, but the COVID-19 emergency also calls our attention to "overlapping crises" (Sultana, 2021). Not only have women, girls and marginalized communities long been sidelined in water and development projects, but such intersecting considerations compound and exacerbate the core challenges of ensuring human well-being, flourishing, or broader socio-environmental sustainability. Something as basic as securing water and sanitation, as called for in the SDGs, is not at all simple, especially if we draw insights and lessons from the diverse body of gender and intersectional work.

References

Abu, T.Z., Bisung, E. and Elliott, S.J. (2019). "What if your husband doesn't feel the pressure?: An exploration of women's involvement in WaSH decision-making in Nyanchwa, Kenya." *International Journal of Environmental Research and Public Health* [online], 16(10). Available from: http://dx.doi.org/10.3390/ijerph16101763.

Adams, E.A., Juran, L. and Ajibade, I. (2018). "Spaces of exclusion' in community water governance: A feminist political ecology of gender and participation in Malawi's urban water user associations." *Geoforum*, 95, 133–142.

Ahlers, R. (2005). "Gender dimensions of neoliberal water policy in Mexico and Bolivia: Empowering or disempowering." In V. Bennet, S. Dávila-Poblete, and M. Nieves Rico (eds.) *Opposing currents: The politics of water and gender in Latin America*. Pittsburgh, PA: University of Pittsburgh Press. pp. 53–71.

Ahlers, R. and Zwarteveen, M. (2009). "The water question in feminism: Water control and gender inequities in a neo-liberal era." *Gender, Place and Culture*, 16(4), 409–426.

Aladuwaka, S. and Momsen, J. (2010). "Sustainable development, water resources management and women's empowerment: The Wanaraniya water project in Sri Lanka." *Gender & Development*, 18(1), 43–58.

Arora-Jonsson, S. (2011). "Virtue and vulnerability: Discourses on women, gender and climate change." *Global Environmental Change*, 21(2), 744–751.

Arora-Jonsson, S. (2014). "Forty years of gender research and environmental policy: Where do we stand?" Women's *Studies International Forum*, 47(B), 295–308.

Barker, J. (2019). "Confluence: Water as an Analytic of Indigenous Feminisms." *American Indian Culture and Research Journal*, 43(3), 1–40.

Barnes, J. (2014). *Cultivating the Nile: The everyday politics of water in Egypt*. Durham, NC: Duke University Press.

Bennett, V., Dávila-Poblete, S. and Nieves Rico, M. eds. (2005). *Opposing currents: The politics of water and gender in Latin America*. Pittsburgh, PA: University of Pittsburgh Press.

Bisung, E. and Elliott, S.J. (2018). "Improvement in access to safe water, household water insecurity, and time savings: A cross-sectional retrospective study in Kenya." *Social Science & Medicine*, 200, 1–8.

Boyce, P., Brown, S., Cavill, S., Chaukekar, S., Chisenga, B., Dash, M., Dasgupta, R., De La Brosse, N., Dhall, P. and Fisher, J. (2018). "Transgender inclusive sanitation-insights from South Asia." *Waterlines*, 37(2), 102–117.

Braun, Y.A. (2011). "The reproduction of inequality: Race, class, gender, and the social organization of work at sites of large-scale development projects." *Social Problems*, 58(2). 281–303.

Brewis, A., L. Zachary DuBois, Amber Wutich, Ellis Adams, Sarah Dickin, Susan Elliott, Vanessa Empinotti, Leila Harris, Elisabeth Ilboudo Nébié, Marina Korzenevica (Forthcoming). "Gender identities, water insecurity, and risk: Re-theorizing the connections for a gender-inclusive toolkit for water insecurity research" WIREs Water.

Buechler, S. (2009). "Gender, water, and climate change in Sonora, Mexico: Implications for policies and programmes on agricultural income-generation." *Gender & Development*, 17(1), 51–66.

Bustamante, R., Peredo, E. and Udaeta, M.E. (2005). "Women in the 'Water War in the Cochabamba Valleys." In V. Bennett, S. Dávila-Poblete and M. Nieves Rico (eds.) *Opposing currents: The politics of water and gender in Latin America*. Pittsburgh, PA: University of Pittsburgh Press. pp. 72–90.

Butler, J. (1990). *Gender trouble: Feminism and the subversion of identity*. New York, NY: Routledge.

Caretta, M.A., Zaragocin, S., Turley, B. and Orellana, K T. (2020). "Women's organizing against extractivism: Toward a decolonial multi-sited analysis." *Human Geography*, 13(1), 49–59.

Cole, S.K., Mullor, E.C., Ma, Y. and Sandang, Y. (2020). "Tourism, water, and gender: An international review of an unexplored nexus." *Wiley Interdisciplinary Reviews: Water*, 7(4), e1442.

Coles, A., Gray, L. and Momsen, J.H. eds. (2015). *The Routledge handbook of gender and development.* London, UK: Routledge.

Coles, A. and Wallace, T. eds. (2005). *Gender, water and development*. New York, NY: Routledge.

Connelly, M.P., Li, T., MacDonald, M. and Parpart, J.L. (2000). "Feminism and development: Theoretical perspectives." In J.L. Parpart, M.P. Connelly, and V.E. Barriteau (eds.) *Theoretical perspectives on gender and development.* Ottawa, ON: IDRC. pp. 51–160.

Costa, J., Hailu, D., Silva, E. and Tsukada, R. (2009). *The implications of water and electricity supply for the time allocation of women in rural Ghana* Working Paper No. 59 [online]. Brasilia, Brazil: International Policy Centre for Inclusive Growth. [Accessed February 10, 2023]. Available from: https://www.econstor.eu/handle/10419/71815

Crenshaw, K. (1990). "Mapping the margins: Intersectionality, identity politics, and violence against women of color." *Stanford Law Review*, 43, 1241.

de Jong, S. (2016). "Mainstream(ing) has never run clean, perhaps never can: Gender in the main/stream of development." In W. Harcourt (ed.) *The Palgrave handbook of gender and development: Critical engagements in feminist theory and practice*. New York, NY: Palgrave Macmillan. pp. 92–105.

Dickin, S., Bisung, E., Nansi, J. and Charles, K. (2021). Empowerment in water, sanitation and hygiene index. *World Development*, 137, 105158.

Drew, G. (2014). Mountain women, dams, and the gendered dimensions of environmental protest in the Garhwal Himalaya. *Mountain Research and Development*, 34(3), 235–242.

Ennis-McMillan, M. (2005). "Women, equity, and household water management in the valley of Mexico." In V. Bennett, S. Dávila-Poblete, and M. Nieves Rico (eds.) *Opposing currents: The politics of water and gender in Latin America*. Pittsburgh, PA: University of Pittsburgh Press. pp. 137–153.

Escobar, A. (2012). *Encountering development: The making and unmaking of the Third World*. Princeton, NJ: Princeton University Press.

Esquivel, V. and Sweetman, C. (2016). "Gender and the sustainable development goals." *Gender & Development*, 24(1), 1–8.

Fisher, J., Cavill, S. and Reed, B. (2017). "Mainstreaming gender in the WASH sector: Dilution or distillation?" *Gender & Development*, 25(2), 185–204.

Georgeson, R. and Hallenbeck, J. (2018). "We have stories: Five generations of Indigenous women in water." *Decolonization-Indigeneity Education & Society*, 7(1), 20–38.

Harcourt, W. ed. (2016). *The Palgrave handbook of gender and development: Critical engagements in feminist theory and practice*. New York, NY: Palgrave Macmillan.

Harris, L.M. (2006). "Irrigation, gender, and social geographies of the changing waterscapes of southeastern Anatolia." *Environment and Planning D: Society and Space*, 24(2), 187–213.

Harris, L.M. (2008). "Water rich, resource poor: Intersections of gender, poverty, and vulnerability in newly irrigated areas of southeastern Turkey." *World Development*, 36(12), 2643–2662.

Harris, L.M. (2009). "Gender and emergent water governance: Comparative overview of neoliberalized natures and gender dimensions of privatization, devolution and marketization." *Gender, Place and Culture*, 16(4), 387–408.

Hayman, E., in collaboration with Wedge, M. Gooshú, Aan, Colleen, J. Tláa, Gooch. (2015). "Storytelling water north of the future Héen Kas'él'ti Xoo 1 (among the ragged lakes): Collaborative water research with Carcross/Tagish First Nation, Yukon Territory, Canada." In S. Buechler and A.M. Hanson

(eds.) *A political ecology of women, water and global environmental change*. New York, NY: Routledge. pp. 185–205.
Ivens, S. (2008). "Does increased water access empower women?" *Development*, 51(1), 63–67.
Kabeer, N. (2005). "Gender equality and women's empowerment: A critical analysis of the third millennium development goal." *Gender & Development*, 13(1), 13–24.
Kilby, P. (2019). *The green revolution: Narratives of politics, technology and gender*. London: Routledge, Taylor & Francis Group.
Leder, S., Clement, F. and Karki, E. (2017). "Reframing women's empowerment in water security programmes in Western Nepal." *Gender & Development*, 25(2), 235–251.
Linton, J. (2010). *What is water?: The history of a modern abstraction*. Vancouver, BC: UBC Press.
MacGregor, S. (2017). *Routledge handbook of gender and environment*. London, UK: Routledge.
Meehan, K. et al. (2020). "Exposing the myths of household water insecurity in the global North: A critical review." *Wiley Interdisciplinary Reviews: Water*, 7(6), e1486.
Mollett, S. and Faria, C. (2013). "Messing with gender in feminist political ecology." *Geoforum*, 45, 116–125.
Morinville, C. and Harris, L.M. (2014). "Participation, politics, and panaceas: Exploring the possibilities and limits of participatory urban water governance in Accra, Ghana. *Ecology and Society*, 19(3), 36.
Mukherjee, S., Sundberg, T. and Schütt, B. (2020). "Assessment of water security in socially excluded areas in Kolkata, India: An approach focusing on water, sanitation and hygiene." *Water*, 12(3), 746.
Mukherjee, S., Sundberg, T. and Schütt, B. (2021). "Issues, dimensions and approaches of assessing urban water security in developing and emerging countries: An inclusive perspective." In P.K. Sikdar (ed.) *Environmental management: Issues and concerns in developing countries*. New Delhi, India: Springer. pp. 151–184.
Nauges, C. and Strand, J. (2013). *Water hauling and girls' school attendance: Some new evidence from Ghana* [online]. Washington, DC: The World Bank. [Accessed February 10, 2023]. Available from: https://elibrary.worldbank.org/doi/abs/10.1596/1813-9450-6443
Neimanis, A. (2013). "Feminist subjectivity, watered." *Feminist Review*, 103, 23–41.
O'Reilly, K. (2010). "Combining sanitation and women's participation in water supply: An example from Rajasthan." *Development in Practice*, 20(1), 45–56.
Özerol, G. and Harris, L.M. (2020). "Gender-sensitive analysis of water governance: Insights for engendering energy transitions." In J. Clancy, G. Özerol, N. Mohlakoana, M. Feenstra and L. Sol Cueva (eds.) *Engendering the energy transition*. Cham, Switzerland: Springer. pp. 59–82.
Parpart, J.L., Connelly, M.P. and Barriteau, V.E. eds. (2000). *Theoretical perspectives on gender and development*. Ottawa, ON: IDRC.
Pauli, B.J. (2019). *Flint fights back: Environmental justice and democracy in the flint water crisis*. Cambridge, MA: MIT Press.
Perkins, P.E. and Figueiredo Walker, P. (2015). "International partnerships of women for sustainable watershed governance in times of climate change." In S. Buechler and A. Hanson (eds.) *A political ecology of women, water and global environmental change*. New York, NY: Routledge.
Porter, F. and Sweetman, C. eds. (2005). *Mainstreaming gender in development: A critical review* [online]. Cowley, Oxford, UK: Oxfam. [Accessed February 10, 2023]. Available from: https://policy-practice.oxfam.org/resources/mainstreaming-gender-in-development-a-critical-review-121089/
Radonic, L. and Jacob, C.E. (2021). "Examining the cracks in universal water coverage: Women document the burdens of household water insecurity." *Water Alternatives*, 14(1), 60–78.
Reddock, R. (2000). "Why gender? Why development?" In J.L. Parpart, M.P. Connelly and V.E. Barriteau (eds.) *Theoretical perspectives on gender and development*. Ottawa, ON: IDRC. pp. 23–50.
Regmi, S.C. (2005). "Gender mainstreaming in the water sector in Nepal: A real commitment or a token?" In A. Coles and T. Wallace (eds.) *Gender, water and development*. New York, NY: Routledge. pp. 95–114.
Sangtin Writers Collective and Nagar, R. (2006). *Playing with fire: Feminist thought and activism through seven lives in India*. Minneapolis, MN: University of Minnesota Press.
Singh, N. and Singh, O.P. (2015). "Climate change, water and gender: Impact and adaptation in North-Eastern Hills of India." *International Social Work*, 58(3), 375–384.
Smith, J.L. (2008). "A critical appreciation of the 'bottom-up' approach to sustainable water management: Embracing complexity rather than desirability." *Local Environment*, 13(4), 353–366.
Sultana, F. (2009). Fluid lives: subjectivities, gender and water in rural Bangladesh. *Gender, Place and Culture*, 16(4), 427–444.
Sultana, F. (2011). Suffering for water, suffering from water: Emotional geographies of resource access, control and conflict. *Geoforum*, 42(2), 163–172.

Sultana, F. (2018). "Gender and water in a changing climate: Challenges and opportunities." In C. Fröhlich, G. Gioli, R. Cremades, and H. Myrttinen (eds.) *Water security across the gender divide*. Cham, Switzerland: Springer. pp. 17–33.

Sultana, F. (2021). "Climate change, COVID-19, and the co-production of injustices: A feminist reading of overlapping crises." *Social & Cultural Geography*, 447–460. https://www.google.com/url?sa=t&rct=j&q=&esrc=s&source=web&cd=&cad=rja&uact=8&ved=2ahUKEwierOnM2tyAAxVnHzQIHcgSDFEQFnoECBMQAQ&url=https%3A%2F%2Fwww.fs.usda.gov%2Fresearch%2Ftreesearch%2F49888&usg=AOvVaw2MAjqAwEb-qJuUW-HJj8ef&opi=89978449

Sweetman, C. and Medland, L. (2017). "Introduction: Gender and water, sanitation and hygiene." *Gender & Development*, 25(2), 153–166.

Truelove, Y. (2019). "Rethinking water insecurity, inequality and infrastructure through an embodied urban political ecology." *Wiley Interdisciplinary Reviews. Water*, 6(3): e1342.

UNEP. (1992). *International conference on water and the environment: Development issues for the 21st century*. Dublin, Ireland: UNEP. [Accessed February 10, 2023]. Available from: https://wedocs.unep.org/handle/20.500.11822/30961

United Nations. (2019). *A Nexus Approach For The SDGs: Interlinkages between the goals and targets*. Geneva, Switzerland. [Accessed February 10, 2023]. Available from: https://sdgtoolkit.org/tool/a-nexus-approach-for-the-sdgs-interlinkages-between-the-goals-and-targets

UN Water. (2021). *Summary progress update 2021 – SDG 6: Water and sanitation for all* [online]. Geneva, Switzerland. [Accessed February 10, 2023]. Available from: https://www.unwater.org/publications/summary-progress-update-2021-sdg-6-water-and-sanitation-for-all

Vinyeta, K., Whyte, K. and Lynn, K. (2016). *Climate change through an intersectional lens: Gendered vulnerability and resilience in indigenous communities in the United States*. Portland, OR: U.S. Department of Agriculture, Forest Service, Pacific Northwest.

Wallace, T. and Wilson, P. (2005). "The challenge to international NGOs of incorporating gender." In Coles, A. and Wallace, T. (eds.) *Gender, water and development*. New York, NY: Routledge.

Whyte, K.P. (2014). "Indigenous women, climate change impacts, and collective action." *Hypatia*, 29(3), 599–616.

Wutich, A., Brewis, A., and Tsai, A. (2020). "Water and mental health". *Wiley Interdisciplinary Reviews: Water*, 7(5), e1461.

Yaka, Ö. (2017). "A feminist-phenomenology of women's activism against hydropower plants in Turkey's Eastern Black Sea region." *Gender, Place & Culture*, 24(6), 869–889.

Yates, J.S., Harris, L.M., and Wilson, N.J. (2017). "Multiple ontologies of water: Politics, conflict and implications for governance." *Environment and Planning D: Society and Space*, 35(5), 797–815.

Zwarteveen, M. (2011). "Questioning Masculinities in Water." *Economic and Political Weekly*, 46(18), 40–48.

29
URBAN WATER

Susan van de Meene

Introduction

In urban areas (towns, cities and megacities), water is influential in supporting human health and economic activity (Grasham et al., 2019; Herslund & Mguni, 2019). The provision of clean and reliable water supply and effective sanitation services reduces the incidence of waterborne disease and supports general hygiene. Economic and industrial activities can consume large quantities of water and also contribute wastewater which may contain pollutants that require specialised treatment; economic and industrial activities support the urban area's economic growth and development. Therefore, providing water for both domestic and industrial consumption is essential in promoting global development. Beyond water supply and sewerage services, effective drainage facilitates all aspects of urban activity: protecting health and life from flooding, reducing flood impacts on transport and other services and reducing damage to domestic, industrial and commercial premises.

This chapter surveys key developments in the literature and transitions in understanding and practice of urban water management and development. First, the conventional approach to urban water management is discussed, followed by the impacts of climate change on cities and how this is affecting and will continue to affect cities. Governance is then considered, providing a broader perspective on urban water management than solely focusing on infrastructure. Lastly, opportunities for future strategies and approaches for urban water management and development are examined. Throughout this chapter, the discussion focuses on developing and improving urban water management to support human societies, and as such it considers the diversity of urban development across a variety of countries, focusing on urban areas in developing countries. Since the challenge to pursue more sustainable urban water management that can mitigate and adapt to climate change is a global one, the discussion seeks insights from both the Global South and North that contribute to this global challenge.

Conventional Water and Urban Development

Water has been essential to supporting urban development across millennia, supporting public health, economic activity, mobility, recreation, social connectivity and ecosystem services (Brown et al., 2009). The infrastructure planned, constructed and operated to support these values has changed over time and varies to best fit the requirements of local conditions.

DOI: 10.4324/9781003095545-33

In most urban areas around the world, responsibility for water supply has been taken on by public authorities. Protecting public health through water supply infrastructure frequently involves planning and constructing large scale, centralised infrastructure which collects, treats and transports water of high quality to households and businesses in the urban area. Surface water, groundwater, desalination or a combination of these water sources have been used. Sanitation services comprise wastewater collection and treatment infrastructure, collecting sewage and industrial discharges from across the city to typically large wastewater treatment plants. Following treatment, the water is then discharged to the environment. Infrastructure used to drain cities of excess rainfall and floodwater is also an important part of the urban fabric. Drainage infrastructure comprises roads, gutters and pipes that convey water away from urban areas and thus reduce the flood risk to lives and property. Drainage infrastructure may discharge to the sewerage infrastructure, thus forming a combined system; in other cities, the drainage (stormwater) and sewerage systems are separate.

These water supply, sanitation and drainage infrastructure are typically what is found across Northern cities. Cities in the Global South have a wider variety of infrastructure that varies in type, extent and reliability (Grasham et al., 2019). For example, water supply infrastructure in urban areas in India includes: piped, public and private wells, hand pumps, open and closed public wells, and tanker truck, and water is typically available for only 4–5 hours per day (McKenzie & Ray, 2009). In Dar es Salaam, Tanzania, groundwater is a local source of water and pit latrines are used which potentially contaminates groundwater supplies (Kyessi, 2005). The intermittent supply, poor water pressure and unpredictable supplies have both financial and health impacts on these populations (McKenzie & Ray, 2009), while underdeveloped sewerage systems leads to poor sewage disposal that contaminates local water supply sources, posing a risk to human health.

Building on well-established concepts such as integrated urban water management, low impact design, sustainable urban drainage and water sensitive urban design, the provision of urban water services has been described in terms of different phases (see Bichai & Cabrera Flamini, 2018 for an overview of these terms in the Global South context). The first phase starts with a focus on water supply infrastructure, then sewerage infrastructure and drainage infrastructure (Brown et al., 2009). The infrastructure, therefore, continues to support development and protect public health, life and property. These authors outline another three phases, which support different and broader societal values: waterways – supporting ecosystem services; water cycle – supporting resource efficiency and the water sensitive city – supporting resilience to climate change and intergenerational equity (Brown et al., 2009). [The authors have identified these specific phases or characteristics of these phases in cities around the world and argue that the different city phases have been observed to transition in a largely linear fashion. However, there are opportunities for cities who may be focusing on water supply or sewerage (for example) to "leapfrog" different phases and thus support diverse and complex societal values (Brodnik et al., 2018; Brown et al., 2009; Evans et al., 2020; Grasham et al., 2019). The process of leapfrogging has been posited to need transdisciplinary science, cross-sector partnerships and innovation experiments that lead to the development of contextualised strategies and absorptive capacity, the ability of a city to learn and implement new technologies that are pre-requisites for a leap-transition (Brodnik et al., 2018).

Taking a closer look at urban areas, it can be seen that multiple city phases (as outlined above) may exist within the one city. Urban water management and urban planning are inextricably linked. The design and spatial arrangement of the city provides space for urban water (and other) infrastructure and services. As cities have developed in numerous different ways and at different rates, so to have the water infrastructure varied. For less developed countries, there are three types of development: "(a) informal settlements (unplanned urban areas commonly referred to as slums), (b) peri-urban areas (unplanned urban areas falling outside of the formal urban boundary and frequently exhibiting rural and urban characteristics) and (c) planned urban areas." (Grasham et al., 2019: 2).

The installation of water supply, sanitation and drainage infrastructure varies across these different types of development. Planned areas have higher quality, more extensive and reliable infrastructure due to the space allocated for different urban features such as roads, water supply areas (e.g. catchments, groundwater pumping areas, etc.), pipes and treatment facilities. This infrastructure is typically large scale, centralised infrastructure that services a large proportion of the urban population. Peri-urban areas have more variable infrastructure and informal settlements have infrastructure that does not provide for much of the population and is the least reliable. The geographic location of different types of urban development also affects their population's exposure to environmental risk, for example, informal areas are often located in areas that are more highly exposed to flooding, have poor drainage, little sanitation (Enqvist & Ziervogel, 2019; Grasham et al., 2019; Kyessi, 2005; Williams et al., 2018). Due to the population density of these areas, implementing change to improve drainage and water infrastructure is difficult (Williams et al., 2018). Residents of these areas are likely to have poorer physical and mental health (Grasham et al., 2019; Narayan et al., 2021). Additionally, residents of informal areas with poor water supply infrastructure or unreliable water supply need to travel distances from their homes to collect water (Kyessi, 2005). In addition to the travel time, collecting water can take time when residents need to wait for a water supply truck to pass by or wait in a queue to fill their water containers (Herslund & Mguni, 2019).

The impact of poor urban water infrastructure is complex and intertwined. Residents without good quality infrastructure take time to collect water and therefore are unable to earn an income; the poor health resulting from inadequate infrastructure also affects the ability to earn an income. Residents with a lower socio-economic status are also less able to install their own water infrastructure on their homes, such as water tanks, gutters in informal and/or peri-urban areas, thus preventing them from improving their water security (Herslund & Mguni, 2019). Providing safe, secure and reliable water supply, sanitation and drainage infrastructure supports better public health and thus enables urban residents to work, study and care for children and other family members, and also actively contribute to economic activity and development of cities.

In general terms, the concept of leapfrogging suggests that undesirable conditions that increase levels of pollution and vulnerability can be bypassed to reach the more desirable sustainable and resilient conditions. Cities in developing countries are particularly well positioned to take these leaps forward because fewer resources have been invested in traditional urban water management infrastructure and institutions which makes them more receptive to water sensitive practices (Brodnik et al., 2018). However, some scholars question the suitability for the water sensitive practices to be implemented in cities in the Global South (Bichai & Cabrera Flamini, 2018; Mishra et al., 2020). Such scholars appreciate the academic contribution of the water sensitive concept and call for greater understanding of the context where these concepts and strategies are being implemented, to avoid negative consequences of the implementation and wasting resources.

Water, climate and urban development

The impacts of climate change are being observed and felt around the world. In some areas, rainfall intensity will increase, in others droughts will increase. These changes to the rainfall and runoff patterns, and changes to groundwater recharge and availability will affect all parts of urban water infrastructure (Narayan et al., 2021): providing water supplies that have been designed on historical rainfall patterns and groundwater availability now need to be supplemented through other means (e.g. increased efficiency to reduce demand; desalination and potable reuse to increase supply). Increases in rainfall intensity will result in more frequent and larger floods due to the drainage infrastructure being overwhelmed. In areas of combined sewer and drainage systems, increased floods can have negative health effects due to the combined system flooding with raw sewage and runoff flowing out of the pipes. Some areas will experience more frequent droughts and more extreme heat events (IPCC, 2023). In urban areas, large populations will be at risk of these climate change hazards (Grasham et al., 2019), and scholars including Hobbie and Grimm (2020: 2) consider that "extreme events are the aspect of climate change that will have the most direct and obvious impact on the greatest number people in their lifetimes."

The impact of climate variability and shocks such as floods and droughts support and deepen poverty (Grasham et al., 2019). The density of cities and close association of people, infrastructure, services and other activities amplify the impact of climate change (Hobbie & Grimm, 2020). For example, during drought, water demand increases for maintenance of vegetation and cooling, which can exacerbate the impacts of poor water supply in areas with intermittent, poor quality and low water pressure in their supply systems. The ecological systems in cities are particularly at risk to climate change due to the interaction of shocks or disturbances, pests and other pathogens and other urban maintenance activities such as construction, soil compaction and de-icing (Hobbie & Grimm, 2020: 2). In summary, challenges facing urban areas that were outlined in section "Conventional Water and Urban Development" are exacerbated by climate change and thus less developed urban areas will need to overcome both the foundational challenges and those of climate change. However, the potential of cities to overcome these climate challenges is significant (Williams et al., 2018) and in the next section, we turn to the governance of urban water to better understand this potential.

Water Governance and Urban Development

When considering urban water and development, infrastructure is often at the forefront, however water governance also has an important impact on urban water and development (Enqvist & Ziervogel, 2021; Muller, 2016; Özerol et al., 2018). Water governance is often defined as "the range of political, social, economic and administrative systems that are in place to regulate development and management of water resources and provisions of water services at different levels of society" (Rogers & Hall, 2003: 88), and thus "urban water governance" focuses on examining these systems in urban areas.

The governance arrangements and systems often reflect the scale and structure of the infrastructure, for example, the governance arrangements that support large scale, centralised water supply and sewerage infrastructure are also centralised, and often publicly owned (Brown et al., 2009). Different ideal types or styles of governance have been identified: hierarchical, market-based and network. These different types of governance have the following characteristics and have been observed together in different settings:

- Hierarchical – large centralised, top-down control arrangements and processes with little public participation;
- Market – allocate resources efficiently and increase public participation; and
- Network – decentralised, small scale arrangements with varied public participation (Kjær, 2004; van de Meene et al., 2011).

Water governance is important in terms of development. With urban populations growing, the focus on cities has increased, with new conceptualisations of cities as places with their own metabolism (Neto, 2016). Challenges facing urban water governance in cities in both the global South and the global North are similar and include: inadequate catchment sources to provide water supplies and ageing pipe and treatment facility infrastructure (Olsson & Head, 2015: 5). In addition, poor urban areas face constraints such as limited financial resources, inadequate operational or maintenance capacities (Nyborg et al., 2016; Olsson & Head, 2015). However, the amount of comparative research into urban water governance in the global South is limited, which means issues particularly relating to development (e.g. justice and equity) are not often investigated (Özerol et al., 2018).

In a number of studies and analyses, Bakker (2003) argued that water governance was critical for development and particularly, ensuring the right to water. The example of Cape Town in South Africa provides insight into the interplay between socio-political conditions and drought, leading to severe water scarcity (Grasham et al., 2019). After building high quality systems models of the Cape Town water supply system, officials ignored the indications requiring future expansion of the supply system, which, when combined with extended drought periods led to severe water shortages (Muller, 2018). The city's leaders did not fully understand the social and financial implications of the stalled supply expansion, which had significant economic and social implications beyond the period of severe water shortages (Muller, 2018).

The response of Cape Town water planners and city leaders has been to expand their top-down, hierarchical governance approach and engage strategically with community groups to diversify and implement multilevel governance (Enqvist & Ziervogel, 2021; Muller, 2018). The demands on an urban water governance system are substantial: the system must plan for the long term, invest in infrastructure, respond to population and climatic changes, reduce inequality, integrate disaster risk reduction and respond to hazards (Grasham et al., 2019; Enqvist & Ziervogel, 2021; Muller, 2018).

As observed in Cape Town, the urban water system is one element of effective water governance and management. A strong enabling environment is needed to support effective and integrated water supply, sanitation (and solid waste management) that comprises: political leadership, integrated planning, shared objectives, monitoring and accountability, effective policy, capacity strengthening, institutional coordination and stakeholder support (Narayan et al., 2021: 18). Developing an effective enabling environment requires partnerships and a focus on "approaches" rather than "solutions" because approaches provide more flexibility to respond to the water governance challenges outlined above (Enqvist & Ziervogel, 2021).

Different forms of governance have been used to more effectively manage urban water and thus support development. In fact, the diversity of governance structures and characteristics aligns with those observed in practice, which have organically emerged to address poor water supply, sewerage and drainage services. Multilevel governance that supports the contributions of both city officials and community members provides opportunities for addressing the complexity of water management in urban areas, and drawing on strengths of these stakeholders to overcome shortages in resources and capacity (Enqvist & Ziervogel, 2021). Indeed, working with multiple stakeholders and developing new approaches to overcome these challenges reflects the diversity of strategies and governance that have existed (often informally) across developing cities. For example, in Dar es Salaam, residents in informal areas form collectives to generate enough capital to connect to the centralised water supply system and households sink boreholes to access water which is then sold to the broader community (Herslund & Mguni, 2019). These novel strategies demonstrate the potential of small scale, community-led water governance and management initiatives to address the challenges facing residents of informal and peri-urban areas, and thus support their development (also see e.g. Kyessi, 2005; Sletto et al., 2019).

Future Opportunities/New Pathways

The importance of water for urban development is clear and it is imperative to work towards effective water governance approaches across water supply, sewerage and drainage infrastructure and services. Engaging with the multiple aspects of water will contribute to resilience than conventional, narrowly focused approaches (Enqvist & Ziervogel, 2019), and also contribute to climate change adaptation (Ziervogel et al., 2010). Initiatives to more effectively engage with multiple stakeholders at a variety of scales can potentially "leapfrog" the errors made in more developed areas, reduce the environmental degradation and move towards sustainable urban water management more quickly (Herslund & Mguni, 2019). These authors argue that compared to "northern" cities, southern cities already undertake practices that are quite "water sensitive," for example, water consumption per household is low due to a lack of water consuming appliances such as washing machines (Herslund & Mguni, 2019: 8). Additionally, by reframing these "less resource intensive practices as desirable opportunities for sustainable urban living," less sustainable practices can be avoided (Herslund & Mguni, 2019: 8). This reframing, or institutional change, requires changes to social institutions (norms, regulations etc) and the technical part of the urban water system to realise more sustainable water management in urban areas and thus support their development (Brown et al., 2009; Pahl-Wostl, 2017). Tapping into residents' creative adaptability will help to overcome the interconnected problems and reduce the exposure to disaster risks (Herslund & Mguni, 2019). Additionally, community involvement incorporates local knowledge and preferences and builds a sense of local ownership (Narayan et al., 2021). Seeking opportunities for concurrently addressing both development and climate change adaptation can decrease opposition to adaptation strategies and thus create incentives to respond to climate change (Ziervogel et al., 2010).

In practical terms, considering what the urban water systems may look like can be undertaken by examining nature-based solutions. Nature-based solutions are also referred to as green infrastructure or low impact development (Askarizadeh et al., 2015). Examples of infrastructure include swales and wetlands, raingardens, green roofs and increase in the tree canopy cover. These tools to retain water within the city, promote infiltration and groundwater recharge and/or evapotranspiration, thereby also reducing the runoff volumes and flow rates; protection of coastal areas from sea inundation is another benefit (Hobbie & Grimm, 2020). Considering climate change impacts, the improvement of infiltration, evapotranspiration and shading can result in urban cooling and a reduction of the urban heat island effect, and also other co-benefits of improved mental health, physical health, reduced violence and crime (Hobbie & Grimm, 2020).

Conclusion

The challenge of effective and sustainable urban water management to support development is complex (Narayan et al., 2021). The conventional approach has been focused on large scale, centralised infrastructure and governance, which has not adequately provided safe, effective and reliable water services to residents across informal settlements and peri-urban areas. This is made even more challenging when adapting to the impacts of climate change: changes to rainfall and runoff patterns, groundwater availability, extreme heat and increased droughts and floods.

Research into these challenges and how to effectively govern and manage urban water has identified that diverse, multilevel governance arrangements and strategies are likely to be effective.

These strategies support community participation, incorporate local community knowledge and preferences and at times, build on the water management strategies that already exist in informal and peri-urban areas. Importantly, implementing these strategies can strengthen the potential of developing cities to leapfrog the challenges and development trajectory observed in many global north cities, thus reducing the environmental impact of conventional urban water management approaches and more effectively supporting the development of urban dwellers.

References

Askarizadeh, A., Rippy, M. A., Fletcher, T. D., Feldman, D. L., Peng, J., Bowler, P., Mehring, A. S., Winfrey, B. K., Vrugt, J. A., AghaKouchak, A., Jiang, S. C., Sanders, B. F., Levin, L. A., Taylor, S. and Grant, S. B. (2015). "From rain tanks to catchments: Use of low-impact development to address hydrologic symptoms of the urban stream syndrome." *Environmental Science & Technology*, 49(11), 264–280. https://doi.org/10.1021/acs.est.5b01635

Bakker, K. (2003). "Archipelagos and networks: Urbanisation and water privatisation in the South." *The Geographical Journal*, 169, 328–341.

Bichai, F. and Cabrera Flamini, A. (2018). "The water-sensitive City: Implications of an urban water management paradigm and its globalization." *Wiley Interdisciplinary Reviews. Water*, 5(3), 1–9.

Brodnik, C., Holden, J., Marino, R., Wright, A., Copa, V., Rogers, B., Arifin, H. S., Brown, R., Djaja11, K., Farrelly, M., Kaswanto, R. L., Marsudiantoro, D., Marthanty, D., Maryonoputri, L., Payne, E., Purwanto, M., Lovering, D. R., Suharnoto, Y., Sumabrata, J., Suwarso, R., Syaukat, Y., Urich, C. and Yuliantoro, D. (2018). "Jumping to the top: Catalysts for leapfrogging to a water sensitive city." In *IOP Conference Series: Earth and Environmental Science* 179(1), 012034. IOP Publishing, Bristol, UK.

Brown, R. R., Keath, N. and Wong, T. H. W. (2009). "Urban water management in cities: Historical, current and future regimes." *Water Science & Technology*, 59(5), 847–855.

Enqvist, J. P. and Ziervogel, G. (2019). "Water governance and justice in cape town: An overview." *Wiley Interdisciplinary Reviews. Water*, 6(4), 1–15.

Enqvist, J. and Ziervogel, G. (2021). "Multilevel Governance for Urban Water Resilience in Bengaluru and Cape Town." In Baird, J., & Plummer, R. (eds.) *Water Resilience*. Springer, Cham.

Evans, D. M., Browne, A. L. and Gortemaker, I. A. (2020). "Environmental leapfrogging and everyday climate cultures: Sustainable water consumption in the Global South." *Climatic Change*. 163, 83–97.

Grasham, C. F., Korzenevica, M. and Charles, K. J. (2019). "On considering climate resilience in urban water security: A review of the vulnerability of the urban poor in sub-Saharan Africa." *WIREs Water*, 6(3), 1–11.

Herslund, L. and Mguni, P. (2019). "Examining urban water management practices – Challenges and possibilities for transitions to sustainable urban water management in Sub-Saharan cities." *Sustainable Cities and Society*, 48, 101573.

Hobbie, S. E. and Grimm, N. B. (2020). "Nature-based approaches to managing climate change impacts in cities." *Philosophical Transactions of the Royal Society B: Biological Sciences*, 375(1794), 20190124.

IPCC (2023) Climate Change 2023: Synthesis Report. A Report of the Intergovernmental Panel on Climate Change. Contribution of Working Groups I, II and III to the Sixth Assessment Report of the Intergovernmental Panel on Climate Change [Core Writing Team, H. Lee and J. Romero (eds.)]. IPCC, Geneva, Switzerland, (in press)

Kjær, A. M. (2004). *Governance*. Polity Press, Cambridge, UK.

Kyessi, A. G. (2005). "Community-based urban water management in fringe neighbourhoods: The case of Dar es Salaam, Tanzania." *Habitat International*, 29(1), 1–25.

McKenzie, D. and Ray, I. (2009). "Urban water supply in India: Status, reform options and possible lessons." *Water Policy*, 11(4), 442–460.

Mishra, G., Acharya, G., Iyer, M. and Doshi, S. (2020). "Deconstructing water sensitivity: Experiences from global cities." *IOP Conference Series: Earth and Environmental Science*, 592(1), 12012.

Muller, M. (2016). "Urban water security in Africa: The face of climate and development challenges." *Development Southern Africa*, 33(1), 67–80.

Muller, M. (2018). "Cape Town's drought: Don't blame climate change." *Nature*, 559(7713), 174–176.

Narayan, A. S., Marks, S. J., Meierhofer, R., Strande, L., Tilley, E., Zurbrügg, C. and Lüthi, C. (2021). "Advancements in and integration of water, sanitation, and solid waste for low- and middle-income countries." *Annual Review of Environment and Resources*, 46(1), 9.1–9.27

Neto, S. (2016). "Water governance in an urban age." *Utilities Policy*, 43, 32–41.
Nyborg, K., Anderies, J. M., Dannenberg, A., Lindahl, T., Schill, C., Schluter, Adger, W. N., Arrow, K. J., Barrett, S., Carpenter, S., Chapin, I. I. I. F. S., Crepin, A.-S., Daily, G., Ehrlich, P., Folke, C., Jager, W., Kautsky, N., Livin, S. A., Madsen, O. J., Polasky, S., Scheffer, M., Walker, B., Weber, E. U., Wilen, J., Xepapadeas, X. and de Zeeuw, A. (2016). "Social norms as solutions." *Science*, 354(6308), 42–43.
Olsson, L. and Head, B. W. (2015). "Urban water governance in times of multiple stressors: An editorial." *Ecology and Society*, 20(1), 4. https://doi.org/10.5751/ES-07300-200127
Özerol, G., Vinke-de Kruijf, J., Brisbois, M. C., Casiano Flores, C., Deekshit, P., Girard, C., Knieper, C., Mirnezami, S. J., Ortega-Reig, M., Ranjan, P., Schröder, N. J. S. and Schröter, B. (2018). "Comparative studies of water governance: A systematic review." *Ecology and Society*, 23(4), 43.
Pahl-Wostl, C. (2017). "An evolutionary perspective on water governance: From understanding to transformation." *Water Resources Management*, 3(10), 2917–2932.
Rogers, P. and Hall, A. W. (2003). "Effective Water Governance." *TEC Background Papers Nr. 7*. Global Water Partnership Technical Committee (TEC).
Sletto, B., Tabory, S. and Strickler, K. (2019). "Sustainable urban water management and integrated development in informal settlements: The contested politics of co-production in Santo Domingo, Dominican Republic." *Global Environmental Change*, 54, 195–202.
van de Meene, S. J., Brown, R. R., & Farrelly, M. A. (2011). "Towards understanding governance for sustainable urban water management." *Global Environmental Change*, 21(3), 1117–1127.
Williams, D. S., Costa, M. M., Celliers, L. and Sutherland, C. (2018). "Informal settlements and flooding: Identifying strengths and weaknesses in local governance for water management." *Water (Basel)*, 10(7), 1–21.
Ziervogel, G., Shale, M. and Du, M. (2010). "Climate change adaptation in a developing country context: The case of urban water supply in Cape Town." *Climate and Development*, 2(2), 94–110.

30
WATER AND HEALTH

Jo Geere, Paul R Hunter, and Bruce Lankford

Introduction

Access to sufficient water for human consumption and domestic use is key to human health and survival (Howard et al., 2020), however, the relationships between WaSH access, water security and human health and development are complex and dynamic. Achieving and maintaining secure WaSH access associated with human health and well-being relies on social and political processes that are negotiated between people with variable levels of resource, voice and influence with which to achieve their aims (Jepson et al., 2017; Linton and Budds, 2014). Communities working and negotiating with other water users and sectors, both across stakeholder groups and upwards in the governance chain and engaging in public-private partnerships, can make an impact in terms of improved WaSH governance. This matters for stronger representation of vulnerable communities who are often excluded from political processes and live with the consequences of poor WaSH access, or when local resources need additional support, for example, during and after extreme weather events (Omasete et al., 2021). The associations between human health and development, WaSH and water security, are described in this chapter, and the complex relationships between them are illustrated as WaSH Multisystems.

Health and Water

In this first section, we outline six subsections that interconnect health and water. This then forms the basis for the second section on WaSH multisystems and water safety planning. We begin by defining what we mean by 'health.'

Health

As early as 1948, the World Health Organisation (WHO) defined health broadly as 'a state of complete physical, mental and social well-being and not merely the absence of disease or infirmity' (WHO, 2018). 'Meaningful occupations' and 'occupational justice' have also been argued as essential to health and well-being, and at the same time, are often compromised by poor WaSH access and water insecurity (Evans et al., 2013; Wrisdale & Geere, 2016). Meaningful occupations

DOI: 10.4324/9781003095545-34

include activities that an individual values and which contribute to their sense of purpose in life. They can include paid, informal or voluntary work, education, leisure, recreation and spiritual practice (Collins, 2018; Wrisdale & Geere, 2016). Furthermore, one of the most important factors enabling a meaningful occupation in adulthood, is access to an education in childhood that is minimally disrupted by ill-health (Abadzi, 2009). In addition, improved maternal education is associated with better health of the next generation (Güneş, 2015). The lack of affordable potable and productive water clearly undermines these health attributes.

Associations between WaSH access, water security and health

Public health protection through prevention of water related disease is a key rationale for Sustainable Development Goal 6 targets 6.1 and 6.2, to achieve safely managed drinking water, improved sanitation facilities and better hygiene practices for all people by 2030 (UN, 2021a). These are collectively labelled 'improved WaSH access' and most research into water and health has focused on water related hazards and routes of infectious disease transmission, as described in the 1972 'Bradley Classification' (Bartram & Hunter, 2015; Luby et al., 2018; Soares et al., 2002; Stewart et al., 2018; Wolf et al., 2018; WHO & UNICEF, 2019). The original classification included four broad disease categories: water-borne, water-washed, water-based and water-related (Bartram & Hunter 2015). Water-borne disease occurs when an infectious organism in contaminated drinking water is ingested. Water-washed diseases are those that are seen more frequently in people who do not have access to a sufficient supply of water to undertake adequate personal hygiene. Water-based diseases are those where the insect vector can only survive in close contact with a water body. Water-related are infections 'spread by insects that breed in water or bite nearby it' (White et. Al., cited by Bartram & Hunter, 2015). However, as Bartram and Hunter (2015) pointed out, this classification no longer adequately covers the range of diseases that are related to water. For example, this classification would struggle to cope with infectious agents that can spread in multiple ways, also it doesn't include non-infectious health risks. Bartram and Hunter (2015) suggested a modification of the original classification system and in Table 30.1 we show an adaptation of their suggested classification.

Whist this classification is helpful it still has its limitations especially as pathogens can spread in multiple ways. Schistosomiasis, for example, is caused by a blood fluke or trematode worm. The worm infects someone's gut or bladder and sheds its eggs in the urine or faeces. If the eggs are shed into water, they hatch, and the miracidia penetrate the bodies of certain freshwater snails where they mature for the next stage. Free swimming cercariae are then released from the snails and penetrate the skin of a new human host and the life cycle continues. In the above classification Schistosomiasis is classed as water-based disease but that classification doesn't give much insight into how the disease can be controlled through WaSH interventions. With comprehensive sanitation provision all eggs would be shed into toilets or latrines and the life cycled would not continue. But even with adequate sanitation there is still the risk of transmission from people defecating or urinating in or near water bodies. If there is an inadequate supply of drinking water then adults may have to be in contact with these compromised water sources. Even with adequate water supplies transmission may occur through children playing in local water sources.

Estimates of the total disease burden associated with inadequate WaSH vary depending on the assumptions made. In a recent analysis by a World Health Organization group, inadequate WaSH was estimated to have caused 1.6 million deaths and 105 million DALYs (Disability Adjusted Life Years – a measure of the prevalence and severity of diseases) (Prüss-Ustün et al., 2019). Of those 1.6 million deaths, 829,000 deaths were estimated to be due to diarrhoeal disease. Much of this high disease burden fell on the children under five years old and was substantially higher in

Table 30.1 Adaptation of Bartram and Hunter (2015) suggested classification.

Class	*Revised sub-class*	*Examples*
Water-borne	Infectious	Typhoid, Diarrhoeal diseases such as cholera, cryptosporidiosis, giardiasis, norovirus, Enterotoxic *E. coli* and Campylobacter Infectious hepatitis
	Toxic chemicals	Arsenic, Chlorination by products, Fluoride
	Inadequate nutrient minerals	Magnesium Fluoride in too high a dose
Water-washed (now) water-access	Superficial	Trachoma
	Intestinal	Several of the diarrhoeal diseases also included as water borne diseases
	Respiratory	Pneumonia, COVID-19
	Hydration	Dehydration
	Water collection diseases	Injury and violence during water collection and carrying, transmission of other infectious diseases whilst collecting water for example, meningococcal meningitis
Water based	Contact	Schistosomiasis, Amoebic encephalitis
	Ingested	Toxic algal poisoning, Guinea worm
Water related		Malaria River blindness, Gambian sleeping sickness
Engineered water systems	Inhaled	Legionella Mycobacterium avium complex – lung disease in immune compromised, Radon – lung cancer
	Contact	Pseudomonas folliculitis associated with hot tubs or ear infections in swimming pools

low-income countries than in middle- and high-income countries. The analysis found the continent with the highest disease burden from inadequate WaSH was Africa.

But inadequate WaSH access is associated with poorer health and adverse health outcomes more broadly than is generally included in estimates of disease burden. An analysis of data from 41 countries found that water access on premises, and near universal use of improved sanitation, was associated with increased rates of access to maternal health care, and associated with improvements to child health, indicated by reduced risk of childhood deaths, stunting, diarrhoea and acute undernutrition (Geere & Hunter, 2020). Also in a recent systematic review, a broad range of health problems were associated with collection of water away from home such as self-reported pain, injury and tiredness or fatigue (Geere et al., 2018). A study of water fetching injuries in 21 low- and middle-income countries showed that increase in distance or queueing time and increased time spent collecting water were all associated with greater risk of physical injury (Venkataramanan et al., 2020). Systematic reviews (Bisung & Elliott 2017; Geere et al., 2018) have also found strong evidence from qualitative data that people report stress or poorer mental health and well-being, reduced perinatal health care access and increased social vulnerability as consequences of collecting water away from home, and that stress or reported injuries can be due to discrimination, confrontation and interpersonal violence at or en route to shared water points. Furthermore, health, climate or conflict related events are likely to exacerbate both communicable and non-communicable health problems associated with poor access to WaSH services and water insecurity (Lankford et al., 2021). None of these issues would traditionally have been considered as contributing to the global burden of disease associated with water supplies.

The 2020 COVID-19 pandemic as an example of a global health shock highlighted how increased need for water for regular handwashing, hygiene and cleaning, and the need to maintain social distancing, create additional challenges for those with poor WaSH access (Macedo Alves et al., 2021). At the same time, the pandemic also challenged the delivery of health (Propper et al., 2020) and WaSH services (Staddon et al., 2020) to marginal groups, created social tensions and exacerbated common mental health problems such as anxiety (Johnson et al., 2021).

WaSH within health care facilities

WaSH in health care facilities refers to the provision of water, sanitation, health care waste management, hygiene and environmental cleaning infrastructure and services across all parts of a health facility. 'Health care facilities' include formally recognised facilities that provide health care. They include primary (health posts and clinics, general medical practitioner practices), secondary (district) and tertiary (national) hospitals; public, private and faith run facilities; and temporary structures for emergency contexts (WHO, 2019a, 2019b).

Worldwide in 2016, one in four health care facilities lacked basic water services, one in five had no sanitation services and two in five had no soap and water or alcohol-based hand rub (WHO, 2019a, 2019b). In low and middle income countries, 38% of health care facilities do not have any water source and up to 90% of health workers do not adhere to recommended hand hygiene practices (WHO, 2019a, 2019b). WaSH services are essential in health care facilities for personal hygiene of staff and patients, for cleaning and sterilising surfaces and instruments and for preparation of medications. Persistent lack of WaSH services and poor hygiene practices in health care facilities compromise patient safety and undermine public health, particularly affecting maternal and child health. For example, in Africa reports indicate that up to 20% of women get a wound infection after a caesarean section, hospital-born babies in low-income settings have neonatal infection rates 3–20 times higher than in high-income settings, and on average 15% of patients will acquire at least one infection in acute care hospitals (WHO, 2019a, 2019b).

In 2017, less than half of the global population were covered by essential health services (UN, 2021b) and this inequality of access combined with poor quality health care services creates a challenge to development by undermining the health and resilience of populations to ongoing adversity and shocks. The most obvious impact of inadequate WASH is on the risk of hospital acquired infections (Watson et al., 2019). However, wider impacts can include low staff morale with poorer care and delayed health care seeking behaviour in potential patients worried about the hygienic state of the hospital (Bouzid et al., 2018).

Education

The impact of the provision of WaSH in schools on absenteeism is somewhat mixed (Trinies et al., 2016). Inadequate WaSH provision in the school environment can have significant negative impacts on child attendance. Provision of free safe drinking water in class could reduce school absenteeism at least during the dry season (Hunter et al., 2014). Furthermore, there is evidence that children who become dehydrated during class, as could be expected in hot low-income countries with inadequate access to drinking water during the day, may have poorer learning (Bar-David et al., 2005). Another area of importance for WaSH provision in schools is services for menstruating girls. Managing menstruation, where girls need water to support hygiene and hand washing whilst at school, can be difficult and may lead to increased absenteeism (McMahon et al., 2011).

Livelihoods

As discussed above 'meaningful occupations' and 'occupational justice' are essential to health and well-being. Livelihoods and WaSH are beneficially connected in three main ways. First, livelihoods and income-earning opportunities are enabled by a predictable and secure supply of water (including potable water) to the household and community (Kayaga et al., 2020). In this instance people use water to produce and sell goods and services. Examples include the irrigation of garden crops; laundry and washing services; cooking food to sell; and keeping small-stock such as chickens. Second, successful livelihoods are supported by an individual's health in turn underwritten by access to water, sanitation and hygiene (Borba et al., 2007). Third, these health and income benefits combine to help inure and strengthen people's resilience to shocks and stresses such floods, drought and disease (Kohlitz et al., 2017, 2019, 2020).

Water security

Water security has been defined as 'the availability of an acceptable quantity and quality of water for health, livelihoods, ecosystems and production, coupled with an acceptable level of water-related risks to people, environments and economies' (Grey & Sadoff, 2007). This focus is on both quantity and quality but also distributional and justice dimensions should feature in definitions of water security (Zeitoun et al., 2016). Water security is also linked to sustainability of services or the source of supply. The World Health Organization estimates that the minimum need for water in an emergency setting is 7.5 to 15 litres per day, the first 2.5 to 3 litres for drinking and the rest for cooking and personal hygiene (Howard et al., 2020). It should also be remembered that there are specific groups for whom water security is particularly challenging because of exclusionary physical, social or political processes (Jepson et al., 2017; Wrisdale & Geere, 2016).

Although there has been debate in the past on the relative importance of water quantity and quality (Evans et al., 2013), both are essential to good health. Water security can be undermined, in a number of ways including the failure of a source of water leading to reduced quantity or by the failure of water treatment facilities leading to reduced quality. Water security is coming under increasing threat as a result of climate change especially following extreme weather events such as flooding and droughts (Stringer et al., 2021). Droughts are probably the biggest threat to water security and health in low-income countries. The adverse health impacts of drought can include malnutrition due to crop failure or increased risks of waterborne disease due to having to rely on alternate poorer quality water sources (Stanke et al., 2013). Interruptions in quality can follow extreme weather or be due to failure in water supply systems. Even where drinking water is not being treated for one day a month, it can undermine the entire benefit of water treatment on diarrhoeal disease reduction (Hunter et al., 2009). Indeed, most outbreaks of water-borne disease whether in low- middle- or high-income countries are associated with some failure in water security. Another source of water insecurity is armed conflict. The destruction of WaSH services during periods of war can add to the huge adverse health impacts that armed conflicts can have on the health of affected civilian populations (Schillinger et al., 2020).

WaSH within WaSH Multisystems

As is discussed in other chapters in this book, systems that deliver WaSH services are complex, comprised of many parts and interconnected with the social and natural environment in which they are situated. To reflect this complexity and variability, we have developed the concept of a WaSH multisystems to facilitate understanding and analysis of different parts and to connect the system to resilience. The concept of WaSH multisystems can be applied both to evaluate and identify key factors that are influencing existing WaSH access and services in a particular setting and situation and to develop contextually appropriate strategies to improve WaSH access so that the resilience of WaSH services to shocks can be strengthened. The model below (Figure 30.1) illustrates the complexity of WaSH systems within a multisystem.

Because the WaSH multisystem is complex, the interactions between the multisystem and human health are also complex. Depending on the constituent parts, design and workings of the multisystem, human health can be adversely affected by failures at any of several points. Alternatively actions taken in some parts of the multisystem may be able to compensate for failings elsewhere and so continue to protect human health. In many otherwise effective multisystems, failures at more than one point may be needed before adverse health impacts are seen. This is enshrined in the World health Organization guidelines for drinking water quality as the multiple barrier approach to drinking water treatment (WHO, 2017).

The most obvious evidence for the impact of failures comes from the investigation of outbreaks of disease. In one typhoid outbreak in the Democratic Republic of the Congo, a study found that the initial focus of the outbreak were military camps that had a generally lower level of hygiene and included latrines that were shared between multiple families (Ali et al., 2017). Subsequently this outbreak spread into the surrounding civilian population primarily through a highly contaminated municipal drinking water network (Brainard et al., 2018). In addition, further local spread was probably associated with unclean domestic latrines. Similarly, threats in one part of the multisystem, say the water source, can be eliminated by action elsewhere, such as improved drinking

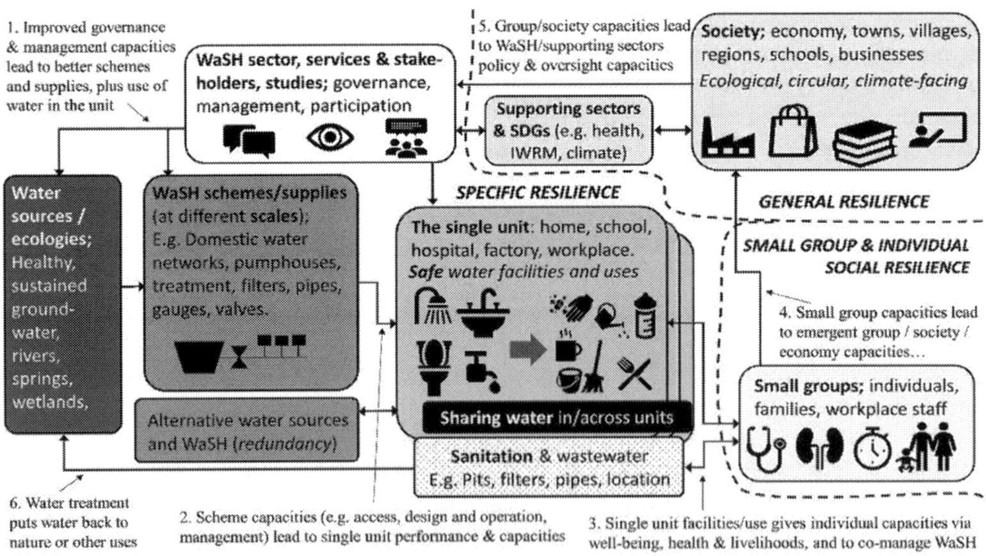

Figure 30.1 A model of the WaSH multisystem.

water treatment at a centralised treatment plant or by the action of people in their own homes such as boiling all water for drinking.

Water safety planning

Water Safety Planning has been developed as a process to safeguard water supplies for human consumption and other uses. Water safety plans (WSPs) provide a framework for assessing water related hazards and risks and appropriate responses (Bartram & Hunter, 2015). Ludi et al (2017) modified a conventional water safety plan to address additional risks posed by climate-related hazards, specifically for small and underserved communities. Their technical brief describes a step-by-step approach and possible interventions to increase climate resilience. Steps within their approach include the engagement of communities, assessment of water resources and WaSH systems, assessment of risks and hazards, the planning and implementation of WaSH interventions, the monitoring of effectiveness and reviewing outcomes before re-engaging with community members (Ludi et al 2017).

Central to WSP is hazard identification and risk prioritisation. These hazards are generally but not exclusively considered to be hazards and risks to public health. Currently risk prioritisation frequently relies on quantitative microbial risk assessment (QMRA) (WHO, 2016). However, as discussed, not all hazards are microbiological. Reliable measures of health, well-being and WaSH multisystem functioning that are relevant to the communities involved (Geere et al., 2010; Venkataramanan et al., 2020; Wutich et al., 2017), should be incorporated into evaluations of WaSH multisystems and Water Safety Planning. These may include measures of non-communicable diseases, such as musculoskeletal injury or psychological distress, access to health care services and experience of interpersonal violence, as well as incidence of communicable diseases.

Conclusion

Understanding the complex relationships between WaSH, water security and human health in development settings is key to achieving and maintaining situations of secure WaSH access that are associated with improved human health, well-being and social development. Resources and evidence-based strategies such as water and sanitation safety planning exist that can also be drawn on to plan, create and maintain WaSH multisystems that address and reduce existing inequalities, build resilience to ongoing adversity and shocks, and support progress towards the SDGs for human health and development.

References

Abadzi, H. (2009). "Instructional time loss in developing countries: Concepts, measurement, and implications." *The World Bank Research Observer*, 24, 267–290.

Ali, E., Van Den Bergh, R., D'hondt, R., Kuma-Kuma, D., De Weggheleire, A., Baudot, Y., Lambert, V., Hunter, P., Zachariah, R. and Maes, P. (2017). "Localised transmission hotspots of a typhoid fever outbreak in the Democratic Republic of Congo." *Pan African Medical Journal*, 28(179). doi: 10.11604/pamj.2017.28.179.10208.

Bar-David, Y. A., Urkin, J. and Kozminsky, E. L. (2005). "The effect of voluntary dehydration on cognitive functions of elementary school children." *Acta Paediatrica*, 94(11), 1667–1673.

Bartram, J. and Hunter, P. (2015). "Bradley classification of disease transmission routes for water-related hazards." In Bartram, J., Baum, R., Coclanis, P A., Gute, D., Kay, D., McFadyen, S., Pond, K., Robertson, W. and Rouse, M. (eds.) *Routledge Handbook of Water and Health*. London, UK and New York, NY, Routledge.

Bisung, E. and Elliott, S. J. (2017). "Psychosocial impacts of the lack of access to water and sanitation in low- and middle-income countries: A scoping review." *Journal of Water & Health*, 15(1), 17–30.

Borba, M., Smet, L. J., Sijbesma, C. and McIntyre, P. (2007). *Enhancing Livelihoods Through Sanitation*." IRC International Water and Sanitation Centre, Delft.

Bouzid, M., Cumming, O. and Hunter, P. R. (2018). "What is the impact of water sanitation and hygiene in healthcare facilities on care seeking behaviour and patient satisfaction? A systematic review of the evidence from low-income and middle-income countries." *BMJ Global Health*, 2018(3), [e000648]. https://doi.org/10.1136/bmjgh-2017-000648.

Brainard, J., D'hondt, R., Ali, E., Van den Bergh, R., De Weggheleire, A., Baudot, Y., Patigny, F., Lambert, V., Zachariah, R., Maes, P. and Kuma-Kuma Kenge, D. (2018). "Typhoid fever outbreak in the Democratic Republic of Congo: Case control and ecological study." *PLoS Neglected Tropical Diseases*, 12(10), e0006795.

Collins, M. (2018). *The Visionary Spirit: Awakening to the Imaginal Realm in the Transformocene Age*. Permanent Publications.

Evans, B., Bartram, J., Hunter, P. R., Rhoderick Williams, A., Geere, J., Majuru, B., Bates, L., Fisher, M., Overbo, A. and Schmidt, W.-P. (2013). *Public Health and Social Benefits of at-House Water Supplies. Final report*. University of Leeds, Leeds.

Geere, J. L., Cortobius, M., Geere, J. H., Hammer, C. and Hunter, P. R. (2018). "Is water carriage associated with the water carrier's health? A systematic review of quantitative and qualitative evidence." *BMJ Global Health*, 3, [e000764]. https://doi.org/10.1136/bmjgh-2018-000764

Geere, J. L. and Hunter, P. R. (2020). "The impacts of water carriage and sanitation provision on maternal and child health. A combined analysis of 49 Multiple Indicator Cluster Surveys from 41 countries." *International Journal of Hygiene and Environmental Health*, 223(1), 238–247.

Geere, J. L., Mokoena, M. M., Jagals, P., Poland, F. and Hartley, S. (2010). "How do children perceive health to be affected by domestic water carrying? Qualitative findings from a mixed methods study in rural South Africa." *Child: Care, Health and Development*, 36(6), 818–826. https://doi.org/10.1111/j.1365-2214.2010.01098.x

Grey, D. and Sadoff, C. W. (2007). "Sink or swim? Water security for growth and development." *Water Policy*, 9(6), 545–571.

Güneş, P. M. (2015). "The role of maternal education in child health: Evidence from a compulsory schooling law." *Economics of Education Review*, 47, 1–6.

Howard, G., Bartam, J., Williams, A., Overbo, A., Fuente, D. and Geere, J.-A. (2020). *Domestic Water Quantity, Service Level and Health*. Geneva, World Health Organization.

Hunter, P. R., Risebro, H., Yen, M., Lefebvre, H., Lo, C., Hartemann, P., Longuet, C. and Jaquenoud, F. (2014). "Impact of the provision of safe drinking water on school absence rates in Cambodia: A quasi-experimental study." *PLoS ONE*, 9(3), e91847

Hunter, P. R., Zmirou-Navier, D. and Hartemann, P. (2009). "Estimating the impact on health of poor reliability of drinking water interventions in developing countries." *Science of the Total Environment*, 407(8), 2621–2624.

Jepson, W., Budds, J., Eichelberger, L., Harris, L., Norman, E., O'Reilly, K., Pearson, A., Shah, S., Shinn, J., Staddon, C., Stoler, J., Wutich, A. and Young, S. (2017). "Advancing human capabilities for water security: A relational approach." *Water Security*, 1, 46–52.

Johnson, S., Dalton-Locke, C., Vera San Juan, N., Foye, U., Oram, S., Papamichail, A., Landau, S., Rowan Olive, R., Jeynes, T., Shah, P., Sheridan Rains, L., Lloyd-Evans, B., Carr, S., Killaspy, H., Gillard, S., Simpson, A., Bell, A., Bentivegna, F., Botham, J., Edbrooke-Childs, J., Goldsmith, L., Grünwald, L., Harju-Seppänen, J., Hatch, S., Henderson, C., Howard, L., Lane, R., Ledden, S., Leverton, M., Lomani, J., Lyons, N., McCrone, P., Ntephe, C. U., Ocloo, J. E., Osborn, D., Pilling, S., Poursanidou, K., Scott, H. R., Steare, T., Stuart, R., Tomlin, A., Turner, K., Tzouvara, V. T. and C.-M. H. P. R. U. Group. (2021). "Impact on mental health care and on mental health service users of the COVID-19 pandemic: A mixed methods survey of UK mental health care staff." *Social Psychiatry and Psychiatric Epidemiology*, 56(1). 25–37.

Kayaga, S., Fisher, J., Goodall, S., Kanyesigye, C., Kaggwa, R., Nambiro, M., Kitakufe, R., Otema, J. B., Mafunguro, R. and Ahabwe, G. (2020). "Enhancing livelihoods of the urban poor through productive uses of utility-supplied water services – Evidence from Kampala." *Cities*, 102(102721). https://doi.org/10.1016/j.cities.2020.102721

Kohlitz, J., Carrard, N. and Foster, T. (2019). Social-ecological system resilience for WASH. In K. Neely (ed.) *Systems Thinking and WASH: Tools and Case Studies for a Sustainable Water Supply*. Rugby, UK, Practical Action Publishing, pp. 79–92.

Kohlitz, J., Chong, J. and Willetts, J. (2020). "Rural drinking water safety under climate change: The importance of addressing physical, social, and environmental dimensions." Resources-*Basel*, 9(6), 77. https://doi.org/10.3390/resources9060077

Kohlitz, J. P., Chong, J. and Willetts, J. (2017). "Climate change vulnerability and resilience of water, sanitation, and hygiene services: A theoretical perspective." *Journal of Water Sanitation and Hygiene for Development*, 7(2), 181–195.

Lankford, B., Geere, J. L., Omasete, J., Newton-Lewis, V., Farr, J. and Casey, V. (2021). *A WaterAid Primer on WASH and Resilience*. London, UK, WaterAid.

Linton, J. and Budds, J. (2014)."The hydrosocial cycle: Defining and mobilizing a relational, dialectical approach to water." *Geoforum*, 57, 170–180.

Luby, S. P., Rahman, M., Arnold, B. D., Unicomb, L., Ashraf, S., Winch, P. J., Stewart, C. P., Begum, F., Hussain, F., Benjamin-Chung, J., Leontsini, E., Naser, A. M., Parvez, S. M., Hubbard, A. E., Lin, A., Nizame, F. A., Jannat, K., Ercumen, A., Ram, P. K., Das, K. K., Abedin, J., Clasen, T. F., Dewey, K. G., Fernald, L. C., Null, C., Ahmed, T. and Colford, J. M. M. (2018). "Effects of water quality, sanitation, handwashing, and nutritional interventions on diarrhoea and child growth in rural Bangladesh: A cluster randomised controlled trial." *Lancet Global Health*, 6, e302–315.

Ludi, E., Mosello, B. and Calow, R. (2017). "Local participatory water supply and climate change risk assessment: Modified water safety plans." *WASH Climate Resilient Development / Technical Brief.* United Nations Children's Fund (UNICEF) and Global Water Partnership (GWP): New York, NY and Stockholm. Available at https://www.gwp.org/globalassets/global/about-gwp/publications/unicef-gwp/gwp_unicef_tech_a_web.pdf [accessed 13/07/2023].

Macedo Alves, E., Geere, J.-A., Gutierres Arteiro da Paz, M., Jacobi, P., Abreu de Castro Grandisoli, E. and Sulaiman, N. S. (2021). "Water security in two megacities: Observations on public actions during 2020 in São Paulo and London." *Water International* 46(6), 883–899.

McMahon, S. A., Winch, P. J., Caruso, B. A., Obure, A. F., Ogutu, E. A., Ochari, I. A. and Rheingans, R. D. (2011). "'The girl with her period is the one to hang her head' Reflections on menstrual management among schoolgirls in rural Kenya." *BMC International Health and Human Rights* 11, 7. https://doi.org/10.1186/1472-698X-11-7

Omasete, J., Forster, J. and Geere, J. (2021). *WASH as the Foundation for Building Climate Resilience of Vulnerable Communities*. London, UK, WaterAid.

Propper, C., Stoye, G. and Zarakano, B. (2020). "The wider impacts of the coronavirus pandemic on the NHS." *Fiscal Studies*, 41(2), 345–356.

Prüss-Ustün, A., Wolf, J., Bartram, J., Clasen, T., Cumming, O., Freeman, M. C., Gordon, B., Hunter, P. R., Medlicott, K. and Johnston, R. (2019). "Burden of disease from inadequate water, sanitation and hygiene for selected adverse health outcomes: An updated analysis with a focus on low-and middle-income countries." *International Journal of Hygiene and Environmental Health*, 222(5), 765–777.

Schillinger, J., Özerol, G., Güven-Griemert, Ş and Heldeweg, M. (2020). "Water in war: Understanding the impacts of armed conflict on water resources and their management." *Wiley Interdisciplinary Reviews,* Water, 7(6), DOI: 10.1002/wat2.1480

Soares, L. C., Griesinger, M. O., Dachs, J. N., Bittner, M. A. and Tavares, S. (2002). "Inequities in access to and use of drinking water services in Latin America and the Caribbean." *Pan American Journal of Public Health*, 11(5–6), 386–396.

Staddon, C., Everard, M., Mytton, M., Octavianti, T., Quinn, P. W. N., Uddin, S. M. N., Young, S. L., Miller, J. D., Budds, J., Geere, J., Meehan, K., Charles, K., Stevenson, E. G. J., Vonk, J. and Mizniak, J. (2020). "Water insecurity compounds the global coronavirus crisis." *Water International*, 45(5), 416–422.

Stanke, C., Kerac, M., Prudhomme, C., Medlock, J. and Murray, V. (2013). "Health effects of drought: A systematic review of the evidence." *PLoS Currents*, 5, ecurrents.dis.7a2cee9e980f91ad7697b570bcc4b004.

Stewart, C. P., Kariger, P., Fernald, L., Pickering, A. J., Arnold, C. D., Arnold, B. F., Hubbard, A. E., Dentz, H. N., Lin, A., Meerkerk, T. J., Milner, E., Swarthout, J., Colford, J. M. and Null, C. (2018). "Effects of water quality, sanitation, handwashing, and nutritional interventions on child development in rural Kenya (WASH Benefits Kenya): A cluster-randomised controlled trial." *Lancet Child and Adolescent Health*, 2(4), 269–280.

Stringer, L. C., Mirzabaev, A., Benjaminsen, T. A., Harris, R. M. B., Jafari, M., Lissner, T. K., Stevens, N. and Tirado-von der Pahlen, C. (2021). "Climate change impacts on water security in global drylands." *One Earth*, 4(6), 851–864.

Trinies, V., Garn, J. V., Chang, H. H. and Freeman, M. C. (2016). "The impact of a school-based water, sanitation, and hygiene program on absenteeism, diarrhea, and respiratory infection: A matched–control trial in Mali." *The American Journal of Tropical Medicine and Hygiene*, 94(6), 1418–1425.

UN. (2021a). "Goal 6: Ensure availability and sustainable management of water and sanitation for all." [Accessed February 10, 2023]. Available from https://sdgs.un.org/goals/goal6

UN. (2021b). "Goal 3: Ensure healthy lives and promote well-being for all at all ages." [Accessed October 10, 2021]. Available from https://sdgs.un.org/goals/goal3

Venkataramanan, V., Geere, J. L., Thomae, B., Stoler, J., Hunter, P. R. and Young, S. L., and H. R. Household Water Insecurity Experiences Research Coordination Network. (2020). "In pursuit of 'safe' water: The burden of personal injury from water fetching in 21 low-income and middle-income countries." *BMJ Global Health*, 5(10), https://doi.org/10.1136/bmjgh-2020-003328.

Watson, J., D'Mello-Guyett, L., Flynn, E., Falconer, J., Esteves-Mills, J., Prual, A., Hunter, P. R., Allegranzi, B., Montgomery, M. and Cumming, O. (2019). "Interventions to improve water supply and quality, sanitation and hand WaSHing facilities in healthcare facilities, and their effect on healthcare-associated infections in low-income and middle-income countries: A systematic review and supplementary scoping review." *BMJ Global Health*, 4(4), e001632.

WHO. (2016). *Quantitative Microbial Risk Assessment: Application for Water Safety Management*. World Health Organization, Geneva.

WHO. (2017). *Guidelines for Drinking-Water Quality: Fourth Edition Incorporating the First Addendum*. World Health Organization, Geneva ISBN 978-92-4-154995-0 https://www.who.int/publications/i/item/9789241549950

WHO. (2018). "Constitution of WHO: Principles." [Accessed May 5, 2018]. Available from http://www.who.int/about/mission/en/

WHO. (2019a). "1 in 4 health care facilities lacks basic water services – UNICEF, WHO." 302021 [Accessed October 30, 2021]. Available from https://www.who.int/news/item/03-04-2019-1-in-4-health-care-facilities-lacks-basic-water-services-unicef-who

WHO. (2019b). Water, sanitation and hygiene in health care facilities: Practical steps to achieve universal access. Geneva, World Health Organization (WHO).

WHO and UNICEF. (2019). *Progress on Household Drinking Water, Sanitation and Hygiene 2000–2017: Special Focus on Inequalities*. Geneva, World Health Organization (WHO) and United Nations Children's Fund (UNICEF).

Wolf, J., Hunter, P. R., Freeman, M. C., Cumming, O., Clasen, T., Bartram, J., Higgins, J. P. T., Johnston, R., Medlicott, K., Boisson, S. and Prüss-Ustün, A. (2018). "Impact of drinking water, sanitation and handwashing with soap on childhood diarrhoeal disease: Updated meta-analysis and meta-regression." *Tropical Medicine and International Health*, 23(5), 508–525.

Wrisdale, L. and Geere, J. (2016). "Factors that impact on access to water and sanitation in rural South Africa: An occupational justice perspective." *Journal of Occupational Science*, 24(3), 259–279.

Wutich, A., Budds, J., Eichelberger, J., Geere, L., Harris, J., Horney, L. M., Jepson, J. A., Norman, W., O'Reilly, E., Pearson, K., Shah, A. L., Shinn, S. H., Simpson, J., Staddon, K., Stoler, C., Teodoro, J. and Young, M. P. (2017). "Advancing methods for research on household water insecurity: Studying entitlements and capabilities, socio-cultural dynamics, and political processes, institutions and governance." *Water Security*, 2, 1–2.

Zeitoun, M., Lankford, B., Krueger, T., Forsyth, T., Carter, R., Hoekstra, A. Y., Taylor, R., Varis, O., Cleaver, F., Boelens, R., Swatuk, L., Tickner, D., Scott, C. A., Mirumachi, N. and Matthews, N. (2016). "Reductionist and integrative research approaches to complex water security policy challenges." *Global Environmental Change*, 39, 143–154.

31

WATER, SANITATION, AND HYGIENE

Nelson Ekane

The Water, Sanitation, and Hygiene (WaSH) Backlog

About 3.6 billion people in the world lack safely managed sanitation[1] services and about 2 billion people lack safely managed drinking water services. Open defecation (OD) is still rife and practiced by about 494 million people. The majority of these people live in rural areas in countries in Central and Southern Asia, Oceania, and sub-Saharan Africa (SSA) (WHO/UNICEF JMP, 2021). Access to water, sanitation, and hygiene (WaSH) is intrinsically important and the benefits are mutually reinforcing (UNDP, 2006). Although the provision of water often takes precedence, provision of sanitation is also of critical importance. Yet, sanitation provision continues to lag behind the provision of water supply (WHO/UNICEF JMP, 2021). One of the consequences of this is the contamination of surface and groundwater sources by poor excreta management practices. This is most easily discernible in Central and Southern Asia, Oceania, and SSA where on-site sanitation is more common than sewer connections even in urban areas (WHO/UNICEF JMP, 2021). For instance, in SSA which is experiencing the slowest rate of progress in terms of improved access to water and sanitation services in the world, only 13% of the population is connected to a sewer network. About 47% of the population depend on onsite sanitation facilities of different standards (ESAWAS, 2022).

The factors contributing to these inadequacies are multifarious, cross-cutting, and complex in nature in terms of technology and design on one hand and social norms and behaviours on the other. Generally, governance of basic WaSH services in some of the regions where the problem is greatest, SSA in particular is characterized by the state acting through non-state actors in supposedly flexible and inclusive state–citizen interactions with non-state actors (private, civil society, and international organisations) filling gaps in resources and service delivery (Oosterveer, 2009; Van Vliet et al., 2011; Ekane et al., 2019). The outcome in terms of inclusive, sustainable, and cost-effective service delivery is mixed in many cases. This is partly explained by the fact that development concerns such as those related to WaSH have predominantly been approached in a linear manner which misses the complexity of the context and the problem (Nordveit, 2010; Van Vliet et al., 2011; Burns & Worsley, 2015). As a result, WaSH programmes and projects have been largely supply or technologically driven (WSP, 2010), and subsidy-driven or donor-dependent (ECA, 2012). These have ingrained different forms of path dependent behaviour

DOI: 10.4324/9781003095545-35

among actors at different levels of society in regions like SSA (Ekane et al., 2014) wherein more emphasis is being placed on infrastructure provision (focusing on numbers of facilities) than on supporting the delivery of functional and sustainable services (GLAAS, 2017). Moreover, WaSH is not getting the same investment commitments as other basic infrastructure projects and programmes. Also, spatial disparity in WaSH service delivery results in lower access and quality in rural areas than in urban areas. This disparity is also glaring between the rich and poor in urban areas (Estache & Wodon, 2014). This situation is exacerbated by the fact that improvement in access to WaSH services is only marginally higher than population growth in regions where the problem is greatest (WHO/UNICEF JMP, 2021). The quality of government in some countries in these regions is also an important contributing factor to the low coverage of WaSH services and slow progress in improving access (Rothstein, 2011). This reflects that some public and private agencies and NGOs delivering basic WaSH services face problems of legitimacy due to low levels of accountability and trust (Ekane, 2018a; Mason et al., 2018; Revilla et al., 2021). This has implications for citizens' willingness-to-pay for services and to actively participate in development (Rothstein, 2011), given that participation partly depends on the type and quality of services citizens receive and their corresponding level of satisfaction (Brinkerhoff et al., 2018). In this chapter, emphasis is placed on improving understanding on the complex societal and behavioural issues surrounding sanitation and hygiene.

Sanitation and Hygiene Enigma

Despite good progress in access to sanitation and hygiene services in some developing countries in recent years, major challenges remain (WHO/UNICEF JMP, 2021). Even when behaviours and practices are successfully changed, new behaviours, and practices are difficult to sustain.What encourages the hygiene behaviours and sanitation practices that experts and practitioners are discouraging? Why do people not always do the things that policy prescribes? Which approaches actually work on the ground in changing sanitation practices and hygiene behaviours? Is the problem the target population, the policies, and strategies, or both? These are some of many complex questions of sociocultural and sociopolitical nature that cannot be tackled with simple answers or trivial solutions (Bouton, 2014; Carter, 2015; Kelly & Barker, 2016; Mara, 2017). This is particularly the case for OD practices, proper use of toilet facilities, and handwashing with water and soap which the SDG for sanitation addresses (WHO/UNICEF JMP, 2021). Handwashing with water and soap and hand disinfection in particular have never been as salient globally as they are presently as a primary barrier to stop transmission of the Coronavirus.

There is increasing interest in improving understanding of why people targeted by public interventions don't do the things that policy prescribes. Also, there is increasing recognition that the problems, context, and complexities of changing behaviours necessitate even more interdisciplinary approaches to improve understanding of the extent to which behavioural interventions affect people's reflection and actions (Carter, 2015). This chapter contributes to this strand of literature and draws on applied behavioural sciences and policy implementation theories to examine why target populations don't do that which experts want them to do, and what it takes to sustainably eliminate OD, promote proper and sustained use of toilet facilities, and instil handwashing with soap at critical times. This is done through the lens of 'carrots' (subsidies, motivation), 'sticks' (fines, penalties), 'sermons' (education, encouragement), and 'nudges' (social or physical environment) which are instruments used to trigger and sustain behaviour change.

Sanitation, Health, and Development

OD and other poor hygiene practices impose costs on society. The human and environmental health costs (Prüss-Ustün et al., 2014; UNDP, 2016) and wider impact on human and economic development (WSP, 2012; UNDP, 2016) are well reported. Poor sanitation and hygiene practices take a toll on human health, particularly among children under five years of age and the vulnerable in society (Prüss-Ustün et al., 2014). Poor health impairs the productive ability of people and keeps them away from school and work. This has implications for human development and undoubtedly exacerbates poverty.

Investing in sanitation and hygiene has been shown to lead to direct health and indirect economic benefits (Wolf et al., 2014; WSP, 2015; Esteves & Cumming, 2016). These are good arguments for universal compliance with proper handwashing and defecation practices, but due to different factors that this chapter examines the potential benefits are yet to trigger complete transformation of undesirable behaviours and practices in communities where they prevail. High prescribed standards for sanitation facilities, high cost of piloted solutions (McGranahan, 2015), structural inequalities and remoteness of rural settlements (O'Reilly et al., 2017), and unavailability of building materials and expertise (Pickering et al., 2015) are some of many factors perpetuating the problem or slowing down progress. Part of the problem is the tendency for experts or practitioners within the sanitation sector to impose their reasoning and rationale on target populations or beneficiaries (Carter, 2015; Bateman & Engel, 2018). This pertains to cognitive biases of experts and even decision-makers in the design and evaluation of policy interventions (Strassheim, 2019).

Policy Approaches to Sanitation

From a public policy point of view, sanitation practices and hygiene behaviours are fostered and managed in different ways to reduce health and social implications. In this chapter, public policy is used to refer to a system of regulatory measures, laws, guidelines, and approaches or interventions used by governments and other non-state actors to change, maintain, or create living conditions that are conducive to human welfare (Shafir, 2013). From a top-down perspective, a combination of different policy instruments are used on target populations to influence their behaviours, and attempt to achieve policy purposes. Target populations are offered reinforcing incentives that invite voluntary exchange through social marketing techniques including motivation or subsidies; coerced to behave in non-voluntary ways through fines, penalties, and other punitive measures within the law; and persuaded to behave voluntarily in a particular manner through education, encouragement, and moral appeals (Bemelmans-Videc et al., 2010). Most of the above strategies like many others employed in several health and hygiene promotion programmes rely mainly on conscious, deliberate, and rational thinking which demands effort in making sense of complex situations (BMJ, 2011; WSP, 2016). These are driven by system 2 mode of thinking (Kahneman, 2011) and constitute the basis for the rational agent model that has hitherto dominated policy-making (Barr et al., 2013). In response to the shortcomings of the assumptions of the rational agent model, applied behavioural scientists advance a more nuanced perspective to policy-making which is intended to protect people from exploitation by market forces by helping them make decisions that serve their best interest (Thaler & Sunstein, 2008; Barr et al., 2013). As one of such strategies, nudge is designed to influence, prompt, or remind targets of appropriate behaviours by altering social or physical environments to make certain behaviours such as handwashing more likely (BMJ, 2011; WSP, 2016). This relies mainly on System 1 mode of thinking which is automatic, affective, and effortless. Both systems 1 and 2 constitute the psychological mechanisms underlying human judgement and decision-making, and in many cases, system 2 endorses what system 1 suggests (Kahneman, 2011). Both systems are important to consider in improving understanding on why people don't do the things that policy prescribes. Thus, a combination of interventions that collectively address both systems is most likely to yield powerful behaviour change strategies (WSP, 2016).

Regarding the extent to which chosen policy instruments contribute to attainment of policy goals in specific contexts, i.e., policy effectiveness, Ingram and Schneider (1991) emphasize that target populations must make decisions and take actions consistent with policy goals for desired outcomes to be attained. Weaver (2014) reiterates that the objectives of government policies can only be achieved if target populations engage in specific behaviours that are consistent with set policy objectives. This is, however, difficult to attain as target populations do not all respond to policy in the same way. Barr et al. (2013) emphasize that understanding and intention do not necessarily lead to action and that behaviour and choices are shaped by contextual nuances. In other words, target populations face multiple barriers to change to the behaviours and practices that policy seeks. This is an important aspect to consider, especially when compliance with policy is low as is the case in some regions with low coverage of sanitation and hygiene services. Individuals may not understand why they have to change certain behaviours and practices or why they are expected to behave in a certain way. This may be because change is imposed on them or because of mistrust in the implementers. This relates to information and cognition constraints (don't want to comply). Even when they acknowledge the need for change, they may be confronted with resource and autonomy constraints ('can't comply'). In other cases, they may simply resist change either because of lack of incentives or weak monitoring and enforcement ('don't need to comply') (Weaver, 2015).

Further, policy instruments do not always deliver expected outcomes and sometimes produce unintended and unwanted effects (Schneider & Ingram, 1990). This is partly attributed to wrong assumptions about cause-effect relationships with resultant poor policy design. The other part has to do with coordination problems and interaction of practitioners and target populations during implementation (Andrews et al., 2017). Pertaining to sanitation and hygiene, these have been reported to contribute to the mismatch between policy intentions, actions, and outcomes on the ground (Ekane, 2018a, 2018b; Ekane et al., 2019).

Choice of Instruments

Different regimes are known to use different instruments even when addressing similar problems (Schneider & Ingram, 1990). Similarly, the choice of approaches depends on the influence the policy community has on the policy agenda (Bateman & Engel, 2018). Burns and Worsley (2015) also argue that the view of experts on development issues occurs within a defined and subjectively bounded domain. This can be said to be the case with some approaches for sanitation provision and hygiene promotion (Engel & Susilo, 2014; Carter, 2015; Bateman & Engel, 2018). Another case in point is the systematic production of evidence through Randomized Controlled Trials (RCTs) which is largely promoted for evaluating policies by virtue of the strength in internal validity. According to Dunlop and Radaelli (2015), heavy reliance on RCTs for evidence-based policy making is a good example of collective illusion of control among experts and policy-makers. This approach of evaluating policy outcomes does not emphasize social and cultural processes and unintended consequences which must be considered when programmes and projects are implemented at scale (Deaton & Cartwright, 2016).

Community Approaches to Sanitation

A variety of participatory community-based approaches to sanitation and hygiene are widely promoted and funded mainly by donor organizations and implemented across developing countries. The aim of these approaches is to create group endorsement for new sanitation and hygiene norms in line with the global development agenda (World Bank, 2015). These approaches are based on principles of social change, triggering of demand, self-realization and collective action, leadership, and innovation. (UNICEF, 2009; Mehta & Movik, 2011; Bateman & Engel, 2018). Some of the most commonly implemented approaches to change sanitation practices and hygiene behaviours in the developing context include Community Health Club (CHC), Community-Led Total Sanitation (CLTS), and Sanitation Marketing (SanMark).

The choice, appropriateness, and effectiveness of the abovementioned approaches and their inherent instruments is a subject of debate in the sanitation sector. From an expert point of view, much is written on the pros and cons of different approaches in changing sanitation practices and hygiene behaviours (Waterkeyn & Cairncross, 2005; Pattanayak et al., 2009; Harvey, 2011; Engel & Susilo, 2014; Freeman et al., 2014; Sigler et al., 2014; Brooks et al., 2015; Burns & Worsley, 2015; Guiteras et al., 2015; Pickering et al., 2015; Kelly & Barker, 2016; WSP, 2016; Watson et al., 2017; Garn et al., 2017; Crocker et al., 2017a, 2017b; Bateman & Engel, 2018; Grover et al., 2018; Tagat & Kapoor, 2018).

Embedded in CHC, CLTS, and SanMark are instruments such as subsidies, motivation ('carrots'); fines, penalties, and even emotional appeals like shame ('sticks'); education, encouragement and moral appeals ('sermons'); and changes in the environment (nudges). CLTS mainly uses a combination of sermons (education, encouragement and moral appeals) and 'sticks' (social shaming, naming) to trigger changes in defecation habits and create open defecation free communities (ODF) (UNICEF, 2009; Sigler et al., 2014; Crocker et al., 2017a, 2017b; Zuin et al., 2019). The CHC approach is mainly built on sermons (education, encouragement and moral appeals) in achieving zero open defecation communities (Waterkeyn & Cairncross, 2005; Brooks et al., 2015; Rosenfeld et al., 2021). SanMark uses mainly carrots (incentives and other social marketing techniques) and sermons (education or encouragement) to create demand for toilets and strengthen supply by building capacity of the local private sector (Barrington et al., 2017). Of the three approaches, CLTS is the only one having some elements of 'sticks' characterized by the evocation of shame and disgust to deter OD practices (Bateman & Engel, 2018).

Appropriateness and Effectiveness of Instruments

The underlying behavioural assumptions of instruments differ, so does their outcomes in different contexts. These assumptions guide the choice of instruments for different policy purposes and, hence, are important for investigating how instruments are implemented and their outcomes in different contexts (Schneider & Ingram, 1990).

Reinforcement – positive incentives ('carrots') and negative incentives including punishment ('sticks') are known to contribute to learning and behaviour change. However, Baumeister et al. (2001) show that punishment leads to faster learning than reinforcement. Along similar lines, Van der Doelen (2010) distinguishes between stimulative or rewarding instruments (subsidies, information programmes) which increase possible alternative types of behaviour and promote legitimacy of interventions, and punishing or repressive instruments (fines) which limit alternatives of behaviour but contribute to desired changes in behaviour and enhance effectiveness of interventions. He concludes that a mix of stimulative and repressive instruments is the key for acceptance and effectiveness of policy.

'Sermons' are considered the most lenient and legitimate instruments (Vedung & Van der Doelen, 2010). Educational appeals recommend and encourage desirable behaviours by showing or promising societal as well as self-interested future benefits (Rothschild, 1999). This is based on the assumption that people will act rationally if given access to education or information (Mehta & Movik, 2011). This is, however, not always effective in a free choice society since self-interests are not often in line with societal interests. As a result, universal compliance is difficult to attain in the short-term (Rothschild, 1999). Moreover, educational appeals often convey long-term benefits for society but may be unable to persuade individuals with short-term expected benefits (Rothschild, 1999). This is a possible shortcoming of CHC which largely depends on this instrument. OD and poor hygiene behaviours present negative externalities to society and, therefore, discouraging such practices and behaviours is for the public good (Ekane, 2018a; Ekane et al., 2019).

The appropriateness of 'carrots', notably direct subsidies in the form of construction materials and even complete latrine structures in sanitation provision and hygiene promotion is well reported. The outcome has been mixed as in the case of India where millions of latrines subsidized by the government are not used for the intended purpose (Mehta & Movik, 2011). Poorly targeted direct subsidies have been reported to undermine community ownership, momentum and ultimately sustainability, especially in communities that: are above the poverty line (Mehta & Movik, 2011; Burns & Worsley, 2015); have previous experience of subsidies (Harvey, 2011); have high baseline sanitation and hygiene coverage (Guiteras et al., 2015). Both CLTS and CHCs discourage the use of direct subsidies even though rewards and encouragement are common with both approaches (Mehta & Movik, 2011). These include certification, prizes, etc. However, Pattanayak et al. (2009) show from studies in Orissa, India that a combination of subsidy and shame during CLTS is more effective than using shame alone to trigger change, particularly for households below the poverty line. Further, recent studies suggest that direct subsidies may be more effective in contexts where building materials and expertise are not available (Pickering et al., 2015).

With respect to 'sticks' or punishment including naming, shaming, bad emotions are known to leave a lasting impression on people, producing thorough and careful cognitive processing and behaviour change as compared to good emotions (Baumeister et al., 2001). This exemplifies how the intuitive processes of system 1 subsequently guides information processing and judgement by system 2 (Kahneman, 2011). This is a plausible explanation for the use of fines, penalties, publicizing names, and evocation of shame and disgust to trigger change in OD practices. However, the enforcement of fines and penalties remains lax in poor contexts in particular, which makes them ineffective as people seldom pay them (Ekane et al., 2019). Naming and shaming have been shown to be effective in triggering behaviour change in some contexts, particularly in households above the poverty line (Pattanayak et al., 2009). Despite the potential effectiveness of naming and shaming, these remain contentious in the sanitation sector (Pattanayak et al., 2009; Engel & Susilo, 2014; Sigler et al., 2014; Pickering et al., 2015; WSP, 2016). The appropriateness and actual effectiveness of CLTS in particular (Pattanayak et al., 2009; Pickering et al., 2015) in poor and vulnerable communities are questioned (WSP, 2016). From a moral point of view, Engel and Susilo (2014) and Bateman and Engel (2018) show that the coercive nature of shame in particular, a pernicious social emotion, undermines effectiveness in promoting long-term behaviour change especially in poor communities. This is consistent with other authors who emphasize that poverty-related policies that evoke shame or stigma are counterproductive (Walker et al., 2013; Dolezal & Lyons, 2017), and undermine human dignity (Chase & Walker, 2012; Bateman & Engel, 2018) and sustainability (Crocker et al., 2017a, 2017b). As a result, the psychosocial implications of eliciting shame in the CLTS process challenges the human rights-based approach to development (Bateman & Engel, 2018). It is for this reason that the Ministry of Rural Development in Cambodia removed the shame component from the CLTS programme in 2013 (MRD/DRHC, 2013; Bateman & Engel, 2018). Following from the above, the assumption that poor, powerless, disadvantaged, vulnerable, or marginalized target populations are less capable of learning or responding positively to educational appeals (Vedung & Van der Doelen, 2010) may justify the use of emotional appeals, sanctions, or force that reinforces stigma and powerlessness particularly among the marginalized groups (Dolezal & Lyons, 2017).

Nudge is designed to influence, prompt, or remind targets of appropriate behaviours by altering social or physical environment to make certain behaviours such as handwashing more likely (BMJ, 2011; Kelly & Barker, 2016; WSP, 2016). However, only few studies on nudging sanitation practices and hygiene behaviours have so far been performed (Dreibelbis et al., 2016; Grover et al., 2018; Tagat & Kapoor, 2018). Dreibelbis et al. (2016) observe that low-cost nudges can prompt significant behaviour change and result in higher rates of handwashing among school-aged children in Bangladesh. Still among school children in Bangladesh, Grover et al. (2018) also observe a sustained improved handwashing behaviour 5 months after a nudge intervention in trials comparing environmental nudges and high-intensity hygiene education intervention on handwashing behaviours. Tagat and Kapoor (2018) acknowledge the influence of religious imagery in public spaces as a deterrent for unhygienic sanitation behaviours in India. These accounts are encouraging but by no means enough to draw conclusions on the effectiveness of nudging in changing sanitation practices and hygiene behaviours. Further, there is limited evidence to suggest that nudges alone will lead to healthier behaviours (BMJ, 2011; Neal et al., 2015). Even more, there is insufficient evidence to recommend nudges as a single approach to handwashing behaviour change (WSP, 2016; Watson et al., 2017). Generally, the evidence base for nudging is still emerging and nudges for handwashing have been so far tested in mainly single contexts or on a limited scale (GHP, 2017) as in Bangladesh (Grover et al., 2018). Thus, more primary research and synthesis of existing evidence is needed to examine the effectiveness and acceptability of nudging interventions in different contexts (BMJ, 2011). Similarly, Crocker et al. (2017a) emphasize the need for further studies to establish the long-term sustainability of CLTS outcomes as well as to evaluate CLTS in combination with interventions having different designs and objectives. Rosenfeld et al. (2021) also recommend studies that compare the CHC approach to other WaSH promotion and behaviour change interventions.

Conclusion

Sanitation service delivery has a strong hygiene behaviour change component which is a very important complement. However, unlike many other social or basic infrastructure services such as water and electricity, sanitation, and hygiene are more complex for a number of reasons, one of which has to do with changing personal or private practices and hygiene behaviours and related sociocultural and traditional norms. Individual sanitation practices and hygiene behaviours including defecation and handwashing habits are generally personal or private issues. Moreover, there are various sociocultural and traditional norms and perceptions of dirt and disease suggesting different kinds of behaviours and practices. This is perhaps most easily discernible in contexts where people hold steadfastly to sociocultural and traditional beliefs, taboos, moral codes, etc.

Following from the above, sanitation and hygiene problems will not be solved by merely building toilets and handwashing facilities. This is because people's behaviours and practices depend to a large extent on context and culture. We know that technological advances have played and will continue to play a key role in providing different options for transforming sanitation and hygiene conditions in different parts of the world. For instance, the latrine is crucial in this process as it helps to physically separate people from their excreta. This has contributed immensely to improvements in human and environmental health and general well-being. However, despite sanitation technology development, diffusion, or adoption of innovations remains slow in some parts of the world.

Pertaining to access and use of toilet facilities, part of the problem is that overemphasis on technology-driven, subsidy-based supply-driven approaches has so far produced sub-optimal results. Undoubtedly, the challenge remains in providing access to context-specific technology solutions and most importantly in getting people to use them.

Universal compliance to recommended practices and behaviours is not achievable in the short-term through educational messages alone which are the most legitimate instruments. Other instruments can trigger short-term desired changes but may not always be legitimate. Education is suitable for targets that 'don't want to comply' due to cognitive, belief, and peer-effect barriers but education may not be enough to overcome self-interest in order to attain the much-needed universal compliance and eradication of OD by 2030. Other instruments are needed to reach individuals that 'don't need to comply' due to monitoring and enforcement problems and those that 'can't comply' due to resource and autonomy problems. Even though the use of coercive instruments such as shaming, especially in poor contexts, is debated, different forms of 'sticks' including shaming, naming and fines are generally perceived as effective in changing sanitation practices and hygiene behaviours such as OD and handwashing but are likely to compromise social relations and values in society. This insight is consistent with previous work on poverty-related shaming.

Note

1 In this chapter, sanitation refers to the provision of services and facilities for the collection, handling, treatment, disposal and/or use of mainly human excreta, and the related health and hygiene behavioural aspects.

References

Andrews, M., Pritchett, L. and Woolcock, M. (2017). *Building State Capability: Evidence, Analysis, Action*. Oxford: Oxford University Press.

Barr, M., Mullainathan, S. and Shafir, E. (2013). "Behavioural informed regulation." In: Shafir, E. (ed.) *The Behavioural Foundations of Public Policy*. New Jersey: Princeton University Press. pp. 440–461.

Barrington, D. J., Sridharan, S., Shields, K. F., Saunders, S. G., Souter, R. T. and Bartram, J. (2017). "Sanitation marketing: A systematic review and theoretical critique using the capability approach." *Social Science and Medicine*, 194, 128–134.

Bateman, M. and Engel, S. (2018). "To shame or not to shame—That is the sanitation question." *Development Policy Review*, 36, 155–173. https://doi.org/10.1111/dpr.12317

Baumeister, R., Bratslavsky, E., Finkenauer, C. and Vohs, K. (2001). "Bad is stronger than good." *Review of General Psychology*, 5(4), 323–370.

Bemelmans-Videc, M.-L., Rist, R. and Vedung, E. (eds.). (2010). *Carrots, Sticks, and Sermons: Policy Instruments and Their Evaluation*. New Brunswick, NJ: Transaction Publishers.

BMJ. (2011). "Judging nudging: Can nudging improve population health? Analysis," *BMJ*, (342), d228.

Bouton, M. (2014). "Why behaviour change is difficult to sustain." *Prevention Medicine*, 68, 29–36.

Brinkerhoff, D., Wetterberg, A. and Wibbels, E. (2018). "Distance, services, and citizen perceptions of the state in rural Africa." *Governance*, 31, 103–124.

Brooks, J., Adams, A. Bendjemil, S. and Rosenfeld, J. (2015). "Putting heads and hands together to change knowledge and behaviours: Community Health Clubs in Port-au-Prince, Haiti." *Waterlines*, 34(4).

Burns, D. and Worsley, S. (2015). *Navigating Complexity in International Development*. Rugby, UK: Practical Action Publishing.
Carter, R. (2015). "What does it take to change behaviour? Editorial," *Waterlines*, 34(4), 277–278.
Chase, E. and Walker, R. (2012). "The co-construction of shame in the context of poverty: beyond a threat to the social bond." *Sociology*, 47(4): 739–754.
Crocker, J., Saywell, D. and Bartram, J. (2017a). "Sustainability of community-led total sanitation outcomes: Evidence from Ethiopia and Ghana." *International Journal of Hygiene and Environmental Health*, 220, 551–557.
Crocker, J., Saywell, D., Shields, K., Kolsky, P. and Bartram, J. (2017b). "The true costs of participatory sanitation: Evidence from community-led total sanitation studies in Ghana and Ethiopia." *Science of the Total Environment*, 601–602, 1075–1083.
Deaton, A. and Cartwright, N. (2016). "Understanding and misunderstanding Randomized Controlled Trials (RCTs)." *NBER Working Paper No. 22595*. Princeton, NJ: Princeton University.
Dolezal, L. and Lyons, B. (2017). "Shaming people about their lifestyle habits does nothing to improve their health." *The Conversation*. [Accessed: June 2018]. Available from http://theconversation.com/shaming-people-about-their-lifestyle-habits-does-nothing-to-improve-their-health-88427
Dreibelbis, R., Kroeger, A., Hossain, K., Venkatesh, M. and Ram, P. (2016). "Behaviour change without behavior change communication: nudging handwashing among primary school students in Bangladesh." *The International Journal of Environmental Research and Public Health*, 13, 129. https://doi.org/10.3390/ijerph13010129.
Dunlop, C. and Radaelli, C. (2015). "Overcoming illusions of control: How to nudge and teach regulatory humility." In: Alemanno, A. and Sibony, A. (eds.). *Nudge and the Law: A European Perspective*. Oxford: Hart Publishing, pp. 139–158.
Ekane, N. (2018a). *Making Sanitation Happen: An Enquiry into Multi-Level Sanitation Governance*. PhD Thesis. KTH.
Ekane, N. (2018b). "How can sanitation policy deliver in Africa? Insights from Rwanda and Uganda. SEI Policy Brief." [Accessed 30 June 2018]. Available from https://www.sei.org/publications/can-sanitation-policy-deliver-africa-insights-rwanda-uganda/
Ekane, N., Kjellén, M., Westlund, H., Ntakarutimana, A. and Mwesige, D. (2019). "From words to action: Linking sanitation policy to service delivery in Rwanda and Uganda." *Development Policy Review*, 38(3), 344–365.
Ekane, N., Nykvist, B., Kjellén, M., Noel, S. and Weitz, N. (2014). "Multi-level sanitation governance: an approach to understanding and overcoming the challenges in the sanitation sector in sub-Saharan Africa." *Waterlines*, 3(33), 242–256. https://doi.org/10.3362/1756-3488.2014.024.
Engel, S. and Susilo, A. (2014). "Shaming and sanitation in Indonesia: A return to colonial public health practices?" *Development and Change*, 45(1), 157–178.
ESAWAS. (2022). *The Water Supply and Sanitation Regulatory Landscape Across Africa*. Eastern and Southern Africa: Water and Sanitation Regulators Association.
Estache, A. and Wodon, Q. (2014). *Infrastructure and Poverty in Sub-Saharan Africa*. NY: Palgrave Macmillan Publishers Limited.
Esteves, M. and Cumming, J. (2016). *The Impact of Water, Sanitation, and Hygiene on Key Health and Social Outcomes: Review of Evidence: Review of Evidence*. DFID Evidence Paper. SHARE and UNICEF. [Accessed 5 January 2021]. Available from https://www.lshtm.ac.uk/sites/default/files/2017-07/WASHEvidencePaper_HighRes_01.23.17_0.pdf
ECA. (2012). "European Union development assistance for drinking water supply and basic sanitation in Sub-Saharan countries," European Court of Auditors (ECA) *Special report* # 13. [Accessed 3 March 2020]. Available from https://op.europa.eu/en/publication-detail/-/publication/8da0f2c6-ea47-4b58-a8fe-c7de33a7c726/language-en
Freeman, M., Stocks, M., Cumming, O., Jeandron, A., Higgins, J., Wolf, J., Prüss-Üstün, A., Bonjour, S., Hunter, P., Fewtrell, L. and Curtis, V. (2014). "Hygiene and health: Systematic review of handwashing practices worldwide and update of health effects." *Tropical Medicine & International Health*, 19, 906–916.
Garn, J., Sclar, G., Freeman, M., Penakalapati, G., Alexander, K., Brooks, P., Rehfuess, E., Boisson, S., Medlicott, K. and Clasen, T. (2017). "The impact of sanitation interventions on latrine coverage and latrine use: A systematic review and meta-analysis." *International Journal of Hygiene and Environmental Health*, 220, 329–340.

GHP. (2017). "Using Nudges to Encourage Handwashing with Soap. Global Handwashing Partnership." [Accessed 16 March 2018]. Available from https://globalhandwashing.org/wp-content/uploads/2017/11/Using-Nudges-to-Encourage-Handwashing-with-Soap.pdf

GLAAS. (2017). "Financing Universal Water, Sanitation and Hygiene under the Sustainable Development Goals (SDGs): UN-Water Global Analysis and Assessment of Sanitation and Drinking-Water." *GLAAS* 2017 Report.

Grover, E., Hossain, M. K., Uddin, S., Venkatesh, M., Ram, P. K. and Dreibelbis, P. K. R. (2018). "Comparing the behavioural impact of a nudge-based handwashing intervention to high-intensity hygiene education: A cluster-randomised trial in rural Bangladesh." *Tropical Medicine and International Health*, 23(1), 10–25. https://doi.org/10.1111/tmi.12999.

Guiteras, R., Levinsohn, J. and Mobarak, A. (2015). "Sanitation subsidies. Encouraging sanitation investment in the developing world: A cluster-randomized trial." *Science*, 348, 903–906.

Harvey, P. (2011). "Zero subsidy strategies for accelerating access to rural water and sanitation services." *Water Science and Technology: A Journal of the International Association on Water Pollution Research*, 63, 1037–1043.

Ingram, H. and Schneider, A. (1991). "The choice of target populations." *Administration and Society*, 23(3), 333–356.

Kahneman, D. (2011). *Thinking Fast and Slow*. New York, NY: Ferrar, Straus and Giroux.

Kelly, M. and Barker, M. (2016). "Why is changing health-related behaviour so difficult?" *Public Health*, 136, 109–116.

Mara, D. (2017). "The elimination of open defecation and its adverse health effects: A moral imperative for governments and development professionals. Review paper," *WASHDev*, 7(1), 1–12.

Mason, N., Oyaya, C. and Boulenouar, J. (2018). "Reforming urban sanitation under decentralisation: Cross-country learning for Kenya and beyond." *Development Policy Review*, 38(1), 42–63.

McGranahan, G. (2015). "Realizing the right to sanitation in deprived urban communities: Meeting the challenges of collective action, coproduction, affordability, and housing tenure." *World Development*, 68, 242–253.

Mehta, L. and Movik, S. (eds.). (2011). *Shit Matters: The Potential of Community-Led Total Sanitation*. Warwickshire, UK: Practical Action Publishing Ltd.

MRD/DRHC. (2013). *National Guidelines on Community-Led Total Sanitation*. Phnom Penh: Royal Government of Cambodia.

Neal, D., Vujcic, J., Hernandez, O. and Wood, W. (2015). "The science of habit: Creating disruptive and sticky behaviour change in handwashing behavior." *WASHplus*. [Accessed 05 June 2017]. Available from http://www.washplus.org/sites/default/files/resource_files/habits-neal2015.pdf

Nordveit, B. (2010). "Development as a complex process of change: Conception and analysis of projects, programmes and policies." *International Journal of Educational Development*, 30, 110–117.

O'Reilly, K., Dhanju, R. and Goel, A. (2017). "Exploring 'the remote' and 'the rural': Open defecation and latrine use in Uttarakhand, India." *World Development*, 93, 193–205.

Pattanayak, S., Yang, J.-C., Dickinson, K., Poulos, C., Patil, S., Mallick, R., Blitstein, J. and Praharaj, P. (2009). "Shame or subsidy revisited: Social mobilization for sanitation in Orissa, India. Bull." *World Health Organization*, 87, 580–587.

Pickering, A., Djebbari, H., Lopez, C., Coulibaly, M. and Alzua, M. (2015). "Effect of a community-led sanitation intervention on child diarrhea and child growth in rural Mali: A cluster-randomised controlled trial." *Lancet Global Health*, 3, e701–e711.

Prüss-Ustün, A., Bartram, J., Clasen, T., Colford, J. T., Curtis, C., Bonjour, V., Dangour, S., De France, A. D., Fewtrell, J., Freeman, L., Gordon, M. C., Hunter, B., Johnston, P. R., Mathers, R. B., Mäusezahl, C., Medlicott, D., Neira, K., Stocks, M., Wolf, M. and Cairncross, J. (2014). "Burden of disease from inadequate water, sanitation, and hygiene in low- and middle-income settings: A retrospective analysis of data from 145 countries." *Tropical Medicine and International Health*, 19(8), 894–905.

Revilla, M. L. D., Qu, F., Seetharam, K. E. and Rao, B. (2021). "Sanitation in the top development journals: A review." *ADBI Working Paper 1253*. Tokyo: Asian Development Bank Institute. [Accessed 28 January 2022]. Available from https://www.adb.org/publications/sanitation-top-development-journals-review

Rosenfeld, J., Berggren, R. and Frerichs, L. (2021). "A review of the community health club literature describing water, sanitation, and hygiene outcomes." *International Journal of Environmental Research and Public Health*, 18(4), 1880.

Rothschild, M. (1999). "Carrots, sticks, and promises: A conceptual framework for the management of public health and social issue behaviours." *Journal of Marketing*, 63(4), 24–37.

Rothstein, B. (2011). *The Quality of Government. Corruption, Social Trust, and Inequality in International Perspective*. Chicago: University of Chicago Press.

Schneider, A. and Ingram, H. (1990). "Behavioural assumptions of policy tools." *Journal of Politics*, 52(2), 510–529.

Shafir, E. (2013). "Introduction." In: Shafir, E. (ed.) *The Behavioral Foundations of Public Policy*. NJ: Princeton University Press. pp. 1–9.

Sigler, R., Mahmoudi, L. and Graham, J. (2014). "Analysis of behavioural change techniques in community-led total sanitation programmes." *Health Promotion International*, 30(1), 16–28.

Strassheim, H. (2019). "Behavioural mechanisms and public policy design: Preventing failures in behavioural public policy." *Public Policy and Administration*, *36*(2), 187–204. https://doi.org/10.1177/0952076719827062.

Tagat, A. and Kapoor, H. (2018). "'Sacred nudging' and sanitation decisions in India." *India Review*, 17(3), 301–319. https://doi.org/10.1080/14736489.2018.1473320.

Thaler, R. H. and Sunstein, C. (2008). *Nudge: Improving Decisions About Health, Wealth, and Happiness*. New Haven, CT: Yale University Press.

UNDP. (2006). "Beyond Scarcity: Power, Poverty, and the Global Water Crisis." *Human Development Report*, NY. 388p.

UNDP. (2016). "UNDP Support to the Implementation of Sustainable Development Goal 6. Sustainable Management of Water and Sanitation." *United Nations Development Programme* (UNDP).

UNICEF. (2009). "Community Approaches to Total Sanitation: Based on case studies from India, Nepal, Sierra Leone, Zambia." *UNICEF Policy and Programing in Practice.*

Van der Doelen, F. (2010). "The 'give and take' packaging of policy instruments: Optimizing legitimacy and effectiveness." In: Bemelmans-Videc, M-L., Rist, R. and Vedung, E. (eds.) *Carrots, Sticks and Sermons: Policy Instruments and Their Evaluation*. New Brunswick, NJ, and London: Transaction Publishers.

Van Vliet, B., Spaargaren, G. and Oosterveer, P. (2011). "Sanitation under challenge: Contributions from the social sciences." *Water Policy*, 13, 797.

Vedung, E. and Van der Doelen, F. (2010). "The sermon: Information programmes in the public policy process – Choice, effects, and evaluation." In: Bemelmans-Videc, M-L., Rist, R. and Vedung, E. (eds.) *Carrots, Sticks and Sermons: Policy Instruments and Their Evaluation*. New Brunswick, NJ, and London: Transaction Publishers.

Walker, R., Bantebya Kyomuhendo, G., Chase, E., Choudhry, S., Gubrium, E., Yongmie Nocola, J., Lodemel, I., Mathew, L., Mwiine, A., Pellissery, S. and Ming, Y. (2013). "Poverty in global perspective: Is shame a common denominator?" *Journal of Social Policy*, 42, 215–233.

Waterkeyn, J. and Cairncross, S. (2005). "Creating demand for sanitation and hygiene through community health clubs: A cost effective intervention in two districts in Zimbabwe." *Social Science and Medicine*, 61, 1958–1970.

Watson, J. A., Ensink, J., Ramos, M., Benelli, P., Holdsworth, E., Dreibelbis, R. and Cumming, O. (2017). "Does targeting children with hygiene promotion messages work? The effect of handwashing promotion targeted at children, on diarrhoea, soil-transmitted helminth infections and behaviour change, in low- and middle-income countries." *Tropical Medicine and International Health*, 22(5), 526–538.

Weaver, K. (2014). "Compliance regimes and barriers to behaviour change." *Governance*, 27(2), 243–265.

Weaver, K. (2015). "Getting people to behave: Research lessons for policy makers." *Public Administration Review*, 75(6), 806–816.

WHO/UNICEF JMP (2021). "Progress on Household Drinking Water, Sanitation, and Hygiene 2000–2020: Five Years into The SDGs." *UNICEF and WHO 2021.*

Wolf, J., Prüss-Ustün, A., Cumming, O., Bartram, J., Bonjour, S., Cairncross, S., Clasen, T., Colford, J. M., Curtis, V., De France, J., Fewtrell, L., Freeman, M. C., Gordon, B., Hunter, P. R., Jeandron, A., Johnston, R. B., Mäusezahl, D., Mathers, C., Neira, M. and Higgins, J. P. T. (2014). "Systematic review: Assessing the impact of drinking water and sanitation on diarrhoeal disease in low-and middle-income settings: Systematic review and meta-regression." *Tropical Medicine and International Health*, 19(8), 928–942.

World Bank. (2015). *World Development Report 2015: Mind, Society, and Behaviour*. Washington, DC: World Bank.

WSP. (2010). *Financing on-Site Sanitation for the Poor: A Six Country Comparative Review and Analysis. WSP Sanitation Global Practice Team*. Washington, DC: WSP, World Bank [Technical Paper].

WSP. (2012). *Promoting Handwashing Behaviour: The Effect of Mass Media and Community Level Interventions in Peru*. Washington, DC: WSP, World Bank.

WSP. (2015). *Investment in Sanitation to Support Economic Growth in Africa: Recommendations to the African Ministers' Council on Water (AMCOW) and Ministers of Finance*. Washington, DC: WSP, World Bank.

WSP. (2016). *Nudging and Habit Change for Open Defecation: New Tactics from Behavioural Science.* Washington, DC: WSP, World Bank.

Zuin, V., Delaire, C., Peletz, R., Cock-Esteb, A., Khush, R. and Albert, J. (2019). "Policy diffusion in the rural sanitation sector: Lessons from community-led total sanitation (CLTS)." *World Development*, 124, 104643.

32
DIGITAL WATER

Karen Bakker, Rosemary Knight, Raymond T. Ng, Alan K. Mackworth, and Max Ritts

We wish to dedicate this chapter to the memory of its lead author, Karen Bakker (1971–2023). Karen passed away tragically during the final stages of its completion to an aggressive form of cancer. Her voice was and remains our project's central guiding force; the one that brought us together to write this piece. Thank you Karen.

Introduction

Digital technologies are spurring a transformation of water management and governance. This chapter examines the application of digital technologies to freshwater management, which may dramatically alter the role that water plays in development and societal evolution. Proponents of these technologies promise that they will accelerate sustainability, enhance hazard mitigation, and incorporate citizens in novel ways in decision-making (Bakker & Ritts, 2018). Sceptics argue that these technologies raise ethical concerns: privacy, transparency, and data colonialism. Additionally, critics point to the water demands of digital infrastructures, in the context of hydrosocial risks (particularly relating to climate change) which are disproportionately borne by underserved communities (Farinosi et al., 2018; Deitz & Meehan, 2019). This debate raises complex questions which merit further scrutiny, given that water is a vital resource for human and environmental health and well-being: a flow resource, essential for life, which sits at the water development nexus.

Our definition of digital includes the socio-technical aspects of digital innovations, grouped into three categories of instrumentation, interconnectedness, and intelligence: (1) Instrumentation refers to digital devices that collect data, often in an automated fashion (e.g., sensors, drones, nano-satellites, smartphone-based interfaces); (2) Interconnectedness refers to cloud-based computing and internet-based databases that enable an exponential increase in data collection and distribution; (3) Intelligence refers to artificial intelligence, data science, and other computational techniques that enable the automation of data analysis and decision-making, and which have transformative potential for social learning and adaptive water management. Combined, these technologies mean that water management and governance will be assisted (or even undertaken) by algorithms. The evolving socio-technical "computational apparatus" that we explore in this chapter does not apply to all forms of water governance. Rather, we focus on examples where digital transformation is articulated with evolution in water governance practices, which is referred to as: "Digital Water."

DOI: 10.4324/9781003095545-36

The Digital Revolution

Over the past several decades, a rapid set of technological innovations has transformed environmental monitoring capabilities (Bakker & Ritts, 2018). These innovations have significant implications for water monitoring and management. As the discussion below illustrates, these technologies create a tension between increasingly centralized forms of digital monitoring, data storage, and analysis; and decentralized yet digitally enabled modalities of citizen science and management. Digital transformation thus creates a constantly shifting dialectic of inclusion and exclusion, which is inflected with socio-economic, political, and intersectional equity issues.

Remote Sensing and Sensor Technologies

In the early 1990s, NASA's "Mission to Planet Earth" launched a global vision: a system of environmental observation satellites that would revolutionize environmental science. Today, several hundred observation satellites orbit Earth, and remote sensing networks are proliferating, driven by nanosatellites and new private sector providers that supply hourly/daily sampling intervals and global coverage. Earth-imaging platforms now combine satellite, airborne, drone, ground-surface, and subsurface data across a range of spatiotemporal scales. Monitoring surface water quantity *and* quality by satellite is technically feasible; with the advent of low-cost nanosatellites, data is also cheaper and more accessible. For example, multispectral satellite data can be used to monitor reservoir water quantity and quality (Bresciani et al., 2019). Landsat thermal imagery can be used to monitor evapotranspiration and drought. Similarly, GRACE satellite monitoring of soil moisture has been correlated with the Palmer Drought Severity Index, with applications for drought and food insecurity forecasts (Anderson et al., 2012; Long et al., 2013).

In the past decade, recent technological advances have also led to the development of cost-effective sensors that are wireless, battery-powered, and "smart" (in the sense of having considerable processing capabilities). The most common sensors typically collect quantitative data (e.g., temperature, salinity, concentrations of certain compounds), but can also capture text streams and videos (such as by drones), and even bioacoustical data expressive of shifting rainfall patterns and riparian ecologies (Ritts & Bakker, 2021). Wireless in-situ sensors for monitoring agricultural water use, for example, are now used in many countries. These sensors can detect water levels, as well as pollutants such as nitrates (Kotamäki et al., 2009; Gartia et al., 2012). The combined application of sensors and data loggers generates real-time information about changing water-related parameters. Environmental sensor networks (ESNs) enable data to be compared and correlated across space and time. The interface between the rapidly proliferating categories of ESNs and remote sensing, supported by internet-based communications and cloud-based data storage, enables unprecedented real-time monitoring and visualization. This has practical applications for water management in the global South, including flood warning alerts for rural areas in Bangladesh and Vietnam, or monitoring the impact of refugee camps on local water systems (Biswas et al., 2020; Nguyen et al., 2020).

Digital Water thus potentially presents a solution to sparse data, slow reporting, and delayed action on water-related threats. However, a new challenge arises: data deluge. Planners have proposed using artificial intelligence (AI) to deal with large datasets (Fang et al., 2019); deep learning may be used to interpret satellite and aerial images. Artificial intelligence is now used in river water quality modelling, the design of novel treatment technologies for water pollution, and in the optimization of process automation for water treatment and desalination plants (Fan et al., 2018; Shafiee et al., 2018; Al Ani et al., 2019; Sun & Scanlon, 2019; Tung & Yaseen, 2020). These AI algorithms can underpin real time decision-support systems for a diverse array of water governance applications, from precision agriculture (reducing water use) to flood hazard response. Automated agricultural techniques (including automation of water and pesticide application, as well as crop rotation) are another example of practical applications.

Even as water governance frameworks undergo digital transformation, water systems remain under-monitored in many regions. Indeed, as a combination of population growth and water scarcity places increasing pressure on water quality and quantity in countries around the world, accurate and continuous monitoring of surface water becomes even more vital. However, water monitoring is consistently underfunded, aggravating risks such as pollution or extreme events. The problem is exacerbated by a shortage of trained personnel. Here, too, digital tools are being tested as a solution. VR training tools provide a method of training personnel for accurate monitoring of water discharge in open channel flows, providing a novel option for quickly training and reassigning staff, and increasing their efficiency in the field (Mirauda et al., 2020). Another emerging development is the use of Augmented Reality to monitor water bodies and water-related landscape features, such as riverbanks. In these schemes, detailed features of existing underground water systems – the component materials, as well as location, age and diameter of water pipes, for instance – can be viewed by technicians above ground, saving expensive and potentially dangerous excavations. However, as explored below, these approaches have yielded uneven results across actual water management regimes and are articulated with socio-economic power structures. The obvious risk is that this deepens the digital divide.

Digital Water and Citizen Science

Digital transformation enables novel forms of user engagement in water monitoring and research in novel ways. New sensor technologies, the increasing availability of smartphone apps, social media, and new data interfaces allowing greater citizen control of data management, are accelerating a shift to citizen-led initiatives and, in some cases, enabling polycentric monitoring and governance (e.g., dynamic, decentralized, and diverse networks or data collection, see also: Jollymore et al., 2017; Pahl-Wostl, 2017; Zheng et al., 2017; Yu et al., 2019).

To give just a few water-related examples, digital crowdsourcing initiatives have been launched for hydrological monitoring in Kenya (Weeser et al., 2018), stream monitoring in the United States (Lowry & Fienen, 2013), and water quality monitoring in China (Zheng et al., 2017). These initiatives differ from earlier citizen science programmes due to their scalability and low cost, as enabled by the mobilization of internet-based interfaces and/or low-cost sensors that allow many untrained observers to collect data on hydrological measurements (e.g. stream stages), facilitated by mobile phone apps, or simple text messaging interfaces that store and display text messages as data on the web (see, e.g., Wong & Kerkez, 2016; Pule et al., 2017; Paul & Buytaert, 2018). The Tsinghua Environment Monitoring Platform (TEMP[1]), for example, deploys WeChat accounts to collect water quality information and monitor water pollution using Volunteered Geographic Information from the public (Zheng et al., 2017). Non-humans, both biotic and abiotic, are also distributed for bio-logging and environmental monitoring, particularly in sites that are challenging for humans to reach, such as deep subsurface water measurements in lakes and oceans; for example, both seals and drones are mounted with sensors in order to track ocean temperatures (Kays et al., 2015).

The Contradictions of Digital Water

In this section, we briefly explore some of the opportunities and challenges posed by the Digital Water agenda at the water-development nexus. First, we discuss the example of groundwater management; an important issue, given that groundwater provides drinking water for over 50% of the world's population, and is vital to agricultural production in many low- and middle-income countries, and rural areas worldwide. Second, we discuss the example of Digital Water technologies applied to urban areas in the global South: home to 3 billion (predicted to be 4 billion by 2030) people, and sites of increasing water stress (UNDESA, 2018).

Groundwater Management in Water-Stressed Agricultural Zones: The Case of California

The advent of modern drilling and pumping technology, combined with the electrification of rural areas, has led to an exponential increase of groundwater exploitation. As the world's population continues to grow, pressures on groundwater are intensifying, notably for agriculture. This is leading to significant water stress and scarcity in groundwater dependent regions, particularly those where groundwater sources are being depleted in an unsustainable manner.

Could digital technologies address this situation, and if so, how? Consider the case of California, which is arguably the region with the most advanced network of Digital Water tools in the world. Following the passage of the Sustainable Groundwater Management Act (SGMA) in 2014, local agencies in California are now charged with both groundwater planning (policy design) and management (policy implementation). This creates a new opportunity for digitally enabled dynamic management. Over the past decade, there have been significant advancements in several technologies that provide information about the state of groundwater systems, with unprecedented coverage. For example, water managers are now able to use technologies such as satellite Interferometric Synthetic Aperture Radar (InSAR) to monitor subsidence and saltwater intrusion in real time.[2] When combined with digital tools that enable "precision agriculture," these data-driven, computationally intensive systems can optimize real-time water use to an unprecedented degree. This is paramount given the well-known increasing threats to groundwater and growing hydrological variability in the context of climate change.

While digital innovation is enabling advances in "top down" remote sensing technologies, it is also enabling advances in "bottom up" citizen science technologies. Researchers are using artificial intelligence-supported citizen science to monitor marine ecosystems and high-tide flooding (Freiwald et al., 2018; Golparvar & Wang, 2020), as well as to promote flood risk communication via the FloodRise project (Cheung & Feldman, 2019). Simultaneously, researchers are attempting to design digital strategies that enable citizen scientists to increase the credibility of their data (which is often a major shortcoming of involving non-experts), and improve water data management (Mayton & Story, 2018). However, Chase and Levine (2018) note the following significant pitfalls: a lack of diversity amongst participants (which tend to be largely white, older, affluent, well-educated); and a concomitant limited impact of citizen science programmes on pre-environmental attitudes. Moreover, Kim et al. (2011) points out that digitally intensive citizen science programmes are only helpful if the data is actually used in decision-making. More data does not automatically lead to more inclusion or better management outcomes.

Moreover, digital technologies have other types of impacts on groundwater. The expansion of wafer production in Silicon Valley in the 1980s, for example, generated significant groundwater contamination (Olivieri et al., 1985). A legacy of contaminated water continues to challenge environmental cleanup efforts in the region and the geographical distribution of water pollution is linked to broader patterns of environmental injustice (Lécuyer, 2017). More recently, the high degree of water (as well as energy) consumption by data centres has come under scrutiny (Siddik et al., 2021); the "data center industrial complex" (as Hogan terms it) intensifies patterns of environmental extraction and injustice (Hogan, 2015, 2018, 2021; Johnson & Hogan, 2017; Jacobson & Hogan, 2019). In the data centre, "Digital Water" assumes a new material expression in the form of "warm water cooling," which is industry-speak for managing computer systems that generate excess heat. A key ethical question emerges: can there be viable forms of digital water governance which do not fuel water-hungry data centres? While tech fixes promising a "cooler cloud" might ameliorate this problem in the long-term (e.g., Staedter, 2019), this creates a contradiction. Technologies that promise to enhance water monitoring and protection are predicated on a set of digital technologies that accelerate groundwater demand and worsen water crises.

The California case mirrors the global challenge facing Digital Water agendas, which largely sidestep structural issues related to political economy (e.g., water rights regimes) and infrastructure development. Moreover, digital technologies pose ethical challenges stemming from their potential environmental impacts. Data derived from an expanded array of sensors, continuously sampling the physical world, requires processing predicated on real-time big-data analytics with high energy use. Furthermore, e-waste is an issue, particularly given the dispersed nature of these distributed networks. The widespread deployment of digital technologies for water monitoring and governance will only be technologically feasible and politically acceptable if these environmental issues are addressed. Analogous to the Green Revolution (which increased yields but also water pollution), the Digital Water agenda optimizes water inputs, but also generates increased climate emissions. It is thus a tool, but not a panacea, at the water-development nexus.

Digital Water and Cities in the Global South

Digital Water technologies have enormous potential to impact water access for the 3 billion people living in urban areas in the Global South (predicted to be more than 4 billion by 2030) (UNDESA, 2018). However, there has only been limited application of digital technologies to water services in these contexts. Could digital technologies enable water services providers to better grapple with the enormous challenges of providing water sustainably and equitably in urban areas, particularly mega-cities? If so, what are the pitfalls? Organizations such as the World Bank have lauded the potential of Digital Water technologies (Ndaw, 2015) whereas scholars have sounded a cautionary note (Amankwaa et al., 2021).

With respect to water supply networks within cities (which tend to have incomplete spatial coverage of the urban population), there is some evidence that enhancing data capture (whether through "distant" technologies such as remote sensing or "proximate" technologies such as smart meters) can enhance efficiency of water infrastructure, thereby reducing the gap between supply and demand. There is also evidence that digital technologies may have positive financial impacts on water services providers by reducing operational costs through the introduction of mobile payments, reducing maintenance costs through faster, more precise detection of leaks and low-pressure zones, and increasing revenue through prepaid meters and water ATMS (water vending machines devices which, analogous to a banking ATM, dispense a volume of water upon payment).

However, although these digital water technologies enhance efficiency and revenue for the water services providers, this does not cover the entirety (or, in the case of many mega-cities) even the majority of the population. Low-income residents of large cities in the global South are the least likely to have access to piped water in their homes; digital water technologies such as water ATMs may further entrench existing inequalities in water supply, whether by imposing fees or substituting pay-for-service water sources for free ones. Occasionally, significant time savings may be achieved for women and children, given their responsibility for sourcing water for households. This benefit often derives from closer proximity of physical infrastructure rather than the digital nature of water meters or vending machines. Moreover, digital water innovations may have the unintended effect of aggregating and financializing water-related costs that low-income households previously paid in small amounts, or "paid" in time rather than money. While, in some cases, water consumers might feel empowered by having greater access to digital information (e.g., insight into their water bills, faster response to complaints, timely notification of water cutoffs in rationed systems), this tends to be a benefit for middle- and higher-income households specifically (Ndaw, 2015; Hanjahanja & Omuto, 2018).

The Digital Water innovations experienced as a positive innovation for these groups may, in contrast, be perceived negatively by lower-income households by increasing the cost of water and enabling faster disconnections. Some users in Kenya, for example, covered the upfront costs of prepaid meters by taking out small loans, which they then found hard to repay (Guma, 2019). Low-income households may also find it challenging to seek assistance from the water supply provider when services move "online." For households without computers or internet access, the costs of seeking assistance using pay-for-use mobile phones can quickly accelerate. The costs associated with adopting Digital Water technologies, particularly where these are passed on to end users, present challenges to both adoption and use for low-income households. The digital divide thus risks exacerbating inequities in water access in cities in the global South.

Digital Water technologies are also often associated with a set of broader changes in the management and political economy of water supply systems. The sociopolitical context in which many water supply systems were developed over the past century emphasized provision of water as a basic service to citizens. Water was often subsidized or even provided free to low-income consumers. In contrast, Digital Water technologies are often aligned with a neoliberal agenda which frames users as customers rather than citizens. Subsidies may be reduced or even eliminated based on a "pay what you use" model that does not account for significant income disparities or a lack of internet access within large cities. Ironically, Digital Water technologies may further entrench the inequalities that their proponents are seeking to address (Amankwaa et al., 2021).

Challenges at the Water-Development Nexus

Regulatory/data challenges

Digital Water offers the potential for a significant enhancement of existing water management regimes, but also for disruption (for the aforementioned water and energy consumption reasons, above all else). Global transparency can be created by expanding access to data from satellites and other sources. For some proponents, robust traceability can be provided by distributed ledgers, such as blockchain, which is enabling monitoring of compliance with environmental and social standards across far-flung supply chains.

New environmental data capabilities also enable non-state actors to take regulation and management into their own hands. The growing ubiquity of smartphones and social media, for example, offers farmers tools to access (and collect) water management information that can help them better manage irrigation, while also enabling regulators to monitor irrigation water use via remote sensing (Kamienski et al., 2018; Foster et al., 2020). These technologies also empower citizens to monitor compliance with environmental protections, and also spur enforcement action. This creates the potential for new modalities of polycentric monitoring and governance (Buytaert et al., 2014).

Big Data-driven management also poses daunting political and institutional challenges for freshwater governance. Efforts to move to more nimble, adaptive management can provoke resistance from those who have vested interests in current systems. Managers and regulators will increasingly operate under the glare of growing public scrutiny and will often have to navigate a proliferating array of competing solutions.

Many of these challenges will centre on data: who collects it, controls it, profits from it. There are several aspects to this challenge. First, sharing data across distributed and varied monitoring networks must be standardized; better meta-data generation and management are essential. DaTaONE is an example of innovative environmental cyber-infrastructure for Earth observational datasets. However, protocols are far from being universal, and considerable challenges attend the implementation of environmental data technologies across scales and borders.

Second, to reduce the amount of raw data to be transmitted across wide geographical areas, and ultimately to conserve power, local data processing within a sensor is essential. Advances in edge computing now enable "smart" sensors to handle vast amounts of data with limited power/storage. Here, a key challenge is sensor network query processing, including query languages, architectures, and query optimization (including de-duplication algorithms) to ensure accuracy and precision. This requires new communication and coordination protocols for sensor networks and new algorithms for data querying. Furthermore, organizing simultaneous temporal attributes (or tracking efforts), including time of acquisition, integration/dwell time, sampling interval, and aggregation time span, can easily overwhelm computing power. Thus, innovation in batteries, power-saving technologies, and backups will be increasingly essential to the functioning and performance of smart grids and app-based conservation.

Third, if sensor data are directly used for real-time decision-making, quality control of sensors is paramount. Detection of a malfunctioning sensor, or a sensor running out of power, needs to be automated with redundant sensors to improve network robustness. Moreover, a requirement essential for public trust is that the sensor networks are highly secured. However, even if it is possible to guarantee full security, it could be very expensive to implement using the currently available technology. This creates a trade-off: security and reliability versus cost. Even if these technical issues are addressed, better data does not necessarily result in better decision-making. Indeed, the belief in data as a panacea for environmental problems is one of the central tropes (and tragedies) in our collective "drama of the commons." Better data is a necessary, but insufficient condition for improved environmental governance if ethical issues are not effectively addressed (Jasanoff, 2007).

Ethical challenges

Digital technologies pose ethical challenges stemming from their potential environmental impacts. Data will be derived from an expanded array of digital monitoring devices, continuously sampling the physical world; its processing will in turn require real-time big-data analytics with greater energy and water demands. Digital Water will only be politically acceptable if these environmental issues are addressed. Given the well-recognized "water data gap," the generation of new data streams has been encouraged by planners and regulators (Josset et al., 2019). But more data does not necessarily lead to better outcomes, and raises ethical concerns (Hogan, 2015, 2021; Johnson, 2019).

Specific ethical concerns associated with Digital Water agendas include: data ownership, intellectual property, algorithmic bias, inclusion/exclusion of local communities and resource users, cybersecurity, and privacy. These issues potentially compromise transparency and trust amongst resource users. Some innovations to address these concerns include data provenance and authentication tools that attempt to trace how specific pieces of data lead to certain decisions, but the broader ethical issues raised above have not been systematically addressed. For example, Digital Water technologies entail social learning through both active and passive data gathering. Passive data gathering creates a risk of "human bycatch" if proper checks are not established. How and when should resource users consent to provide data or surveillance? Moreover, there is also an active role for water users in collecting and responding to real-time data; for example, the involvement of "citizen scientists" through apps and digital platforms, the use of humans as "citizen sensors," creating new opportunities for citizen engagement and distributed social learning. However, active data gathering by citizens implies voluntary submission to surveillance, providing data while having their online actions thoroughly indexed. There is thus a potential for exploitative modes of recruitment and usage if proper checks are not established; privacy breaches may also be an issue.

A second concern relates to governance. The growing ubiquity of smartphones and social media empowers users and civil society organizations to monitor compliance and spur enforcement action, but how and when they should do so is contested. How could Digital Water technologies enable inclusive decision-making and decision-support frameworks, particularly with respect to underrepresented users who are often excluded from formal water decision-making processes? The assessment of the development of these decision-support applications is interdisciplinary, and draws on computer science, data science, ecology, economics, engineering, geography, law, oceanography, and political science. If decision-support systems are not designed to address issues of intersectional equity, the risk is of deepening the digital divide (disenfranchising groups of users who are already often excluded from formal water-related decision-making processes). Ethical issues emerge as citizens interacting with freshwater monitoring technologies become involuntarily exposed to tracking and surveillance, potentiating new "data harms" or perverse forms of "automated inequality" (Eubanks, 2018), or as the enactment of alternative water knowledge systems are unintentionally delimited (an issue with particular relevance to Indigenous peoples). Indeed, "Indigenous Data Sovereignty" may include requests to restrict monitoring from certain waters, placing limitations on management goals or outcomes (Kukutai & Taylor, 2016). More generally, democratizing data access, with associated "translation" algorithms accessible to a broad range of stakeholders, is critical for ethical digital water governance and socio-environmental justice. At a design level, transparency is critical, and should include the full data life cycle, including data provenance, tracking, and data archiving.

This chapter raises new questions. How can involving citizens in Digital Water initiatives create new modes of distributed social learning and responsive water management? What are the implications of Digital Water for local communities and resource users? How could Digital Water technologies enable inclusive decision-making and decision-support frameworks, particularly with respect to underrepresented users who are often excluded from formal water decision-making processes? Digital Water may transform our experience of water abundance and scarcity and create new political possibilities for water governance in which every user is potentially a citizen scientist and a sensor. This enhances the potential for egalitarian inclusion but also increases risks of surveillance eco-capitalism (e.g., mobilizing environmental data to generate capital accumulation).

We echo calls for codes of conduct that guide citizens and institutions to take stock of the proliferation of new social relationships and ethical challenges created by Digital Water technologies. We also call for dialogue between social and natural scientists, bridging the "can-do" innovation of science and engineering with the "should do" critiques of social science analysis (Jasanoff, 2007). Ethics is not an afterthought, but rather a crucial input into the design of the Digital Water agenda, particularly given the need for sustainable management of the water-development nexus, on which we all depend for health and life itself.

Notes

1 See http://www.thuhjjc.com/
2 Airborne electromagnetic (AEM) methods measure electrical resistivity related to sediment type and water quality and have been used to map the large-scale structure of groundwater systems in California's agriculturally intensive Central Valley (Knight et al., 2018) and saltwater intrusion into aquifers along the coast (Gottschalk et al., 2020). Satellite Interferometric Synthetic Aperture Radar (InSAR) has been used to measure subsidence (i.e., compaction of the aquifer system) related to groundwater withdrawal (Farr & Liu, 2014) – a key issue in California that had previously been understudied.

References

Al Aani, S., Bonny, T., Hasan, S. W. and Hilal, N. (2019). "Can machine language and k intelligence revolutionize process automation for water treatment and desalination?" *Desalination*, 458, 84–96.

Amankwaa, G., Heeks, R. and Browne, A. L. (2021). "Digital innovations and water services in global South cities: A systematic literature review." *Water Alternatives*, 14(2), 619–644.

Anderson, M. C., Allen, R. G., Morse, A. and Kustas, W. P. (2012). "Use of Landsat thermal imagery in monitoring evapotranspiration and managing water resources." *Remote Sensing of Environment*, 122, 50–65.

Bakker, K. and Ritts, M. (2018). "Smart earth: A meta-review and implications for environmental governance." *Global Environmental Change*, 52, 201–211.

Biswas, N. K., Hossain, F., Bonnema, M., Aminul Haque, A. M., Biswas, R. K., Bhuyan, A. and Hossain, A. (2020). "A computationally efficient flash flood early warning system for a mountainous and transboundary river basin in Bangladesh." *Journal of Hydroinformatics*, 22(6), 1672–1692.

Bresciani, M., Giardino, C., Stroppiana, D., Dessena, M. A., Buscarinu, P., Cabras, L., Schenk, K., Heege, T., Bernet, H., Bazdanis, G. and Tzimas, A. (2019). "Monitoring water quality in two dammed reservoirs from multispectral satellite data." *European Journal of Remote Sensing*, 52(sup4), 113–122.

Buytaert, W., Zulkafli, Z., Grainger, S., Acosta, L., Alemie, T. C., Bastiaensen, J., de Bièvre, B., Bhusal, J., Clark, J., Dewulf, A., Foggin, M., Hannah, D. M., Hergarten, C., Isaeva, A., Karpouzoglou, T., Pandeya, B., Paudel, D., Keshav, S., Tammo, S., Tilahun, S., van Hecken, G. and Zhumanova, M. (2014). "Citizen science in hydrology and water resources: Opportunities for knowledge generation, ecosystem service management, and sustainable development." *Frontiers in Earth Science*, 2(26), 1–21.

Chase, S. K. and Levine, A. (2018). "Citizen science: Exploring the potential of natural resource monitoring programs to influence environmental attitudes and behaviors." *Conservation Letters*, 11(2), 1–10.

Cheung, W. and Feldman, D. (2019). "Can citizen science promote flood risk communication?" *Water*, 11(10), 1961–1970.

Deitz, S. and Meehan, K. (2019). "Plumbing poverty: Mapping hot spots of racial and geographic inequality in US household water insecurity." *Annals of the American Association of Geographers*, 109(4), 1092–1109.

Eubanks, V. (2018). *Automating Inequality: How High-Tech Tools Profile, Police, and Punish the Poor.* New York City, NY: St. Martin's Press.

Fan, M., Hu, J., Cao, R., Ruan, W. and Wei, X. (2018). "A review on experimental design for pollutants removal in water treatment with the aid of artificial intelligence." *Chemosphere*, 200, 330–343.

Fang, F., Dilkins, B. and Plumptree, A. J. (eds.). (2019). *Artificial Intelligence and Conservation*. Cambridge, UK: Cambridge University Press.

Farinosi, F., Giupponi, C., Reynaud, A., Ceccherini, G., Carmona-Moreno, C., De Roo, A., Gonzalez-Sanchez, D. and Bidoglio, G. (2018). "An innovative approach to the assessment of hydro-political risk: A spatially explicit, data driven indicator of hydro-political issues." *Global Environmental Change*, 52, 286–313.

Farr, T. G. and Liu, Z. (2014). "Monitoring subsidence associated with groundwater dynamics in the Central valley of California using interferometric radar." In: Laksmi, V. et al. (eds.) *Remote Sensing of the Terrestrial Water Cycle*. Washington, DC: American Geophysical Union. pp. 397–406.

Foster, T., Mieno, T. and Brozović, N. (2020). "Satellite-based monitoring of irrigation water use: Assessing measurement errors and their implications for agricultural water management policy." *Water Resources Research*, 56(11), 1–19.

Freiwald, J., Meyer, R., Caselle, J. E., Blanchette, C. A., Hovel, K., Neilson, D., Dugan, J., Altstatt, J., Nielsen, K. and Bursek, J. (2018). "Citizen science monitoring of marine protected areas: Case studies and recommendations for integration into monitoring programs." *Marine Ecology*, 39, 1–11.

Gartia, M. R., Braunschweig, B., Chang, T.-W., Moinzadeh, P., Minsker, B. F., Agha, G., Wieckowski, A., Keeferh, L. L. and Liu, G. L. (2012). "The microelectronic wireless nitrate sensor network for environmental water monitoring." *Journal of Environmental Monitoring*, 14(12), 3068–3075.

Golparvar, B. and Wang, R. Q. (2020). "AI-supported citizen science to monitor high-tide flooding in Newport Beach, California." *Proceedings of the 3rd ACM SIGSPATIAL International Workshop on Advances in Resilient and Intelligent Cities*. Seattle, Washington, 3 November, 66–69. https://doi.org/10.1145/3423455.3430315

Gottschalk, I., Knight, R., Asch, T., Abraham, J. and Cannia, J. (2020). "Using an airborne electromagnetic method to map saltwater intrusion in the Northern Salinas Valley, California." *Geophysics*, 85(4), B119–B131.

Guma, P. K. (2019). "Smart urbanism? ICTs for water and electricity supply in Nairobi." *Urban Studies*, 56(11), 2333–2352.

Hanjahanja, R. and Omuto, C. (2018). "Do prepaid water meters improve the quality of water service delivery? The case of Nakuru, Kenya." *Smart Water*, 3(1), 1–12.

Hogan, M. (2015). "Data flows and water woes: The Utah data center." *Big Data & Society*, 2(2), 1–12.

Hogan, M. (2018). "Big data ecologies." *Ephemera*, 18(3), 631–657.

Hogan, M. (2021). "The data center industrial complex." In: Ruiz, R. and Jue, M. (eds.) *Saturation: An Elemental Politics*. Durham, NC: Duke UP. pp. 283–306.

Jacobson, K. and Hogan, M. (2019). "Retrofitted data centres: A new world in the shell of the old." *Work Organisation, Labour & Globalisation*, 13(2), 78–94.

Jasanoff, S. (2007). "Technologies of humility." *Nature*, 450(7166), 33–33.

Johnson, A. (2019). "Data centers as infrastructural in-betweens: Expanding connections and enduring marginalities in Iceland." *American Ethnologist*, 41(1), 75–88.

Johnson, A. and Hogan, M. (2017). "Introducing location and dislocation: Global geographies of digital data." *Imaginations: Journal of Cross-Cultural Image Studies*, 8(2), 4–7.

Jollymore, A., Haines, M. J., Satterfield, T. and Johnson, M. S. (2017). "Citizen science for water quality monitoring: Data implications of citizen perspectives." *Journal of Environmental Management*, 200, 456–467.

Josset, L., Allaire, M., Hayek, C., Rising, J., Thomas, C. and Lall, U. (2019). "The US water data gap—A survey of state-level water data platforms to inform the development of a national water portal." *Earth's Future*, 7(4), 433–449.

Kamienski, C., Taumberger, M., Dantas, R., Toscano, A., Cinotti, T. S., Maia, R. F. and Neto, A. T. (2018). "Swamp: An IoT-based smart water management platform for precision irrigation in agriculture." *2018 Global Internet of Things Summit (GIoTS)*. Bilbao, Spain, 4–7 June, 1–6.

Kays, R., Crofoot, M. C., Jetz, W. and Wikelski, M. (2015). "Terrestrial animal tracking as an eye on life and planet." *Science*, 348(6240), 1–10.

Kim, S., Robson, C., Zimmerman, T., Pierce, J. S. and Haber, E. M. (2011). "Creek Watch: Pairing usefulness and usability for successful citizen science." *Proceedings of the SIGCHI Conference on Human Factors in Computing Systems*. Montreal, QC, 22–27 April, 2125–2134.

Knight, R., Smith, R., Asch, T., Abraham, J., Cannia, J., Viezzoli, A. and Fogg, G. (2018). "Mapping aquifer systems with airborne electromagnetics in the Central Valley of California." *Groundwater*, 56(6), 893–908.

Kotamäki, N., Thessler, S., Koskiaho, J., Hannukkala, A. O., Huitu, H., Huttula, T., Havento, J. and Järvenpää, M. (2009). "Wireless in-situ sensor network for agriculture and water monitoring on a river basin scale in southern Finland: Evaluation from a data user's perspective." *Sensors*, 9(4), 2862–2883.

Kukutai, T. and Taylor, J. (2016). *Indigenous Data Sovereignty: Toward an Agenda*. Canberra: Australian National University Press.

Lécuyer, C. (2017). "From clean rooms to dirty water: Labor, semiconductor firms, and the struggle over pollution and workplace hazards in Silicon Valley." *Information & Culture*, *52*(3), 304–333.

Long, D., Scanlon, B. R., Longuevergne, L., Alexander, Y., Sun, D., Fernando, N. and Save, H. (2013). "GRACE satellite monitoring of large depletion in water storage in response to the 2011 drought in Texas." *Geophysical Research Letters*, 40(13), 3395–3401.

Lowry, C. S. and Fienen, M. N. (2013). "CrowdHydrology: Crowdsourcing hydrologic data and engaging citizen scientists." *GroundWater*, 51(1), 151–156.

Mayton, H. M. and Story, S. D. (2018). "Identifying common ground for sustainable water data management: The case of California." *Water Policy*, 20(6), 1191–1207.

Mirauda, D., Capece, N. and Erra, U. (2020). "Sustainable water management: Virtual reality training for open-channel flow monitoring." *Sustainability*, 12(3), 1–18.

Ndaw, M. F. (2015). *Unlocking the Potential of Information Communications Technology to Improve Water and Sanitation Services: Summary of Findings and Recommendations*. Washington, DC: Water and Sanitation Program, World Bank.

Nguyen, H. T., Duong, T. Q., Nguyen, L. D., Vo, T. Q. M., Tran, N. T., Dang, P. D. N., Nguyen, L. D., Dang, C. K. and Nguyen, L. K. (2020). "Development of a spatial decision support system for real-time flood early warning in the Vu Gia-Thu Bon River Basin, Quang Nam Province, Vietnam." *Sensors*, *20*(6), 1–16.

Olivieri, A., Eisenberg, D., Kurtovich, M. and Pettegrew, L. (1985). "Ground-water contamination in Silicon Valley." *Journal of Water Resources Planning and Management*, 111(3), 346–358.

Pahl-Wostl, C. (2017). "An evolutionary perspective on water governance: From understanding to transformation." *Water Resources Management*, 31(10), 2917–2932.

Paul, J. D. and Buytaert, W. (2018). "Citizen science and low-cost sensors for integrated water resources management." *Advances in Chemical Pollution, Environmental Management and Protection*, (3), 1–33.

Pule, M., Yahya, A. and Chuma, J. (2017). "Wireless sensor networks: A survey on monitoring water quality." *Journal of Applied Research and Technology*, 15(6), 562–570.

Ritts, M. and Bakker, K. (2021). "Conservation acoustics: Animal sounds, audible natures, cheap nature." *Geoforum*, 124, 144–155.

Shafiee, M. E., Barker, Z. and Rasekh, A. (2018). "Enhancing water system models by integrating big data." *Sustainable Cities and Society*, 37, 485–491.

Siddik, M. A. B., Shehabi, A. and Marston, L. (2021). "The environmental footprint of data centers in the United States." *Environmental Research Letters*, 16(6), 1–12.

Staedter, T. (2019). "A cooler cloud: A new type of thermal conductor cools data centers without water, refrigerants, chillers, or server fans-[News." *IEEE Spectrum*, 56(2), 9–10.

Sun, A. Y. and Scanlon, B. R. (2019). "How can Big Data and machine learning benefit environment and water management: A survey of methods, applications, and future directions." *Environmental Research Letters*, 14(7), 1–29.

Tung, T. M. and Yaseen, Z. M. (2020). "A survey on river water quality modelling using artificial intelligence models: 2000–2020." *Journal of Hydrology*, 585, 1–62.

UNDESA. (2018). *World Urbanization Prospects*. New York, NY: UN Department for Economic and Social Affairs.

Weeser, B., Stenfert Kroese, J., Jacobs, S. R., Njue, N., Kemboi, Z., Ran, A., Rufino, M. C. and Breuer, L. (2018). "Citizen science pioneers in Kenya–A crowdsourced approach for hydrological monitoring." *Science of the Total Environment*, 631, 1590–1599.

Wong, B. P. and Kerkez, B. (2016). "Real-time environmental sensor data: An application to water quality using web services." *Environmental Modelling & Software*, 84, 505–517.

Yu, Y., Zhu, Y., Wan, D., Zhao, Q., Shu, K. and Liu, H. (2019). "Applications of social media in hydroinformatics: A survey." *arXiv:1905.03035*, 1–37.

Zheng, H., Hong, Y., Long, D. and Jing, H. (2017). "Monitoring surface water quality using social media in the context of citizen science." *Hydrology and Earth System Sciences*, 21(2), 949–961.

INDEX

Note: Page references in *italics* denote figures, in **bold** tables and with "n" endnotes.

FSC
www.fsc.org
MIX
Papier | Fördert gute Waldnutzung
FSC® C083411

Kolibri 360 GmbH
Ferdinand-Jühlke-Straße 7
99095 Erfurt, Deutschland
produktsicherheit@kolibri360.de